VOLUME ONE HUNDRED AND TWENTY TWO

PROGRESS IN MOLECULAR BIOLOGY AND TRANSLATIONAL SCIENCE

Molecular Basis of Memory

VOLUME ONE HUNDRED AND TWENTY TWO

PROGRESS IN MOLECULAR BIOLOGY AND TRANSLATIONAL SCIENCE

Molecular Basis of Memory

Edited by

ZAFAR U. KHAN
Universidad de Málaga, Málaga, Spain

E. CHRIS MULY
Emory University, Atlanta, GA, USA

ELSEVIER

AMSTERDAM • BOSTON • HEIDELBERG • LONDON
NEW YORK • OXFORD • PARIS • SAN DIEGO
SAN FRANCISCO • SINGAPORE • SYDNEY • TOKYO
Academic Press is an imprint of Elsevier

Academic Press is an imprint of Elsevier
225 Wyman Street, Waltham, MA 02451, USA
525 B Street, Suite 1800, San Diego, CA 92101-4495, USA
Radarweg 29, PO Box 211, 1000 AE Amsterdam, The Netherlands
The Boulevard, Langford Lane, Kidlington, Oxford, OX5 1GB, UK
32 Jamestown Road, London NW1 7BY, UK

First edition 2014

Library of Congress Cataloging-in-Publication Data
A catalog record for this book is available from the Library of Congress

British Library Cataloguing in Publication Data
A catalogue record for this book is available from the British Library

ISBN: 978-0-12-420170-5
ISSN: 1877-1173

Printed and bound by CPI Group (UK) Ltd, Croydon, CR0 4YY

CONTENTS

CONTRIBUTORS

Daniel L. Alkon
Blanchette Rockefeller Neurosciences Institute, Morgantown, West Virginia, USA

Raül Andero
Department of Psychiatry and Behavioral Sciences, Emory University School of Medicine, and Yerkes National Primate Research Center, Atlanta, Georgia, USA

Amy F.T. Arnsten
Department of Neurobiology, Yale University School of Medicine, New Haven, Connecticut, USA

Fabricio Ballarini
Instituto de Biología Celular y Neurociencias "Prof. E. De Robertis", Consejo Nacional de Investigaciones Científicas y Técnicas de Argentina, Facultad de Medicina, Universidad de Buenos Aires, Buenos Aires, Argentina

P.J. Banks
School of Physiology and Pharmacology, University of Bristol, Bristol, United Kingdom

Z.I. Bashir
School of Physiology and Pharmacology, University of Bristol, Bristol, United Kingdom

Antoine Besnard
Harvard Stem Cell Institute, Harvard Medical School, Center for Regenerative Medicine, Massachusetts General Hospital, Boston, Massachusetts, USA

Anna Beyeler
Department of Brain and Cognitive Sciences, Picower Institute for Learning and Memory, Massachusetts Institute of Technology, Cambridge, Massachusetts, USA

Johannes Bohacek
Brain Research Institute, Medical Faculty of the University of Zürich, and Department of Health Sciences and Technology, Swiss Federal Institute of Technology, Brain Research Institute Zürich, Switzerland

M.W. Brown
School of Physiology and Pharmacology, University of Bristol, Bristol, United Kingdom

Jocelyne Caboche
INSERM, UMRS 952; CNRS, UMR7224, Physiopathologie des Maladies du Système Nerveux Central, and UPMC University Paris 6, Paris, France

Victoria S. Cavener
Vanderbilt Brain Institute, Vanderbilt University School of Medicine, Nashville, Tennessee, USA

Dennis C. Choi
Department of Psychiatry and Behavioral Sciences, Emory University School of Medicine, and Yerkes National Primate Research Center, Atlanta, Georgia, USA

Roger J. Colbran
Department of Molecular Physiology and Biophysics; Vanderbilt Brain Institute, and Vanderbilt-Kennedy Center for Research on Human Development, Vanderbilt University School of Medicine, Nashville, Tennessee, USA

Sabrina Davis
CNRS, Centre de Neurosciences Paris-Sud, and Centre de Neurosciences Paris-Sud, Univ Paris-Sud, UMR 8195, Orsay, France

Christine A. Eckhardt
Department of Brain and Cognitive Sciences, Picower Institute for Learning and Memory, Massachusetts Institute of Technology, Cambridge, Massachusetts, USA

Wambura C. Fobbs
Laboratory of Neural Systems, Decision Science, Learning and Memory, Neurobiology and Behavior Program, Psychology Department, University of Washington, Seattle, Washington, USA

Katharina Gapp
Brain Research Institute, Medical Faculty of the University of Zürich, and Department of Health Sciences and Technology, Swiss Federal Institute of Technology, Brain Research Institute Zürich, Switzerland

Thu N. Huynh
Center for Neural Science, New York University, New York, USA

Nidhi Jalan-Sakrikar
Department of Molecular Physiology and Biophysics, Vanderbilt University School of Medicine, Nashville, Tennessee, USA

Lu E. Jin
Department of Neurobiology, Yale University School of Medicine, New Haven, Connecticut, USA

Zafar U. Khan
Laboratory of Neurobiology at CIMES and Department of Medicine at Faculty of Medicine, University of Málaga, Málaga, Spain

Eric Klann
Center for Neural Science, New York University, New York, USA

Serge Laroche
CNRS, Centre de Neurosciences Paris-Sud, and Centre de Neurosciences Paris-Sud, Univ Paris-Sud, UMR 8195, Orsay, France

Joseph E. LeDoux
Center for Neuroscience, New York University, and Emotional Brain Institute, Nathan Kline Institute for Psychiatric Research, Orangeburg, New York, USA

Isabelle M. Mansuy
Brain Research Institute, Medical Faculty of the University of Zürich, and Department of Health Sciences and Technology, Swiss Federal Institute of Technology, Brain Research Institute Zürich, Switzerland

Elisa Martín-Montañez
Laboratory of Neurobiology at CIMES, and Department of Pharmacology at Faculty of Medicine, University of Málaga, Málaga, Spain

María Cecilia Martínez
Instituto de Biología Celular y Neurociencias "Prof. E. De Robertis", Consejo Nacional de Investigaciones Científicas y Técnicas de Argentina, Facultad de Medicina, Universidad de Buenos Aires, Buenos Aires, Argentina

Sheri J.Y. Mizumori
Laboratory of Neural Systems, Decision Science, Learning and Memory, Neurobiology and Behavior Program, Psychology Department, University of Washington, Seattle, Washington, USA

Diego Moncada
Instituto de Biología Celular y Neurociencias "Prof. E. De Robertis", Consejo Nacional de Investigaciones Científicas y Técnicas de Argentina, Facultad de Medicina, Universidad de Buenos Aires, Buenos Aires, Argentina, and Neurophysiology of Learning and Memory Research Group, Leibniz-Institut für Neurobiologie, Magdeburg, Germany

E. Chris Muly
Atlanta Department of Veterans Affairs Medical Center, Decatur; Department of Psychiatry and Behavioral Sciences, Emory University, Atlanta, and Division of Neuropharmacology and Neurological Diseases, Yerkes National Primate Research Center, Atlanta, Georgia, USA

Irene Navarro-Lobato
Laboratory of Neurobiology at CIMES, and Department of Medicine at Faculty of Medicine, University of Málaga, Málaga, Spain

Kerry J. Ressler
Department of Psychiatry and Behavioral Sciences, Emory University School of Medicine; Yerkes National Primate Research Center, Atlanta, Georgia, and Howard Hughes Medical Institute, Chevy Chase, Maryland, USA

Emanuela Santini
Center for Neural Science, New York University, New York, USA

Hillary C. Schiff
Center for Neuroscience, New York University, New York, USA

Robert M. Sears
Center for Neuroscience, New York University, New York, USA

Brian C. Shonesy
Department of Molecular Physiology and Biophysics, Vanderbilt University School of Medicine, Nashville, Tennessee, USA

Miao-Kun Sun
Blanchette Rockefeller Neurosciences Institute, Morgantown, West Virginia, USA

Kay M. Tye
Department of Brain and Cognitive Sciences, Picower Institute for Learning and Memory, Massachusetts Institute of Technology, Cambridge, Massachusetts, USA

Alexandra Veyrac
CNRS, Centre de Neurosciences Paris-Sud, and Centre de Neurosciences Paris-Sud, Univ Paris-Sud, UMR 8195, Orsay, France

Haydée Viola
Instituto de Biología Celular y Neurociencias "Prof. E. De Robertis", Consejo Nacional de Investigaciones Científicas y Técnicas de Argentina, Facultad de Medicina, and Departamento de Fisiología, Biología Molecular y Celular, Facultad de Ciencias Exactas y Naturales, Universidad de Buenos Aires, Buenos Aires, Argentina

E.C. Warburton
School of Physiology and Pharmacology, University of Bristol, Bristol, United Kingdom

Bisrat T. Woldemichael
Brain Research Institute, Medical Faculty of the University of Zürich, and Department of Health Sciences and Technology, Swiss Federal Institute of Technology, Brain Research Institute Zürich, Switzerland

PREFACE

Life is all memory, except for the one present moment that goes by you so quickly you hardly catch it going.

Tennessee Williams

Memory is central to our conceptualization of who we are as individuals. Memories of our childhood, of our experiences, and of our accomplishments are, for all intents and purposes, our identity. Memory allows us to place ourselves in the context of a family and society, and without a memory of our past, we could not effectively navigate our present nor plan for our future. Modern society places increased demands on memory as people are now called upon to have more information memorized than in centuries past. Furthermore, modern hygiene and health care have dramatically lengthened life expectancy over the past century, resulting in an increase in the number of people with memory impairment. Thus, there is a pressing and increasing need to identify strategies to optimize and preserve memory function.

In this volume, recent research is presented and reviewed which has pushed our understanding of memory processes beyond circuit activity to the underlying molecular mechanisms. It begins with a review of memory impairment in aging and disease (Chapter 1) and then examines the role of individual kinases and transcription factors shown to play a role in memory processes (Chapters 2–4). The molecular mechanisms of synaptic plasticity and its role in memory are then considered (Chapters 5–6). The next section considers the interface between molecules and circuits in different types of memory processes (Chapters 7–10). The volume concludes with considerations of the contributions of epigenetic studies and optogenetic approaches to the study of memory and a proposed hypothesis to link molecular synaptic changes to changes integrated into complex neural circuits (Chapters 11–13). It is our hope that this volume will provide a useful overview to the current state of knowledge in a rapidly evolving field with tremendous potential to translate into clinical benefits.

E. Chris Muly and Zafar U. Khan

CHAPTER ONE

Memory Deficits in Aging and Neurological Diseases

Zafar U. Khan[*,†], Elisa Martín-Montañez[*,‡], Irene Navarro-Lobato[*,†], E. Chris Muly[§,¶,||]

[*]Laboratory of Neurobiology at CIMES, University of Málaga, Málaga, Spain
[†]Department of Medicine at Faculty of Medicine, University of Málaga, Málaga, Spain
[‡]Department of Pharmacology at Faculty of Medicine, University of Málaga, Málaga, Spain
[§]Atlanta Department of Veterans Affairs Medical Center, Decatur, Georgia, USA
[¶]Department of Psychiatry and Behavioral Sciences, Emory University, Atlanta, Georgia, USA
[||]Division of Neuropharmacology and Neurological Diseases, Yerkes National Primate Research Center, Atlanta, Georgia, USA

Contents

Abstract

Memory is central to our ability to perform daily life activities and correctly function in society. Improvements in public health and medical treatment for a variety of diseases have resulted in longer life spans; however, age-related memory impairments have been significant sources of morbidity. Loss in memory function is not only associated with aging population but is also a feature of neurodegenerative diseases such as Alzheimer's disease and other psychiatric and neurological disorders. Here, we focus

Progress in Molecular Biology and Translational Science, Volume 122
ISSN 1877-1173
http://dx.doi.org/10.1016/B978-0-12-420170-5.00001-5

on current understanding of the impact of normal aging on memory and what is known about its mechanisms, and further review pathological mechanisms behind the cause of dementia in Alzheimer's disease. Finally, we discuss schizophrenia and look into abnormalities in circuit function and neurotransmitter systems that contribute to memory impairment in this illness.

1. INTRODUCTION

Intact memory function is critical to carry out daily life activities, such as managing finances, remembering to take medications, driving, remembering a grandchild's birthday, and learning to use a new computer. Deficits in memory function are seen in aging and Alzheimer's disease but are also features of other psychiatric and neurological disorders. Mental retardation, autism, attention deficit disorder, learning disability, schizophrenia, and depression all have memory components, as do Parkinson's, Huntington's, and other neurodegenerative diseases. It is believed that memory deficits arise from aberrant activity in neuronal networks and diminished synaptic activity.[1] Active synaptic remodeling is essential for the formation of memories and is thought to be regulated by many biological events. These include gene expression, protein translation, protein degradation, cytoskeletal dynamics, extracellular matrix interactions, second-messenger signaling, and neurotransmitter receptor trafficking and function[2] (Fig. 1.1). Compromise of one or more of these interrelated biological events can hamper cognitive performance and produce memory loss.

2. MEMORY LOSS IN AGING

Aging differentially affects cognitive and neurological functioning.[3,4] In particular, abilities reliant on processing speed such as problem solving, inhibitory function, working memory, long-term memory, and spatial ability decline with age.[5] Aging can also affect the encoding of new information, particularly when effortful or strategic processes are required.[6] For example, younger adults benefit from a higher degree of detailed information processing than older adults, suggesting that older individuals encode the detail of new information less thoroughly.[7] Neuroimaging studies have shown an age-related decrease in functional activity in the medial temporal lobes and the prefrontal cortex while encoding new information and that has subsequently been linked to poorer memory performance.[8–10] In general,

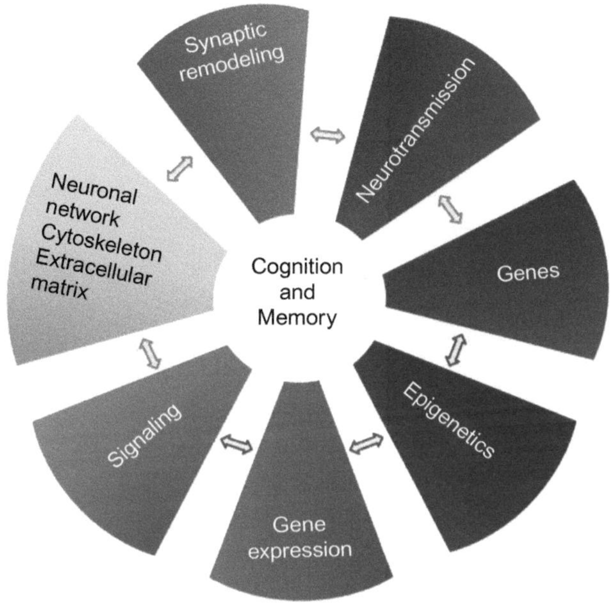

Figure 1.1 Interrelated biological functions that mediate cognition and memory. (See the color plate.)

episodic memory (conscious recollection of experienced events) declines with age[11–13]; however, semantic memory (facts, meanings, concepts, and knowledge) does not show age-related decline. In fact, semantic memory increases with age. It has been argued that the difficulties older individuals experience in word finding and name retrieval are associated more with problems retrieving this information rather than an actual semantic memory storage deficit. Also, like semantic memory, implicit or procedural memory (unconscious automated actions and movements, such as playing a guitar or riding a bike) is relatively unaffected by aging.[11,12,14,15]

Recognition memory, which is one of the most studied examples of episodic memory, is supported by recollection and familiarity. Recollection is recovery of qualitative information with specific details about a prior event. However, familiarity is when specific details are not recovered but people believe that the event has occurred previously and seems very familiar. In aging, retrieval of information is affected by the degree to which brain effort is required during processing. Retrieval of familiar recognition memory, which is relatively simple and lacks contextual details, is less affected by normal aging than recollection.[16,17] Perhaps, because recollection requires more effortful processing during access to specific details stored in brain. Typically, familiarity and recollection work together to support recognition memory; however, older individuals may tend to over rely on familiarity when experiencing recollection failures. Neuroimaging studies have shown that reliance on familiarity is a compensatory mechanism for poor recollection.[18] Although performance on tasks of both types of recognition memory can be sensitive to normal aging, older individuals fare much better on yes–no or forced-choice simple recognition tasks than on recollection tasks, suggesting that the information is stored in memory but it is not easily accessible.[16,19,20] Thus, older individuals may properly encode and store information but struggle when asked to retrieve this information. When recollection demands are eliminated with a recognition task during performance, age effects are attenuated. It is thought that deficits in effortful memory processes arise primarily due to inefficiency of the frontal lobes to engage in the self-initiation of strategic processes at both encoding and retrieval. In the following sections, underlying biological mechanisms associated with memory decline during aging are described.

2.1. Altered epigenetic regulation

Aging leads to the transcriptional downregulation of genes involved in synaptic maintenance and function.[21–24] This change in gene expression can alter neuronal integrity and also destabilize dendritic spines and their synaptic organization, which is vital for normal memory processing. Chromatin remodeling via histone modification and DNA methylation represents two major mechanisms of epigenetic regulation of gene expression. Chromatin regulation is critical for controlling the expression of genes that are required for memory processes,[25] as demonstrated by animal studies and the association of these modifications with human disorders characterized by intellectual disability.[26] It has been shown that epigenetic mechanisms contribute to reduced gene expression in the brain and that this alteration correlates very well with cognitive decline in aging.[27] For example, in the human prefrontal cortex, a noticeable downregulation of genes implicated in synaptic plasticity was observed.[24] Similar observations were made in the hippocampus of human and rodent brain.[28–30] Therefore, downregulation in the epigenetic control of gene expression in brain areas implicated in memory processing during aging can alter synaptic plasticity and memory function.[31,32] The posttranslational modifications of histone chromatin protein, such as acetylation, ADP-ribosylation, and ubiquitination,[33] regulate chromatin remodeling processes to facilitate the genetic DNA availability for transcription.[33] Most studies of posttranslational modifications are focused on histone acetylation, which has been shown to play a crucial role in regulating synaptic function and long-term memory formation.[34–36] Age-dependent decline in histone acetylation and resulting disruption of memory-related target genes are key to the deterioration of memory in rodents.[32,37,38] Consistent with these observations, increasing histone acetylation via treatments with histone deacetylase (HDAC) inhibitors rescues memory deficits associated with cognitive aging and other memory disorders.[35,39] On the other hand, methylation modifications of DNA have also been shown to play important role in regulating gene expression. In a genome-wide study of aging individuals, substantial decline in DNA methylation was observed and that this change coincided well with the memory deficit.[33,40]

2.2. Impaired cytoskeletal dynamics

The neuronal cytoskeleton is composed of microtubules and neurofilaments that are found throughout dendrites, axons, and cell bodies,[41] and of actin filaments that are specially enriched in spines and growth cones. It is widely believed that information in the brain can be stored by strengthening or weakening existing synapses, as well as appearance or disappearance of dendritic spines and their synapses. Functional and structural changes at spines and synapses are considered to be the basis of learning and memory in the brain.[42,43] Numerous studies have demonstrated that the actin cytoskeleton plays a pivotal role in the formation and elimination, motility and stability, and size and shape of dendritic spines.[44–48] In addition, modulation of actin dynamics drives the morphological changes in dendritic spines that are associated with alteration in synaptic strength.[49,50] At synapses, the actin cytoskeleton not only contributes to overall structure of synapses but also plays important roles in synaptic activities that range from organizing the postsynaptic density[51] and anchoring postsynaptic receptors[52] to facilitating the trafficking of synaptic cargos[53] and localizing the translation machinery.[54] Thus actin dynamics are fundamental for normal memory processing, and during aging, the regulation of the actin cytoskeleton becomes defective.[55,56] It has also been shown that dysregulation of the actin cytoskeleton is associated with memory disorders.[56] Furthermore, impaired synaptic plasticity and memory deficits in the aging brain have been linked to failure of actin polymerization in the spine head[57] and to elevated breakdown of cytoskeleton proteins, such as microtubule-associated protein 2A/B.[58]

2.3. Dysregulated extracellular matrix activity

The extracellular matrix occupies approximately 20% of the normal adult brain[59] and is composed of mainly unbranched polysaccharide hyaluronic acid and chondroitin sulfate proteoglycans of the lectican family, such as aggrecan, versican, brevican, and neurocan,[60] where hyaluronic acid acts as a backbone to recruit proteoglycans and glycoproteins into extracellular matrix structures.[61–63] In addition, a variety of other glycoproteins and proteoglycans contribute to extracellular matrix of the brain. These include laminins, tenascins, pentraxins, phosphocan, reelin, thrombospondins, and heparan-sulfate proteoglycans. Matrix-shaping enzymes, like proteases and hyaluronidases, are also found in the brain matrix structures.[61,63–66] The components of the extracellular matrix form a condensed and stable lattice-like network structure known as perineuronal nets. These perineuronal nets wrap tightly around the synapses on the cell soma and proximal dendrites,[67–70] serving as ground for various secreted glycoproteins, lectins, growth factors, and enzymes to modulate dendritic spines through interactions with receptors and other proteins. Extracellular matrix organization influences normal physiology of dendritic spines and synapses, and alterations in the extracellular matrix composition may underlie abnormal dendritic spine development and synaptic plasticity.[71,72] Indeed, a large family of hyaluronic acid-binding proteoglycans has been shown to play an important role in regulating synaptic function and plasticity,[73] and in the aging brain, the extracellular matrix activity was found altered.[74] Abnormal dendritic spine morphology and/or altered spine density is a hallmark of many neurological disorders associated with learning deficits, memory loss, and other cognitive deficits. Many components of the extracellular matrix can affect dendritic spines and synapses. Therefore, it is not surprising that the changes in extracellular matrix composition as a result of gene regulation or posttranslational modifications of several extracellular genes can affect cognitive functions as well. Several extracellular proteins, including reelin, elastin, thrombospondins, and some proteoglycans, have been implicated in cognitive deficits seen in several neurological disorders and aging.[75–80]

2.4. Dysfunction of neural networks

Age-dependent behavioral and cognitive decline is well documented,[81–83] and neurophysiological signatures have also been shown to slow with age.[84] Biological explanations for this slowing have generally focused on processes, such as age-related reductions in axon myelination,[85,86] degradation of the synaptic network, and slower interactions between networks of neurons, which would increase the processing time to complete a task.[87]

2.4.1 Imaging of brain structure and function

Studies on normal aging-related changes show that the brain shrinks in volume. Accumulated data indicate an annual reduction of 0.5–1% in most brain areas.[88] However, the pattern of changes is highly heterogeneous. The largest reduction was observed in frontal and temporal association cortex, areas whose functions are typically associated with age-related cognitive decline.[89–91] The decline in brain volume is probably not related to neuronal loss, rather, shrinkage of neurons and synaptic spines has been observed during aging. In addition, there is a reduction of up to 50% in the length of myelinated axons.[88] Together, these results indicate an age-related reduction in the connections between neurons.

In behavioral studies, aged individuals typically showed decreases in the activation of left-hemisphere prefrontal cortex during encoding of an episodic memory; however, during retrieval, a decrease in the activation of right-hemisphere prefrontal cortex together with increase in left-hemisphere prefrontal cortex was observed.[92] Given that during retrieval, prefrontal cortex activity of young individuals is mainly associated with right hemisphere, age-related decrease in the activity of right-hemisphere prefrontal cortex coupled with age-related increase in left-hemisphere prefrontal cortex activity leads to a more symmetric pattern of prefrontal cortex activity in aged individuals. Moreover, most of the imaging studies are consistent with the notion that neural activity in prefrontal cortex is more symmetric with aging.

Aging is accompanied by cognitive decline, including processing speed and episodic memory.[93–95] However, there exists substantial heterogeneity in cognitive aging[96,97] with some individuals showing minimal or no effect.[98] It is thought that individuals resisting age-related cognitive decline have maintained brain integrity better[99] and are spared from the structural and functional alterations of brain structures, such as the prefrontal cortex and the medial temporal lobe, which are associated with cognitive decline.[89–91] The better cognitive performance in aging-resistant individuals is correlated with increased brain activation, particularly in the prefrontal cortex,[100] which has been interpreted as a compensatory process to maintain performance. However, prefrontal activity appears not to be sufficient for the maintenance of cognitive ability during aging. A recent study has found that preservation of both hippocampus and prefrontal function are important for the maintenance of good memory function in individuals resisting age-related cognitive decline.[101]

2.4.2 Synaptic dynamics

Normal brain aging is associated with a decline in cognitive performance.[102,103] This age-related functional impairment does not involve a loss of neurons.[104,105] Instead, it likely depends on the impairment of synaptic contacts between axonal boutons and dendritic spines.[103,106] Both size and number of dendritic spines on pyramidal neurons of the cerebral cortex and hippocampus are affected by aging.[103,107–109] It has been shown that the spine loss in prefrontal cortex is highly selective to long-necked thin spines and the number of mushroom-shaped spines is stable throughout the lifetime of the animal.[110–114] Given that thin spines are highly plastic and dynamic compared with the mushroom spines, cognitive decline in aging is likely to be associated with impairment in the plasticity of network activity that this group of spines supports. Furthermore, alterations in the size and stability of dendritic spines and axonal boutons in the cortex are related to weaker synapses in aging brain.[115] These weaker synapses are less capable of short-term plasticity, and therefore the neural circuitry is less adaptable. By contrast, memory processing in the hippocampus is associated with synaptic stability and the conversion of simple synapses into complex synaptic structures such as perforated synapses and multisynaptic boutons. Indeed, an increase in the number of perforated synapses has been associated with the induction and maintenance of long-term potentiation,[116,117] and multisynaptic boutons are increased in the hippocampal CA1 area following associative learning.[118] On the other hand, aging cortical circuits display higher rates of axonal bouton formation, elimination, and destabilization.[119] In addition, persistent axonal boutons which do not disappear with age and are believed to encode long-term memory are not affected. These findings indicate that memory deficits in the aged brain arise not through an inability to form new synapses but rather through decreased synaptic activity.

It has been shown that despite an age-dependent dendritic spine loss, patch-clamp recording of synaptic activity revealed an increase in miniature EPSC frequency restricted to aged animals with preserved memory. In contrast, a strong positive relationship between miniature IPSC frequency and the occurrence of both exploratory behavior and novelty-related memory deficits was observed.[120] The enhanced inhibitory tone was accompanied by a deficit in activity-driven inhibition, also suggesting an impaired dynamic range for modulation of inhibition in the cognitively impaired aged animals. Therefore, functional synaptic imbalance toward inhibitory activity as a substrate of cognitive impairment in the aged cohort while, in contrast, an increase in excitatory synaptic tone is observed in aged unimpaired animals. The balance between basal inhibitory and excitatory synaptic tone may thus define unimpaired from impaired cognitive aging.

2.4.3 Gamma frequency

The current view is that age-dependent cognitive slowing is caused by a reduction in the speed of neural signaling and slower interactions between networks of neurons.[121] In the brain, neural communication is organized by gamma oscillations at 30–100 Hz and this gamma frequency activity is the result of the action of fast-spiking inhibitory neurons,[122,123] as well as their interactions with local excitatory neurons.[124,125] The activity of neurons that are engaged in the same information computation can become dynamically synchronized within the gamma cycle.[126–128] Consistent with this idea, treatment with barbiturates, which widen the gamma period by increasing the inhibitory $GABA_A$ receptor time constant,[129] also increases the behavioral reaction times.[130] In aging rats, a reduction in 40–70 Hz gamma oscillation and an increase of the cortical gamma period have been shown.[121] These reductions were observed in the form of local field potentials, later peaks in fast-spiking neuron autocorrelations, and delays in the spiking of inhibitory neurons following local excitatory signals. Thus, the reductions in the strength and efficiency of gamma oscillations may contribute to the slowing of behavior and cognition during aging.

2.5. Impaired neurotransmitter systems

From patient studies, animal research, pharmacological intervention, and molecular genetics, brain neurotransmitter systems have long been recognized for their fundamental role in memory and cognitive functions. The cognitive decline with age is associated with altered function of neurotransmitter systems, including the dopaminergic, glutamatergic, cholinergic, and serotonergic systems.[131]

2.5.1 Dopamine

Dopamine is one of the most important neurotransmitters in the brain and dopaminergic pathways are critical for cognitive functions.[132] Postmortem studies of human aging show that dopaminergic neurons in the substantia nigra are lost 2–6% per decade,[133] a figure that correlates well with the decrease in striatal dopamine neurotransmitter availability.[134] These changes are accompanied by age-related decrease in dopamine D1[135] and D2[136] receptors at a rate of 3% per decade. There is also evidence for age-related loss of other proteins related to dopamine neurotransmission. Molecular imaging studies show loss in the expression of dopamine transporter in aging striatum,[137,138] and further reveal age-related loss of both striatal D1 and D2 receptor densities to a similar degree.[139] Both dopamine synthesis and levels of the synthetic enzyme tyrosine hydroxylase decline with age.[140] In older adults, behavioral deficits in episodic memory are better accounted for by D2 receptor binding than D1.[141] On the other hand, a dopamine D1 receptor agonist-mediated long-term improvement in working memory of aged rhesus monkeys suggests that increasing D1 activation may be beneficial during aging.[142] A decline in the function of the dopaminergic system in aging has been linked to age-related deficits in memory processing speed, episodic memory, working memory, and cognitive control.[141,143] It has been shown that aging impacts the mesolimbic dopaminergic system, a pathway linked to cognitive, emotional, and motivational processing, more than the nigrostriatal dopaminergic system[144] which is primarily involved in motor functions. In addition, dopamine is critical for hippocampus-dependent consolidation of memory for novel events.[145] The magnitude of dopamine release in mesolimbic brain areas, such as hippocampus, regulates long-term plasticity[146] and consolidation of memory.[145] Dopamine has shown a strong correlation with age-related cognitive deficits that are especially pronounced in tasks that are executively demanding.[141,147] These observations reflect that for successful performance on such demanding tasks greater dopamine release is required.[148] Therefore, age-related dopamine deficit may particularly be detrimental for more demanding tasks.

2.5.2 Glutamate

Glutamate is the major excitatory neurotransmitter in the brain and has been known to play a central role in learning and memory. Glutamate facilitates release of brain-derived neurotrophic factor and is essential for long-term potentiation[149] and long-term depression,[150] neurogenesis, activity-dependent neuronal survival, and neural outgrowth during development.[151] The stimulation of ionotropic NMDA receptors allows calcium ions to enter into the cell, leading to activation of several intracellular signaling pathways and promotion of receptor trafficking, gene expression, and long-term potentiation.[149,152] Changes in calcium homeostasis have been associated with age-related memory impairments.[107] In fact, the induction and maintenance of long-term potentiation is compromised during aging.[153,154] It has been shown that while young rats rely on NMDA receptor-dependent long-term potentiation to build new memories, aged rats without cognitive deficit use an NMDA receptor-independent strategy.[155] This could be due to the decrease in expression of ionotropic receptor subunits that generally occurs in aging.[156] Similar to long-term potentiation, NMDA receptor-independent long-term depression has also been correlated with successful cognitive aging in rats.[157] Aged Rhesus monkeys have a decreased density of synaptic GluR2 receptor in dendritic spines, and this decrease correlates well with impaired recognition memory.[158]

2.5.3 Acetylcholine

Acetylcholine is used as a neurotransmitter by neurons located in the brainstem and basal forebrain which project to brain regions, such as the cerebral cortex and the hippocampus, involved in cognitive function and memory processing. Basal forebrain cholinergic neurons have been shown to undergo moderate degeneration during aging, giving rise to cholinergic hypofunction and subsequently progressive memory deficits.[159,160] In contrast to the basal forebrain, the cholinergic cells of the pontine cholinergic system appear to resist neurodegeneration.[161] A selective vulnerability of basal forebrain cholinergic neurons during normal aging is thought to be due to an altered regulation of nerve growth factor function.[162] Several studies have shown a correlation between the reduced activity of choline acetyltransferase, an enzyme involved in the synthesis of acetylcholine neurotransmitter, and age-related memory impairment.[163–165] Aged rats impaired in spatial learning have lower choline acetyltransferase activity relative to both young rats and aged animals with preserved cognitive function.[163,165–167] There is an increase in the frontal cholinergic activity in aging, which has been suggested as an act to compensate age-related decline in the activity of cholinergic system.[168] This functional compensation requires the recruitment of additional frontal regions to maintain adequate performance on cognitive tasks.

2.5.4 Serotonin

Both the level of the neurotransmitter serotonin and its receptor function are severely affected in aging,[169–171] and it has been suggested that this alteration is responsible for some of the frontal cognitive deficits. A receptor imaging study estimated about 25% drop in frontal lobe serotonin receptors in aged individuals,[172] amounting to a decline of approximately 5% per decade. However, more recent studies reported cross-sectional estimates of 2–11% decline per decade for serotonin receptors in various brain regions, including the hippocampus, midbrain, thalamus, and hypothalamus.[173–175] Similarly, the binding potential in key brain areas of the serotonergic system, such as limbic structures and raphe nuclei, declined with a rate of about

3–4% per decade in older adults.[176] An association of lower serotonin level has been shown with impaired cognitive performance in older individuals,[177] and the depletion of tryptophan, a precursor of serotonin, was shown to be associated with lower performance on episodic memory.[178] In addition, both serotonin transporter and serotonin receptor are implicated in memory function.[179] It has been shown that serotonin plays a critical role in the adaptation to changes in stimulus-reward contingencies during learning and the regulation of impulsive behavior.[180,181] Dysfunction in serotonergic systems in aging is particularly pronounced when reward information is uncertain or the task context requires flexible adaptations to changing stimulus-reward contingencies.

2.5.5 γ-Aminobutyric acid

γ-Aminobutyric acid (GABA) is the principal inhibitory neurotransmitter in the brain and inhibitory interneurons that mediate GABAergic function compose 20% of cortical neurons. GABA is implicated in cortical maturation, synaptic plasticity, and cognition. Attenuation of GABAergic neurotransmission by treatment with nonselective benzodiazepine receptor inverse agonists improved cognition in animals[182,183] and in healthy human volunteers.[184] The α5 subunit of $GABA_A$ receptors is believed to be responsible for the GABA-mediated cognitive effects.[185,186] This notion is supported by the preferential localization of α5 subunits in the hippocampus[187] and the facilitation in cognitive performance after the administration of selective inverse agonist in both rodents[185,188] and nonhuman primates.[189]

2.5.6 Norepinephrine

Norepinephrine is a catecholamine produced by dopamine β-hydroxylase.[190] In the brain, it is primarily synthesized in neurons of the locus coeruleus and to a lesser extent in the lateral tegmental field.[191] Norepinephrine binds to adrenergic receptors to exert its effect. The adrenergic system is involved in the consolidation of memory[192] and norepinephrine levels are significantly reduced during aging.[193] Considering that norepinephrine drives phosphorylation of the GluR1 subunit facilitating AMPA receptor trafficking to synaptic sites and long-term potentiation induction in the hippocampus,[194] its deficit during aging could hamper experience-driven synaptic modifications and memory formation. In aged monkeys, the reduction in persistent firing during working memory in prefrontal cortex was rescued by treatment with guanfacine,[195] an agonist of α2 adrenergic receptor and the administration of

$\alpha 2$ antagonist selectively impaired memory-guided performance that was associated with error perseveration.

3. MEMORY DEFICIT IN ALZHEIMER'S DISEASE

Alzheimer's disease is an irreversible, progressive neurodegenerative disorder and is clinically characterized by memory loss and cognitive decline.[196] Based on current understanding of the pathology and clinical phenotype, Alzheimer's disease comprises three main phases: (i) a presymptomatic stage in which subjects are cognitively normal but pathology (plaques and tangles) has started to accumulate; (ii) a variable period of up to 5 years of prodromal decline in cognition known as mild cognitive impairment that is characterized by a relatively isolated impairment in short-term memory, occasionally accompanied by working memory deficit; and (iii) a third stage when patients manifest clinical dementia with impairments in multiple cognitive domains and loss of function in daily activities.[197] Deficits in episodic memory are the clinical hallmark of Alzheimer's disease. The loss in episodic memory is believed to be due to disruption in the neuron-to-neuron communications in the brain areas critical for the memory system and is thought to commence from the neuronal loss, loss of synapses, and pruning of dendrites in temporal lobe and diencephalon. In following sections, the underlying causes of the memory loss in Alzheimer's disease are discussed.

3.1. High amyloid-β peptide

Although physiological amyloid-β concentrations are essential for normal memory function and synaptic plasticity,[198] high levels of the same peptide, as is seen in Alzheimer's disease, are associated with synaptic dysfunction and memory loss.[199–202] Studies in transgenic mice point to soluble intraneuronal amyloid-β as the initial mediator of tau pathology and cognitive decline[203,204] and results from patients with Alzheimer's disease further show that this soluble pool of amyloid-β correlates better with cognitive decline than amyloid-β plaques.[205,206] Amyloid-β causes endocytic removal of AMPA receptors from the synaptic cell surface, leading to loss of dendritic spines and NMDA responses,[207] activities considered critical for memory processing in the brain. However, it has been argued that amyloid-β while important is not the only cause of memory loss seen in Alzheimer's disease[208] because approaches aimed at reducing amyloid load to clear the brain from amyloid-β deposits failed to improve memory.[209,210]

3.2. Altered tau protein phosphorylation

Tau promotes assembly and maintenance of microtubule structures, and this activity is regulated by the degree of its phosphorylation. In Alzheimer's disease, tau phosphorylation is at least threefold greater than normal. This hyperphosphorylation causes aggregation of tau protein into paired helical filaments forming neurofibrillary tangles, a morphological aberration considered as a histopathological hallmark of the disease.[211] Hyperphosphorylation suppresses the ability of tau to stimulate microtubule assembly,[212] an activity essential for normal cellular morphology, viability, and function. Tau aggregates block intracellular trafficking of neurotrophins and other functional proteins and decrease the synaptic availability of molecular components crucial for memory processing.[213]

3.3. Degeneration of cholinergic neurons in Alzheimer's disease

Degeneration of cholinergic neurons is one of the fundamental steps in the pathogenesis of Alzheimer's disease. Several studies have shown a disruption in cholinergic system and have suggested a link between this cholinergic loss of function and cognitive decline in Alzheimer's disease.[214] Currently, most prescribed medication against Alzheimer's disease is designed to restore the cholinergic function through regulation of acetylcholinestrase activity. In addition, long-term treatment with M1 agonist, AF267B, reverses cognitive impairment and decreases amyloid-β and tau pathology in transgenic mice.[215] Further, M1 deletion in a transgenic mouse model of Alzheimer's disease exacerbates cognitive impairment through dysregulation of CREB signaling and accelerated synaptotoxicity.[216] Recent study found that selective α7nACh agonist, A-582941, restores cognitive function while not affecting amyloid-β and tau pathology in transgenic mice.[217] Several reports have shown that blocking or eliminating α7nACh receptors could alleviate some symptoms of Alzheimer's disease. For instance, deletion of the α7nACh receptor gene ameliorates certain behavioral deficits in a transgenic mouse model of Alzheimer's disease.[218]

4. MEMORY IMPAIRMENT IN SCHIZOPHRENIA

While the most prominent aspect of the neurodevelopmental disorder schizophrenia is psychosis, it is also characterized by memory impairments. The memory component of schizophrenia is a complex syndrome that includes other cognitive functions as well. This memory impairment includes both short- and long-term memory processes. The impairment of the short-term memory process known also as working memory is well established in schizophrenia.[219,220] Working memory is the process by which information is held "online" for short periods of time to allow behavior to be guided by an internal representation of a stimulus rather than the stimulus itself. Impaired working memory has been argued to be a core feature of schizophrenia.[221] In addition to working memory, patients with schizophrenia show impairment on long-term memory including episodic memory and semantic memory.[222–224] Implicit memory is relatively spared in schizophrenia.[225] The nature of the long-term memory deficits in schizophrenia does not appear to be a problem with storage *per se*. A meta-analysis conducted by Aleman and colleagues[226] suggested no storage deficit in schizophrenia with no difference between immediate and delayed recall. Rather, they found evidence for a difference between recall and recognition performance and suggested a deficit in the retrieval of stored information. A similar pattern was reported in a study by Holthausen and colleagues and they suggested that a decrease in processing speed could be the cause of decreased long-term memory performance.[224] Another review suggested that the impairment in verbal declarative memory seen in schizophrenia is due largely to problems in encoding new information.[227] Thus, patients with schizophrenia appear to store information adequately but are impaired in their ability to encode and retrieve that information. In the following sections, the underlying causes of the memory dysfunction in schizophrenia are discussed.

4.1. Altered brain circuits

Classically, long-term memory is thought to depend on the hippocampus and other medial temporal lobe structures while working memory function is seen as a property of the prefrontal cortex. However, functional imaging studies suggest distributed circuits including both areas are involved in both long- and short-term memories,[228] though medial temporal lobe structures may only be used in short-term memory when working memory capacity is exceeded.[229] Schizophrenia is associated with pathology in both the prefrontal cortex[230–233] and medial temporal lobe.[234–236]

Functional imaging studies in patients with schizophrenia have found evidence for medial temporal lobe involvement in both long-term and working memory function in schizophrenia. Patients show less activation of medial temporal lobe while maintaining information online during a working memory task, and unlike control subjects, in patients with schizophrenia task performance is not correlated with medial temporal signal.[237] Patients showed impaired activation of hippocampal/parahippocampal regions with both working memory and long-term memory tasks.[238] Similarly, functional imaging studies have implicated prefrontal dysfunction in both types of memory impairment in schizophrenia. Patients were found to have impaired activation of right dorsolateral prefrontal cortex during the performance of both working and long-term memory tasks.[238] Another study found working memory tasks activated prefrontal activity, but this activation was reduced and less focal in patients.[239] This study also found that certain long-term memory tasks activate prefrontal cortex in controls but not in patients. These results have led to the suggestion that memory deficits in schizophrenia share a common mechanism marked by prefrontal cortical dysfunction and impaired working memory function that disrupts long-term memory[238,239]; however, given the evidence for medial temporal lobe involvement in the pathology of schizophrenia, multiple mechanisms seem likely to play a role in memory impairment in this condition.

4.2. Abnormal neurotransmitter systems

In addition to a general reduction of prefrontal and perhaps medial temporal circuitry in schizophrenia,[230,240–242] the dopaminergic and GABAergic neurotransmitter systems seem likely to be involved in memory impairment in schizophrenia. Dopamine action at D1 family receptors in the prefrontal cortex is critical for working memory function.[243] In patients with schizophrenia, there is an increase in D1 receptor binding in prefrontal cortex and this is inversely correlated with working memory function.[244] A role for dopamine in long-term memory has been inferred from the finding that individuals with catechol-*O*-methyltransferase containing the methionine allele, which increases prefrontal dopamine availability, show reduced interference from other memories.[245] GABAergic mechanisms also play a role in working memory.[246,247] Within the cortex, GABAergic interneurons are particularly affected in schizophrenia.[248]

Thus, schizophrenia is characterized by impaired working memory and episodic and semantic long-term memory. This impairment is associated with altered activation of both prefrontal and medial temporal areas known to be involved in these processes and while the pathology of schizophrenia has not been fully elucidated, these memory impairments appear to stem from altered prefrontal dopaminergic and GABAergic systems in these patients.

5. CONCLUDING REMARKS

Memory is supported by large distributed networks in the brain, and these networks are sustained through many interrelated biological events that are associated with gene expression, protein translation and degradation, cytoskeleton and extracellular matrix structures, second-messenger signaling, and neurotransmitter receptor trafficking and function. Evidence in the literature suggests that an aberration in communication of any of the interconnected brain areas of these neural networks results in memory dysfunction and this is a common feature of conditions where memory loss have regularly been observed, including aging and many neurological and psychiatric diseases. This implies that memory deficits arise as a consequence of slowed synaptic remodeling and diminished synaptic activity. However, aging, Alzheimer's disease, and schizophrenia involve distinct mechanisms to reach to this stage. During aging, decrements in a variety of interrelated biological functions are thought to be implicated in slowing synaptic activity and memory deficit. However, in contrast to aging, in Alzheimer's disease, dysfunction of neuronal network communication is due to neuronal degeneration, while in schizophrenia, it is associated with the altered communication between prefrontal and medial temporal areas and abnormal prefrontal dopaminergic and GABAergic systems.

ACKNOWLEDGMENTS

Research in the Khan laboratory on cognitive dysfunction is supported by Grants from the Ministerio de Ciencia e Innovación (BFU2010-16500) and Junta de Andalucia (CTS 586). Dr. Muly's research is supported by a Merit Award from the Office of Research and Development, Department of Veterans Affairs, and by an NIH/NCRR base Grant (Grant P51RR000165) to Yerkes National Primate Research Center.

REFERENCES

1. Wilson IA, Gallagher M, Eichenbaum H, Tanila H. Neurocognitive aging: prior memories hinder new hippocampal encoding. *Trends Neurosci.* 2006;29(12):662–670.
2. Bibb JA, Mayford MR, Tsien JZ, Alberini CM. Cognition enhancement strategies. *J Neurosci.* 2010;30(45):14987–14992.
3. Christensen H. What cognitive changes can be expected with normal ageing? *Aust N Z J Psychiatry.* 2001;35(6):768–775.
4. Ruffman T, Henry JD, Livingstone V, Phillips LH. A meta-analytic review of emotion recognition and aging: implications for neuropsychological models of aging. *Neurosci Biobehav Rev.* 2008;32(4):863–881.

5. Park DC, Reuter-Lorenz P. The adaptive brain: aging and neurocognitive scaffolding. *Annu Rev Psychol.* 2009;60:173–196.
6. Daum I, Graber S, Schugens MM, Mayes AR. Memory dysfunction of the frontal type in normal ageing. *Neuroreport.* 1996;7(15–17):2625–2628.
7. Craik FIM, Lockhart R. Levels of processing: a framework for memory research. *J Verb Learn Verb Behav.* 1972;11(6):671–684.
8. Cabeza R, Grady CL, Nyberg L, et al. Age-related differences in neural activity during memory encoding and retrieval: a positron emission tomography study. *J Neurosci.* 1997;17(1):391–400.
9. Stebbins GT, Carrillo MC, Dorfman J, et al. Aging effects on memory encoding in the frontal lobes. *Psychol Aging.* 2002;17(1):44–55.
10. Daselaar SM, Veltman DJ, Rombouts SA, Raaijmakers JG, Jonker C. Neuroanatomical correlates of episodic encoding and retrieval in young and elderly subjects. *Brain.* 2003;126(Pt. 1):43–56.
11. Craik FIM, Jennings J. Human memory. In: Craik FIM, Salthouse T, eds. *The Handbook of Aging and Cognition.* Hillsdale: Lawrence Erlbaum Associates, Inc.; 1992:51–110.
12. Nilsson LG. Memory function in normal aging. *Acta Neurol Scand Suppl.* 2003;179:7–13.
13. Park D. The basic mechanisms accounting for age-related decline in cognitive function. In: Park D, Schwarz N, eds. *Cognitive Aging: A Primer.* New York, NY: Psychology Press; 2000:3–21.
14. Churchill JD, Stanis JJ, Press C, Kushelev M, Greenough WT. Is procedural memory relatively spared from age effects? *Neurobiol Aging.* 2003;24(6):883–892.
15. Monti L, Gabrieli J, Reminger S, Rinaldi JA, Wilson RS, Fleischman DA. Differential effects of aging and Alzheimer's disease on conceptual implicit and explicit memory. *Neuropsychology.* 1996;10(1):101–112.
16. Craik FIM, McDowd J. Age differences in recall and recognition. *J Exp Psychol Learn Mem Cogn.* 1987;13(3):474–479.
17. Perlmutter M. Age differences in adults' free recall, cued recall, and recognition. *J Gerontol.* 1979;34(4):533–539.
18. Daselaar SM, Fleck MS, Dobbins IG, Madden DJ, Cabeza R. Effects of healthy aging on hippocampal and rhinal memory functions: an event-related fMRI study. *Cereb Cortex.* 2006;16(12):1771–1782.
19. Kemps E, Newson R. Comparison of adult age differences in verbal and visuo-spatial memory: the importance of 'pure', parallel and validated measures. *J Clin Exp Neuropsychol.* 2006;28(3):341–356.
20. Nyberg L, Maitland SB, Ronnlund M, et al. Selective adult age differences in an age-invariant multifactor model of declarative memory. *Psychol Aging.* 2003;18(1):149–160.
21. Blalock EM, Chen KC, Sharrow K, et al. Gene microarrays in hippocampal aging: statistical profiling identifies novel processes correlated with cognitive impairment. *J Neurosci.* 2003;23(9):3807–3819.
22. Lee CK, Weindruch R, Prolla TA. Gene-expression profile of the ageing brain in mice. *Nat Genet.* 2000;25(3):294–297.
23. Loerch PM, Lu T, Dakin KA, et al. Evolution of the aging brain transcriptome and synaptic regulation. *PLoS One.* 2008;3(10):e3329.
24. Lu T, Pan Y, Kao SY, et al. Gene regulation and DNA damage in the ageing human brain. *Nature.* 2004;429(6994):883–891.
25. Crick F. Memory and molecular turnover. *Nature.* 1984;312(5990):101.
26. Levenson JM, Sweatt JD. Epigenetic mechanisms in memory formation. *Nat Rev Neurosci.* 2005;6(2):108–118.
27. Yankner BA, Lu T, Loerch P. The aging brain. *Annu Rev Pathol.* 2008;3:41–66.

28. Berchtold NC, Cribbs DH, Coleman PD, et al. Gene expression changes in the course of normal brain aging are sexually dimorphic. *Proc Natl Acad Sci USA*. 2008;105(40):15605–15610.
29. Verbitsky M, Yonan AL, Malleret G, Kandel ER, Gilliam TC, Pavlidis P. Altered hippocampal transcript profile accompanies an age-related spatial memory deficit in mice. *Learn Mem*. 2004;11(3):253–260.
30. Xu X, Zhan M, Duan W, et al. Gene expression atlas of the mouse central nervous system: impact and interactions of age, energy intake and gender. *Genome Biol*. 2007;8(11):R234.
31. Kandel ER. The molecular biology of memory storage: a dialogue between genes and synapses. *Science*. 2001;294(5544):1030–1038.
32. Peleg S, Sananbenesi F, Zovoilis A, et al. Altered histone acetylation is associated with age-dependent memory impairment in mice. *Science*. 2010;328(5979):753–756.
33. Liu L, van Groen T, Kadish I, Tollefsbol TO. DNA methylation impacts on learning and memory in aging. *Neurobiol Aging*. 2009;30(4):549–560.
34. Sweatt JD. Experience-dependent epigenetic modifications in the central nervous system. *Biol Psychiatry*. 2009;65(3):191–197.
35. Stilling RM, Fischer A. The role of histone acetylation in age-associated memory impairment and Alzheimer's disease. *Neurobiol Learn Mem*. 2011;96(1):19–26.
36. Peixoto L, Abel T. The role of histone acetylation in memory formation and cognitive impairments. *Neuropsychopharmacology*. 2013;38(1):62–76.
37. Zeng Y, Tan M, Kohyama J, et al. Epigenetic enhancement of BDNF signaling rescues synaptic plasticity in aging. *J Neurosci*. 2011;31(49):17800–17810.
38. Castellano JF, Fletcher BR, Kelley-Bell B, Kim DH, Gallagher M, Rapp PR. Age-related memory impairment is associated with disrupted multivariate epigenetic coordination in the hippocampus. *PLoS One*. 2012;7(3):e33249.
39. Fischer A, Sananbenesi F, Mungenast A, Tsai LH. Targeting the correct HDAC(s) to treat cognitive disorders. *Trends Pharmacol Sci*. 2010;31(12):605–617.
40. Penner MR, Roth TL, Chawla MK, et al. Age-related changes in Arc transcription and DNA methylation within the hippocampus. *Neurobiol Aging*. 2010;32(12):2198–2210.
41. Goldman JE. Immunocytochemical studies of actin localization in the central nervous system. *J Neurosci*. 1983;3(10):1952–1962.
42. Holtmaat A, Svoboda K. Experience-dependent structural synaptic plasticity in the mammalian brain. *Nat Rev Neurosci*. 2009;10(9):647–658.
43. Kasai H, Fukuda M, Watanabe S, Hayashi-Takagi A, Noguchi J. Structural dynamics of dendritic spines in memory and cognition. *Trends Neurosci*. 2010;33(3):121–129.
44. Halpain S. Actin and the agile spine: how and why do dendritic spines dance? *Trends Neurosci*. 2000;23(4):141–146.
45. Luo L. Actin cytoskeleton regulation in neuronal morphogenesis and structural plasticity. *Annu Rev Cell Dev Biol*. 2002;18:601–635.
46. Ethell IM, Pasquale EB. Molecular mechanisms of dendritic spine development and remodeling. *Prog Neurobiol*. 2005;75(3):161–205.
47. Tada T, Sheng M. Molecular mechanisms of dendritic spine morphogenesis. *Curr Opin Neurobiol*. 2006;16(1):95–101.
48. Schubert V, Dotti CG. Transmitting on actin: synaptic control of dendritic architecture. *J Cell Sci*. 2007;120(Pt. 2):205–212.
49. Matus A. Actin-based plasticity in dendritic spines. *Science*. 2000;290(5492):754–758.
50. Cingolani LA, Goda Y. Actin in action: the interplay between the actin cytoskeleton and synaptic efficacy. *Nat Rev Neurosci*. 2008;9(5):344–356.
51. Sheng M, Hoogenraad CC. The postsynaptic architecture of excitatory synapses: a more quantitative view. *Annu Rev Biochem*. 2007;76:823–847.

52. Renner M, Specht CG, Triller A. Molecular dynamics of postsynaptic receptors and scaffold proteins. *Curr Opin Neurobiol.* 2008;18(5):532–540.
53. Schlager MA, Hoogenraad CC. Basic mechanisms for recognition and transport of synaptic cargos. *Mol Brain.* 2009;2:25.
54. Bramham CR. Local protein synthesis, actin dynamics, and LTP consolidation. *Curr Opin Neurobiol.* 2008;18(5):524–531.
55. Ramakers GJ. Rho proteins, mental retardation and the cellular basis of cognition. *Trends Neurosci.* 2002;25(4):191–199.
56. Newey SE, Velamoor V, Govek EE, Van Aelst L. Rho GTPases, dendritic structure, and mental retardation. *J Neurobiol.* 2005;64(1):58–74.
57. Rex CS, Lin CY, Kramar EA, Chen LY, Gall CM, Lynch G. Brain-derived neurotrophic factor promotes long-term potentiation-related cytoskeletal changes in adult hippocampus. *J Neurosci.* 2007;27(11):3017–3029.
58. Bernath E, Kupina N, Liu MC, Hayes RL, Meegan C, Wang KK. Elevation of cytoskeletal protein breakdown in aged Wistar rat brain. *Neurobiol Aging.* 2006;27(4):624–632.
59. Sykova E, Nicholson C. Diffusion in brain extracellular space. *Physiol Rev.* 2008;88(4):1277–1340.
60. Bradbury EJ, Moon LD, Popat RJ, et al. Chondroitinase ABC promotes functional recovery after spinal cord injury. *Nature.* 2002;416(6881):636–640.
61. Bandtlow CE, Zimmermann DR. Proteoglycans in the developing brain: new conceptual insights for old proteins. *Physiol Rev.* 2000;80(4):1267–1290.
62. Rauch U. Extracellular matrix components associated with remodeling processes in brain. *Cell Mol Life Sci.* 2004;61(16):2031–2045.
63. Frischknecht R, Seidenbecher CI. The crosstalk of hyaluronan-based extracellular matrix and synapses. *Neuron Glia Biol.* 2008;4(3):249–257.
64. Dityatev A, Schachner M. Extracellular matrix molecules and synaptic plasticity. *Nat Rev Neurosci.* 2003;4(6):456–468.
65. Christopherson KS, Ullian EM, Stokes CC, et al. Thrombospondins are astrocyte-secreted proteins that promote CNS synaptogenesis. *Cell.* 2005;120(3):421–433.
66. Dityatev A, Fellin T. Extracellular matrix in plasticity and epileptogenesis. *Neuron Glia Biol.* 2008;4(3):235–247.
67. Hartig W, Brauer K, Bruckner G. Wisteria floribunda agglutinin-labelled nets surround parvalbumin-containing neurons. *Neuroreport.* 1992;3(10):869–872.
68. Carulli D, Rhodes KE, Fawcett JW. Upregulation of aggrecan, link protein 1, and hyaluronan synthases during formation of perineuronal nets in the rat cerebellum. *J Comp Neurol.* 2007;501(1):83–94.
69. Carulli D, Pizzorusso T, Kwok JC, et al. Animals lacking link protein have attenuated perineuronal nets and persistent plasticity. *Brain.* 2010;133(Pt. 8):2331–2347.
70. Dityatev A, Schachner M, Sonderegger P. The dual role of the extracellular matrix in synaptic plasticity and homeostasis. *Nat Rev Neurosci.* 2010;11(11):735–746.
71. Berardi N, Pizzorusso T, Maffei L. Critical periods during sensory development. *Curr Opin Neurobiol.* 2000;10(1):138–145.
72. Kwok JC, Dick G, Wang D, Fawcett JW. Extracellular matrix and perineuronal nets in CNS repair. *Dev Neurobiol.* 2011;71(11):1073–1089.
73. Galtrey CM, Fawcett JW. The role of chondroitin sulfate proteoglycans in regeneration and plasticity in the central nervous system. *Brain Res Rev.* 2007;54(1):1–18.
74. Kurihara D, Yamashita T. Chondroitin sulfate proteoglycans down-regulate spine formation in cortical neurons by targeting tropomyosin-related kinase B (TrkB) protein. *J Biol Chem.* 2012;287(17):13822–13828.
75. Madhusudan A, Sidler C, Knuesel I. Accumulation of reelin-positive plaques is accompanied by a decline in basal forebrain projection neurons during normal aging. *Eur J Neurosci.* 2009;30(6):1064–1076.

76. Knuesel I, Nyffeler M, Mormede C, et al. Age-related accumulation of Reelin in amyloid-like deposits. *Neurobiol Aging*. 2009;30(5):697–716.
77. Irwin SA, Galvez R, Greenough WT. Dendritic spine structural anomalies in fragile-X mental retardation syndrome. *Cereb Cortex*. 2000;10(10):1038–1044.
78. Kaufmann WE, Moser HW. Dendritic anomalies in disorders associated with mental retardation. *Cereb Cortex*. 2000;10(10):981–991.
79. Barnes AP, Milgram SL. Signals from the X: signal transduction and X-linked mental retardation. *Int J Dev Neurosci*. 2002;20(3–5):397–406.
80. Fiala JC, Spacek J, Harris KM. Dendritic spine pathology: cause or consequence of neurological disorders? *Brain Res Brain Res Rev*. 2002;39(1):29–54.
81. Salthouse TA. Speed of behavior and its implications for cognition. In: Birren JE, ed. *Handbook of the Psychology of Aging*. 2nd ed. New York: Van Nostrand Reinhold; 1985:400–426.
82. Salthouse TA. The processing-speed theory of adult age differences in cognition. *Psychol Rev*. 1996;103(3):403–428.
83. Birren JE, Fisher LM. Aging and speed of behavior: possible consequences for psychological functioning. *Annu Rev Psychol*. 1995;46:329–353.
84. Bashore TR. Age-related changes in mental processing revealed by analyses of event-related brain potentials. In: Rohrbaugh JW, Parasuraman R, Johnson R, eds. *Event-Related Brain Potentials: Basic Issues and Applications*. New York: Oxford UP; 1990:242–274.
85. Peters A. The effects of normal aging on myelinated nerve fibers in monkey central nervous system. *Front Neuroanat*. 2009;3:11.
86. Penke L, Munoz Maniega S, Murray C, et al. A general factor of brain white matter integrity predicts information processing speed in healthy older people. *J Neurosci*. 2010;30(22):7569–7574.
87. Cerella J. Aging and information processing rate. In: Birren JE, Schaie KW, eds. *The Handbook of the Psychology of Aging*. 3rd ed. San Diego: Academic; 1990:201–221.
88. Fjell AM, Walhovd KB. Structural brain changes in aging: courses, causes and cognitive consequences. *Rev Neurosci*. 2010;21(3):187–221.
89. Golomb J, Kluger A, de Leon MJ, et al. Hippocampal formation size in normal human aging: a correlate of delayed secondary memory performance. *Learn Mem*. 1994;1(1):45–54.
90. Rajah MN, D'Esposito M. Region-specific changes in prefrontal function with age: a review of PET and fMRI studies on working and episodic memory. *Brain*. 2005;128(Pt. 9):1964–1983.
91. Persson J, Pudas S, Lind J, Kauppi K, Nilsson LG, Nyberg L. Longitudinal structure-function correlates in elderly reveal MTL dysfunction with cognitive decline. *Cereb Cortex*. 2012;22(10):2297–2304.
92. Cabeza R. Cognitive neuroscience of aging: contributions of functional neuroimaging. *Scand J Psychol*. 2001;42(3):277–286.
93. Schaie KW. The course of adult intellectual development. *Am Psychol*. 1994;49(4):304–313.
94. Verhaeghen P, Salthouse TA. Meta-analyses of age-cognition relations in adulthood: estimates of linear and nonlinear age effects and structural models. *Psychol Bull*. 1997;122(3):231–249.
95. Ronnlund M, Nyberg L, Backman L, Nilsson LG. Stability, growth, and decline in adult life span development of declarative memory: cross-sectional and longitudinal data from a population-based study. *Psychol Aging*. 2005;20(1):3–18.
96. Christensen H, Mackinnon AJ, Korten AE, et al. An analysis of diversity in the cognitive performance of elderly community dwellers: individual differences in change scores as a function of age. *Psychol Aging*. 1999;14(3):365–379.

97. Wilson RS, Beckett LA, Barnes LL, et al. Individual differences in rates of change in cognitive abilities of older persons. *Psychol Aging*. 2002;17(2):179–193.
98. Habib R, Nyberg L, Nilsson LG. Cognitive and non-cognitive factors contributing to the longitudinal identification of successful older adults in the betula study. *Neuropsychol Dev Cogn B Aging Neuropsychol Cogn*. 2007;14(3):257–273.
99. Nyberg L, Lovden M, Riklund K, Lindenberger U, Backman L. Memory aging and brain maintenance. *Trends Cogn Sci*. 2012;16(5):292–305.
100. Eyler LT, Sherzai A, Kaup AR, Jeste DV. A review of functional brain imaging correlates of successful cognitive aging. *Biol Psychiatry*. 2011;70(2):115–122.
101. Pudas S, Persson J, Josefsson M, de Luna X, Nilsson LG, Nyberg L. Brain characteristics of individuals resisting age-related cognitive decline over two decades. *J Neurosci*. 2013;33(20):8668–8677.
102. Grady C. The cognitive neuroscience of ageing. *Nat Rev Neurosci*. 2012;13(7):491–505.
103. Morrison JH, Baxter MG. The ageing cortical synapse: hallmarks and implications for cognitive decline. *Nat Rev Neurosci*. 2012;13(4):240–250.
104. Tigges J, Herndon JG, Peters A. Neuronal population of area 4 during the life span of the rhesus monkey. *Neurobiol Aging*. 1990;11(3):201–208.
105. Freeman SH, Kandel R, Cruz L, et al. Preservation of neuronal number despite age-related cortical brain atrophy in elderly subjects without Alzheimer disease. *J Neuropathol Exp Neurol*. 2008;67(12):1205–1212.
106. Hof PR, Morrison JH. The aging brain: morphomolecular senescence of cortical circuits. *Trends Neurosci*. 2004;27(10):607–613.
107. Burke SN, Barnes CA. Senescent synapses and hippocampal circuit dynamics. *Trends Neurosci*. 2010;33(3):153–161.
108. Geinisman Y, deToledo-Morrell L, Morrell F, Persina IS, Rossi M. Age-related loss of axospinous synapses formed by two afferent systems in the rat dentate gyrus as revealed by the unbiased stereological dissector technique. *Hippocampus*. 1992;2(4):437–444.
109. Foster TC, Barnes CA, Rao G, McNaughton BL. Increase in perforant path quantal size in aged F-344 rats. *Neurobiol Aging*. 1991;12(5):441–448.
110. Matsuzaki M, Ellis-Davies GC, Nemoto T, Miyashita Y, Iino M, Kasai H. Dendritic spine geometry is critical for AMPA receptor expression in hippocampal CA1 pyramidal neurons. *Nat Neurosci*. 2001;4(11):1086–1092.
111. Kasai H, Matsuzaki M, Noguchi J, Yasumatsu N, Nakahara H. Structure-stability-function relationships of dendritic spines. *Trends Neurosci*. 2003;26(7):360–368.
112. Zuo Y, Lin A, Chang P, Gan WB. Development of long-term dendritic spine stability in diverse regions of cerebral cortex. *Neuron*. 2005;46(2):181–189.
113. Bourne J, Harris KM. Do thin spines learn to be mushroom spines that remember? *Curr Opin Neurobiol*. 2007;17(3):381–386.
114. Holtmaat A, Wilbrecht L, Knott GW, Welker E, Svoboda K. Experience-dependent and cell-type-specific spine growth in the neocortex. *Nature*. 2006;441(7096):979–983.
115. Mostany R, Anstey JE, Crump KL, Maco B, Knott G, Portera-Cailliau C. Altered synaptic dynamics during normal brain aging. *J Neurosci*. 2013;33(9):4094–4104.
116. Geinisman Y, deToledo-Morrell L, Morrell F. Induction of long-term potentiation is associated with an increase in the number of axospinous synapses with segmented postsynaptic densities. *Brain Res*. 1991;566(1–2):77–88.
117. Geinisman Y, Detoledo-Morrell L, Morrell F, Persina IS, Beatty MA. Synapse restructuring associated with the maintenance phase of hippocampal long-term potentiation. *J Comp Neurol*. 1996;368(3):413–423.
118. Geinisman Y, Berry RW, Disterhoft JF, Power JM, Van der Zee EA. Associative learning elicits the formation of multiple-synapse boutons. *J Neurosci*. 2001;21(15):5568–5573.

119. Grillo FW, Song S, Teles-Grilo Ruivo LM, et al. Increased axonal bouton dynamics in the aging mouse cortex. *Proc Natl Acad Sci USA*. 2013;110(16):E1514–E1523.
120. Bories C, Husson Z, Guitton MJ, De Koninck Y. Differential balance of prefrontal synaptic activity in successful versus unsuccessful cognitive aging. *J Neurosci*. 2013;33(4):1344–1356.
121. Insel N, Patron LA, Hoang LT, et al. Reduced gamma frequency in the medial frontal cortex of aged rats during behavior and rest: implications for age-related behavioral slowing. *J Neurosci*. 2012;32(46):16331–16344.
122. Hasenstaub A, Shu Y, Haider B, Kraushaar U, Duque A, McCormick DA. Inhibitory postsynaptic potentials carry synchronized frequency information in active cortical networks. *Neuron*. 2005;47(3):423–435.
123. Cardin JA, Carlen M, Meletis K, et al. Driving fast-spiking cells induces gamma rhythm and controls sensory responses. *Nature*. 2009;459(7247):663–667.
124. Economo MN, White JA. Membrane properties and the balance between excitation and inhibition control gamma-frequency oscillations arising from feedback inhibition. *PLoS Comput Biol*. 2012;8(1):e1002354.
125. Wang XJ. Neurophysiological and computational principles of cortical rhythms in cognition. *Physiol Rev*. 2010;90(3):1195–1268.
126. Womelsdorf T, Fries P, Mitra PP, Desimone R. Gamma-band synchronization in visual cortex predicts speed of change detection. *Nature*. 2006;439(7077):733–736.
127. Harris KD, Csicsvari J, Hirase H, Dragoi G, Buzsaki G. Organization of cell assemblies in the hippocampus. *Nature*. 2003;424(6948):552–556.
128. Fries P. Neuronal gamma-band synchronization as a fundamental process in cortical computation. *Annu Rev Neurosci*. 2009;32:209–224.
129. Whittington MA, Traub RD, Jefferys JG. Synchronized oscillations in interneuron networks driven by metabotropic glutamate receptor activation. *Nature*. 1995;373(6515):612–615.
130. Meador KJ, Loring DW, Moore EE, et al. Comparative cognitive effects of phenobarbital, phenytoin, and valproate in healthy adults. *Neurology*. 1995;45(8):1494–1499.
131. Dickstein DL, Kabaso D, Rocher AB, Luebke JI, Wearne SL, Hof PR. Changes in the structural complexity of the aged brain. *Aging Cell*. 2007;6(3):275–284.
132. Jackson DM, Westlind-Danielsson A. Dopamine receptors: molecular biology, biochemistry and behavioural aspects. *Pharmacol Ther*. 1994;64(2):291–370.
133. Fearnley JM, Lees AJ. Ageing and Parkinson's disease: substantia nigra regional selectivity. *Brain*. 1991;114(Pt. 5):2283–2301.
134. Snow BJ, Tooyama I, McGeer EG, et al. Human positron emission tomographic [18 F] fluorodopa studies correlate with dopamine cell counts and levels. *Ann Neurol*. 1993;34(3):324–330.
135. Rinne JO, Lonnberg P, Marjamaki P. Age-dependent decline in human brain dopamine D1 and D2 receptors. *Brain Res*. 1990;508(2):349–352.
136. Seeman P, Bzowej NH, Guan HC, et al. Human brain dopamine receptors in children and aging adults. *Synapse*. 1987;1(5):399–404.
137. Erixon-Lindroth N, Farde L, Wahlin TB, Sovago J, Halldin C, Backman L. The role of the striatal dopamine transporter in cognitive aging. *Psychiatry Res*. 2005;138(1):1–12.
138. Mozley LH, Gur RC, Mozley PD, Gur RE. Striatal dopamine transporters and cognitive functioning in healthy men and women. *Am J Psychiatry*. 2001;158(9):1492–1499.
139. Kaasinen V, Rinne JO. Functional imaging studies of dopamine system and cognition in normal aging and Parkinson's disease. *Neurosci Biobehav Rev*. 2002;26(7):785–793.
140. McCormack AL, Di Monte DA, Delfani K, et al. Aging of the nigrostriatal system in the squirrel monkey. *J Comp Neurol*. 2004;471(4):387–395.

141. Backman L, Ginovart N, Dixon RA, et al. Age-related cognitive deficits mediated by changes in the striatal dopamine system. *Am J Psychiatry*. 2000;157(4):635–637.
142. Castner SA, Goldman-Rakic PS. Enhancement of working memory in aged monkeys by a sensitizing regimen of dopamine D1 receptor stimulation. *J Neurosci*. 2004;24(6):1446–1450.
143. Volkow ND, Logan J, Fowler JS, et al. Association between age-related decline in brain dopamine activity and impairment in frontal and cingulate metabolism. *Am J Psychiatry*. 2000;157(1):75–80.
144. Cruz-Muros I, Afonso-Oramas D, Abreu P, et al. Aging of the rat mesostriatal system: differences between the nigrostriatal and the mesolimbic compartments. *Exp Neurol*. 2007;204(1):147–161.
145. O'Carroll CM, Martin SJ, Sandin J, Frenguelli B, Morris RG. Dopaminergic modulation of the persistence of one-trial hippocampus-dependent memory. *Learn Mem*. 2006;13(6):760–769.
146. Frey U, Morris RG. Synaptic tagging: implications for late maintenance of hippocampal long-term potentiation. *Trends Neurosci*. 1998;21(5):181–188.
147. Volkow ND, Gur RC, Wang GJ, et al. Association between decline in brain dopamine activity with age and cognitive and motor impairment in healthy individuals. *Am J Psychiatry*. 1998;155(3):344–349.
148. Mattay VS, Goldberg TE, Fera F, et al. Catechol O-methyltransferase val158-met genotype and individual variation in the brain response to amphetamine. *Proc Natl Acad Sci USA*. 2003;100(10):6186–6191.
149. Baudry M, Bi X, Gall C, Lynch G. The biochemistry of memory: the 26 year journey of a 'new and specific hypothesis'. *Neurobiol Learn Mem*. 2011;95(2):125–133.
150. Collingridge GL, Peineau S, Howland JG, Wang YT. Long-term depression in the CNS. *Nat Rev Neurosci*. 2010;11(7):459–473.
151. Mattson MP. Glutamate and neurotrophic factors in neuronal plasticity and disease. *Ann N Y Acad Sci*. 2008;1144:97–112.
152. Lee YS, Silva AJ. The molecular and cellular biology of enhanced cognition. *Nat Rev Neurosci*. 2009;10(2):126–140.
153. Barnes CA. Long-term potentiation and the ageing brain. *Philos Trans R Soc Lond B Biol Sci*. 2003;358(1432):765–772.
154. Kumar A. Long-term potentiation at CA3-CA1 hippocampal synapses with special emphasis on aging, disease, and stress. *Front Aging Neurosci*. 2011;3:7.
155. Boric K, Munoz P, Gallagher M, Kirkwood A. Potential adaptive function for altered long-term potentiation mechanisms in aging hippocampus. *J Neurosci*. 2008;28(32):8034–8039.
156. Zhao X, Rosenke R, Kronemann D, et al. The effects of aging on N-methyl-D-aspartate receptor subunits in the synaptic membrane and relationships to long-term spatial memory. *Neuroscience*. 2009;162(4):933–945.
157. Lee HK, Min SS, Gallagher M, Kirkwood A. NMDA receptor-independent long-term depression correlates with successful aging in rats. *Nat Neurosci*. 2005;8(12):1657–1659.
158. Hara Y, Punsoni M, Yuk F, et al. Synaptic distributions of GluA2 and PKMzeta in the monkey dentate gyrus and their relationships with aging and memory. *J Neurosci*. 2012;32(21):7336–7344.
159. Hartig W, Bauer A, Brauer K, et al. Functional recovery of cholinergic basal forebrain neurons under disease conditions: old problems, new solutions? *Rev Neurosci*. 2002;13(2):95–165.
160. Schliebs R, Arendt T. The significance of the cholinergic system in the brain during aging and in Alzheimer's disease. *J Neural Transm*. 2006;113(11):1625–1644.
161. McKinney M, Jacksonville MC. Brain cholinergic vulnerability: relevance to behavior and disease. *Biochem Pharmacol*. 2005;70(8):1115–1124.

162. Henriksson BG, Soderstrom S, Gower AJ, Ebendal T, Winblad B, Mohammed AH. Hippocampal nerve growth factor levels are related to spatial learning ability in aged rats. *Behav Brain Res.* 1992;48(1):15–20.
163. Gallagher M, Burwell RD, Kodsi MH, et al. Markers for biogenic amines in the aged rat brain: relationship to decline in spatial learning ability. *Neurobiol Aging.* 1990;11(5):507–514.
164. Lee JM, Ross ER, Gower A, Paris JM, Martensson R, Lorens SA. Spatial learning deficits in the aged rat: neuroanatomical and neurochemical correlates. *Brain Res Bull.* 1994;33(5):489–500.
165. Dunbar GL, Rylett RJ, Schmidt BM, Sinclair RC, Williams LR. Hippocampal choline acetyltransferase activity correlates with spatial learning in aged rats. *Brain Res.* 1993;604(1–2):266–272.
166. Colombo PJ, Gallagher M. Individual differences in spatial memory and striatal ChAT activity among young and aged rats. *Neurobiol Learn Mem.* 1998;70(3):314–327.
167. Luine V, Hearns M. Spatial memory deficits in aged rats: contributions of the cholinergic system assessed by ChAT. *Brain Res.* 1990;523(2):321–324.
168. Dumas JA, Newhouse PA. The cholinergic hypothesis of cognitive aging revisited again: cholinergic functional compensation. *Pharmacol Biochem Behav.* 2011;99(2):254–261.
169. Eppinger B, Hammerer D, Li SC. Neuromodulation of reward-based learning and decision making in human aging. *Ann N Y Acad Sci.* 2011;1235:1–17.
170. Murakami H, Bessinger K, Hellmann J, Murakami S. Manipulation of serotonin signal suppresses early phase of behavioral aging in Caenorhabditis elegans. *Neurobiol Aging.* 2008;29(7):1093–1100.
171. Petrascheck M, Ye X, Buck LB. A high-throughput screen for chemicals that increase the lifespan of Caenorhabditis elegans. *Ann N Y Acad Sci.* 2009;1170:698–701.
172. Wong DF, Wagner Jr HN, Dannals RF, et al. Effects of age on dopamine and serotonin receptors measured by positron tomography in the living human brain. *Science.* 1984;226(4681):1393–1396.
173. Costes N, Merlet I, Ostrowsky K, et al. A 18 F-MPPF PET normative database of 5-HT1A receptor binding in men and women over aging. *J Nucl Med.* 2005;46(12):1980–1989.
174. Hesse S, Barthel H, Murai T, et al. Is correction for age necessary in neuroimaging studies of the central serotonin transporter? *Eur J Nucl Med Mol Imaging.* 2003;30(3):427–430.
175. Yamamoto M, Suhara T, Okubo Y, et al. Age-related decline of serotonin transporters in living human brain of healthy males. *Life Sci.* 2002;71(7):751–757.
176. Moller M, Jakobsen S, Gjedde A. Parametric and regional maps of free serotonin 5HT1A receptor sites in human brain as function of age in healthy humans. *Neuropsychopharmacology.* 2007;32(8):1707–1714.
177. Topic B, Willuhn I, Palomero-Gallagher N, Zilles K, Huston JP, Hasenohrl RU. Impaired maze performance in aged rats is accompanied by increased density of NMDA, 5-HT1A, and alpha-adrenoceptor binding in hippocampus. *Hippocampus.* 2007;17(1):68–77.
178. Riedel WJ, Klaassen T, Deutz NE, van Someren A, van Praag HM. Tryptophan depletion in normal volunteers produces selective impairment in memory consolidation. *Psychopharmacology (Berl).* 1999;141(4):362–369.
179. Buhot MC, Martin S, Segu L. Role of serotonin in memory impairment. *Ann Med.* 2000;32(3):210–221.
180. Cools R, Lewis SJ, Clark L, Barker RA, Robbins TW. L-DOPA disrupts activity in the nucleus accumbens during reversal learning in Parkinson's disease. *Neuropsychopharmacology.* 2007;32(1):180–189.

181. Tanaka SC, Doya K, Okada G, Ueda K, Okamoto Y, Yamawaki S. Prediction of immediate and future rewards differentially recruits cortico-basal ganglia loops. *Nat Neurosci.* 2004;7(8):887–893.
182. Jensen LH, Stephens DN, Sarter M, Petersen EN. Bidirectional effects of beta-carbolines and benzodiazepines on cognitive processes. *Brain Res Bull.* 1987;19(3):359–364.
183. Venault P, Chapouthier G, de Carvalho LP, et al. Benzodiazepine impairs and beta-carboline enhances performance in learning and memory tasks. *Nature.* 1986;321(6073):864–866.
184. Duka T, Ott H, Rohloff A, Voet B. The effects of a benzodiazepine receptor antagonist beta-carboline ZK-93426 on scopolamine-induced impairment on attention, memory and psychomotor skills. *Psychopharmacology (Berl).* 1996;123(4):361–373.
185. Collinson N, Atack JR, Laughton P, Dawson GR, Stephens DN. An inverse agonist selective for alpha5 subunit-containing GABAA receptors improves encoding and recall but not consolidation in the Morris water maze. *Psychopharmacology (Berl).* 2006;188(4):619–628.
186. Crestani F, Keist R, Fritschy JM, et al. Trace fear conditioning involves hippocampal alpha5 GABA(A) receptors. *Proc Natl Acad Sci USA.* 2002;99(13):8980–8985.
187. Friston KJ, Grasby PM, Bench CJ, et al. Measuring the neuromodulatory effects of drugs in man with positron emission tomography. *Neurosci Lett.* 1992;141(1):106–110.
188. Savic MM, Clayton T, Furtmuller R, et al. PWZ-029, a compound with moderate inverse agonist functional selectivity at GABA(A) receptors containing alpha5 subunits, improves passive, but not active, avoidance learning in rats. *Brain Res.* 2008;1208:150–159.
189. Ballard TM, Knoflach F, Prinssen E, et al. RO4938581, a novel cognitive enhancer acting at GABAA alpha5 subunit-containing receptors. *Psychopharmacology (Berl).* 2009;202(1–3):207–223.
190. Coyle JT. Biochemical aspects of neurotransmission in the developing brain. *Int Rev Neurobiol.* 1977;20:65–103.
191. Smythies J, Section III. The norepinephrine system. *Int Rev Neurobiol.* 2005;64:173–211.
192. Schultz W. Getting formal with dopamine and reward. *Neuron.* 2002;36(2):241–263.
193. Marcyniuk B, Mann DM, Yates PO. Loss of nerve cells from locus coeruleus in Alzheimer's disease is topographically arranged. *Neurosci Lett.* 1986;64(3):247–252.
194. Hu H, Real E, Takamiya K, et al. Emotion enhances learning via norepinephrine regulation of AMPA-receptor trafficking. *Cell.* 2007;131(1):160–173.
195. Wang M, Gamo NJ, Yang Y, et al. Neuronal basis of age-related working memory decline. *Nature.* 2011;476(7359):210–213.
196. Khachaturian ZS. Diagnosis of Alzheimer's disease. *Arch Neurol.* 1985;42(11):1097–1105.
197. Jack Jr CR, Knopman DS, Jagust WJ, et al. Hypothetical model of dynamic biomarkers of the Alzheimer's pathological cascade. *Lancet Neurol.* 2010;9(1):119–128.
198. Puzzo D, Privitera L, Leznik E, et al. Picomolar amyloid-beta positively modulates synaptic plasticity and memory in hippocampus. *J Neurosci.* 2008;28(53):14537–14545.
199. Yankner BA, Duffy LK, Kirschner DA. Neurotrophic and neurotoxic effects of amyloid beta protein: reversal by tachykinin neuropeptides. *Science.* 1990;250(4978):279–282.
200. Kamenetz F, Tomita T, Hsieh H, et al. APP processing and synaptic function. *Neuron.* 2003;37(6):925–937.
201. Shankar GM, Li S, Mehta TH, et al. Amyloid-beta protein dimers isolated directly from Alzheimer's brains impair synaptic plasticity and memory. *Nat Med.* 2008;14(8):837–842.

202. Puzzo D, Privitera L, Palmeri A. Hormetic effect of amyloid-beta peptide in synaptic plasticity and memory. *Neurobiol Aging*. 2012;33(7):1484.e15–1484.e24.
203. Takahashi RH, Capetillo-Zarate E, Lin MT, Milner TA, Gouras GK. Co-occurrence of Alzheimer's disease ss-amyloid and tau pathologies at synapses. *Neurobiol Aging*. 2010;31(7):1145–1152.
204. Oddo S, Caccamo A, Shepherd JD, et al. Triple-transgenic model of Alzheimer's disease with plaques and tangles: intracellular Abeta and synaptic dysfunction. *Neuron*. 2003;39(3):409–421.
205. Lue LF, Kuo YM, Roher AE, et al. Soluble amyloid beta peptide concentration as a predictor of synaptic change in Alzheimer's disease. *Am J Pathol*. 1999;155(3):853–862.
206. McLean CA, Cherny RA, Fraser FW, et al. Soluble pool of Abeta amyloid as a determinant of severity of neurodegeneration in Alzheimer's disease. *Ann Neurol*. 1999;46(6):860–866.
207. Hsieh H, Boehm J, Sato C, et al. AMPAR removal underlies Abeta-induced synaptic depression and dendritic spine loss. *Neuron*. 2006;52(5):831–843.
208. Villemagne VL, Pike KE, Darby D, et al. Abeta deposits in older non-demented individuals with cognitive decline are indicative of preclinical Alzheimer's disease. *Neuropsychologia*. 2008;46(6):1688–1697.
209. Holmes C, Boche D, Wilkinson D, et al. Long-term effects of Abeta42 immunisation in Alzheimer's disease: follow-up of a randomised, placebo-controlled phase I trial. *Lancet*. 2008;372(9634):216–223.
210. Green RC, Schneider LS, Amato DA, et al. Effect of tarenflurbil on cognitive decline and activities of daily living in patients with mild Alzheimer disease: a randomized controlled trial. *JAMA*. 2009;302(23):2557–2564.
211. Alonso A, Zaidi T, Novak M, Grundke-Iqbal I, Iqbal K. Hyperphosphorylation induces self-assembly of tau into tangles of paired helical filaments/straight filaments. *Proc Natl Acad Sci USA*. 2001;98(12):6923–6928.
212. Lindwall G, Cole RD. Phosphorylation affects the ability of tau protein to promote microtubule assembly. *J Biol Chem*. 1984;259(8):5301–5305.
213. Salehi A, Delcroix JD, Mobley WC. Traffic at the intersection of neurotrophic factor signaling and neurodegeneration. *Trends Neurosci*. 2003;26(2):73–80.
214. Mufson EJ, Counts SE, Perez SE, Ginsberg SD. Cholinergic system during the progression of Alzheimer's disease: therapeutic implications. *Expert Rev Neurother*. 2008;8(11):1703–1718.
215. Caccamo A, Oddo S, Billings LM, et al. M1 receptors play a central role in modulating AD-like pathology in transgenic mice. *Neuron*. 2006;49(5):671–682.
216. Medeiros R, Kitazawa M, Caccamo A, et al. Loss of muscarinic M1 receptor exacerbates Alzheimer's disease-like pathology and cognitive decline. *Am J Pathol*. 2011;179(2):980–991.
217. Medeiros R, Chabrier MA, LaFerla FM. Elucidating the triggers, progression, and effects of Alzheimer's disease. *J Alzheimers Dis*. 2013;33(Suppl. 1):S195–S210.
218. Dziewczapolski G, Glogowski CM, Masliah E, Heinemann SF. Deletion of the alpha 7 nicotinic acetylcholine receptor gene improves cognitive deficits and synaptic pathology in a mouse model of Alzheimer's disease. *J Neurosci*. 2009;29(27):8805–8815.
219. Park S, Holzman PS. Schizophrenics show spatial working memory deficits. *Arch Gen Psychiatry*. 1992;49(12):975–982.
220. Gold JM, Carpenter C, Randolph C, Goldberg TE, Weinberger DR. Auditory working memory and Wisconsin Card Sorting Test performance in schizophrenia. *Arch Gen Psychiatry*. 1997;54(2):159–165.
221. Silver H, Feldman P, Bilker W, Gur RC. Working memory deficit as a core neuropsychological dysfunction in schizophrenia. *Am J Psychiatry*. 2003;160(10):1809–1816.

222. McKay AP, McKenna PJ, Bentham P, Mortimer AM, Holbery A, Hodges JR. Semantic memory is impaired in schizophrenia. *Biol Psychiatry*. 1996;39(11):929–937.
223. Rushe TM, Woodruff PW, Murray RM, Morris RG. Episodic memory and learning in patients with chronic schizophrenia. *Schizophr Res*. 1999;35(1):85–96.
224. Holthausen EA, Wiersma D, Sitskoorn MM, Dingemans PM, Schene AH, van den Bosch RJ. Long-term memory deficits in schizophrenia: primary or secondary dysfunction? *Neuropsychology*. 2003;17(4):539–547.
225. Clare L, McKenna PJ, Mortimer AM, Baddeley AD. Memory in schizophrenia: what is impaired and what is preserved? *Neuropsychologia*. 1993;31(11):1225–1241.
226. Aleman A, Hijman R, de Haan EH, Kahn RS. Memory impairment in schizophrenia: a meta-analysis. *Am J Psychiatry*. 1999;156(9):1358–1366.
227. Cirillo MA, Seidman LJ. Verbal declarative memory dysfunction in schizophrenia: from clinical assessment to genetics and brain mechanisms. *Neuropsychol Rev*. 2003;13(2):43–77.
228. Andreasen NC, O'Leary DS, Arndt S, et al. Short-term and long-term verbal memory: a positron emission tomography study. *Proc Natl Acad Sci USA*. 1995;92(11):5111–5115.
229. Jeneson A, Squire LR. Working memory, long-term memory, and medial temporal lobe function. *Learn Mem*. 2012;19(1):15–25.
230. Selemon LD, Rajkowska G, Goldman-Rakic PS. Abnormally high neuronal density in the schizophrenic cortex. A morphometric analysis of prefrontal area 9 and occipital area 17. *Arch Gen Psychiatry*. 1995;52(10):805–818, discussion 819–820.
231. Johnston-Wilson NL, Sims CD, Hofmann JP, et al. Disease-specific alterations in frontal cortex brain proteins in schizophrenia, bipolar disorder, and major depressive disorder. The Stanley Neuropathology Consortium. *Mol Psychiatry*. 2000;5(2):142–149.
232. Selemon LD, Kleinman JE, Herman MM, Goldman-Rakic PS. Smaller frontal gray matter volume in postmortem schizophrenic brains. *Am J Psychiatry*. 2002;159(12):1983–1991.
233. Black JE, Kodish IM, Grossman AW, et al. Pathology of layer V pyramidal neurons in the prefrontal cortex of patients with schizophrenia. *Am J Psychiatry*. 2004;161(4):742–744.
234. Arnold SE, Trojanowski JQ. Recent advances in defining the neuropathology of schizophrenia. *Acta Neuropathol*. 1996;92(3):217–231.
235. Dean B, Opeskin K, Pavey G, Hill C, Keks N. Changes in protein kinase C and adenylate cyclase in the temporal lobe from subjects with schizophrenia. *J Neural Transm*. 1997;104(11–12):1371–1381.
236. McCarley RW, Wible CG, Frumin M, et al. MRI anatomy of schizophrenia. *Biol Psychiatry*. 1999;45(9):1099–1119.
237. Luck D, Danion JM, Marrer C, Pham BT, Gounot D, Foucher J. Abnormal medial temporal activity for bound information during working memory maintenance in patients with schizophrenia. *Hippocampus*. 2010;20(8):936–948.
238. Barch DM, Csernansky JG, Conturo T, Snyder AZ. Working and long-term memory deficits in schizophrenia: is there a common prefrontal mechanism? *J Abnorm Psychol*. 2002;111(3):478–494.
239. Ragland JD, Blumenfeld RS, Ramsay IS, et al. Neural correlates of relational and item-specific encoding during working and long-term memory in schizophrenia. *Neuroimage*. 2012; 59(2):1719–1726.
240. Garey LJ, Ong WY, Patel TS, et al. Reduced dendritic spine density on cerebral cortical pyramidal neurons in schizophrenia. *J Neurol Neurosurg Psychiatry*. 1998;65(4):446–453.
241. Glantz LA, Lewis DA. Decreased dendritic spine density on prefrontal cortical pyramidal neurons in schizophrenia. *Arch Gen Psychiatry*. 2000;57(1):65–73.

242. Pierri JN, Volk CL, Auh S, Sampson A, Lewis DA. Decreased somal size of deep layer 3 pyramidal neurons in the prefrontal cortex of subjects with schizophrenia. *Arch Gen Psychiatry*. 2001;58(5):466–473.
243. Goldman-Rakic PS, Muly 3rd EC, Williams GV. D(1) receptors in prefrontal cells and circuits. *Brain Res Brain Res Rev*. 2000;31(2–3):295–301.
244. Abi-Dargham A, Mawlawi O, Lombardo I, et al. Prefrontal dopamine D1 receptors and working memory in schizophrenia. *J Neurosci*. 2002;22(9):3708–3719.
245. Wimber M, Schott BH, Wendler F, et al. Prefrontal dopamine and the dynamic control of human long-term memory. *Transl Psychiatry*. 2011;1:e15.
246. Sawaguchi T, Matsumura M, Kubota K. Delayed response deficit in monkeys by locally disturbed prefrontal neuronal activity by bicuculline. *Behav Brain Res*. 1988;31(2):193–198.
247. Rao SG, Williams GV, Goldman-Rakic PS. Destruction and creation of spatial tuning by disinhibition: GABA(A) blockade of prefrontal cortical neurons engaged by working memory. *J Neurosci*. 2000;20(1):485–494.
248. Lewis DA, Pierri JN, Volk DW, Melchitzky DS, Woo TU. Altered GABA neurotransmission and prefrontal cortical dysfunction in schizophrenia. *Biol Psychiatry*. 1999;46(5):616–626.

CHAPTER TWO

The "Memory Kinases": Roles of PKC Isoforms in Signal Processing and Memory Formation

Miao-Kun Sun, Daniel L. Alkon
Blanchette Rockefeller Neurosciences Institute, Morgantown, West Virginia, USA

Contents

Abstract

The protein kinase C (PKC) isoforms, which play an essential role in transmembrane signal conduction, can be viewed as a family of "memory kinases." Evidence is emerging that they are critically involved in memory acquisition and maintenance, in addition to their involvement in other functions of cells. Deficits in PKC signal cascades in neurons are one of the earliest abnormalities in the brains of patients suffering from Alzheimer's disease. Their dysfunction is also involved in several other types of memory impairments, including those related to emotion, mental retardation, brain injury, and vascular dementia/ischemic stroke. Inhibition of PKC activity leads to a reduced capacity of many types of learning and memory, but may have therapeutic values in treating substance abuse or aversive memories. PKC activators, on the other hand, have been shown to

Progress in Molecular Biology and Translational Science, Volume 122
ISSN 1877-1173
http://dx.doi.org/10.1016/B978-0-12-420170-5.00002-7

possess memory-enhancing and antidementia actions. PKC pharmacology may, therefore, represent an attractive area for developing effective cognitive drugs for the treatment of many types of memory disorders and dementias.

1. INTRODUCTION

The protein kinase C (PKC) family is composed of serine/threonine protein kinases of at least 12 isoforms in mammals and is part of an essential signaling network in the brain. They are involved in a variety of signal pathways regulating cell growth, differentiation, apoptosis, transformation, tumorigenicity, synaptic function, behavior, and cognition.[1–3] These isoforms are commonly classified into three subgroups, based on their protein structure and second messenger requirements: classical PKC (cPKC; α, βI, βII, and γ), novel PKC (nPKC; δ, ε, η, θ, and μ), and atypical PKC (aPKC; ζ, ι, and λ, including PKMζ, which consists of a PKC catalytic domain without the regulatory domain). The number of isoforms may differ in other species. In *Aplysia*, for example, only three isoforms, Apls I, II, and III, have been identified so far.

Interest has been raised in these "memory kinases," since some PKC isoforms, particularly the PKCα, γ, ε, and ζ, are highly and selectively involved in the molecular cascades that are critical parts of the signaling processes in memory and memory disorders, while PKCδ may underlie ischemic brain injury and vascular dementia. Targeting these kinases thus holds great potential in developing effective therapeutics for the treatment of memory disorders.

2. DISTRIBUTION OF PKC ISOFORMS

Unlike the majority of enzymes, PKC isoforms can be rather nonspecific toward their substrates between different isoforms. Their substrate selectivity thus depends on where they are when activated, that is, their intracellular distribution.

The distribution of PKC isoforms can be summarized as follows. The PKCα and PKCδ are expressed in all tissues,[4,5] particularly prominent in the brain. PKCζ is present in most tissues, particularly the lung (maximally expressed), brain, and liver. Other isoforms are expressed more in a tissue-specific manner: ε predominantly in the brain and, to a lesser degree, kidney, βI in the spleen, βII in the spleen and brain, η in the keratinocytes, and θ in skeletal muscle, T cells, and epidermis. Although brains are found to contain the most PKCγ immunoreactivity, some immunostaining has also been seen in the adrenal tissue.[6] PKCι/λ can be found in the testis and insulin-secreting cells.

Interestingly, most cells express more than one PKC isoform. The presence of multiple isoforms in the same tissue and within the same cell suggests that individual isoforms might mediate distinct or even opposite cellular functions. One example is PKCδ and PKCε in the mitochondria, where they regulate energy homeostasis by transmitting opposing signals to the pyruvate dehydrogenase complex (PDHC). Activated PKCε transmits an inhibitory signal to the PDHC, whereas PKCδ opposes this signal and stimulates the PDHC.[7] Their functional balance thus determines the flux of fuel entering the Krebs cycle.

3. STRUCTURE OF PKC ISOFORMS

The PKC isoforms consist of a regulatory domain (C1 and C2) and a catalytic domain (C3, the ATP-binding domain; and C4, the protein substrate-binding domain), separated by isozyme-specific variable regions (as V; Fig. 2.1). The catalytic region is highly conserved, while the regulatory region is less so.

The cPKC isoforms require Ca^{2+} and diacylglycerol (DAG)/or other PKC activators for activation. Activity of the cPKC is regulated by three phosphorylation residues in the activation loop, turn motif (TM), and hydrophobic motif (HM). The activation loop of PKCα and βII, for

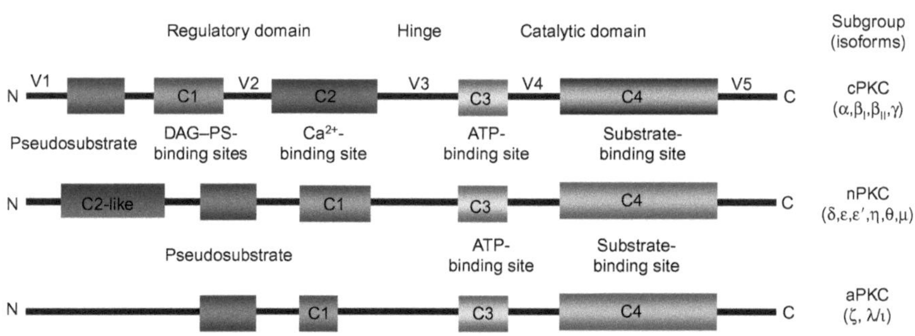

Figure 2.1 The PKC isoforms. The homologous domains (C1, C2, C3, and C4) are separated by isoform-unique (variable or V) regions. The C1 domain contains binding sites for diacylglycerol (DAG) and phosphatidylserine (PS). (See the color plate.)

instance, is phosphorylated by 3-phosphoinositide-dependent kinase 1 (PDK-1),[8–10] which then allows them to autophosphorylate the HM and TM.[11] TM and HM phosphorylation of PKCα and βII can also be mediated by the rapamycin complex 2 in mammalian.[12,13] Phosphorylation of these three residues stabilizes the cPKCs, enabling them to readily receive signals from second messengers. Dephosphorylation of TM and HM is critical for the inactivation and degradation of PKCα and βII.[14–16] PP2A, PP1, and PH domain and leucine-rich repeat protein phosphatase (PHLPP), a Ser/Thr phosphatase, can dephosphorylate PKC.[17–19] Activation and inactivation of cPKCs are mainly driven by the DAG–C1a-binding kinetics, which are rate-limiting,[20,21] while the C2–membrane interaction is the major determinant for the localization of cPKCs to the plasma/nuclear membrane.[22] In the Ca^{2+}-bound form, the surface charge changes to positive from negative and the Ca^{2+}–C2 domain becomes very much attracted to phospholipids.[23,24]

The nPKCs require only DAG/or other activators for activation and contain a C2 domain that lacks the acidic Ca^{2+}-binding pocket. The structure determines that the Ca^{2+}-binding affinity of nPKCs is very low and Ca^{2+} is not required for activation. Drugs targeting the nPKC isoforms are thus more reliable in producing the expected pharmacological impact.

The aPKC isoforms lack both the Ca^{2+}-binding site in the C2 domain and one-half of the C1 homologous domain (atypical C1 domain).[25–27] aPKCs are thus insensitive to Ca^{2+} and not activated by DAG or phorbol esters. But they can be activated by arachidonic acid and ceramide.[28] In addition, direct protein–protein interactions through their PB1 domain can also modulate their activity.[29,30]

All the PKC isoforms, except PKCμ (human) and its murine homologue, PKD, have an N-terminal pseudosubstrate motif near the C1 domain (Fig. 2.1). The pseudosubstrate acts as an autoinhibitory domain that binds to the PKC catalytic domain, thereby maintaining an inactive state. Upon cleavage of the autoregulatory region, the PKC isoforms can be transformed into a persistently active kinase (PKM). For instance, PKCδ can be cleaved by caspase-3 to generate a catalytically active kinase.[31,32] Cleavage of the autoregulatory region is not the only way to generate a PKM. PKMζ can also be generated in neurons by a transcriptional and translational mechanism unique to the ζ gene, *PRKCZ*, which produces the ζ catalytic domain directly through new protein synthesis.[33]

4. PKC ISOFORMS IN SIGNAL TRANSDUCTION

Three major steps are involved in the activation of the PKC isoforms: the binding of required activators, membrane association and translocation, and binding to specific anchoring molecules. The phosphoinositide signaling pathway is one of the major cascades that activate PKC isoforms. Stimulation of certain G-protein-coupled receptors activates phospholipase C (PLC), which hydrolyzes phosphatidylinositol-4,5-bisphosphate to form inositol triphosphate (IP3) and DAG. IP3 binds to intracellular receptors, releasing Ca^{2+} from the endoplasmic reticulum. The combination of the Ca^{2+} wave and DAG simulates the cPKC isoforms, while DAG alone activates the nPKC isoforms. It has been shown that oscillations in Ca^{2+} are directly translated into oscillations of substrate phosphorylation,[34] with a time delay only 10 s. Activation of PKC isoforms is characterized by a rapid translocation (<250 ms)[24] from the cytosol to the plasma/nuclear membrane or cytoskeleton, involving binding to their regulatory domains of intracellular receptors for inactivated kinases (RICKs) and intracellular receptors for activated kinases (RACKs) in the specific translocation of PKCs from the cytosol to different compartments.[35] RACKs bind only activated PKC, but are not necessarily substrates of the enzyme, and the binding does not involve the catalytic domain of the kinase. The RACK binding is, however, required for PKC isoforms to mediate their cellular responses. Inhibition of PKC binding to RACKs *in vivo* has been shown to inhibit PKC translocation and PKC-mediated functions.[36,37] Peptides that mimic either the PKC-binding site on RACKs or the RACK-binding site on PKC isozymes are isozyme-specific translocation inhibitors of PKCs.

PKC may phosphorylate various proteins and enzymes and alter their properties and activities. Mitogen-activated protein kinases (MAPKs), such as ERK, p38, and JNK, are common downstream effectors of PKCs. Activation of PKC modulates Akt signaling and activates cell migration-related molecules, such as focal adhesion kinase, paxillin, and vinculin. Other examples include the plasma membrane ATPase (PMCA) by PKCγ,[38] PLCβs by KCα,[39] phospholipases D (PLD) by cPKCs,[40–43] downregulation of DGKδ and DGKζ by PKCα,[44,45] and activation of adenylate cyclases (AC)2 and AC5 by PKCα.[46,47]

4.1. Synaptic transmission

Synaptic transmission plays an essential role in signal processes in the brain. Functional neuronal synapses require a variety of molecular players. PKC is a well-known regulator of synaptic functions, including the synthesis, vesicle-refilling, and release of neurotransmitters in the cholinergic, γ-aminobutyric acid (GABA)-ergic, dopaminergic, and glutamatergic (Glu) systems.[48–50] PKC activation leads to phosphorylation of GluR 2/3 (at serine 880) in the Purkinje cells. GluR 2/3 phosphorylation appears to be the critical step for parallel fiber long-term depression.[51] Phosphorylation of GluR1 on serine 818 by PKC controls synaptic incorporation of GluR1-containing AMPA receptors into the synapses during long-term potentiation (LTP).[52] Inhibition of PKC activity holds AMPARs at the perisynaptic regions, thus making both LTP and spine expansion labile.[53] PKC changes GABAR functions and their allosteric modulations through phosphorylating several GABAR subunits within their major intracellular domains. Muscarinic activation of G-protein-coupled receptors leads to stimulation of PLC, which cleaves the membrane phospholipids phosphatidylinositol-4,5-bisphosphate to form IP3 and DAG. IP3 and DAG release Ca^{2+} and activate PKC, respectively. PKC activation, on the other hand, enhances acetylcholine release from the hippocampus.[54] Activation of the presynaptic α7 acetylcholine receptors on the Glu terminals in the CA1 region of the rat hippocampus facilitates glutamate release via an action on PKC.[55] In PKCε knockout mice, nicotinic regulation of dopamine release is reduced in the brain reward network,[56] most likely due to a downregulation of α6 nAChR subunit mRNA in the ventral midbrain and striatum.[57] PKC also regulates the expression of genes in mature neurons,[58] as well as activity and cell surface expression of several plasma membrane proteins, such as neurotransmitter transporters (serotonin, dopamine, norepinephrine, glutamate, and GABA), G-protein-coupled receptors, and the Na^+/H^+ antiporter.

4.2. Ion channels

PKC activity affects the operation of many ionic channels. Activation of PKC enhances Ca^{2+} action potentials and decreases voltage-gated Na^+ currents[59–63] through enhancement of intrinsic slow inactivation gating.[62] Several K^+ channels are inhibited by PKC activation. These include the voltage-dependent K^+ currents,[64,65] the Ca^{2+}-activated K^+ currents in the hippocampus, and the delayed rectifier K^+ channel (PKCε).[66]

4.3. Neurite outgrowth

Activity of the PKC isoforms influences the process of neurite outgrowth or the induction of apoptosis. PKC may play an important role in neuronal differentiation and synaptogenesis via stimulating HuD-mediated mRNA stability in the hippocampal neurons,[67] while PKC inhibitors block neurite outgrowth in hippocampal organotypic cultures.[68] Activation of PKC with 12-myristate 13-acetate, an analogue of DAG, induces rapid morphological plasticity and formation of dendritic lamellae in dendrites of hippocampal neurons.[69] PKCε activation leads to an increased expression of brain-derived neurotrophic factor (BDNF), which initiates complex signaling pathways that modify/repair synaptic structure and function.[70] PKCε has been found to be involved in green tea polyphenols precondition-induced protection of neurons against oxygen–glucose deprivation-induced cell death.[71] Apolipoprotein E3 can increase expression of PKCε and thereby protect synapses from Aβ oligomers-induced loss in primary hippocampal neurons in culture.[72] Neurons overexpressing PKMζ, an independent C-terminal domain of PKCζ, exhibit shorter spines, primarily the stubby type.[73] Dynamic changes in spine morphology, which accompany functional plasticity and contribute to the permanence of the changes in synaptic strength, are achieved by F-actin stabilization or destabilization leading to spine enlargement or spine loss, respectively. The brain-specific PACSIN 1 (protein kinase C and casein kinase 2 substrate in neurons 1) is enriched at synapses and has been proposed to affect the formation and maturation of dendritic spines.[74] There is also evidence that astrocytes are active participants in synaptic formation and modification.[75] Local astrocytic contact with cultured rat hippocampal neurons via integrin receptors promotes global synaptogenesis.[76] The astrocyte–neuron contact activates PKC through an arachidonic acid cascade in neurons, triggering excitatory synaptogenesis, a process that can be blocked by inhibitors of both integrins and PKC *in vitro*.[76] Synaptogenesis induced by training and PKC activation has also been demonstrated *in vivo* with rat spatial maze learning.[77,78]

PKC isoforms appear to have a major role in regulating survival signals in a variety of cell types, including neurons. In general, PKCα, β, ε, and ζ function as suppressors of apoptosis,[79] while PKC δ and θ are mainly pro-apoptotic.[80] Plasmalemmal repair/sealing is necessary for survival of damaged neurons, and involves nPKC isoforms.[9] An inhibitor of an nPKC (an nPKCη, pseudosubstrate fragment) has been found to decrease the frequency and rate of plasmalemmal sealing in the B104 hippocampal cells.[81]

Nuclear translocation of PKCζ is believed to mediate excitotoxic death of the cortical neurons, involving caspase-3-dependent PKCζ degradation. Aspirin directly inhibits PKCζ activity, thereby protecting against excitotoxic death of the cortical neurons.[82]

Translocation of PKCε to the mitochondria likely plays an important role in preventing oxygen–glucose deprivation/reperfusion-induced apoptosis or necrosis.[83–85] PKC oxidation is essential in cell signaling and pathology.[86,87] PKCε directly phosphorylates and regulates mitochondrial ATP-sensitive K^+ channels (mitoK^+_{ATP} channels), which are crucial in preserving mitochondrial membrane potential.[88,89]

4.4. Synaptic plasticity

The most widely studied form of synaptic plasticity in mammals is LTP, first characterized in the hippocampus by Lømo[90] and Bliss and Lømo.[91] Activation of PKCs facilitates LTP.[2,92,93] Although enormous efforts have been made to define LTP's role in memory acquisition and long-term maintenance, it remains to be studied whether LTP is necessarily related to memory formation, including the well-established hippocampus-dependent memories, for many observations of a dissociation between them.[94–96] In 2002, however, PKMζ was found to be both necessary and sufficient for maintaining the late-phase LTP in hippocampal slices.[97,98] PKMζ is translated within only 10 min in response to LTP-inducing stimuli,[33,99–103] suggesting its local synthesis. Immediately after translation, the nascent PKMζ is rapidly phosphorylated by PDK-1, which locks the PKMζ in a maximally activated state.[101,104] Once formed, the protein remains active until it is degraded. At the synapse, PKMζ increases the aggregation of postsynaptic density protein 95 (PSD-95)[105,106] and the GluA2 subunit of the AMPAR,[107,108] by a process involving inhibition of endocytosis.[107–109] PSD-95 at synapses is crucial to spine maturation.[110,111] However, when PKMζ inhibitors, such as ZIP (a synthetic, cell-permeable zeta inhibitory peptide of the amino acid sequence of pseudosubstrate), are applied experimentally, the additional postsynaptic AMPARs are actively eliminated and the synapse returns to its naive, basal state.

4.5. Learning and memory

Learning and lasting memories involve activity-generated cellular and synaptic signals that promote gene transcription and protein synthesis in the brain. PKC isoforms play crucial roles in the acquisition of many types of learning and memory. These include Pavlovian conditioning of *Hermissenda*,[1,2] rodent spatial learning,[112–114] conditioned avoidance,[96] learning and memory of eye blink conditioning,[1,115,116] olfactory discrimination learning,[117] contextual fear memory,[118] and drug-associated reward memory.[119–121] PKCα has been proposed to be genetically linked to memory capacity and risk for post-traumatic stress disorder (PTSD) in humans.[122] PKCε activation leads to an enhanced BDNF activity, resulting in an increase expression of the Ca^{2+} release channel isoforms ryanodine receptor 2 and 3 and PKMζ in the hippocampus.[70] Memory task learning is associated with PKC immunoreactivity in the principal hippocampal neurons[123] and stimulation of muscarinic cholinergic receptors is associated with an increase in PKCγ immunoreactivity.[124] The mouse knockout of the PKCγ gene has various CNS defects, including impaired learning and memory,[125,126] reduced neuropathic pain,[127] and decreased anxiety.[128] Aged rats exhibit lower retention of spatial memory, associated with lower densities of mushroom spines and synapses in the apical dendrites of CA1 pyramidal cells of the hippocampus.[129] PKC activation with bryostatin-1 or DCP-LA for 15 days during training can restore mushroom spine synapses and memory in the aged rats to the levels of young ones, thus reversing the synaptic and memory decline in aging.[129] Similar effects of PKC activation on memory performance have also been observed in animals who suffered memory impairments due to neurotoxic amyloid, cerebrovascular deficits (see below), or mental retardation (Sun M-K *et al.*, unpublished observations).

It has been proposed that the persistent activity of PKMζ is the molecular mechanism of lasting memory.[130–132] PKMζ is synthesized *de novo* as a constitutively active kinase. Overexpressing PKMζ enhances memory in flies and rodents.[21,130] Inhibiting PKMζ disrupts long-term memory. The mechanisms involved include persistent inhibition of GluR2-AMPAR removal from the surface of postsynaptic sites[107,108] and/or alterations in spine structure.[73] ZIP effectively blocks the action of PKMζ both biochemically *in vitro* and within neurons[133] and reverses LTP *in vivo* 1 day after induction and erases the memory for long-term sensitization of the siphon-withdrawal reflex in *Aplysia* as late as 7 days after training.[134] It has been shown that locally inhibiting PKMζ rapidly reverses long-term memories, including spatial long-term memory,[135–137] trace memories in the hippocampus,[138] olfactory avoidance learning in *Drosophila*,[21] active memories in the basolateral amygdala,[108,136,139–141] appetitive memories in the nucleus accumbens,[119,142,143] habit memory in the dorsal lateral striatum,[144] and elementary associations,[129,145–147] extinction memories of reward-related and aversive cues,[148] and skilled sensorimotor memories[149] in the neocortex, several days or weeks after training,[147,150] or for at least 1 month in rats.[145] If one considers that the half-life of intracranially injected ZIP is ~2 h and that it is virtually cleared from the brain within a day,[141] it is rather interesting that the disruption of previously stored memory by ZIP lasts far longer and appears permanent.[109] Even after ZIP is eliminated, the information encoded as which synapses originally contained PKMζ cannot be recovered. After ZIP has cleared, new memories can again be formed,[135,149,150] and even erased again by ZIP.[146] These observations suggest that transiently inhibiting PKMζ erases the long-term memory trace but does not damage the neural networks. Some forms of memory are, however, found not to be erased by inhibiting PKMζ. These include short-term memories mediated by the hippocampus[150] and neocortex,[146] and certain long-term memories, such as latent inhibition and attenuation of neophobia.[145]

The downregulation of PKMζ has recently been observed during memory extinction and reconsolidation blockade.[151] Decreases in PKMζ have also been observed in the hippocampus with spatial familiarity.[152] Abnormal aggregations of PKMζ are also observed in and near neurofibrillary tangles in the brains of individuals with Alzheimer's disease (AD).[153]

If information is encoded as the presence or absence of PKMζ at specific synapses, interrupting the activity of PKMζ should effectively erase this information.[109,131] Recent studies, however, cast doubt on the importance of PKMζ in the maintenance of long-term memory. First, PKMζ is not required for hippocampal synaptic plasticity at Schaffer collateral-CA1 synapses, and learning and memory.[154] The mice lacking both PKCζ and PKMζ (*Prkcz*$^{-/-}$ mice) do not show deficits in learning or memory in tests of cued fear conditioning, novel object recognition, object location recognition, conditioned place preference for cocaine, or motor learning, when compared with their wild-type littermates.[155] The same studies also show that ZIP can erase reward learning even when PKMζ is not present,[154,155] suggesting that effects of ZIP on memory maintenance occur through PKMζ-independent mechanisms, such as decreased synaptic transmission in nontetanized control slices by ZIP.[154]

5. PKC INHIBITORS

Because PKC isoforms have unique and sometimes opposing roles, the use of isoform-selective inhibitors is essential. Like PKC activators, developing PKC inhibitors that affect the function of only one PKC isozyme has proven to be difficult, because of the substantial homology among PKC isoforms.[156] Nevertheless, several types of PKC inhibitors are currently available.[157]

The kinase ATP-binding site is a narrow hydrophobic pocket between two lobes and linked by a flexible hinge region. ATP-binding site inhibitors bind in this hydrophobic pocket and usually form a number of hydrogen bonds to the hinge region. This hydrophobic pocket is conserved and as such poor selectivity is a common feature with this class of inhibitors. They can inhibit a number of other protein kinases at higher concentrations.[158] Staurosporine, for instance, is a highly potent PKC inhibitor and has poor selectivity. Chemical modification of the indolocarbazole core structure has lead to the next generation of synthetic bisindole maleimides, which show some improved selectivity within the PKC isoforms. For example, ruboxistaurin (LY333531) and enzastaurin (LY317615) are selective low nanomolar PKCβ inhibitors. Other available ATP-binding site inhibitors of PKCβ include LY379196 and CGP53353. Rottlerin is used as an ATP-binding site competitive inhibitor of PKCδ.[159–161] It has been shown to block other kinase and nonkinase proteins *in vitro*, to activate multiple Ca^{2+}-sensitive K^+ channels with high potency, and to uncouple mitochondria.[162] Twenty millimolars rottlerin produced a substantial inhibition of c-Jun N-terminal kinase 1 a1 (JNK1 a1), mitogen- and stress-activated protein kinase 1 (MSK-1), cAMP-dependent protein kinase (PKA), 3-phosphoinositide-dependent protein kinase 1 (PDK-1), protein kinase B a (PKB a, also called Akt), and glycogen synthase kinase 3β (GSK3β).[163] Spheciosterol sulfate C is a small molecule, which can selectively inhibit PKCζ.[164]

Chelerythrine is one small molecule inhibitor, competitive for the protein substrate-binding pocket.[165]

There is one class of functional inhibitor that act via the C1a/b binding domains. Calphostin C[166] is a C1 domain antagonist that requires light activation for its function.[167]

It is well established that PKC isoform inhibition, like PKC deficits, impairs learning and memory.[96,168] Since PKC is involved in mood regulation and adult cell proliferation, several inhibitors, such as tamoxifen and chelerythrine, have been found to cause depressive-like behavior in forced swim test.[169] However, PKC inhibitors are more than just research tools. These compounds may find their use in mania or certain phase of bipolar disorder.[170] Inhibiting PKC isoforms could provide a strategy for the treatment of comorbid nicotine and alcohol addictions, substance abuse, fear, and PTSD.

PKCδ inhibitors may also be used for the treatment of vascular dementia. Cerebral ischemia increases PKCδ mRNA and protein levels in the cortex and hippocampus.[171,172] Activation of PKCδ by cerebral ischemia results in cytochrome *c* release from the mitochondria and apoptosis.[173] The increased PKCδ expression in the penumbral area may be responsible for delayed neuronal damage.[174] A selective PKCδ peptide inhibitor, for example, has been found to reduce cellular injury in a rat hippocampal slice model of cerebral ischemia. The inhibitor decreased infarct size *in vivo* in rats with transient middle cerebral artery occlusion when administered at the onset, at 1 h, or at 6 h of reperfusion.[175]

6. PKC ACTIVATORS

PKC activators, such as DAG, arachidonic acid, phorbol esters, aplysiatoxins, teleocidins, and bryostatins, bind to a hydrophilic cleft in a largely hydrophobic surface of the C1 domains, resulting in an enhanced hydrophobicity of the surface and promoting the interaction between the C1 domain and the phospholipid bilayer of the cell membranes. 8-[2-(2-Pentylcyclopropyl-methyl)-cyclopropyl]-octanoic acid (DCP-LA), on the other hand, has been shown to selectively activate PKCε, through binding to the phosphatidylserine binding site.[176,177] The interaction drives removal of the pseudosubstrate region from the catalytic site of the enzyme. When dosed appropriately, these activators may produce memory-enhancing and antidementia effects.

6.1. Memory-enhancing

Activation of PKCε and/or PKCα enhances learning and memory.[77,113] Bryostatin-1 (100 and 200 nM) has been found to strongly facilitate LTP induction in Schaffer-collateral fibers of the hippocampus to single-theta burst stimulation.[93] The effects are blocked by specific inhibitors of PKCα (Ro 32-0432 at 20 nM) and PKCε (TAT-εV1-2 peptide at 500 nM), not of PKCδ (rottlerin at 1 μM).[93] Increasing PKMζ activity has also been found to enhancing memory in flies and rodents.[21,130]

6.2. Restoring synaptic and network functions

Activation of PKCε and PKCα isoforms increases the expression and release of BDNF in the hippocampus.[78,178] The activation is associated with synaptogenesis, synaptic remodel and repair, resulting in a restoration of hippocampus-dependent cognitive functions.[78,179]

Synaptogenesis, synaptic remodel and repair, requires integrity of mitochondria. A major component of the neuroprotective mechanism of PKCε is mediated by regulation of mitochondrial function. PKCε phosphorylates cytochrome *cIV* in rat hippocampal synaptosomes.[85] Selective activation of PKCε in mitochondria with either general PKCε activator peptide, TAT-ΨεRACK, or mitochondrial-selective PKCε activator, TAT-ΨεHSP90, has been shown neuroprotective in primary brain cell cultures and reducing focal ischemic brain injury in mice with TAT-ΨεHSP90.[180]

6.3. Antidementia effects

Dementia includes AD, vascular dementia, frontotemporal dementia, and dementia with Lewy bodies and currently has no curative therapeutics.

AD, characterized as a disease of cognitive deficits, is the most common form of dementia among the elderly population. It is associated with an abnormal processing of amyloid precursor protein (APP) and accumulation of neurite plaques containing amyloid β-peptide (Aβ) in the brain. Reduction in amyloid plaque load in the brain is accompanied by an improvement in LTP and cognitive functions.[181] Aβ reduces PKC isozyme levels[182,183] (Fig. 2.2). Aβ contains a putative PKC pseudosubstrate domain and can directly inhibit PKC isoforms, including PKCα and PKCε.[184] Through binding to PKC, Aβ blocks PKC activation and induces PKC degradation,[185] reduces PKC-mediated phosphorylation,[183] and decreases PKC membrane translocation.[183] Synaptotoxic Aβ oligomers inhibit PKC isoforms, decrease RyR2 protein expression, and block BDNF-induced RyR-dependent spine remodeling in hippocampal neurons.[186,243] Aβ oligomers also decrease RACK1 distribution in the membrane fraction of cortical neurons.[187]

APP is mainly cleaved by α-secretase,[188,189] which is most likely the ADAM10 (a disintegrin and metalloprotease 10).[190,191] The cleavage produces a large extracellular soluble APP fragment (sAPPα) and a smaller membrane-bound intracellular fragment, C83, which is soluble and non-toxic. α-Secretase cleavage occurs at the plasma membrane,[192] but can also occur in the trans-Golgi network.[193] PKC and MAP kinase play a central role in stimulating α-secretase cleavage of APP. However, when APP is cleaved by β- and γ-secretase enzymes, it leads to the formation of Aβ1–40 and Aβ1–42, which, unlike sAPP, are insoluble and prone to aggregation, forming oligomers and plaques. β-Secretase cleavage of wild-type APP occurs mainly in the endosome and to a lower extent in the trans-Golgi network.[194–196]

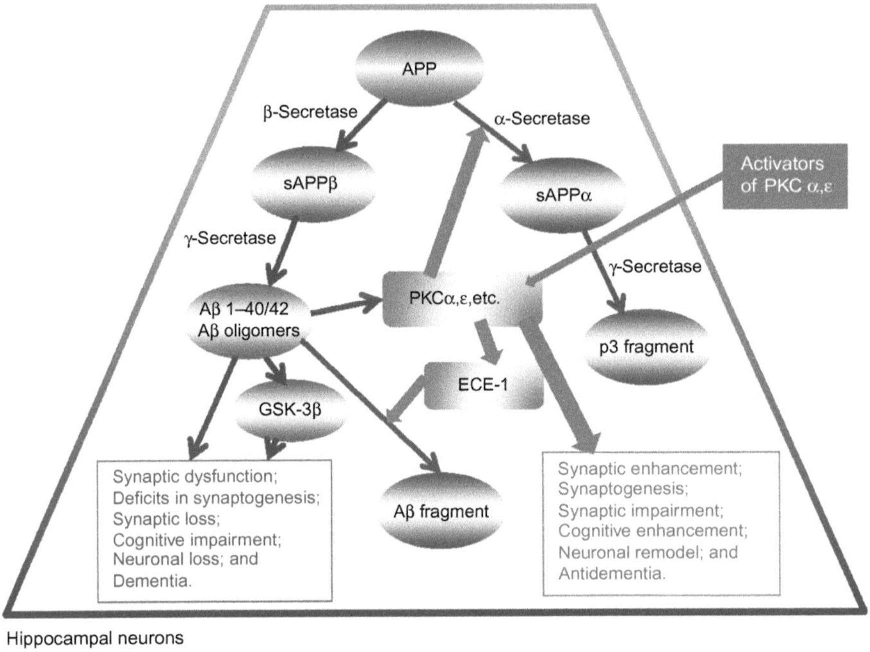

Figure 2.2 Schematic summary of interaction between Aβ production/clearance and PKC isoform-specific activators. PKC isoform-specific activators produce antidementia effects through antagonism of the neurotoxic effects of amyloids on PKC, activating α-secretase and endothelin-converting enzyme (ECE). (See the color plate.)

The Aβ hypothesis of AD pathogenesis has facilitated a strong hope that the ability to halt or reverse AD should be possible. As the generation of Aβ is at the beginning of the cascade, the mechanisms of Aβ generation and Aβ prevention are the target of intensive research both in academia and in pharmaceutical companies. However, results from large clinical trials have been disappointing thus far, since patients with dramatic clearance of amyloid showed no clear change in clinical course.[197–199] Patients who received a γ-secretase inhibitor in a recent clinical trial also showed apparently worse cognitive functions.[200]

PKCα and ε isoforms, on the other hand, activate α-secretase[201–205] and Aβ degradation,[177,206] thus promoting the production of sAPPα and reducing the accumulation of pathogenic Aβ levels in the brain. PKCε activates the Aβ degrading activity of endothelin-converting enzyme type 1 (ECE-1), which might be mediated via the MAPK pathway as well. Overexpression of human PKCε reduces Aβ levels significantly in the brain.[204] In APP transgenic mice, overexpression of PKCε selectively increases the activity of ECE,[205,206] which plays a key role in degrading Aβ in the brains[207,208] and is expressed in several populations of neurons including the hippocampal cells.[207,209] Bryostatin-1 has been found to reduce Aβ levels.[210,211] Furthermore, PKC activation inhibits glycogen synthase 3 kinase,[212] thereby reducing tau protein hyperphosphorylation.[213] Tau is a microtubule-associated protein typically found in the axon of neurons and involved in microtubule assembly and the stabilization of growth axons.[214] The hyperphosphorylation of tau prevents its binding to taxol-stabilized microtubules and disrupts microtubule assembly from tau and tubulin.[215]

PKC activators may also find their therapeutic values in vascular dementia, including those memory impairments from acute stroke. PKCε mediates ischemic tolerance.[216] Activation of PKCε, as a vital part of adenosine/NMDA-activated signal transduction pathway, protects neurons from ischemia–reperfusion injury.[39,217,218] PKCε phosphorylates and inhibits GSK3β, the inhibition of which during reperfusion promotes glycogen synthesis, thus decreasing glycolysis and associated harmful H^+ production during reperfusion.[219] Ischemic preconditioning is associated with PKCε-mediated phosphorylation of the mitochondrial K^+_{ATP} channels[218] and increased synaptosomal mitochondrial respiration.[85] PKCε knockout mice lose the preconditioning effect of ischemia–reperfusion.[218] Electroacupuncture pretreatment has been shown to produce PKCε-mediated antiapoptosis and rapid tolerance to focal cerebral ischemia (MCAO) in rats.[220] Postischemic activation in rats with intermittent doses of bryostatin-1, which is a relatively specific activator of PKCε, has been found to restore neurotrophic activity and synaptogenesis in the hippocampus and spatial learning and memory performance after global cerebral ischemia.[178,179] Thus, an appropriate activation of PKC isoforms with targeted PKC activators may represent an effective therapeutic approach to stroke/ischemia–reperfusion injury and associated memory impairment, through activation of ischemic preconditioning responses and enhancement of neurotrophic activity, synaptogenesis, and synaptic remodeling. The effects of PKCε activators on ischemic impairment of cognition may partially result from an indirect action through its role as an upstream regulator of endothelial cytoprotection, exerting combined anti-apoptotic and anti-inflammatory effects.[221] PKCε inhibits GSK3β.[222]

7. CONCLUSIONS

Our understanding of how memories are maintained is still relatively primitive. Impacts of PKCε,α activation on memory performance are derived through an enhanced memory efficiency, while antidementia effects are achieved via synaptogenesis, synaptorepair, and restriction on Aβ accumulation in AD brains. The downside of such agents also needs consideration. ADAM10, for instance, sheds over 30 different membrane proteins besides APP, such as Notch, betacellulin, klotho, and N-cadherin.[223–225] The discovery that hippocampal neurogenesis in AD relies on Aβ oligomers, not monomers or fibrils, raises the questions how better to treat AD.[226]

In addition, enhancing memory may not always be beneficial. Memory is also involved in fear and traumatic events. Enhancing memory mechanisms may lead to increased risk for PTSD and fear memories. PKC activity is increased in the prefrontal cortex following exposure to even mild, uncontrollable stress. PKCζ is a determinant of metabolic plasticity deficiency, allowing glucose-addicted human cancer cells to reprogram their metabolism in response to glucose deprivation.[227] There is compelling evidence that the amygdala plays an important role in enhancing the formation of memory for emotionally arousing information[228] and that the hippocampus is crucial for binding memory items to their context and for the successful transfer of this information into long-term memory.[229,230] Emotional memory involves PKC.[231] PKCα, for instance, is proposed to be involved in aversive memories, underlying a genetic link between memory capacity and the risk for PTSD.[122] PKMζ, on the other hand, has been found to maintain aversion memory in the bilateral amygdala.[146]

Another issue with PKC activation is persistent pain. Full-length PKCs contribute to the induction of pain hypersensitivity and hyperalgesia.[232–234] PKCε-mediated phosphorylation of $Na_v1.8$ at S1452 increases the sodium channel $Na_v1.8$ currents in nociceptors and produces mechanical hyperalgesia in mice.[235] PKMζ is essential for its maintenance by sustaining spinal nociceptive plasticity.[236] Thus, spinal PKMζ inhibition prevented the enhanced response to subsequent stimuli.

A third issue is substance abuse, that is, formation and maintenance of substance reward-related memory.[121,146,237] Morphine-induced reward memory, however, may involve the PKCγ isoform.[121] The protein levels of PKCγ, but not PKCα, βI, βII, and/or ε, were significantly upregulated in membrane functions of the limbic forebrain obtained from morphine-conditioned mice.[238] An intracerebroventricular injection of calphostin C, a PKC inhibitor, produced a dose-dependent reduction in the morphine-induced place preference.[238] PKC isoforms, especially the γ isoform, are essential for the acquisition and maintenance of morphine-associated reward memory.[121] PKMζ level is increased by single or five consecutive cocaine injections and PKMζ inhibition by chelerythrine or myristoylated ζ inhibitory peptide significantly attenuated the cocaine exposure-induced enhancement of AMPAR/NMDAR ratio.[237] Intracranial infusion of myristoylated ζ inhibitory peptide in the VTA disrupted cocaine-conditioned place preference.[237] Mice lacking PKCε show decreased nicotine self-administration and fail to develop a conditioned place preference for nicotine.[56]

Nevertheless, for several reasons, PKC isoforms may be targeted for achieving a better antidementia effect than other factors. First, PKC is activated by synaptic inputs and intracellular signals that are involved in information processing in cognition, including Glu inputs,[239] cholinergic inputs,[61] serotonergic inputs,[59,60] dopaminergic inputs,[240] intracellular calcium and DAG elevation, and other hormonal stimulation.[241] Second, functional deficits of PKC isoforms are involved in memory disorders.[186,242] Third, PKCε,α isoforms promote synaptic repair and synaptogenesis and reduce/block memory-related pathogenesis.

REFERENCES

1. Alkon DL, Nelson TJ, Zhao WQ, Cavallaro S. Time domains of neuronal Ca^{2+} signaling and associative memory: steps through a calexcitin, ryanodine receptor, K^+ channel cascade. *Trends Neurosci.* 1998;21:529–537.
2. Alkon DL, Sun MK, Nelson TJ. PKC signaling deficits: a mechanistic hypothesis for the origins of Alzheimer's disease. *Trends Pharmacol Sci.* 2007;28:51–60.
3. Lipp P, Reither G. Protein kinase C: the "master" of calcium and lipid. *Cold Spring Harb Perspect Biol.* 2011;3:a004556.
4. Ohno S, Kawasaki H, Imajoh S, Suzuki K. Tissue-specific expression of three distinct types of rabbit protein kinase C. *Nature.* 1987;325:161–166.
5. Breitkreutz D, Braiman-Wiksman L, Daum N, Denning MF, Tennenbaum T. Protein kinase C family: on the crossroads of cell signaling in skin and tumor epithelium. *J Cancer Res Clin Oncol.* 2007;11:793–808.
6. Wetsel WC, Khan WA, Merchenthaler I, et al. Tissue and cellular distribution of the extended family of protein kinase C isoenzymes. *J Cell Biol.* 1992;117:121–133.
7. Gong JL, Hoyos B, Acin-Perez R, et al. Two protein kinase C isoforms, δ and ε, regulate energy homeostasis in mitochondria by transmitting opposing signals to the pyruvate dehydrogenase complex. *FASEB J.* 2012;26:3537–3549.
8. Dutil EM, Toker A, Newton AC. Regulation of conventional protein kinase C isozymes by phosphoinositide-dependent kinase 1 (PDK-1). *Curr Biol.* 1998;8:1366–1375.
9. Le Good JA, Ziegler WH, Parekh DB, Alessi DR, Cohen P, Parker PJ. Protein kinase C isotypes controlled by phosphoinositide 3-kinase through the protein kinase PDK1. *Science.* 1998;281:2042–2045.
10. Toker A, Newton AC. Cellular signaling: pivoting around PDK-1. *Cell.* 2000;2:185–188.
11. Edwards AS, Newton AC. Phosphorylation at conserved carboxyl-terminal hydrophobic motif regulates the catalytic and regulatory domains of protein kinase C. *J Biol Chem.* 1997;272:18382–18390.

12. Facchinetti V, Ouyang W, Wei H, et al. The mammalian target of rapamycin complex 2 controls folding and stability of Akt and protein kinase C. *EMBO J.* 2008;27:1932–1943.
13. Ikenoue T, Inoki K, Yang Q, Zhou X, Guan KL. Essential function of TORC2 in PKC and Akt turn motif phosphorylation, maturation and signaling. *EMBO J.* 2008;27:1919–1931.
14. Hansra G, Bornancin F, Whelan R, Hemmings BA, Parker PJ. 12-*O*-tetradecanoylphorbol-13-acetate-induced dephosphorylation of protein kinase Cα correlates with the presence of a membrane-associated protein phosphatase 2A heterotrimer. *J Biol Chem.* 1996;271:32785–32788.
15. Hansra G, Garcia-Paramio P, Prevostel C, Whelan RD, Bornancin F, Parker PJ. Multisite dephosphorylation and desensitization of conventional protein kinase C isotypes. *Biochem J.* 1999;342:337–344.
16. Newton AC. Protein kinase C. Poised to signal. *Am J Physiol Endocrinol Metab.* 2010;298:E395–E402.
17. Keranen LM, Dutil EM, Newton AC. Protein kinase C is regulated *in vivo* by three functionally distinct phosphorylations. *Curr Biol.* 1995;5:1394–1403.
18. Gysin S, Imber R. Phorbol-ester-activated protein kinase C-α lacking phosphorylation at Ser-657 is down-regulated by a mechanism involving dephosphorylation. *Eur J Biochem.* 1997;249:156–160.
19. Gao T, Brognard J, Newton AC. The phosphatase PHLPP controls the cellular levels of protein kinase C. *J Biol Chem.* 2008;283:6300–6311.
20. Nalefski EA, Newton AC. Membrane binding kinetics of protein kinase C bII mediated by the C2 domain. *Biochemistry.* 2001;40:13216–13229.
21. Dries DR, Newton AC. Kinetic analysis of the interaction of the C1 domain of protein kinase C with lipid membranes by stopped-flow spectroscopy. *J Biol Chem.* 2008;283:7885–7893.
22. Scott AM, Antal CE, Newton AC. Electrostatic and hydrophobic interactions differentially tune membrane binding kinetics of the C2 domain of protein kinase Cα. *J Biol Chem.* 2013;288:16905–16915.
23. Kohout SC, Corbalan-Garcia S, Torrecillas A, Gomez-Fernandez JC, Falke JJ. C2 domains of protein kinase C isoforms α, β, and γ: activation parameters and calcium stoichiometries of the membrane-bound state. *Biochemistry.* 2002;41:11411–11424.
24. Reither G, Schaefer M, Lipp P. PKCα: a versatile key for decoding the cellular calcium toolkit. *J Cell Biol.* 2006;174:521–533.
25. Mott HR, Carpenter JW, Zhong S, Ghosh S, Bell RM, Campbell SL. The solution structure of the Raf-1 cysteine-rich domain: a novel ras and phospholipid binding site. *Proc Natl Acad Sci USA.* 1996;93:8312–8317.
26. Zhou M, Horita DA, Waugh DS, Byrd RA, Morrison DK. Solution structure and functional analysis of the cysteine-rich C1 domain of kinase suppressor of Ras (KSR). *J Mol Biol.* 2002;315:435–446.
27. Colon-Gonzalez F, Kazanietz MG. C1 domains exposed: from diacylglycerol binding to protein–protein interactions. *Biochim Biophys Acta.* 2006;1761:827–837.
28. Hirai T, Chida K. Protein kinase Cz (PKCz): activation mechanisms and cellular functions. *J Biochem.* 2003;133:1–7.
29. Puls A, Schmidt S, Grawe F, Stabel S. Interaction of protein kinase C zeta with ZIP, a novel protein kinase C-binding protein. *Proc Natl Acad Sci USA.* 1997;94:6191–6196.
30. Qiu RG, Abo A, Steven Martin G. A human homolog of the *C. elegans* polarity determinant Par-6 links Rac and Cdc42 to PKCz signaling and cell transformation. *Curr Biol.* 2000;10:697–707.
31. Emoto Y, Manome Y, Meinhardt G, et al. Proteolytic activation of protein kinase C delta by an ICE-like protease in apoptotic cells. *EMBO J.* 1995;14:6148–6156.

32. Kanthasamy AG, Kitazawa M, Kanthasamy A, Anantharam V. Role of proteolytic activation of protein kinase Cδ in oxidative stress-induced apoptosis. *Antioxid Redox Signal.* 2003;5:609–620.
33. Hernandez AI, Blace N, Crary JF, et al. Protein kinase Mζ synthesis from a brain mRNA encoding an independent protein kinase Cζ catalytic domain. Implications for the molecular mechanism of memory. *J Biol Chem.* 2003;278(41):40305–40316.
34. Violin JD, Zhang J, Tsien RY, Newton AC. A genetically encoded fluorescent reporter reveals oscillatory phosphorylation by protein kinase C. *J Cell Biol.* 2003;161:899–909.
35. Mochly-Rosen D, Gordon AS. Anchoring proteins for protein kinase C: a means for isozyme selectivity. *FASEB J.* 1998;1:35–42.
36. Johnson JA, Gray MO, Chen CH, Mochly-Rosen D. A protein kinase C translocation inhibitor as an isozyme-selective antagonist of cardiac function. *J Biol Chem.* 1996;271:24962–24966.
37. Smith BL, Mochly-Rosen D. Inhibition of protein kinase C function by injection of intracellular receptor for the enzyme. *Biochem Biophys Res Commun.* 1992;188:1235–1240.
38. Wang KK, Wright LC, Machan CL, Allen BG, Conigrave AD, Roufogalis BD. Protein kinase C phosphorylates the carboxyl terminus of the plasma membrane Ca^{2+}-ATPase from human erythrocytes. *J Biol Chem.* 1991;266:9078–9085.
39. Xu J, Lan D, Yang G, Li T, Liu LM. Hemorrhagic preconditioning improves vascular reactivity after hemorrhagic shock by activation of PKCα and PKCε via the adenosine A_1 receptors in rats. *J Trauma Acute Care Surg.* 2013;74:1266–1274.
40. Hornia A, Lu Z, Sukezane T, et al. Antagonistic effects of protein kinase C α and δ on both transformation and phospholipase D activity mediated by the epidermal growth factor receptor. *Mol Cell Biol.* 1999;19:7672–7680.
41. Han JM, Kim Y, Lee JS, et al. Localization of phospholipase D1 to caveolin-enriched membrane via palmitoylation: implications for epidermal growth factor signaling. *Mol Biol Cell.* 2002;13:3976–3988.
42. Chen JS, Exton JH. Regulation of phospholipase D2 activity by protein kinase Ca. *J Biol Chem.* 2004;279:22076–22083.
43. Lee CS, Kim KL, Jang JH, Choi YS, Suh PG, Ryu SH. The roles of phospholipase in EGFR signaling. *Biochim Biophys Acta.* 2009;1791:862–868.
44. Luo B, Prescott SM, Topham MK. Protein kinase C α phosphorylates and negatively regulates diacylglycerol kinase ζ. *J Biol Chem.* 2003;278:39542–39547.
45. Imai S, Kai M, Yamada K, Kanoh H, Sakane F. The plasma membrane translocation of diacylglycerol kinase δ1 is negatively regulated by conventional protein kinase C-dependent phosphorylation at Ser-22 and Ser-26 within the pleckstrin homology domain. *Biochem J.* 2004;382:957–966.
46. Kawabe J, Iwami G, Ebina T, et al. Differential activation of adenylyl cyclase by protein kinase C isoenzymes. *J Biol Chem.* 1994;269:16554–16558.
47. Zimmermann G, Taussig R. Protein kinase C alters the responsiveness of adenylyl cyclases to G protein α and βγ subunits. *J Biol Chem.* 1996;271:27161–27166.
48. Stevens CF, Sullivan JM. Regulation of the readily releasable vesicle pool by protein kinase C. *Neuron.* 1998;21:885–893.
49. Dobransky T, Doherty-Kirby A, Kim A, Brewer D, Lajoie G, Rylett RJ. Protein kinase C isoforms differentially phosphorylate human choline acetyltransferase regulating its catalytic activity. *J Neurochem.* 2004;279:52059–52068.
50. Okada M, Zhu G, Yoshida S, Hirose S, Kaneko S. Protein kinase associated with gating and closing transmission mechanisms in temporoammonic pathway. *Neuropharmacology.* 2004;47:485–504.

51. Rekart JL, Meiri K, Routtenberg A. Hippocampal-dependent memory is impaired in heterozygous GAP-43 knockout mice. *Hippocampus.* 2005;15:1–7.
52. Boehm J, Kang M-G, Johnson RC, Esteban J, Huganir RL, Malinow R. Synaptic incorporation of AMPA receptors during LTP is controlled by a PKC phosphorylation site on GluR1. *Neuron.* 2006;51:213–225.
53. Yang Y, Wang XB, Zhou Q. Perisynaptic GluR2-lacking AMPA receptors control the reversibility of synaptic and spine modifications. *Proc Natl Acad Sci USA.* 2010;107:11999–112004.
54. Chaki S, Muramatsu M, Otomo S. Involvement of protein kinase C activation of acetylcholine release from rat hippocampal slices by minaprine. *Neurochem Int.* 1994;24:37–41.
55. Yamamoto S, Kanno T, Nagata T, Yaguchi T, Tanaka A, Nishizaki T. The linoleic acid derivative FR236924 facilitates hippocampal synaptic transmission by enhancing activity of presynaptic alpha7 acetylcholine receptors on the glutamatergic terminals. *Neuroscience.* 2005;130:207–213.
56. Lee AM, Messing RO. Protein kinase C epsilon modulates nicotine consumption and dopamine reward signals in the nucleus accumbens. *Proc Natl Acad Sci USA.* 2011;108:16080–16085.
57. Exley R, Clements MA, Hartung H, McIntosh JM, Cragg SJ. α6-Containing nicotinic acetylcholine receptors dominate the nicotine control of dopamine neurotransmission in nucleus accumbens. *Neuropsychopharmacology.* 2008;33:2158–2166.
58. Roberson ED, English JD, Adams JP, Selcher JC, Kondratick C, Sweatt JD. The mitogen-activated protein kinase cascade couples PKA and PKC to cAMP response element binding protein phosphorylation in area CA1 of hippocampus. *J Neurosci.* 1999;19:4337–4348.
59. Carr DB, Cooper DC, Ulrich SL, Spruston N, Surmeier DJ. Serotonin receptor activation inhibits sodium current and dendritic excitability in prefrontal cortex via a protein kinase C-dependent mechanism. *J Neurosci.* 2002;22:6846–6855.
60. Carr DB, Day M, Cantrell AR, et al. Transmitter modulation of slow, activity-dependent alterations in sodium channel availability endows neurons with a novel form of cellular plasticity. *Neuron.* 2003;39:793–806.
61. Chen Y, Cantrell AR, Messing RO, Scheuer T, Catterall WA. Specific modulation of Na^+ channels in hippocampal neurons by protein kinase C epsilon. *J Neurosci.* 2005;25:507–513.
62. Chen Y, Yu FH, Surmeier J, Scheuer T, Catterall WA. Neuromodulation of Na^+ channel slow inactivation via cAMP-dependent protein kinase and protein kinase C. *Neuron.* 2006;49:409–420.
63. González M, Kazanietz MG, Robinson MB. Regulation of the neuronal glutamate transporter excitatory amino acid carrier-1 (EAAC1) by different protein kinase C subtypes. *Mol Pharmacol.* 2002;62:901–910.
64. Alkon DL, Neary JT, Naito S, Coulter D, Kubota M, Rasmussen H. C-kinase activation prolongs CA-dependent inactivation of K currents. *Biochem Biophys Res Commun.* 1986;134:1245–1253.
65. Farley J, Auerbach S. Protein kinase C activation induces conductance changes in Hermissenda photoreceptors like those seen in associative learning. *Nature.* 1986;319:220–223.
66. Song CY, Xi HJ, Yang L, et al. Propofol inhibited the delayed rectifier potassium current (Ik) via activation of protein kinase C epsilon in rat parietal cortical neurons. *Eur J Pharmacol.* 2011;653:16–20.
67. Lim CS, Alkon DL. Protein kinase C stimulates HuD-mediated mRNA stability and protein expression of neurotrophic factors and enhances dendritic maturation of hippocampal neurons in culture. *Hippocampus.* 2012;22:2303–2319.

68. Toni N, Stoppini L, Muller D. Staurosporine but not chelerythrine inhibits regeneration in hippocampal organotypic cultures. *Synapse*. 1997;27:199–207.
69. Pilpel Y, Segal M. Activation of PKC induces rapid morphological plasticity in dendrites of hippocampal neurons via Rac and Rho-dependent mechanisms. *Eur J Neurosci*. 2004;19:3151–3164.
70. Adasme T, Haeger P, Paula-Lima AC, et al. Involvement of ryanodine receptors in neurotrophin-induced hippocampal synaptic plasticity and spatial memory formation. *Proc Natl Acad Sci USA*. 2011;108:3029–3034.
71. Gundimeda U, McNeill TH, Elhiani AA, Schiffman JE, Hinton DR, Gopalakrishna R. Green tea polyphenols precondition against cell death induced by oxygen–glucose deprivation via stimulation of laminin receptor, generation of reactive oxygen species, and activation of protein kinase Cε. *J Biol Chem*. 2012;287:34694–34708.
72. Sen A, Alkon DL, Nelson TJ. Apolipoprotein E3 (ApoE3) but not ApoE4 protects against synaptic loss through increased expression of protein kinase Cε. *J Biol Chem*. 2012;287:15947–15958.
73. Ron S, Dudai Y, Segal M. Overexpression of PKMζ alters morphology and function of dendritic spines in cultured cortical neurons. *Cereb Cortex*. 2012;22:2519–2528.
74. Schael S, Nuchel J, Muller S, Petermann P, Kormann J, Perez-Otano I. Casein kinase 2 phosphorylation of protein kinase C and casein kinase 2 substrate in neurons (PACSIN) 1 protein regulates neuronal spine formation. *J Biol Chem*. 2013;288:9303–9312.
75. Haydon PG. Glia: listening and talking. *Nat Rev Neurosci*. 2001;2:185–193.
76. Hama H, Hara C, Yamaguchi K, Miyawaki A. PKC signaling mediates global enhancement of excitatory synaptogenesis in neurons triggered by local contact with astrocytes. *Neuron*. 2004;41:405–415.
77. Hongpaisan J, Alkon DL. A structural basis for enhancement of long-term associative memory in single dendritic spines regulated by PKC. *Proc Natl Acad Sci USA*. 2007;104:19571–19576.
78. Hongpaisan J, Sun M-K, Alkon DL. PKCε activation prevents synaptic loss, Aβ elevation, and cognitive deficits in Alzheimer's disease transgenic mice. *J Neurosci*. 2011;31:630–643.
79. Khadra N, Bresson-Bepoldin L, Penna A, et al. CD95 triggers orai1-mediated localized Ca^{2+} entry, regulates recruitment of protein kinase C (PKC) β2, and prevents death-inducing signaling complex formation. *Proc Natl Acad Sci USA*. 2011;108:19072–19077.
80. Basu A, Pal D. Two faces of protein kinase Cδ: the contrasting roles of PKCδ in cell survival and cell death. *Sci World J*. 2010;10:2272–2284.
81. Spaeth CS, Boydston EA, Figard LR, Zuzek A, Bittner GD. A model for sealing plasmalemmal damage in neuron and other eukaryotic cells. *J Neurosci*. 2010;30:15790–15800.
82. Crisanti P, Leon A, Lim DM, Omri B. Aspirin prevention of NMDA-induced neuronal death by direct protein kinase Cζ inhibition. *J Neurochem*. 2005;93:1587–1593.
83. Baines CP, Zhang J, Wang GW, et al. Mitochondrial PKCepsilon and MAPK form signaling modules in the murine heart: enhanced mitochondrial PKCepsilon-MAPK interactions and differential MAPK activation in PKCepsilon-induced cardioprotection. *Circ Res*. 2002;8:390–397.
84. Kim E, Raval AP, Defazio RA, Perez-Pinzon MA. Ischemic preconditioning via ε protein kinase C activation requires cyclooxygenase-2 activation *in vitro*. *Neuroscience*. 2007;145:931–941.
85. Dave KR, DeFazio RA, Raval AP, et al. Ischemic preconditioning targets the respiration of synaptic mitochondria via protein kinase Cε. *J Neurosci*. 2008;28:4172–4182.
86. Cosentino-Gomes D, Rocco-Machado N, Meyer-Fernandes JR. Cell signaling through protein kinase C oxidation and activation. *Int J Mol Sci*. 2012;13:10697–10721.

87. Huang YZ, McNamara JO. Neuroprotective effects of reactive oxygen species mediated by BDNF-independent activation of TrkB. *J Neurosci*. 2012;32:15521–15532.
88. Ardehali H. Signaling mechanisms in ischemic preconditioning. Interaction of PKCε and MitoK(ATP) in the inner membrane of mitochondria. *Circ Res*. 2006;99:798–800.
89. Raval AP, Dave KR, DeFazio RA, Perez-Pinzon MA. εPKC phosphorylates the mitochondrial K^+ (ATP) channel during induction of ischemic preconditioning in the rat hippocampus. *Brain Res*. 2007;1184:345–353.
90. Lømo T. Frequency potentiation of excitatory synaptic activity in the dentate area of the hippocampal formation. *Acta Physiol Scand*. 1966;68(Suppl.):128.
91. Bliss TVP, Lømo T. Long-lasting potentiation of synaptic transmission in the dentate area of the anaesthetized rabbit following stimulation of the perforant path. *J Physiol*. 1973;232:331–356.
92. Ramakers GM, Pasinelli P, Hens JJ, Gispen WH, De Graan PN. Protein kinase C in synaptic plasticity: changes in the in situ phosphorylation state of identified pre- and postsynaptic substrates. *Prog Neuropsychopharmacol Biol Psychiatry*. 1997;21:455–486.
93. Kim H, Han SH, Quan HY, et al. Bryostatin-1 promotes long-term potentiation via activation of PKCα and PKCε in the hippocampus. *Hippocampus*. 2012;226:348–355.
94. Meiri N, Sun M-K, Segal Z, Alkon DL. Memory and long-term potentiation (LTP) dissociated: normal spatial memory despite CA1 LTP elimination with Kv1.4 antisense. *Proc Natl Acad Sci USA*. 1998;95:15037–15042.
95. Zamanillo D, Sprengel R, Hvalby O, et al. Importance of AMPA receptors for hippocampal synaptic plasticity but not for spatial learning. *Science*. 1999;284:1805–1811.
96. Weeber EJ, Atkins CM, Selcher JC, et al. A role for the beta isoform of protein kinase C in fear conditioning. *J Neurosci*. 2000;20:5906–5914.
97. Ling DS, Benardo LS, Serrano PA, et al. Protein kinase Mζ is necessary and sufficient for LTP maintenance. *Nat Neurosci*. 2002;5(4):295–296.
98. Pang PT, Teng HK, Zaitsev E, et al. Cleavage of proBDNF by tPA/plasmin is essential for long-term hippocampal plasticity. *Science*. 2004;306:487–491.
99. Sacktor TC, Osten P, Valsamis H, Jiang X, Naik MU, Sublette E. Persistent activation of the ζ isoform of protein kinase C in the maintenance of long-term potentiation. *Proc Natl Acad Sci USA*. 1993;90:8342–8346.
100. Osten P, Valsamis L, Harris A, Sacktor TC. Protein synthesis-dependent formation of protein kinase Mζ in long-term potentiation. *J Neurosci*. 1996;16:2444–2451.
101. Kelly MT, Crary JF, Sacktor TC. Regulation of protein kinase Mζ synthesis by multiple kinases in long-term potentiation. *J Neurosci*. 2007;27(13):3439–3444.
102. Kelly MT, Yao Y, Sondhi R, Sacktor TC. Actin polymerization regulates the synthesis of PKMζ in LTP. *Neuropharmacology*. 2006;52(1):41–45.
103. Sacktor TC. PKMzeta, LTP maintenance, and the dynamic molecular biology of memory storage. *Prog Brain Res*. 2008;169:27–40.
104. Yao Y, Shao C, Jothianandana D, Tcherepanova A, Shouval H, Sacktor T. Matching biochemical and functional efficacies confirm ZIP as a potent competitive inhibitor of PKMζ in neurons. *Neuropharmacology*. 2012;64:37–44.
105. Shao CY, Sondhi R, van de Nes PS, Sacktor TC. PKMζ is necessary and sufficient for synaptic clustering of PSD-95. *Hippocampus*. 2011;22(7):1501–1507.
106. Yoshii A, Murata Y, Kim J, Zhang C, Shokat KM, Constantine-Paton M. TrkB and protein kinase Mζ regulate synaptic localization of PSD-95 in developing cortex. *J Neurosci*. 2011;31:11894–11904.
107. Yao Y, Kelly MT, Sajikumar S, et al. PKMζ maintains late long-term potentiation by *N*-ethylmaleimide-sensitive factor/GluR2-dependent trafficking of postsynaptic AMPA receptors. *J Neurosci*. 2008;28:7820–7827.
108. Migues PV, Hardt O, Wu DC, et al. PKMζ maintains memories by regulating GluR2-dependent AMPA receptor trafficking. *Nat Neurosci*. 2010;13(5):630–634.

109. Sacktor TC. How does PKMζ maintain long-term memory? *Nat Rev Neurosci.* 2011;12:9–15.
110. Kim E, Sheng M. PDZ domain proteins of synapses. *Nat Rev Neurosci.* 2004;5:771–781.
111. van Zundert B, Yoshii A, Constantine-Paton M. Receptor compartmentalization and trafficking at glutamate synapses: a developmental proposal. *Trends Neurosci.* 2004;27:428–437.
112. Colombo PJ, Wetsel WC, Gallagher M. Spatial memory is related to hippocampal subcellular concentrations of calcium-dependent protein kinase C isoforms in young and aged rats. *Proc Natl Acad Sci USA.* 1997;94:14195–14199.
113. Sun MK, Alkon DL. Dual effects of bryostatin-1 on spatial memory and depression. *Eur J Pharmacol.* 2005;512:43–51.
114. Rochefort C, Arabo A, Andre M, Poucet B, Save E, Rondi-Reig L. Cerebellum shapes hippocampal spatial code. *Science.* 2011;334:385–389.
115. Schreurs BG, Oh MM, Alkon DL. Pairing-specific long-term depression of Purkinje cell excitatory postsynaptic potentials results from a classical conditioning procedure in the rabbit cerebellar slice. *J Neurophysiol.* 1996;75:1051–1060.
116. Schreurs BG, Tomsic D, Gusev PA, Alkon DL. Dendritic excitability microzones and occluded long-term depression after classical conditioning of the rabbit's nictitating membrane response. *J Neurophysiol.* 1997;77:86–92.
117. Olds JL, Bhalla US, Mcphie DL, Lester DS, Bower JM, Alkon DL. Lateralization of membrane-associated protein-kinase-C in rat piriform cortex—specific to operant training cues in the olfactory modality. *Behav Brain Res.* 1994;61:37–46.
118. Ahi J, Radulovic J, Spiess J. The role of hippocampal signaling cascades in consolidation of fear memory. *Behav Brain Res.* 2004;149:17–31.
119. Li YQ, Xue YX, He YY, et al. Inhibition of PKMzeta in nucleus accumbens core abolishes long-term drug reward memory. *J Neurosci.* 2011;31(14):5436–5446.
120. Nimitvilai S, Arora DS, Brodie MS. Reversal of dopamine inhibition of dopaminergic neurons of the ventral terminal area is mediated by protein kinase C. *Neuropharmacology.* 2012;37:543–556.
121. Ping XJ, Ma YY, Li YJ, et al. Essential role of protein kinase C in morphine-induced rewarding memory. *Neuropharmacology.* 2012;62:959–966.
122. de Quervain DJF, Kolassa IT, Ackermann S, et al. PKCα is genetically linked to memory capacity in healthy subjects and to risk for posttraumatic stress disorder in genocide survivors. *Proc Natl Acad Sci USA.* 2012;109:8746–8751.
123. Van der Zee EA, Compaan JC, Bohus B, Luiten PG. Alterations in the immunoreactivity for muscarinic acetylcholine receptors and colocalized PKC gamma in mouse hippocampus induced by spatial discrimination learning. *Hippocampus.* 1995;5:349–362.
124. Van der Zee EA, Compaan JC, de Beer M, Luiten PG. Changes in PKC gamma immunoreactivity in mouse hippocampus induced by spatial discrimination learning. *J Neurosci.* 1992;12:4808–4815.
125. Abeliovich A, Chen C, Goda Y, Silva AJ, Stevens CF, Tonegawa S. Modified hippocampal long-term potentiation in PKC gammamutant mice. *Cell.* 1993;75:1253–1262.
126. Abeliovich A, Paylor R, Chen C, Kim JJ, Wehner JM, Tonegawa S. PKC gamma mutant mice exhibit mild deficits in spatial and contextual learning. *Cell.* 1993;75:1263–1271.
127. Malmberg AB, Chen C, Tonegawa S, Basbaum AI. Preserved acute pain and reduced neuropathic pain in mice lacking PKCgamma. *Science.* 1997;278:279–283.
128. Bowers BJ, Collins AC, Tritto T, Wehner JM. Mice lacking PKC gamma exhibit decreased anxiety. *Behav Genet.* 2000;30:111–121.
129. Hongpaisan J, Xu C, Sen A, Nelson TJ, Alkon DL. PKC activation during training restores mushroom spine synapses and memory in the aged rat. *Neurobiol Dis.* 2013;55:44–62.

130. Shema R, Haramati S, Ron S, et al. Enhancement of consolidated long-term memory by overexpression of protein kinase Mzeta in the neocortex. *Science.* 2011;331(6021):1207–1210.
131. Sacktor TC. Memory maintenance by PKMzeta—an evolutionary perspective. *Mol Brain.* 2012;5:31.
132. Glanzman DI. PKM and the maintenance of memory. *Biol Rep.* 2013;5:4. http://f1000.com/prime/report/b/5/4.
133. Serrano P, Yao Y, Sacktor TC. Persistent phosphorylation by protein kinase Mζ maintains late-phase long-term potentiation. *J Neurosci.* 2005;25(8):1979–1984.
134. Cai D, Pearce K, Chen SP, Glanzman DL. Protein kinase M maintains long-term sensitization and long-term facilitation in *Aplysia. J Neurosci.* 2011;31:6421–6431.
135. Serrano P, Friedman EL, Kenney J, et al. PKMζ maintains spatial, instrumental, and classically conditioned long-term memories. *PLoS Biol.* 2008;6(12):2698–2706.
136. Hardt O, Migues PV, Hastings M, Wong J, Nader K. PKMzeta maintains 1-day- and 6-day-old long-term object location but not object identity memory in dorsal hippocampus. *Hippocampus.* 2010;20(6):691–695.
137. Barry J, Rivard B, Fox S, Fenton A, Sacktor T, Muller R. Inhibition of PKMζ disrupts the stable spatial discharge of hippocampal place cells in a familiar environment. *J Neurosci.* 2012;32:13753–13762.
138. Madronal N, Gruart A, Sacktor TC, Delgado-Garcia JM. PKMζ inhibition reverses learning-induced increases in hippocampal synaptic strength and memory during trace eyeblink conditioning. *PLoS One.* 2010;5(4):e10400.
139. Kwapis JL, Jarome TJ, Lonergan ME, Helmstetter FJ. Protein kinase Mzeta maintains fear memory in the amygdala but not in the hippocampus. *Behav Neurosci.* 2009;123(4):844–850.
140. Gamiz F, Gallo M. Intra-amygdala ZIP injections impair the memory of learned active avoidance responses and attenuate conditioned taste-aversion acquisition in rats. *Learn Mem.* 2011;18(8):529–533.
141. Kwapis JL, Jarome TJ, Gilmartin MR, Helmstetter FJ. Intra-amygdala infusion of the protein kinase Mzeta inhibitor ZIP disrupts foreground context fear memory. *Neurobiol Learn Mem.* 2012;98(2):148–153.
142. Crespo JA, Stockl P, Ueberall F, Jenny M, Saria A, Zernig G. Activation of PKCzeta and PKMzeta in the nucleus accumbens core is necessary for the retrieval, consolidation and reconsolidation of drug memory. *PLoS One.* 2012;7(2):e30502.
143. Shabashov D, Shohami E, Yaka R. Inactivation of PKMzeta in the NAc shell abolished cocaine-conditioned reward. *J Mol Neurosci.* 2012;47(3):546–553.
144. Pauli WM, Clark AD, Guenther HJ, O'Reilly RC, Rudy JW. Inhibiting PKMzeta reveals dorsal lateral and dorsal medial striatum store the different memories needed to support adaptive behavior. *Learn Mem.* 2012;19(7):307–314.
145. Shema R, Sacktor TC, Dudai Y. Rapid erasure of long-term memory associations in cortex by an inhibitor of PKMζ. *Science.* 2007;317(5840):951–953.
146. Shema R, Hazvi S, Sacktor TC, Dudai Y. Boundary conditions for the maintenance of memory by PKMζ in neocortex. *Learn Mem.* 2009;16(2):122–128.
147. Sacco T, Sacchetti B. Role of secondary sensory cortices in emotional memory storage and retrieval in rats. *Science.* 2010;329(5992):649–656.
148. He YY, Xue YX, Wang JS, et al. PKMzeta maintains drug reward and aversion memory in the basolateral amygdala and extinction memory in the infralimbic cortex. *Neuropsychopharmacology.* 2011;36(10):1972–1981.
149. von Kraus LM, Sacktor TC, Francis JT. Erasing sensorimotor memories via PKMzeta inhibition. *PLoS One.* 2010;5(6):e11125.
150. Pastalkova E, Serrano P, Pinkhasova D, Wallace E, Fenton AA, Sacktor TC. Storage of spatial information by the maintenance mechanism of LTP. *Science.* 2006;313(5790):1141–1144.

151. Xue YX, Luo YX, Wu P, et al. A memory retrieval-extinction procedure to prevent drug craving and relapse. *Science*. 2012;336:241–245.
152. Moncada D, Viola H. PKMzeta inactivation induces spatial familiarity. *Learn Mem*. 2008;15(11):810–814.
153. Crary JF, Shao CY, Mirra SS, Hernandez AI, Sacktor TC. Atypical protein kinase C in neurodegenerative disease I: PKMζ aggregates with limbic neurofibrillary tangles and AMPA receptors in Alzheimer disease. *J Neuropathol Exp Neurol*. 2006;65(4):319–326.
154. Volk LJ, Bachman JL, Johnson R, Yu Y, Huganir RL. PKM-ζ is not required for hippocampal synaptic plasticity, learning and memory. *Nature*. 2013;493:420–423.
155. Lee AM, Kanter BR, Wang D, et al. *PrKcz* null mice show normal learning and memory. *Nature*. 2013;493:416–420.
156. Mochly-Rosen D, Das K, Grimes KV. Protein kinase C, an elusive therapeutic target? *Nat Rev Drug Discov*. 2012;11:937–957.
157. Roffey J, Rosse C, Linch M, Hibbert A, McDonald NQ, Parker PJ. Protein kinase C intervention—the state of play. *Curr Opin Cell Biol*. 2009;21:268–279.
158. Fabian MA, Biggs 3rd WH, Treiber DK, et al. A small molecule-kinase interaction map for clinical kinase inhibitors. *Nat Biotechnol*. 2005;23:329–336.
159. Kang JW, Park YS, Lee DH, et al. Intracellular interaction of interleukin (IL)-32α with protein kinase Cε (PKCε) and STAT3 protein augments IL-6 production in THP-1 promonocytic cells. *J Biol Chem*. 2012;287:35556–35564.
160. Okuwa H, Kanno T, Fujita Y, et al. Sphingosine suppresses mesothelioma cell proliferation by inhibiting PKC-δ and inducing cell cycle arrest at the G(0)/G(1) phase. *Cell Physiol Biochem*. 2012;30:995–1004.
161. Leppänen T, Korhonen R, Laavola M, Nieminen R, Tuominen RK, Moilanen E. Down-regulation of protein kinase Cδ inhibits inducible nitric oxide synthase expression through IRF1. *PLoS One*. 2013;8:e52741.
162. Soltoff SP. Rottlerin: an inappropriate and ineffective inhibitor of PKCδ. *Trends Pharmacol Sci*. 2007;28:453–458.
163. Davies SP, Reddy H, Caivano M, Cohen P. Specificity and mechanism of action of some commonly used protein kinase inhibitors. *Biochem J*. 2000;351:95–105.
164. Whitson EL, Bugni TS, Chockalingam PS, et al. Spheciosterol sulfates, PKCzeta inhibitors from a Philippine sponge Spheciospongia sp. *J Nat Prod*. 2008; 71:1213–1217.
165. Herbert JM, Augereau JM, Gleye J, Maffrand JP. Chelerythrine is a potent and specific inhibitor of protein kinase C. *Biochem Biophys Res Commun*. 1990;172:993–999.
166. Kobayashi E, Nakano H, Morimoto M, Tamaoki T. Calphostin C (UCN-1028C), a novel microbial compound, is a highly potent and specific inhibitor of protein kinase C. *Biochem Biophys Res Commun*. 1989;159:548–553.
167. Bruns RF, Miller FD, Merriman RL, et al. Inhibition of protein kinase C by calphostin C is light-dependent. *Biochem Biophys Res Commun*. 1991;176:288–293.
168. Bonini JS, Da Silva WC, Bevilaqua LRM, Medina JH, Izquierdo I, Cammarota M. On the participation of hippocampal PKC in acquisition, consolidation and reconsolidation of spatial memory. *Neuroscience*. 2007;147:37–45.
169. Abrial E, Etievant A, Betry C, et al. Protein kinase C regulates mood-related behaviors and adult hippocampal cell proliferation in rats. *Prog Neuropsychopharmacol Biol Psychiatry*. 2013;43:40–48.
170. Zarate CA, Manji HK. Proetin kinase C inhibitors: rationale for use and potential in the treatment of bipolar disorder. *CNS Drugs*. 2009;23:569–582.
171. Koponen S, Goldsteins G, Keinanen R, Koistinaho J. Induction of protein kinase Cdelta subspecies in neurons and microglia after transient global brain ischemia. *J Cereb Blood Flow Metab*. 2000;20:93–102.

172. Miettinen S, Roivainen R, Keinanen R, Hokfelt T, Koistinaho J. Specific induction of protein kinase C delta subspecies after transient middle cerebral artery occlusion in the rat brain: inhibition by MK-801. *J Neurosci.* 1996;16:6236–6245.
173. Dave KR, Bhattacharya SK, Saul I, et al. Activation of protein kinase C delta following cerebral ischemia leads to release of cytochrome C from the mitochondria via bad pathway. *PLoS One.* 2011;6:e22057.
174. Phan TG, Wright PM, Markus R, Howells DW, Davis SM, Donnan GA. Salvaging the ischaemic penumbra: more than just reperfusion? *Clin Exp Pharmacol Physiol.* 2002;29:1–10.
175. Bright R, Raval AP, Dembner JM, et al. Protein kinase C delta mediates cerebral reperfusion injury *in vivo*. *J Neurosci.* 2004;24:6880–6888.
176. Kanno T, Yamamoto H, Yaguchi T, et al. The linoleic acid derivative DCP-LA selectively activates PKC-epsilon, possibly binding to the phosphatidylserine binding site. *J Lipid Res.* 2006;47:1146–1156.
177. Nelson TJ, Alkon DL. Neuroprotective versus tumorigenic protein kinase C activators. *Trends Biochem Sci.* 2009;34:136–145.
178. Sun M-K, Hongpaisan J, Nelson TJ, Alkon DL. Poststroke neuronal rescue and synaptogenesis mediated *in vivo* by protein kinase C in adult brains. *Proc Natl Acad Sci USA.* 2008;105:13620–13625.
179. Sun M-K, Hongpaisan J, Alkon DL. Postischemic PKC activation rescues retrograde and anterograde long-term memory. *Proc Natl Acad Sci USA.* 2009; 106:14676–14689.
180. Sun X, Budas GR, Xu L, Barreto GE, Mochly-Rosen D, Giffard RG. Selective activation of protein kinase Cε in mitochondria is neuroprotective *in vitro* and reduces focal ischemic brain injury in mice. *J Neurosci Res.* 2013;91:799–807.
181. Postina R, Schroeder A, Dewachter I, et al. A disintegrin-metalloproteinase prevents amyloid plaque formation and hippocampal defects in an Alzheimer disease mouse model. *J Clin Invest.* 2004;113:1456–1464.
182. Desdouits F, Buxbaum JD, Desdouits-Magnen J, Nairn AC, Greengard P. Amyloid alpha peptide formation in cell-free preparations. Regulation by protein kinase C, calmodulin, and calcineurin. *J Biol Chem.* 1996;271:24670–24674.
183. Pakask M, Balaspiri L, Checler F, Kasa P. Human amyloid-beta causes changes in the levels of endothelial protein kinase C and its alpha isoform *in vitro*. *Neurochem Int.* 2001;41:409–414.
184. Lee W, Boo JH, Jung MW, et al. Amyloid beta peptide directly inhibits PKC activation. *Mol Cell Neurosci.* 2004;26:222–231.
185. Cordey M, Gundimeda U, Gopalakrishna R, Pike CJ. Estrogen activates protein kinase C in neurons: role in neuroprotection. *J Neurochem.* 2003;84:1340–1348.
186. Govoni S, Bergamaschi S, Racchi M, et al. Cytosol protein kinase C downregulation in fibroblasts from Alzheimer's disease patients. *Neurology.* 1993;43:2581–2586.
187. Liu W, Dou F, Feng J, Yan Z. RACK1 is involved in beta-amyloid impairment of muscarinic regulation of GABAergic transmission. *Neurobiol Aging.* 2011;32:1818–1826.
188. Fahrenholz F. Alpha-secretase as a therapeutic target. *Curr Alzheimer Res.* 2007;4:412–417.
189. Lichtenthaler SF. Alpha-secretase in Alzheimer's disease: molecular identity, regulation and therapeutic potential. *J Neurochem.* 2011;116:10–21.
190. Jorissen E, Prox J, Bernreuther C, et al. The disintegrin/metalloproteinase ADAM10 is essential for the establishment of the brain cortex. *J Neurosci.* 2010;30:4833–4844.
191. Kuhn PH, Wang H, Dislich B, et al. ADAM10 is the physiologically relevant, constitutive alpha-secretase of the amyloid precursor protein in primary neurons. *EMBO J.* 2010;29:3020–3032.
192. Sisodia SS. Beta-amyloid precursor protein cleavage by a membrane-bound protease. *Proc Natl Acad Sci USA.* 1992;89:6075–6079.

193. Skovronsky DM, Moore DB, Milla ME, Doms RW, Lee VM. Protein kinase C-dependent alpha-secretase competes with beta-secretase for cleavage of amyloid-beta precursor protein in the trans-golgi network. *J Biol Chem.* 2000;275:2568–2575.
194. Asiedu MN, Tillu DV, Melemedjian OK, et al. Spinal protein kinase Mζ underlies the maintenance mechanism of persistent nociceptive sensitization. *J Neurosci.* 2011; 31:6646–6653.
195. Koo EH, Squazzo SL. Evidence that production and release of amyloid beta-protein involves the endocytic pathway. *J Biol Chem.* 1994;269:17386–17389.
196. Vassar R, Bennett BD, Babu-Khan S, et al. Beta-secretase cleavage of Alzheimer's amyloid precursor protein by the transmembrane aspartic protease BACE. *Science.* 1999;286:735–741.
197. Holmes C, Boche D, Wilkinson D, et al. Long-term effects of Abeta42 immunisation in Alzheimer's disease: follow-up of a randomised, placebo-controlled phase I trial. *Lancet.* 2008;372:216–223.
198. Schenk DB, Seubert P, Grundman M, Black R. A beta immunotherapy: lessons learned for potential treatment of Alzheimer's disease. *Neurodegener Dis.* 2005;2:255–260.
199. Serrano-Pozo A, William CM, Ferrer I, et al. Beneficial effect of human anti-amyloid-beta active immunization on neurite morphology and tau pathology. *Brain.* 2010;133:1312–1327.
200. Lilly. Lilly Halts Development of Semagacestat for Alzheimer's Disease Based on Preliminary Results of Phase III Clinical Trials. http://newsroom.lilly.com/releasedetail.cfm?ReleaseID=499794; 2011. Accessed March 13, 2012.
201. Etcheberrigaray R, Tan M, Dewachter I, et al. Therapeutic effects of PKC activators in Alzheimer's disease transgenic mice. *Proc Natl Acad Sci USA.* 2004;101:11141–11146.
202. Khan TK, Nelson TJ, Verma VA, Wender PA, Alkon DL. A cellular model of Alzheimer's disease therapeutic efficacy: PKC activation reverses Abeta-induced biomarker abnormality on cultured fibroblasts. *Neurobiol Dis.* 2009;34:332–339.
203. Nelson TJ, Cui C, Luo Y, Alkon DL. Reduction of beta-amyloid levels by novel PKC epsilon activators. *J Biol Chem.* 2009;284:34514–34521.
204. Kim T, Hinton DJ, Choi DS. Protein kinase C-regulated Aβ production and clearance. *Int J Alzheimers Dis.* 2011;2011:857368.
205. Zhu G, Wang D, Lin YH, McMahon T, Koo EH, Messing RO. Protein kinase Cε suppresses Aβ production and promotes activation of α-secretase. *Biochem Biophys Res Commun.* 2001;285:997–1006.
206. Choi DS, Wang D, Yu GQ, et al. PKCε increases endothelin converting enzyme activity and reduces amyloid plaque pathology in transgenic mice. *Proc Natl Acad Sci USA.* 2006;103:8215–8220.
207. Eckman EA, Reed DK, Eckman CB. Degradation of the Alzheimer's amyloid β peptide by endothelin-converting enzyme. *J Biol Chem.* 2001;276:24540–24548.
208. Eckman EA, Watson M, Marlow L, Sambamurti K, Eckman CB. Alzheimer's disease β-amyloid peptide is increased in mice deficient in endothelin-converting enzyme. *J Biol Chem.* 2003;278:2081–2084.
209. Funalot B, Quimet T, Claperon A, et al. Endothelin-converting enzyme-1 is expressed in human cerebral cortex and protects against Alzheimer's disease. *Mol Psychiatry.* 2004;9:1122–1128.
210. Kozikowski AP, Nowak I, Petukhov PA, et al. New amide-bearing benzolactam-based protein kinase C modulators induce enhanced secretion of the amyloid precursor protein metabolic sAPPalpha. *J Med Chem.* 2003;46:364–373.
211. Lahn M, Sundell K, Gleave M, et al. Protein kinase C-alpha in prostate cancer. *BJU Int.* 2004;93:1076–1081.
212. Lavoie L, Band CJ, Kong M, Bergeron JJM, Posner BI. Regulation of glycogen synthase in rat hepatocytes. Evidence for multiple signaling pathway. *J Biol Chem.* 1999;274:28279–28285.

213. Cho JH, Johnson GV. Glycogen synthase kinase 3 beta induces caspase-cleaved tau aggregation in situ. *J Biol Chem.* 2004;279:54716–54723.
214. Mailliot C, Bussiere T, Hamdane M, et al. Pathological tau phenotypes. The weight of mutations, polymorphisms, and differential neuronal vulnerabilities. *Ann NY Acad Sci.* 2000;920:107–114.
215. Mandelkow EM, Mandelkow E. Tau in Alzheimer's disease. *Trends Cell Biol.* 1998;8:425–427.
216. Liu C, Peng Z, Zhang N, et al. Identification of differentially expressed microRNAs and their PKC-isoform specific gene network prediction during hypoxic preconditioning and focal cerebral ischemia of mice. *J Neurochem.* 2012;120:830–841.
217. Di-Capua N, Sperling O, Zoref-Shani E. Protein kinase C-ε is involved in the adenosine-activated signal transduction pathway conferring protection against ischemia–reperfusion injury in primary rat neuronal cultures. *J Neurochem.* 2003;84:409–412.
218. Raval AP, Dave KR, Mochly-Rosen D, Sick TJ, Pérez-Pinzón MA. εPKC is required for the induction of tolerance by ischemic and NMDA-mediated preconditioning in the organotypic hippocampal slice. *J Neurosci.* 2003;23:384–391.
219. Takeishi Y, Ping P, Bolli R, Kirkpatrick DL, Hoit BD, Walsh RA. Transgenic overexpression of constitutively active protein kinase C epsilon causes concentric cardiac hypertrophy. *Circ Res.* 2000;86:1218–1223.
220. Wang Q, Li XY, Chen YK, et al. Activation of epsilon protein kinase C-mediated anti-apoptosis is involved in rapid tolerance induced by electroacupuncture pretreatment through cannabinoid receptor type 1. *Stroke.* 2011;42:389–396.
221. Dumont O, Mylroie H, Bauer A, et al. Protein kinase Cε activity induces anti-inflammatory and anti-apoptotic genes via an ERK1/2- and NF-κB-dependent pathway to enhance vascular protection. *Biochem J.* 2012;447:193–204.
222. McCarthy J, Lochner A, Opie LH, Sack MN, Essop MF. PKCε promotes cardiac mitochondrial and metabolic adaptation to chronic hypobaric hypoxia by GSK3β inhibition. *J Cell Physiol.* 2011;226:2457–2468.
223. Janes PW, Saha N, Barton WA, et al. Adam meets Eph: an ADAM substrate recognition module acts as a molecular switch for ephrin cleavage in trans. *Cell.* 2005;123:291–304.
224. Pruessmeyer J, Ludwig A. The good, the bad and the ugly substrates for ADAM10 and ADAM17 in brain pathology, inflammation and cancer. *Semin Cell Dev Biol.* 2009;20:164–174.
225. Reiss K, Saftig P. The "a disintegrin and metalloprotease" (ADAM) family of sheddases: physiological and cellular functions. *Semin Cell Dev Biol.* 2009;20:126–137.
226. Meneghini V, Bortolotto V, Francese MT, et al. High-mobility group box-1 protein and β-amyloid oligomers promote neuronal differentiation of adult hippocampal neural progenitors via receptor for advanced glycation end product/nuclear factor-κB axis: relevance for Alzheimer's disease. *J Neurosci.* 2013;33:6047–6059.
227. Ma L, Tao Y, Duran A, et al. Control of nutrient stress-induced metabolic reprogramming by PKCζ in tumorigenesis. *Cell.* 2013;152:599–611.
228. McGaugh JL. Memory consolidation and the amygdala: a systems perspective. *Trends Neurosci.* 2002;25:456–461.
229. McGaugh JL. Memory—a century of consolidation. *Science.* 2000;287:248–251.
230. Ranganath C. A unified framework for the functional organization of the medial temporal lobes and the phenomenology of episodic memory. *Hippocampus.* 2010; 20:1263–1290.
231. Rodrigues SM, Schafe GE, LeDoux JE. Molecular mechanisms underlying emotional learning and memory in the lateral amygdala. *Neuron.* 2004;44:75–91.

232. Mao J, Price DD, Phillips LL, Lu J, Mayer DJ. Increases in protein kinase C gamma immunoreactivity in the spinal cord dorsal horn of rats with painful mononeuropathy. *Neurosci Lett.* 1995;198:75–78.
233. Miletic V, Bowen KK, Miletic G. Loose ligation of the rat sciatic nerve is accompanied by changes in the subcellular content of protein kinase C beta II and gamma in the spinal dorsal horn. *Neurosci Lett.* 2000;288:199–202.
234. Zhou Y, Li GD, Zhao ZQ. State-dependent phosphorylation of epsilon-isozyme of protein kinase C in adult rat dorsal root ganglia after inflammation and nerve injury. *J Neurochem.* 2003;85:571–580.
235. Wu DF, Chandra D, McMahon T, et al. PKCε phosphorylation of the sodium channel $Na_v1.8$ increases channel function and produces mechanical hyperalgesia in mice. *J Clin Invest.* 2012;122:1306–1315.
236. Laferriere A, Pitcher MH, Haldane A, et al. PKMζ is essential for spinal plasticity underlying the maintenance of persistent pain. *Mol Pain.* 2011;7:99.
237. Ho SY, Chen CH, Liu TH, Chang HF, Liou JC. Protein kinase Mζ is necessary for cocaine-induced synaptic potentiation in the ventral tegmental area. *Biol Psychiatry.* 2012;71:706–713.
238. Narita M, Aoki T, Ozaki S, Yajima Y, Suzuki T. Involvement of protein kinase Cgamma isoform in morphine-induced reinforcing effects. *Neuroscience.* 2001;103:309–314.
239. Hasham MI, Pelech SL, Krieger C. Glutamate-mediated activation of protein kinase C in hippocampal neurons. *Neurosci Lett.* 1997;228:115–118.
240. Maurice N, Tkatch T, Meisler M, Sprunger LK, Surmeier DJ. D1/D5 dopamine receptor activation differentially modulates rapidly inactivating and persistent sodium currents in prefrontal cortex. *J Neurosci.* 2001;21:2268–2277.
241. Sato T, Tanaka K, Teramoto T, et al. Facilitative effect of a novel AVP fragment analog, NC-1900, on memory retention and recall in mice. *Peptides.* 2004;25:1139–1146.
242. Cole G, Dobkins KR, Hansen LA, Terry RD, Saitoh T. Decreased levels of protein kinase C in Alzheimer brain. *Brain Res.* 1988;452:165–174.
243. Paula-Lima AC, Adasme T, SanMartín C, Sebollela A, Hetz C, Carrasco MA. Amyloid β-peptide oligomers stimulate RyR-mediated Ca^{2+} release inducing mitochondrial fragmentation in hippocampal neurons and prevent RyR-mediated dendritic spine remodeling produced by BDNF. *Antioxid Redox Signal.* 2011;14:1209–1223.

CHAPTER THREE

CaMKII: A Molecular Substrate for Synaptic Plasticity and Memory

Brian C. Shonesy*, Nidhi Jalan-Sakrikar*, Victoria S. Cavener†, Roger J. Colbran*,†,‡

*Department of Molecular Physiology and Biophysics, Vanderbilt University School of Medicine, Nashville, Tennessee, USA

†Vanderbilt Brain Institute, Vanderbilt University School of Medicine, Nashville, Tennessee, USA

‡Vanderbilt-Kennedy Center for Research on Human Development, Vanderbilt University School of Medicine, Nashville, Tennessee, USA

Contents

Abstract

Learning and memory is widely believed to result from changes in connectivity within neuronal circuits due to synaptic plasticity. Work over the past two decades has shown that Ca^{2+} influx during LTP induction triggers the activation of CaMKII in dendritic spines. CaMKII activation results in autophosphorylation of the kinase rendering it constitutively active long after the Ca^{2+} dissipates within the spine. This "molecular switch"[1]

Progress in Molecular Biology and Translational Science, Volume 122
ISSN 1877-1173
http://dx.doi.org/10.1016/B978-0-12-420170-5.00003-9

mechanism is essential for LTP and learning and memory. Here, we discuss this key regulatory mechanism and the diversity of downstream targets that can be modulated by CaMKII to exert dynamic control of synaptic structure and function.

1. MEMORY AND SYNAPTIC PLASTICITY

Memories are generally thought to be encoded by long-lasting changes in synaptic connectivity. Therefore, deciphering the mechanism of memory formation can only be accomplished through understanding the molecular basis for synaptic regulation. Excitatory synapses can undergo long-lasting increases in synaptic strength in response to short periods of elevated activity. This process, known as long-term potentiation (LTP), is thought to underlie the formation of memories at the cellular level. Another form of plasticity that can occur at glutamatergic synapses is long-term depression (LTD), which as the name suggests is a long-lasting decrease in the strength of synaptic transmission. Both LTP and LTD can occur in the same cell and are thought to oppose one another, thereby allowing the dynamic regulation of synaptic strength and precise control on the storage of information within neural circuits. Synaptic plasticity can occur at synapses throughout the brain; however, the hippocampal CA1 glutamatergic synapse has been the dominant model for studying this phenomenon.

LTP is typically observed by recording postsynaptic responses to electrical stimulation of axonal inputs in the cell body layer of hippocampal CA1 region. Although LTP can be triggered rapidly (within seconds), it can lasts for hours *in vitro* and much longer *in vivo*. Also in the CA1, LTP is normally input specific; LTP at one synapse does not increase the strength of neighboring synapses. This suggests that mechanisms normally supporting LTP are compartmentalized.[2] Postsynaptic Ca^{2+} signaling plays a central role in LTP, as well as in many other forms of plasticity, including several forms of LTD. Thus, the postsynaptic signaling mechanisms must allow Ca^{2+} to elicit LTP under one condition while triggering LTD in another. The timescale of synaptic Ca^{2+} signals is ~100 ms; therefore, another requirement for LTP expression is that the transient Ca^{2+} signal must initiate a sustained pathway to create the long-lasting changes in synaptic activity. These changes are typically associated with both ultrastructural changes in synapses and functional enhancements of glutamate receptors. Calcium/calmodulin-dependent protein kinase II (CaMKII) is widely believed to be a key enzyme that can mediate both of these changes, due to its unique molecular properties which will be the focus of this chapter.

2. MOLECULAR MECHANISMS OF LTP

LTP can be induced using several stimulation paradigms, including high-frequency tetanus trains of 100 Hz, or "theta burst" stimulation (e.g., four stimuli at 100 Hz repeated several times at 200-ms time intervals), which may better simulate normal physiological hippocampal firing patterns.[3] During elevated synaptic activity, glutamate activates AMPA-type glutamate receptors (AMPARs), leading to depolarization of the postsynaptic spine. This depolarization in combination with the presence of glutamate facilitates the opening of the Ca^{2+}-permeable NMDA-type of glutamate receptor (NMDAR), triggering a Ca^{2+}-mediated signaling cascade, ultimately resulting in the potentiation of AMPAR-mediated synaptic current, and an enlargement of dendritic spines.

3. SYNAPTIC ARCHITECTURE

The anatomical and morphological organization of excitatory synapses plays a key role in controlling synaptic plasticity. Excitatory glutamatergic terminals usually form synapses with postsynaptic morphological specializations called dendritic spines, $\sim 1\ \mu m^3$ protrusions from the main dendritic shaft. The neck of the spines provides an important restriction on the diffusion of Ca^{2+} between the spine head and the dendritic shaft, contributing to the synaptic specificity of LTP.[4] Much of the signaling that underlies LTP occurs within individual spines at the postsynaptic density (PSD) or its neighboring perisynaptic and extrasynaptic regions. The PSD is an electron-dense thickening underneath the postsynaptic membrane containing a conglomeration of 400–1000 proteins.[5] Among the large number of PSD proteins, the core functional elements are glutamate-gated ion channels, AMPARs and NMDARs, and metabotropic glutamate receptors (mGluRs). These channels/receptors are supported by direct interactions with a network of scaffolding proteins, such as postsynaptic density-95 (PSD-95), synapse-associated protein-97 (SAP97), and homer, and other scaffolding proteins anchored to the actin cytoskeleton, such as spinophilin/neurabin, α-actinin, and CaMKIIβ, and/or associated with other proteins such as A-kinase-anchoring protein 79/150 (AKAP79/150),

Shank, and densin-180. Both glutamate receptors and scaffolding proteins are dynamically regulated by a variety of PSD-associated kinases including protein kinase A (PKA), protein kinase Cγ, casein kinase 2 (CK2), and CaMKIIα/β. Opposing these kinases are phosphatases like protein phosphatase 1 (PP1) and protein phosphatase 2B. Other signaling molecules that are present at the PSD are GTPases like Arf and Ras and their regulators SynGAP, TIAM1, and kalirin-7 (for a detailed review of the PSD, see Ref. 152). Alterations in the expression, localization, interactions, and/or activation of these PSD proteins are believed to underlie the structural and functional changes at excitatory synapses that are associated with synaptic plasticity. Specific molecular disruptions of many of these processes can cause dysfunction in synaptic plasticity, learning, and/or memory (see below). Moreover, several neurological and psychiatric diseases are associated with changes in number and or morphology of dendritic spines.

4. CALCIUM/CALMODULIN-DEPENDENT PROTEIN KINASE

Multiple molecular approaches have been used to alter CaMKII expression, activity, or subcellular targeting, with profound effects on synaptic plasticity, learning, and memory. In order to fully understand the role of the kinase in these processes, the impact of these manipulations on the functional properties of the kinase and its ability to target key downstream molecules must be ascertained.

Enhanced AMPAR responses to glutamate following LTP induction result from an increased number and intrinsic activity of AMPARs in the postsynaptic membrane.[6,7] Introduction of exogenous-activated CaMKII into the postsynaptic neuron can induce very similar alterations of synaptic AMPAR responses, occluding the increase in response to LTP induction.[8–11] Moreover, increased CaMKII activity can cause changes in spine morphology. Additionally, there is an increase of CaMKII levels, and CaMKII activity, in the PSD following LTP.[12–14] Similarly, if CaMKII activity is inhibited using CaMKII antagonists[15] or by genetic deletion of CaMKIIα,[16,17] LTP is either significantly reduced or absent entirely. These data establish a key role for CaMKII during LTP induction. Ongoing kinase activity may not be required to maintain the enhanced synaptic transmission because CaMKII inhibition after LTP induction generally fails to reverse the increase of synaptic activity. However, this remains a somewhat controversial issue (see below).

5. CaMKII ISOFORMS

There are four homologous isoforms of CaMKII, α, β, γ, and δ, varying in molecular weight from 50 to 60 kDa.[18] The primary mRNA transcripts are subjected to alternative splicing, generating more than 20 variants of the kinase with distinct cellular and tissue distribution. All four isoforms are expressed in the brain, but the α and β isoforms are most abundant.[19] Interestingly, the CaMKIIα mRNA is transported into neuronal dendrites where it is locally translated following LTP induction, whereas CaMKIIβ mRNAs are translated in the soma. While CaMKIIβ splice variants are differentially expressed during embryonic and postnatal development in most rodent brain regions, CaMKIIα is first expressed beginning at postnatal day 5 and is expressed at very high levels in neurons in certain parts of the adult forebrain. Thus, the α:β isoform ratio is about 3:1 in adult rodent forebrain, but about 1:3 in the cerebellum[20–22]; in the hippocampus, these two isoforms together constitute ~2% of the total protein.[22,23]

6. CaMKII HOLOENZYME

CaMKII subunits assemble to form a dodecameric holoenzyme which is arranged as two stacked hexameric rings as determined by low-resolution transmission electron microscopy and more recently by atomic resolution crystallography.[24–26] The C-terminal association domains of each subunit combine to form the central core of each ring with the N-terminal catalytic domains extending radially outward (Fig. 3.1). Inside the cell, subunits of CaMKII can mix to form heteromeric CaMKII holoenzymes composed of varying ratios of different isoforms,[27,28] adding to the complexity of structural configurations and diversity of its biological functions. The isoform composition of holoenzymes is believed to be dictated stochastically by the relative rates of their translation.

7. CaMKII REGULATION

Almost all CaMKII isoforms share a similar domain organization and basic regulatory functions. In the inactive conformation, the kinase activity of each subunit is suppressed by interaction of the central autoinhibitory

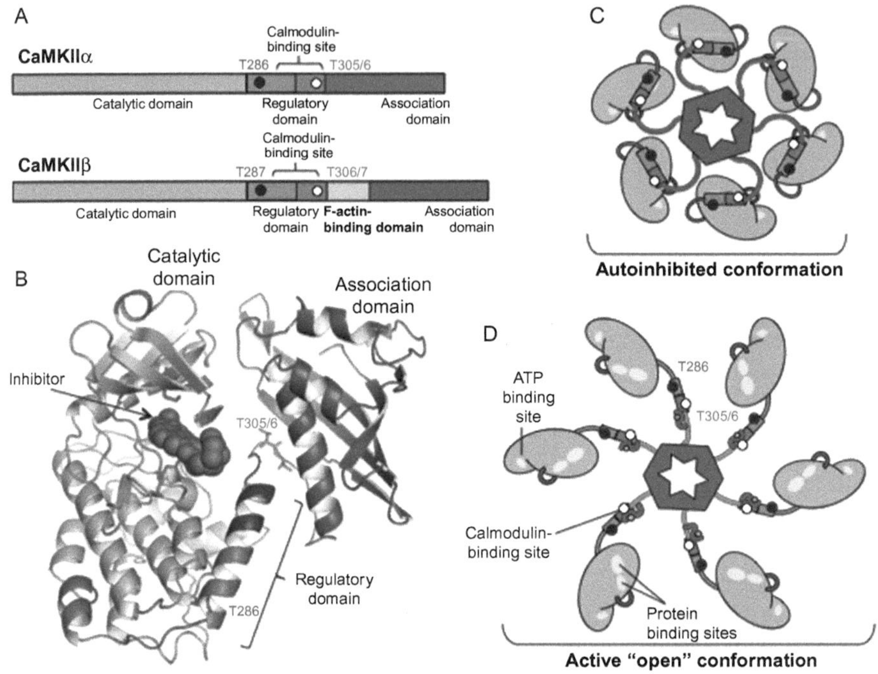

Figure 3.1 Domain organization and structure of CaMKII. (A) CaMKIIα and CaMKIIβ have a similar overall domain organization (as labeled) with the exception of an F-actin-binding domain inserted in CaMKIIβ. (B) Structure of an individual "autoinhibited" CaMKII subunit. This individual subunit structure was "excised" from a structure (PDB:3SOA) of an intact inactive CaMKIIα holoenzyme structure using PyMol (DeLano Scientific). Domains and key residues in CaMKII were colored as in (A). The yellow structure in the center is a CaMKII inhibitor bound in the kinase active site in this structure. (C) Cartoon of the compact inactive CaMKII holoenzyme structure. The kinase catalytic domains (pink) decorate the outside of a hexameric central hub formed by the association domains (teal), linked by the regulatory domains. Autoinhibitory interactions of the regulatory and catalytic domains hold the kinase domains in a compact closed conformation. For clarity, the illustration includes a single ring, whereas the intact dodecameric holoenzyme consists of a stacked pair of rings. (D) Cartoon illustrating conformational changes associated with CaMKII activation. Binding of Ca^{2+}/CaM disrupts interactions of the regulatory (blue) and catalytic (pink) domains (see B); the kinase domains swing away from the hub of the holoenzyme in an open conformation such that the active site is accessible to ATP and protein substrates. (See the color plate.)

domain (AID) with the catalytic domain. Structures of various CaMKII isoforms[24,29] show how the regulatory segment makes extensive contacts with the kinase domain. Ca^{2+}/CaM binding to a region overlapping with the AID disrupts these interactions to expose the catalytic site for binding of ATP and substrate (see Fig. 3.1). Interestingly, the sensitivity of CaMKII to

Ca^{2+}/CaM activation is dictated by the subunit composition; CaMKIIβ has a higher affinity for Ca^{2+}/CaM compared to CaMKIIα.[30]

7.1. Thr286 autophosphorylation

While the basic mechanism for Ca^{2+}/CaM-dependent activation of CaMKII is similar to that for activation of the other members of the CaM kinase family, the holoenzyme structure endows CaMKII with a unique regulatory mechanism for sensing and transducing intracellular Ca^{2+} signals. Ca^{2+}/CaM independently activates each subunit in a holoenzyme, but activation of adjacent subunits results in transautophosphorylation of Thr286 in the AID[31–34] (Fig. 3.1). Importantly, Thr286 autophosphorylation increases the affinity of CaMKII for Ca^{2+}/CaM by more than 1000-fold[35] and also interferes with the AID interaction with the catalytic domain, resulting in autonomous (or Ca^{2+}/CaM independent) CaMKII activity.[36] This unique ability to remain constitutively active beyond the window of the initiating Ca^{2+} signal allows CaMKII to serve as a "memory molecule."[1,37,38]

Thr286 autophosphorylation of CaMKII is critical for certain physiological functions of CaMKII.[39,40] Mice harboring a Thr286 to Ala mutation, which prevents Thr286 autophosphorylation (T286A-KI mice), have reduced hippocampal LTP,[39] and have reduced CaMKII targeting to the synapse.[40] Consistent with these physiological alterations, these mice show impairments in a variety of spatial and working memory tasks, and other behavioral alterations.[39,40]

7.2. Thr305/6 autophosphorylation

The self-regulated mechanism of CaMKII also includes autophosphorylation at Thr305 and/or Thr306, in the Ca^{2+}/CaM-binding domain.[41] Slow autophosphorylation at Thr306 can occur in the basal state (basal autophosphorylation), blocking Ca^{2+}/CaM binding and hence rendering the kinase insensitive to Ca^{2+} transients.[42] Alternatively, removal of Ca^{2+}/CaM from Thr286 autophosphorylated CaMKII (autonomously active CaMKII) promotes rapid autophosphorylation at either Thr305 or Thr306 (sequential autophosphorylation), again preventing Ca^{2+}/CaM binding, although the kinase remains constitutively active (until Thr286 is dephosphorylated).[43] Autophosphorylation at Thr305/6 also regulates synaptic plasticity by modulating CaMKII localization and/or activity in neurons.[44–46] Mice harboring a knock-in mutation mimicking autophosphorylation of the inhibitory Thr305 site (T305D) have reduced CaMKII targeting to PSDs and impaired LTP and learning.[44] In contrast, a double knock-in mutation of Thr305 and Thr306 to prevent inhibitory autophosphorylation (TT305/306VA) resulted in enhanced hippocampus-dependent learning and a decreased threshold for LTP induction.[44] Interestingly, the T305D mice show enhanced LTD, possibly reflecting a shift in the kinase–phosphatase balance.

7.3. Regulation by other posttranslational modifications

Although Ca^{2+}/CaM activation of CaMKIIα and autophosphorylation at Thr286, Thr305, and Thr306 are by far the most well-studied means of CaMKII regulation to date, several other regulatory mechanisms may contribute to the functional modulation. Although the roles of the equivalent site in CaMKIIβ (Thr287, Thr306, and Thr307) have not been rigorously examined, it seems likely that they will also affect the physiological functions of CaMKIIβ. Moreover, other sites of CaMKII phosphorylation are also poorly understood. For example, Thr253 in CaMKIIα appears to be phosphorylated *in vivo* and may enhance targeting to the PSD, while having no effect on kinase activity.[47] Ser314 phosphorylation of CaMKIIα also has been detected *in vitro*, but does not appear to substantially affect Ca^{2+}/CaM binding,[41,48] which might be predicted due to its proximity to Thr305/6, and no other function of this site has been shown to date. In addition, Thr382 in an alternatively spliced F-actin-binding domain of CaMKIIβ was found to be rapidly autophosphorylated; although the functional relevance is unclear, the fact that this site is not conserved in CaMKIIα suggests that it may modulate CaMKIIβ-specific functions.[49]

Recent studies have shown that CaMKIIδ can be oxidized at Met281 and Met282 in the regulatory domain, imparting Ca^{2+}/CaM-independent activity to the kinase, similar to the effects of autophosphorylation at Thr286.[50] Oxidation of these sites has multiple pathological roles in cardiomyocytes and smooth muscle.[51–54] Met281 and Met282 are identical in CaMKIIβ and are conserved as Cys280 and Met281 in CaMKIIα, but little is known about the role of CaMKII oxidation in neurons and how this might affect synaptic plasticity.

7.4. Modulation of CaMKII by dephosphorylation

CaMKII Thr286 can be dephosphorylated by PP1, PP2A, and PP2C *in vitro*.[55–57] PP1 and PP2A appear to be the primary phosphatases acting on Thr286 in the brain, and PP1 likely accounts for the bulk of activity on PSD-localized CaMKII. Transgenic mice overexpressing an inhibitor of PP1, inhibitor-1, have increased levels of phospho-CaMKII Thr286 in the hippocampus and displayed enhanced learning and memory.[58] There is evidence that translocation of activated Thr286-autophosphorylated CaMKII during LTP induction may saturate PP1 in the PSD; thus, PP1 acts to "gate" CaMKII autophosphorylation at the synapse.[59–61] However, a more recent study has suggested that PP1 targets other sites in CaMKII, but does not target Thr286 within the PSD.[38] This emphasizes the generally poor understanding of the important dynamic interplay between CaMKII autophosphorylation and protein phosphatases.

7.5. Modulation of CaMKII activity by protein–protein interactions

In addition to posttranslational modifications, CaMKII may be regulated by interactions with a number of other neuronal proteins. For example, CaMKII can be directly activated by α-actinin in a Ca^{2+}-independent manner, likely due to the similarity between the CaMKII-binding domain on α-actinin and the C-lobe of calmodulin.[62] This interaction selectively activates CaMKII interactions with GluN2B-containing NMDARs (see below), enhancing GluN2B phosphorylation at Ser1303.[62] In contrast, α-actinin can antagonize Ca^{2+}/CaM-dependent activation of CaMKII toward other substrates, such as Ser831 in GluA1. In a related vein, a recent study found that densin is a potent inhibitor of GluA1 Ser831 phosphorylation, but has little effect on GluN2B Ser1303 phosphorylation.[63] These findings suggest that α-actinin and densin have the capability to modify the downstream substrate selectivity of CaMKII, providing another potential mechanism for fine-tuning CaMKII signaling *in vivo*.

8. CaMKII TARGETING

Targeting of CaMKII can have diverse effects on signaling: (1) by localizing the kinase to specific sites where it can be activated by discrete Ca^{2+} signals; (2) by colocalizing the kinase with specific substrates, thus effectively directing kinase activity; and (3) by localizing the kinase close to or away from phosphatases, thus modulating CaMKII autophosphorylation and associated kinase activity.[64]

8.1. Spatial and temporal neuronal localization

Studies several years ago first found that CaMKII transiently translocates to synapses following neuronal activation. This translocation required CaMKII activation, and could be prolonged by autophosphorylation at Thr286, or destabilized by autophosphorylation at Thr305/6.[14,65,66] Recent development of fluorescence resonance energy transfer (FRET)-based imaging approaches has allowed the visualization of CaMKII activation in the spine and dendrite during synaptic stimulation.[13,67–70] The probe used, Camui, is a full-length fusion protein of CaMKII that contains donor and acceptor fluorophores at its N- and C-termini. When CaMKII is activated, the resulting changes in conformation (see Fig. 3.1) cause a decreased amount of FRET due to the increased distance between the acceptor and donor fluorophores. This allows Camui to detect conformational changes associated with either Ca^{2+}/CaM binding or Thr286 autophosphorylation.[67] Using this probe in combination with two-photon glutamate uncaging in hippocampal organotypic cultures, Lee *et al.* were able to demonstrate that during the induction of LTP, CaMKII activation was restricted to stimulated spines, suggesting that CaMKII is highly compartmentalized.[13] Activation in spines was preceded by activation in the nearby dendrite and decayed after ~2 min. When Camui-T286A is expressed, CaMKII activation decays at much faster rate, reflecting the rapid decay of Ca^{2+} in the spine.[71] While the role of CaMKII in LTP maintenance is still under debate (see Refs. 15,72–74), these studies showing the synapse specificity and rapid activation of Camui in spines clearly support the long-held view for its role in synapse-specific LTP induction.

8.2. NMDA-type glutamate receptor

The high-affinity CaMKII-binding domain in GluN2B (residues 1290–1309) bears striking similarity with the AID of CaMKII and binds to the T-site on CaMKII[75] (the region in the catalytic domain of CaMKII where the AID containing Thr286 docks). As a result, binding to GluN2B keeps the AID displaced, thus locking the kinase in a Ca^{2+}/CaM-independent state even in the absence of T286 autophosphorylation.[76] However, GluN2B can inhibit Ca^{2+}/CaM dependent and autonomous activity of CaMKII. This inhibition also involved GluN2B binding to the T-site and was uncompetitive for ATP[77]; thus, interaction with GluN2B modulates the kinetic parameters for ATP binding to CaMKII and catalysis.[78,79]

CaMKII targeting to the NMDAR is dependent on receptor activation, Ca^{2+} influx, and is stabilized by Thr286 autophosphorylation.[76,80–82] Conditions that promote Thr286 autophosphorylation increase the coimmunoprecipitation of CaMKII with GluN2B, including high-frequency tetanic stimulation and ischemia.[83–86] In HEK293 cells, GluN2B, but not GluN2A subunit supports the redistribution of CaMKII upon NMDA stimulation,[82] suggesting that GluN2B plays a dominant role in CaMKII association with NMDA receptors. Interference with CaMKII–GluN2B interaction by overexpression of the GluN2B C-terminal domain or by mutation of GluN2B strongly reduces CaMKII targeting and LTP induction.[87,88] Recent studies indicate that this complex also plays a role in LTP maintenance.[74,89] Furthermore, mice with mutations that prevent CaMKII–GluN2B-binding have reduced LTP, impaired GluA1/TARP trafficking to synapses, and lower GluA1 Ser831 phosphorylation levels. Interestingly, these mice are able to learn normally, but their ability to consolidate memory was impaired.[90]

Interactions with GluN2B may also facilitate modulation of other synaptic signaling pathways that have not been historically associated with CaMKII. For example, ERK1/2 plays an important role in AMPAR trafficking and spine growth[91] and transcriptional activation[92] during LTP induction. ERK1/2 is activated by Ca^{2+} influx through NMDARs during LTP induction,[93] and recent studies suggest that this ERK1/2 activation requires CaMKIIα interaction with GluN2B.[94] CaMKII forms a complex with CK2 and GluN2B, which facilitates CK2 phosphorylation of GluN2B-S1480.[89] Phosphorylation of S1480 disrupts GluN2B binding to PSD-95, thus allowing for its removal from the synapse by lateral diffusion.[95,96]

Taken together, these data make the GluN2B subunit of NMDAR an excellent candidate responsible for stimulus-induced CaMKII translocation to the PSD. This is conceptually supported by the fact that mutations in the CaMKII catalytic domain that interfere with binding to GluN2B also disrupt synaptic targeting of CaMKII in neuron.[80] However, the impact of these mutations on CaMKII interactions with other key proteins is poorly understood. Significantly, interaction with GluN2B alone cannot quantitatively account for all the CaMKII found at the PSD. Recent proteomics studies have shown that there are ~20-fold more CaMKII holoenzymes in the PSD than NMDAR subunits on a molar basis.[23] Consistent with these estimates of a high CaMKII:NMDAR ratio in the PSD, another recent study suggests that only 2% of CaMKII in dendritic spines may be bound to the NMDARs in the PSD and of that only 10% may be held active by GluN2B subunit.[21] Also autophosphorylation at Thr305/6 promotes CaMKII dissociation from the PSD yet appear to have modest effects on direct interactions with GluN2B.[81,97] Hence, it is likely that CaMKII targeting to the PSD is a reflection of GluN2B interactions along with additional interactions with other PSD proteins.

8.3. Actin

Actin is a highly dynamic and prominent cytoskeletal protein that serves as a framework for mechanical stability of dendritic spine structure.[98,99] It also serves as a scaffold for recruiting several postsynaptic proteins. Actin exists in a dynamic equilibrium between two forms, the monomeric globular form (G-actin) and the filamentous form (F-actin), which is modulated during LTP induction.[100]

CaMKIIβ contains an F-actin-binding domain, which specifically allows for direct association with the actin cytoskeleton. This interaction occurs in the basal state and is antagonized by Ca^{2+}/CaM binding to CaMKIIβ.[101] Recent studies indicate that mice harboring an A303R mutation of CaMKIIβ, in which Ca^{2+}/CaM-dependent activation is prevented, do not have deficits in LTP or learning and memory, and do not have altered targeting of holoenzymes to the PSD. However, in this same study, deletion of CaMKIIβ disrupts targeting of CaMKIIα to the synapse, LTP, and learning and memory in CaMKIIβ-KO mice.[102] This suggests that the role of CaMKIIβ may be to target CaMKIIα to synapses, with CaMKIIα performing the kinase function of the holoenzyme. Moreover, overexpression of CaMKIIβ, but not CaMKIIα, increases the neurite extensions and formation of new synapses in dissociated neuronal cultures.[103] RNAi-mediated knockdown of CaMKIIβ leads to a reduction in dendritic spine head volume, whereas overexpression of CaMKIIβ reduced the actin turnover rate in the spine head.[72] In combination, these findings indicate that direct CaMKIIβ interactions with F-actin are important for some isoform-specific responses.

Recent studies have shown that the well-established F-actin-binding proteins spinophilin and α-actinin can bind to CaMKIIα, thereby indirectly targeting CaMKII holoenzymes, perhaps to different pools of F-actin.[62,104,105] The functional relevance of these interactions is unclear. However, it is tempting to speculate that they may play a role in regulating CaMKII actions on multiple proteins that play important roles in modulating F-actin dynamics and spine morphology (see Table 3.1 and Fig. 3.2).

9. AMPAR-MEDIATED POTENTIATION

The enhancement of AMPAR-mediated synaptic transmission is the defining process of LTP. This can be achieved through increased single channel conductance of AMPARs[136] and/or through an increased number of AMPARs at the synapse, specifically homomeric GluA1-AMPARs.[6]

Table 3.1 CaMKII substrates involved in plasticity

Substrate	Phos site	Effect on function	References
Ion channels and auxiliary subunits			
AMPARs	See text		
NMDARs	See text		
GABA-A β1	S384, S409	S384, ↑ channel current	106
GABA-A β2	S410	S410, ↑ surface expression	107
GABA-A β3	S383, S409	S383, ↑ channel current; S409, no known effect	
GABA-A γ2S	S348, T350	No known effect	
GABA-A γ2L	S343, S348, S350	No known effect	
Kv4.2	S438, S459	↑ total expression of Kv4.2	108
		↑ channel current	
Stargazin	C-tail (between residues 229 and 253)[1]	Phosphomimic mutation of nine serines between residues 229 and 235 or overexpression of constitutively active CaMKII enhances the trapping of stargazin (and therefore GluA1-AMPARs) in the PSD[2]	1. 109 2. 110
Scaffolding, cytoskeletal, and motor proteins			
MAP2	18 sites[1]	Inhibits microtubule assembly by MAP2.[2] Depolarization of cultured neurons results in increased phosphorylation of CaMKII as well as PKC sites.[3] MAP2 phosphorylation increases after LTP induction,[4] but whether this is CaMKII dependent has not been reported	1. 111 2. 112 3. 113 4. 114
MUPP1		Binds to association domain of CaMKII and dissociates when Ca^{2+}/CaM binds to CaMKII. Bridges the indirect interaction of SynGAPα with CaMKII (see above)	115

Continued

Table 3.1 CaMKII substrates involved in plasticity—cont'd

Substrate	Phos site	Effect on function	References
Myosin Va	S1650	Can bind to CaMKII and deliver CaM to activate the kinase.[1] Phosphorylation by CaMKII inhibits myosin-V-mediated organelle transport in Xenopus melanosomes.[2] Myosin Va is required for LTP and CaMKII-mediated GluA1 synaptic delivery[3]; however, it is not clear if CaMKII phosphorylation is involved in this role	1. 116 2. 117 3. 118
SAP97	S232	Disrupts SAP97 interaction with NR2A *in vitro* and in heterologous cells	119
Signaling proteins			
CK2		CaMKII–CK2–GluN2B complex formation facilitates CK2 phosphorylation of GluN2B S1480,[1] which disrupts GluN2B binding to PSD-95, thus allowing for its removal from the synapse by lateral diffusion[1,2]	1. 89 2. 95
DGLα	S782, S808	Decreases DGLα-mediated synthesis of the endocannabinoid 2-AG, which mediates STD and LTD in many brain regions. Inhibition of CaMKII causes enhanced STD in the mouse striatum	120
Kalirin-7	Thr95	Rho-GEF that is phosphorylated by CaMKII in response to NMDA activation in spines, and is required for activity-dependent synaptic insertion of GluA1, as well as spine enlargement during LTP	121
SynGAPα/β	S750/1/6, S764/5, S1058, S1123[1]	Inhibits GAP activity, thereby increasing Ras activity[1]. CaMKII can only phosphorylate the α isoform through an indirect interaction via MUPP1. Ca^{2+}/CaM dissociates CaMKII from the SynGAPα–MUPP1 complex, resulting in dephosphorylation. Dissociation of these complexes causes decreased p38 MAPK activity and potentiates AMPAR-mediated synaptic transmission.[2] SynGAPβ directly binds to CaMKII[3], but the role of the CaMKII–SynGAPβ interaction in synaptic function has not been investigated	1. 122 2. 115 3. 123

Tiam1		Phosphorylation stimulates GDP/GTP exchange activity of Tiam1.[1] Tiam1 activates Rac1 which ultimately leads to activity dependent spine remodeling[2]	1. 124 2. 125 3. 126
Regulators of protein turnover			
CREB	S142	Prevents dimerization, thus preventing its effect on transcription	127
CPEB	T171[1]	Stimulates CPEB-dependent protein synthesis during LTP induction[2] which is essential for the maintenance of LTP[3]	1. 128 2. 129 3. 130
ΔFosB	S27	Stabilizes ΔFosB resulting in accumulation. ΔFosB signals to upregulate CaMKII transcription in the nucleus accumbens; thus CaMKII phosphorylation of ΔFosB generates a circular feedforward loop that may have important implications in plasticity	131
MeCP2	S421	Mediates MeCP2-dependent spine growth, increased dendritic branching and transcription upregulation of activity dependent genes including BDNF	132
Proteosome subunit Rpt6	S120	↑ proteosomal activity CaMKII also acts to target the entire proteosome in a kinase-independent manner[1,2,3]	1. 133 2. 134 3. 135

CK2, casein kinase 2; CPEB, cytoplasmic polyadenylation element-binding protein; CREB, cAMP response element-binding protein; DGLα, diacylglycerol lipase α; Kv4.2, potassium voltage-gated channel subunit Kv4.2; MAP2, microtubule-associated protein 2; MeCP2, methyl-CpG-binding protein 2; MUPP1, multiple PDZ domain protein 1; Tiam1, T-lymphoma invasion and metastasis-inducing protein 1; SAP97, synapse-associated protein 97; SynGAP, synaptic Ras GTPase-activating protein.

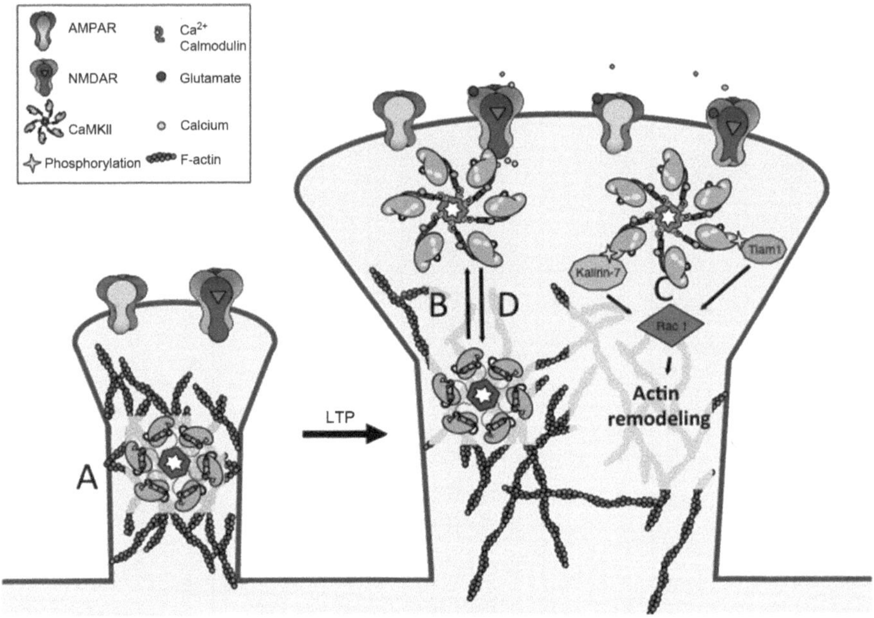

Figure 3.2 The structural role of CaMKII in spine growth during LTP. (A) In the basal synapse, inactive CaMKIIβ bundles F-actin filaments to stabilize the structure of the spine.[101] (B) During LTP induction, CaMKIIβ is activated and uncouples it from F-actin, allowing for the reorganization of actin through other signaling pathways. (C) CaMKII can also promote reorganization of F-actin via the activation of Kalirin-7[121] and/or Tiam1[124–126] which signal through Rac1 to cause spine enlargement. (D) After the induction phase, inactive CaMKIIIβ can rebind and bundle the F-actin to stabilize the new structure during the maintenance of LTP. (See the color plate.)

Although initial studies suggested that Ser831 phosphorylation enhanced the conductance of only homomeric GluA1-AMPARs,[137] a recent study found that association of the AMPAR auxiliary subunit stargazin permitted enhanced conductance in GluA1/GluA2 heteromers following GluA1 phosphorylation at Ser831 by CaMKII.[138–140] Moreover, several studies suggest that Ser831 phosphorylation does in fact increase during LTP[141,142] and that AMPAR conductance increases following LTP induction.[143] However, preventing Ser831 phosphorylation alone in S831A-KI mice does not prevent LTP unless Ser845, a target of PKA, is also mutated to alanine.[144]

LTP is also substantially dependent on the synaptic insertion of new AMPARs. Indeed, it is this immobilization of AMPARs within the PSD that likely allows for long-term enhancement of transmission during the maintenance of LTP. The current model for GluA1-AMPAR insertion during LTP involves a three-step process: (1) AMPARs are first inserted from intracellular vesicular pools into the plasma membrane at extrasynaptic sites; (2) plasma membrane AMPARs diffuse laterally into the synapse; and (3) synaptic AMPARs are then stabilized within the PSD (for review, see Ref. 145). Each of these steps can be modulated by several mechanisms, and CaMKII has been implicated at multiple steps. For example, CaMKII may regulate targets like myosin Va, SAP97, stargazin, as well as others, which are summarized in Table 3.1 and Fig. 3.3.

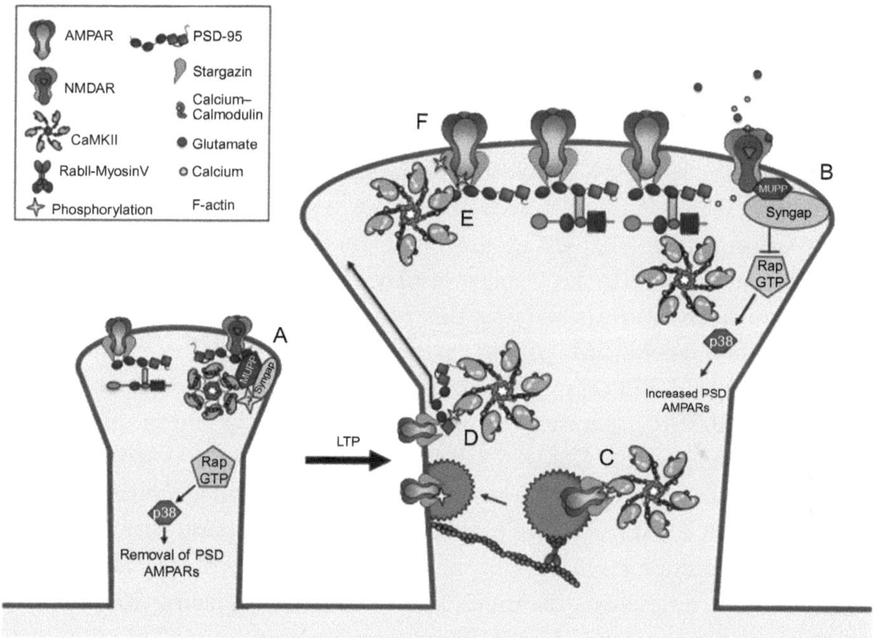

Figure 3.3 The role of CaMKII in potentiation of AMPAR-mediated transmission during LTP induction. (A) In the basal state, CaMKII forms a complex with MUPP1 and SynGAPα.[115] This permits the CaMKII-mediated phosphorylation of SynGAP, inhibiting its GAP activity, thus disinhibiting RapGTP and p38 MAPK. P38 promotes the removal of AMPARs from the synapses,[146] thereby maintaining a low-basal state of transmission. (B) Activation of CaMKII disrupts its binding to the MUPP1–SynGAP complex, after which SynGAP is dephosphorylated allowing for the inhibition of p38 MAPK[115] signaling via RapGTP, thus promoting synaptic accumulation of AMPARs. (C) Activated CaMKII is known to be important for the trafficking of AMPARs; however, the exact mechanism is not clear and may involve multiple mechanisms. (D) After myosin V-dependent insertion at extrasynaptic sites, AMPARs are trafficked by lateral diffusion into the PSD.[145] (E) Stargazin phosphorylation by CaMKII is required to trap AMPARs in the PSD by permitting the binding of stargazin to PSD-95.[110] (F) Phosphorylation of synaptic GluA1 subunits at S831 increases AMPAR conductance after the induction of LTP.[141–143] (See the color plate.)

10. CaMKII AND LTD

In addition to the aforementioned prototypical roles of CaMKII in LTP, several recent studies have implicated CaMKII in synaptic depression. In the hippocampus, mGluR1/5-mediated LTD can occur through the internalization of AMPARs, in a protein-synthesis-dependent mechanism.[147] Recent work by Mockett *et al.* showed that acute inhibition of CaMKII prevented the induction of mGluR-LTD in the hippocampus.[148] Furthermore, CaMKII inhibition also prevented the mGluR-mediated induction of protein synthesis occurring after acute activation of mGluRs, but did not affect basal protein synthesis. This was supported by the fact that CaMKII-Thr286 phosphorylation was increased after activation of mGluRs by the agonist DHPG, suggesting that CaMKII may play a central role in this form of LTD. CaMKIIα is also essential for LTD at parallel fiber–Purkinje cell synapses in the cerebellum.[149] Moreover, this study showed that CaMKIIα-KO mice display deficits in cerebellar learning tasks. The mechanism underlying this process was recently explored in a study by Kawaguchi and Hirano, which suggests that during induction of this form of LTD, CaMKII may indirectly signal through the nitric oxide (cGMP/PKG) pathway by negatively regulating phosphodiesterase 1.[150] Interestingly, the CaMKIIα-KO mice do not have deficits in LTP at Purkinje cell synapses[149]; however, CaMKIIβ-KO mice actually exhibit an inversion in the direction of plasticity such that protocols that elicit LTP in WT mice induce LTD in KO mice and LTD protocols trigger LTP.[151] These studies emphasize the nuances of CaMKII signaling mechanisms such that these conserved CaMKII isoforms apparently mediate quite distinct downstream signals in some cellular contexts.

Our lab has recently demonstrated that CaMKII plays a role in modulating endocannabinoid plasticity in the striatum.[120] The endocannabinoid 2-arachidonoyl glycerol (2-AG) mediates both short- and long-term depression through its retrograde action on presynaptic cannabinoid type 1 receptors (CB1R). 2-AG is synthesized on demand in the spine by diacylglycerol lipase α (DGLα) in response to mGluR activation, increased spine Ca^{2+}, or a synergistic pathway involving both. These studies revealed that CaMKIIα directly binds to DGLα *in vitro* and coimmunoprecipitates with DGLα in mouse striatum. Moreover, CaMKIIα phosphorylates DGLα at Ser782 and Ser808, and phosphomimetic mutations of these sites decrease DGLα activity *in vitro*. CaMKII modulation of 2-AG synthesis appears to play a tonic role in restraining 2-AG signaling in striatal neurons since acute inhibition of postsynaptic CaMKII, or Thr286 to Ala mutation in CaMKIIα, enhances endocannabinoid-mediated short-term depression.

11. CONCLUDING REMARKS

Although much has been learned about the importance of CaMKII in plasticity as well as learning and memory, there is still much that remains unclear. For example, carefully detailing how CaMKII is regulated by specific protein–protein interactions will likely generate significant advances in our understanding of how plasticity is regulated at different synapses and under different conditions. Moreover, a large number of phosphorylation sites on CaMKII remain unexplored and may contribute greatly to synaptic function by either autophosphorylation or phosphorylation by other kinases. Elucidating how CaMKII is specifically and differentially regulated at different synapses or possibly during the storage of different forms of memory would greatly enhance our understanding of memory formation throughout the central nervous system.

REFERENCES

1. Lisman JE. A mechanism for memory storage insensitive to molecular turnover: a bistable autophosphorylating kinase. *Proc Natl Acad Sci USA*. 1985;82(9):3055–3057.
2. Malenka RC, Nicoll RA. Long-term potentiation—a decade of progress? *Science*. 1999;285(5435):1870–1874.
3. Albensi BC, Oliver DR, Toupin J, Odero G. Electrical stimulation protocols for hippocampal synaptic plasticity and neuronal hyper-excitability: are they effective or relevant? *Exp Neurol*. 2007;204(1):1–13.
4. Higley MJ, Sabatini BL. Calcium signaling in dendritic spines. *Cold Spring Harb Perspect Biol*. 2012;4(4):a005686.
5. Kennedy MB. The postsynaptic density at glutamatergic synapses. *Trends Neurosci*. 1997;20(6):264–268.
6. Shi SH, Hayashi Y, Petralia RS, et al. Rapid spine delivery and redistribution of AMPA receptors after synaptic NMDA receptor activation. *Science*. 1999;284(5421):1811–1816.
7. Shi S, Hayashi Y, Esteban JA, Malinow R. Subunit-specific rules governing AMPA receptor trafficking to synapses in hippocampal pyramidal neurons. *Cell*. 2001;105(3):331–343.
8. Lledo PM, Hjelmstad GO, Mukherji S, Soderling TR, Malenka RC, Nicoll RA. Calcium/calmodulin-dependent kinase II and long-term potentiation enhance synaptic transmission by the same mechanism. *Proc Natl Acad Sci USA*. 1995;92(24):11175–11179.
9. Poncer JC, Esteban JA, Malinow R. Multiple mechanisms for the potentiation of AMPA receptor-mediated transmission by alpha-Ca2+/calmodulin-dependent protein kinase II. *J Neurosci*. 2002;22(11):4406–4411.
10. Thiagarajan TC, Piedras-Renteria ES, Tsien RW. alpha- and betaCaMKII. Inverse regulation by neuronal activity and opposing effects on synaptic strength. *Neuron*. 2002;36(6):1103–1114.

11. Hayashi Y, Shi SH, Esteban JA, Piccini A, Poncer JC, Malinow R. Driving AMPA receptors into synapses by LTP and CaMKII: requirement for GluR1 and PDZ domain interaction. *Science.* 2000;287(5461):2262–2267.
12. Fukunaga K, Stoppini L, Miyamoto E, Muller D. Long-term potentiation is associated with an increased activity of Ca2+/calmodulin-dependent protein kinase II. *J Biol Chem.* 1993;268(11):7863–7867.
13. Lee SJ, Escobedo-Lozoya Y, Szatmari EM, Yasuda R. Activation of CaMKII in single dendritic spines during long-term potentiation. *Nature.* 2009;458(7236):299–304.
14. Strack S, Choi S, Lovinger DM, Colbran RJ. Translocation of autophosphorylated calcium/calmodulin-dependent protein kinase II to the postsynaptic density. *J Biol Chem.* 1997;272(21):13467–13470.
15. Chen HX, Otmakhov N, Strack S, Colbran RJ, Lisman JE. Is persistent activity of calcium/calmodulin-dependent kinase required for the maintenance of LTP? *J Neurophysiol.* 2001;85(4):1368–1376.
16. Silva AJ, Stevens CF, Tonegawa S, Wang Y. Deficient hippocampal long-term potentiation in alpha-calcium-calmodulin kinase II mutant mice. *Science.* 1992;257(5067):201–206.
17. Silva AJ, Paylor R, Wehner JM, Tonegawa S. Impaired spatial learning in alpha-calcium-calmodulin kinase II mutant mice. *Science.* 1992;257(5067):206–211.
18. Colbran RJ, Soderling TR. Calcium/calmodulin-dependent protein kinase II. *Curr Top Cell Regul.* 1990;31:181–221.
19. Kennedy MB, McGuinness T, Greengard P. A calcium/calmodulin-dependent protein kinase from mammalian brain that phosphorylates Synapsin I: partial purification and characterization. *J Neurosci.* 1983;3(4):818–831.
20. Hudmon A, Schulman H. Neuronal CA2+/calmodulin-dependent protein kinase II: the role of structure and autoregulation in cellular function. *Annu Rev Biochem.* 2002;71:473–510.
21. Feng B, Raghavachari S, Lisman J. Quantitative estimates of the cytoplasmic, PSD, and NMDAR-bound pools of CaMKII in dendritic spines. *Brain Res.* 2011;1419:46–52.
22. Erondu NE, Kennedy MB. Regional distribution of type II Ca2+/calmodulin-dependent protein kinase in rat brain. *J Neurosci.* 1985;5(12):3270–3277.
23. Cheng D, Hoogenraad CC, Rush J, et al. Relative and absolute quantification of postsynaptic density proteome isolated from rat forebrain and cerebellum. *Mol Cell Proteomics.* 2006;5(6):1158–1170.
24. Chao LH, Stratton MM, Lee IH, et al. A mechanism for tunable autoinhibition in the structure of a human Ca2+/calmodulin-dependent kinase II holoenzyme. *Cell.* 2011;146(5):732–745.
25. Kanaseki T, Ikeuchi Y, Sugiura H, Yamauchi T. Structural features of Ca2+/calmodulin-dependent protein kinase II revealed by electron microscopy. *J Cell Biol.* 1991;115(4):1049–1060.
26. Kolodziej SJ, Hudmon A, Waxham MN, Stoops JK. Three-dimensional reconstructions of calcium/calmodulin-dependent (CaM) kinase IIalpha and truncated CaM kinase IIalpha reveal a unique organization for its structural core and functional domains. *J Biol Chem.* 2000;275(19):14354–14359.
27. Bennett MK, Erondu NE, Kennedy MB. Purification and characterization of a calmodulin-dependent protein kinase that is highly concentrated in brain. *J Biol Chem* 1983;258(20):12735–12744.
28. Miller SG, Kennedy MB. Distinct forebrain and cerebellar isozymes of type II Ca2+/calmodulin-dependent protein kinase associate differently with the postsynaptic density fraction. *J Biol Chem.* 1985;260(15):9039–9046.
29. Rellos P, Pike AC, Niesen FH, et al. Structure of the CaMKIIdelta/calmodulin complex reveals the molecular mechanism of CaMKII kinase activation. *PLoS Biol.* 2010;8(7):e1000426.

30. Brocke L, Chiang LW, Wagner PD, Schulman H. Functional implications of the subunit composition of neuronal CaM kinase II. *J Biol Chem.* 1999;274(32):22713–22722.
31. Mukherji S, Brickey DA, Soderling TR. Mutational analysis of secondary structure in the autoinhibitory and autophosphorylation domains of calmodulin kinase II. *J Biol Chem.* 1994;269(32):20733–20738.
32. Bradshaw JM, Hudmon A, Schulman H. Chemical quenched flow kinetic studies indicate an intraholoenzyme autophosphorylation mechanism for Ca2+/calmodulin-dependent protein kinase II. *J Biol Chem.* 2002;277(23):20991–20998.
33. Chao LH, Pellicena P, Deindl S, Barclay LA, Schulman H, Kuriyan J. Intersubunit capture of regulatory segments is a component of cooperative CaMKII activation. *Nat Struct Mol Biol.* 2010;17(3):264–272.
34. Hanson SK, Grotta JC, Waxham MN, Aronowski J, Ostrow P. Calcium/calmodulin-dependent protein kinase II activity in focal ischemia with reperfusion in rats. *Stroke.* 1994;25(2):466–473.
35. Meyer T, Hanson PI, Stryer L, Schulman H. Calmodulin trapping by calcium-calmodulin-dependent protein kinase. *Science.* 1992;256(5060):1199–1202.
36. Ohsako S, Nakazawa H, Sekihara S, Ikai A, Yamauchi T. Role of threonine-286 as autophosphorylation site for appearance of Ca2(+)-independent activity of calmodulin-dependent protein kinase II alpha subunit. *J Biochem.* 1991;109(1):137–143.
37. Griffith LC. Calcium/calmodulin-dependent protein kinase II: an unforgettable kinase. *J Neurosci.* 2004;24(39):8391–8393.
38. Mullasseril P, Dosemeci A, Lisman JE, Griffith LC. A structural mechanism for maintaining the 'on-state' of the CaMKII memory switch in the post-synaptic density. *J Neurochem.* 2007;103(1):357–364.
39. Giese KP, Fedorov NB, Filipkowski RK, Silva AJ. Autophosphorylation at Thr286 of the alpha calcium-calmodulin kinase II in LTP and learning. *Science.* 1998;279(5352):870–873.
40. Gustin RM, Shonesy BC, Robinson SL, et al. Loss of Thr286 phosphorylation disrupts synaptic CaMKIIalpha targeting, NMDAR activity and behavior in pre-adolescent mice. *Mol Cell Neurosci.* 2011;47(4):286–292.
41. Hanson PI, Schulman H. Inhibitory autophosphorylation of multifunctional Ca2+/calmodulin-dependent protein kinase analyzed by site-directed mutagenesis. *J Biol Chem.* 1992;267(24):17216–17224.
42. Colbran RJ. Inactivation of Ca2+/calmodulin-dependent protein kinase II by basal autophosphorylation. *J Biol Chem.* 1993;268(10):7163–7170.
43. Hashimoto Y, Schworer CM, Colbran RJ, Soderling TR. Autophosphorylation of Ca2+/calmodulin-dependent protein kinase II. Effects on total and Ca2+-independent activities and kinetic parameters. *J Biol Chem.* 1987;262(17):8051–8055.
44. Elgersma Y, Fedorov NB, Ikonen S, et al. Inhibitory autophosphorylation of CaMKII controls PSD association, plasticity, and learning. *Neuron.* 2002;36(3):493–505.
45. Pi HJ, Otmakhov N, El Gaamouch F, Lemelin D, De Koninck P, Lisman J. CaMKII control of spine size and synaptic strength: role of phosphorylation states and nonenzymatic action. *Proc Natl Acad Sci USA.* 2010;107(32):14437–14442.
46. Pi HJ, Otmakhov N, Lemelin D, De Koninck P, Lisman J. Autonomous CaMKII can promote either long-term potentiation or long-term depression, depending on the state of T305/T306 phosphorylation. *J Neurosci.* 2010;30(26):8704–8709.
47. Migues PV, Lehmann IT, Fluechter L, et al. Phosphorylation of CaMKII at Thr253 occurs in vivo and enhances binding to isolated postsynaptic densities. *J Neurochem.* 2006;98(1):289–299.
48. Colbran RJ, Soderling TR. Calcium/calmodulin-independent autophosphorylation sites of calcium/calmodulin-dependent protein kinase II. Studies on the effect of phosphorylation of threonine 305/306 and serine 314 on calmodulin binding using synthetic peptides. *J Biol Chem.* 1990;265(19):11213–11219.

49. Miller SG, Patton BL, Kennedy MB. Sequences of autophosphorylation sites in neuronal type II CaM kinase that control Ca2(+)-independent activity. *Neuron.* 1988;1(7):593–604.
50. Erickson JR, Joiner ML, Guan X, et al. A dynamic pathway for calcium-independent activation of CaMKII by methionine oxidation. *Cell.* 2008;133(3):462–474.
51. Luo M, Guan X, Luczak ED, et al. Diabetes increases mortality after myocardial infarction by oxidizing CaMKII. *J Clin Invest.* 2013;123(3):1262–1274.
52. Scott JA, Xie L, Li H, et al. The multifunctional Ca2+/calmodulin-dependent kinase II regulates vascular smooth muscle migration through matrix metalloproteinase 9. *Am J Physiol Heart Circ Physiol.* 2012;302(10):H1953–H1964.
53. He BJ, Joiner ML, Singh MV, et al. Oxidation of CaMKII determines the cardiotoxic effects of aldosterone. *Nat Med.* 2011;17(12):1610–1618.
54. Swaminathan PD, Purohit A, Soni S, et al. Oxidized CaMKII causes cardiac sinus node dysfunction in mice. *J Clin Invest.* 2011;121(8):3277–3288.
55. Shields SM, Ingebritsen TS, Kelly PT. Identification of protein phosphatase 1 in synaptic junctions: dephosphorylation of endogenous calmodulin-dependent kinase II and synapse-enriched phosphoproteins. *J Neurosci.* 1985;5(12):3414–3422.
56. Dosemeci A, Reese TS. Inhibition of endogenous phosphatase in a postsynaptic density fraction allows extensive phosphorylation of the major postsynaptic density protein. *J Neurochem.* 1993;61(2):550–555.
57. Fukunaga K, Kobayashi T, Tamura S, Miyamoto E. Dephosphorylation of autophosphorylated Ca2+/calmodulin-dependent protein kinase II by protein phosphatase 2C. *J Biol Chem.* 1993;268(1):133–137.
58. Genoux D, Haditsch U, Knobloch M, Michalon A, Storm D, Mansuy IM. Protein phosphatase 1 is a molecular constraint on learning and memory. *Nature.* 2002;418(6901):970–975.
59. Lisman JE, Zhabotinsky AM. A model of synaptic memory: a CaMKII/PP1 switch that potentiates transmission by organizing an AMPA receptor anchoring assembly. *Neuron.* 2001;31(2):191–201.
60. Blitzer RD, Connor JH, Brown GP, et al. Gating of CaMKII by cAMP-regulated protein phosphatase activity during LTP. *Science.* 1998;280(5371):1940–1942.
61. Brown GP, Blitzer RD, Connor JH, et al. Long-term potentiation induced by theta frequency stimulation is regulated by a protein phosphatase-1-operated gate. *J Neurosci.* 2000;20(21):7880–7887.
62. Jalan-Sakrikar N, Bartlett RK, Baucum 2nd AJ, Colbran RJ. Substrate-selective and calcium-independent activation of CaMKII by alpha-actinin. *J Biol Chem.* 2012;287(19):15275–15283.
63. Jiao Y, Jalan-Sakrikar N, Robison AJ, Baucum 2nd AJ, Bass MA, Colbran RJ. Characterization of a central Ca2+/calmodulin-dependent protein kinase IIalpha/beta binding domain in densin that selectively modulates glutamate receptor subunit phosphorylation. *J Biol Chem.* 2011;286(28):24806–24818.
64. Colbran RJ. Targeting of calcium/calmodulin-dependent protein kinase II. *Biochem J.* 2004;378(Pt 1):1–16.
65. Shen K, Teruel MN, Connor JH, Shenolikar S, Meyer T. Molecular memory by reversible translocation of calcium/calmodulin-dependent protein kinase II. *Nat Neurosci.* 2000;3(9):881–886.
66. Shen K, Meyer T. Dynamic control of CaMKII translocation and localization in hippocampal neurons by NMDA receptor stimulation. *Science.* 1999;284(5411):162–166.
67. Kwok S, Lee C, Sanchez SA, Hazlett TL, Gratton E, Hayashi Y. Genetically encoded probe for fluorescence lifetime imaging of CaMKII activity. *Biochem Biophys Res Commun.* 2008;369(2):519–525.
68. Takao K, Okamoto K, Nakagawa T, et al. Visualization of synaptic Ca2+/calmodulin-dependent protein kinase II activity in living neurons. *J Neurosci.* 2005;25(12):3107–3112.

69. Thaler C, Koushik SV, Puhl 3rd HL, Blank PS, Vogel SS. Structural rearrangement of CaMKIIalpha catalytic domains encodes activation. *Proc Natl Acad Sci USA*. 2009;106(15):6369–6374.
70. Mower AF, Kwok S, Yu H, et al. Experience-dependent regulation of CaMKII activity within single visual cortex synapses in vivo. *Proc Natl Acad Sci USA*. 2011;108(52):21241–21246.
71. Sabatini BL, Oertner TG, Svoboda K. The life cycle of Ca(2+) ions in dendritic spines. *Neuron*. 2002;33(3):439–452.
72. Okamoto K, Narayanan R, Lee SH, Murata K, Hayashi Y. The role of CaMKII as an F-actin-bundling protein crucial for maintenance of dendritic spine structure. *Proc Natl Acad Sci USA*. 2007;104(15):6418–6423.
73. Malinow R, Schulman H, Tsien RW. Inhibition of postsynaptic PKC or CaMKII blocks induction but not expression of LTP. *Science*. 1989;245(4920):862–866.
74. Sanhueza M, Fernandez-Villalobos G, Stein IS, et al. Role of the CaMKII/NMDA receptor complex in the maintenance of synaptic strength. *J Neurosci*. 2011;31(25):9170–9178.
75. Strack S, McNeill RB, Colbran RJ. Mechanism and regulation of calcium/calmodulin-dependent protein kinase II targeting to the NR2B subunit of the N-methyl-D-aspartate receptor. *J Biol Chem*. 2000;275(31):23798–23806.
76. Bayer KU, De Koninck P, Leonard AS, Hell JW, Schulman H. Interaction with the NMDA receptor locks CaMKII in an active conformation. *Nature*. 2001;411(6839):801–805.
77. Robison AJ, Bartlett RK, Bass MA, Colbran RJ. Differential modulation of Ca2+/calmodulin-dependent protein kinase II activity by regulated interactions with N-methyl-D-aspartate receptor NR2B subunits and alpha-actinin. *J Biol Chem*. 2005;280(47):39316–39323.
78. Cheriyan J, Kumar P, Mayadevi M, Surolia A, Omkumar RV. Calcium/calmodulin dependent protein kinase II bound to NMDA receptor 2B subunit exhibits increased ATP affinity and attenuated dephosphorylation. *PLoS One*. 2011;6(3):e16495.
79. Pradeep KK, Cheriyan J, Suma Priya SD, et al. Regulation of Ca2+/calmodulin-dependent protein kinase II catalysis by N-methyl-D-aspartate receptor subunit 2B. *Biochem J*. 2009;419(1):123–132, 124 pp following 132.
80. Bayer KU, LeBel E, McDonald GL, O'Leary H, Schulman H, De Koninck P. Transition from reversible to persistent binding of CaMKII to postsynaptic sites and NR2B. *J Neurosci*. 2006;26(4):1164–1174.
81. Leonard AS, Bayer KU, Merrill MA, et al. Regulation of calcium/calmodulin-dependent protein kinase II docking to N-methyl-D-aspartate receptors by calcium/calmodulin and alpha-actinin. *J Biol Chem*. 2002;277(50):48441–48448.
82. Strack S, Colbran RJ. Autophosphorylation-dependent targeting of calcium/calmodulin-dependent protein kinase II by the NR2B subunit of the N-methyl-D-aspartate receptor. *J Biol Chem*. 1998;273(33):20689–20692.
83. Gardoni F, Schrama LH, Kamal A, Gispen WH, Cattabeni F, Di Luca M. Hippocampal synaptic plasticity involves competition between Ca2+/calmodulin-dependent protein kinase II and postsynaptic density 95 for binding to the NR2A subunit of the NMDA receptor. *J Neurosci*. 2001;21(5):1501–1509.
84. Leonard AS, Lim IA, Hemsworth DE, Horne MC, Hell JW. Calcium/calmodulin-dependent protein kinase II is associated with the N-methyl-D-aspartate receptor. *Proc Natl Acad Sci USA*. 1999;96(6):3239–3244.
85. Meng F, Guo J, Zhang Q, Song B, Zhang G. Autophosphorylated calcium/calmodulin-dependent protein kinase II alpha (CaMKII alpha) reversibly targets to and phosphorylates N-methyl-D-aspartate receptor subunit 2B (NR2B) in cerebral ischemia and reperfusion in hippocampus of rats. *Brain Res*. 2003;967(1–2):161–169.
86. Meng F, Zhang G. Autophosphorylated calcium/calmodulin-dependent protein kinase II alpha induced by cerebral ischemia immediately targets and phosphorylates

N-methyl-D-aspartate receptor subunit 2B (NR2B) in hippocampus of rats. *Neurosci Lett.* 2002;333(1):59–63.
87. Barria A, Malinow R. NMDA receptor subunit composition controls synaptic plasticity by regulating binding to CaMKII. *Neuron.* 2005;48(2):289–301.
88. Zhou Y, Takahashi E, Li W, et al. Interactions between the NR2B receptor and CaMKII modulate synaptic plasticity and spatial learning. *J Neurosci.* 2007;27(50):13843–13853.
89. Sanz-Clemente A, Gray JA, Ogilvie KA, Nicoll RA, Roche KW. Activated CaMKII couples GluN2B and casein kinase 2 to control synaptic NMDA receptors. *Cell Rep.* 2013;3(3):607–614.
90. Halt AR, Dallapiazza RF, Zhou Y, et al. CaMKII binding to GluN2B is critical during memory consolidation. *EMBO J.* 2012;31(5):1203–1216.
91. Patterson MA, Szatmari EM, Yasuda R. AMPA receptors are exocytosed in stimulated spines and adjacent dendrites in a Ras-ERK-dependent manner during long-term potentiation. *Proc Natl Acad Sci USA.* 2010;107(36):15951–15956.
92. Chwang WB, O'Riordan KJ, Levenson JM, Sweatt JD. ERK/MAPK regulates hippocampal histone phosphorylation following contextual fear conditioning. *Learn Mem.* 2006;13(3):322–328.
93. Bading H, Greenberg ME. Stimulation of protein tyrosine phosphorylation by NMDA receptor activation. *Science.* 1991;253(5022):912–914.
94. El Gaamouch F, Buisson A, Moustie O, et al. Interaction between alphaCaMKII and GluN2B controls ERK-dependent plasticity. *J Neurosci.* 2012;32(31):10767–10779.
95. Chung HJ, Huang YH, Lau LF, Huganir RL. Regulation of the NMDA receptor complex and trafficking by activity-dependent phosphorylation of the NR2B subunit PDZ ligand. *J Neurosci.* 2004;24(45):10248–10259.
96. Sanz-Clemente A, Matta JA, Isaac JT, Roche KW. Casein kinase 2 regulates the NR2 subunit composition of synaptic NMDA receptors. *Neuron.* 2010;67(6):984–996.
97. Robison AJ, Bass MA, Jiao Y, et al. Multivalent interactions of calcium/calmodulin-dependent protein kinase II with the postsynaptic density proteins NR2B, densin-180, and alpha-actinin-2. *J Biol Chem.* 2005;280(42):35329–35336.
98. Cingolani LA, Goda Y. Actin in action: the interplay between the actin cytoskeleton and synaptic efficacy. *Nat Rev Neurosci.* 2008;9(5):344–356.
99. Dillon C, Goda Y. The actin cytoskeleton: integrating form and function at the synapse. *Annu Rev Neurosci.* 2005;28:25–55.
100. Okamoto K, Bosch M, Hayashi Y. The roles of CaMKII and F-actin in the structural plasticity of dendritic spines: a potential molecular identity of a synaptic tag? *Physiology.* 2009;24:357–366.
101. Shen K, Teruel MN, Subramanian K, Meyer T. CaMKIIbeta functions as an F-actin targeting module that localizes CaMKIIalpha/beta heterooligomers to dendritic spines. *Neuron.* 1998;21(3):593–606.
102. Borgesius NZ, van Woerden GM, Buitendijk GH, et al. betaCaMKII plays a nonenzymatic role in hippocampal synaptic plasticity and learning by targeting alphaCaMKII to synapses. *J Neurosci.* 2011;31(28):10141–10148.
103. Fink CC, Bayer KU, Myers JW, Ferrell Jr JE, Schulman H, Meyer T. Selective regulation of neurite extension and synapse formation by the beta but not the alpha isoform of CaMKII. *Neuron.* 2003;39(2):283–297.
104. Baucum 2nd AJ, Jalan-Sakrikar N, Jiao Y, et al. Identification and validation of novel spinophilin-associated proteins in rodent striatum using an enhanced ex vivo shotgun proteomics approach. *Mol Cell Proteomics.* 2010;9(6):1243–1259.
105. Baucum 2nd AJ, Brown AM, Colbran RJ. Differential association of postsynaptic signaling protein complexes in striatum and hippocampus. *J Neurochem.* 2013;124(4):490–501.
106. McDonald BJ, Moss SJ. Differential phosphorylation of intracellular domains of gamma-aminobutyric acid type A receptor subunits by calcium/calmodulin type

2-dependent protein kinase and cGMP-dependent protein kinase. *J Biol Chem.* 1994;269(27):18111–18117.
107. Houston CM, Lee HH, Hosie AM, Moss SJ, Smart TG. Identification of the sites for CaMK-II-dependent phosphorylation of GABA(A) receptors. *J Biol Chem.* 2007;282(24):17855–17865.
108. Varga AW, Yuan LL, Anderson AE, et al. Calcium-calmodulin-dependent kinase II modulates Kv4.2 channel expression and upregulates neuronal A-type potassium currents. *J Neurosci.* 2004;24(14):3643–3654.
109. Tomita S, Stein V, Stocker TJ, Nicoll RA, Bredt DS. Bidirectional synaptic plasticity regulated by phosphorylation of stargazin-like TARPs. *Neuron.* 2005;45(2):269–277.
110. Opazo P, Labrecque S, Tigaret CM, et al. CaMKII triggers the diffusional trapping of surface AMPARs through phosphorylation of stargazin. *Neuron.* 2010;67(2):239–252.
111. Walaas SI, Nairn AC. Multisite phosphorylation of microtubule-associated protein 2 (MAP-2) in rat brain: peptide mapping distinguishes between cyclic AMP-, calcium/calmodulin-, and calcium/phospholipid-regulated phosphorylation mechanisms. *J Mol Neurosci.* 1989;1(2):117–127.
112. Yamamoto H, Fukunaga K, Tanaka E, Miyamoto E. Ca2+- and calmodulin-dependent phosphorylation of microtubule-associated protein 2 and tau factor, and inhibition of microtubule assembly. *J Neurochem.* 1983;41(4):1119–1125.
113. Diaz-Nido J, Montoro RJ, Lopez-Barneo J, Avila J. High external potassium induces an increase in the phosphorylation of the cytoskeletal protein MAP2 in rat hippocampal slices. *Eur J Neurosci.* 1993;5(7):818–824.
114. Fukunaga K, Muller D, Miyamoto E. Increased phosphorylation of Ca2+/calmodulin-dependent protein kinase II and its endogenous substrates in the induction of long-term potentiation. *J Biol Chem.* 1995;270(11):6119–6124.
115. Krapivinsky G, Medina I, Krapivinsky L, Gapon S, Clapham DE. SynGAP-MUPP1-CaMKII synaptic complexes regulate p38 MAP kinase activity and NMDA receptor-dependent synaptic AMPA receptor potentiation. *Neuron.* 2004;43(4):563–574.
116. Costa MC, Mani F, Santoro Jr W, Espreafico EM, Larson RE. Brain myosin-V, a calmodulin-carrying myosin, binds to calmodulin-dependent protein kinase II and activates its kinase activity. *J Biol Chem.* 1999;274(22):15811–15819.
117. Karcher RL, Roland JT, Zappacosta F, et al. Cell cycle regulation of myosin-V by calcium/calmodulin-dependent protein kinase II. *Science.* 2001;293(5533):1317–1320.
118. Correia SS, Bassani S, Brown TC, et al. Motor protein-dependent transport of AMPA receptors into spines during long-term potentiation. *Nat Neurosci.* 2008;11(4):457–466.
119. Gardoni F, Mauceri D, Fiorentini C, et al. CaMKII-dependent phosphorylation regulates SAP97/NR2A interaction. *J Biol Chem.* 2003;278(45):44745–44752.
120. Shonesy BC, Wang X, Rose KL, et al. CaMKII regulates diacylglycerol lipase-alpha and striatal endocannabinoid signaling. *Nat Neurosci.* 2013;16(4):456–463.
121. Xie Z, Srivastava DP, Photowala H, et al. Kalirin-7 controls activity-dependent structural and functional plasticity of dendritic spines. *Neuron.* 2007;56(4):640–656.
122. Chen HJ, Rojas-Soto M, Oguni A, Kennedy MB. A synaptic Ras-GTPase activating protein (p135 SynGAP) inhibited by CaM kinase II. *Neuron.* 1998;20(5):895–904.
123. Li W, Okano A, Tian QB, et al. Characterization of a novel synGAP isoform, synGAP-beta. *J Biol Chem.* 2001;276(24):21417–21424.
124. Fleming IN, Elliott CM, Exton JH. Phospholipase C-gamma, protein kinase C and Ca2+/calmodulin-dependent protein kinase II are involved in platelet-derived growth factor-induced phosphorylation of Tiam1. *FEBS Lett.* 1998;429(3):229–233.
125. Tolias KF, Bikoff JB, Burette A, et al. The Rac1-GEF Tiam1 couples the NMDA receptor to the activity-dependent development of dendritic arbors and spines. *Neuron.* 2005;45(4):525–538.
126. Carlisle HJ, Manzerra P, Marcora E, Kennedy MB. SynGAP regulates steady-state and activity-dependent phosphorylation of cofilin. *J Neurosci.* 2008;28(50):13673–13683.

127. Wu X, McMurray CT. Calmodulin kinase II attenuation of gene transcription by preventing cAMP response element-binding protein (CREB) dimerization and binding of the CREB-binding protein. *J Biol Chem*. 2001;276(3):1735–1741.
128. Atkins CM, Nozaki N, Shigeri Y, Soderling TR. Cytoplasmic polyadenylation element binding protein-dependent protein synthesis is regulated by calcium/calmodulin-dependent protein kinase II. *J Neurosci*. 2004;24(22):5193–5201.
129. Atkins CM, Davare MA, Oh MC, Derkach V, Soderling TR. Bidirectional regulation of cytoplasmic polyadenylation element-binding protein phosphorylation by Ca2+/calmodulin-dependent protein kinase II and protein phosphatase 1 during hippocampal long-term potentiation. *J Neurosci*. 2005;25(23):5604–5610.
130. Alarcon JM, Malleret G, Touzani K, et al. Chromatin acetylation, memory, and LTP are impaired in CBP+/− mice: a model for the cognitive deficit in Rubinstein-Taybi syndrome and its amelioration. *Neuron*. 2004;42(6):947–959.
131. Robison AJ, Vialou V, Mazei-Robison M, et al. Behavioral and structural responses to chronic cocaine require a feedforward loop involving DeltaFosB and calcium/calmodulin-dependent protein kinase II in the nucleus accumbens shell. *J Neurosci*. 2013;33(10):4295–4307.
132. Zhou Z, Hong EJ, Cohen S, et al. Brain-specific phosphorylation of MeCP2 regulates activity-dependent Bdnf transcription, dendritic growth, and spine maturation. *Neuron*. 2006;52(2):255–269.
133. Bingol B, Wang CF, Arnott D, Cheng D, Peng J, Sheng M. Autophosphorylated CaMKIIalpha acts as a scaffold to recruit proteasomes to dendritic spines. *Cell*. 2010;140(4):567–578.
134. Hamilton AM, Oh WC, Vega-Ramirez H, et al. Activity-dependent growth of new dendritic spines is regulated by the proteasome. *Neuron*. 2012;74(6):1023–1030.
135. Djakovic SN, Schwarz LA, Barylko B, DeMartino GN, Patrick GN. Regulation of the proteasome by neuronal activity and calcium/calmodulin-dependent protein kinase II. *J Biol Chem*. 2009;284(39):26655–26665.
136. Luthi A, Wikstrom MA, Palmer MJ, et al. Bi-directional modulation of AMPA receptor unitary conductance by synaptic activity. *BMC Neurosci*. 2004;5:44.
137. Derkach V. Zooming in on AMPA receptor regulation by CaMKII. *Nat Neurosci*. 2011;14(6):674–675.
138. Kristensen AS, Jenkins MA, Banke TG, et al. Mechanism of Ca2+/calmodulin-dependent kinase II regulation of AMPA receptor gating. *Nat Neurosci*. 2011;14(6):727–735.
139. Barria A, Derkach V, Soderling T. Identification of the Ca2+/calmodulin-dependent protein kinase II regulatory phosphorylation site in the alpha-amino-3-hydroxyl-5-methyl-4-isoxazole-propionate-type glutamate receptor. *J Biol Chem*. 1997;272(52):32727–32730.
140. Roche KW, O'Brien RJ, Mammen AL, Bernhardt J, Huganir RL. Characterization of multiple phosphorylation sites on the AMPA receptor GluR1 subunit. *Neuron*. 1996;16(6):1179–1188.
141. Lee HK, Barbarosie M, Kameyama K, Bear MF, Huganir RL. Regulation of distinct AMPA receptor phosphorylation sites during bidirectional synaptic plasticity. *Nature*. 2000;405(6789):955–959.
142. Barria A, Muller D, Derkach V, Griffith LC, Soderling TR. Regulatory phosphorylation of AMPA-type glutamate receptors by CaM-KII during long-term potentiation. *Science*. 1997;276(5321):2042–2045.
143. Benke TA, Luthi A, Isaac JT, Collingridge GL. Modulation of AMPA receptor unitary conductance by synaptic activity. *Nature*. 1998;393(6687):793–797.

144. Lee HK, Takamiya K, He K, Song L, Huganir RL. Specific roles of AMPA receptor subunit GluR1 (GluA1) phosphorylation sites in regulating synaptic plasticity in the CA1 region of hippocampus. *J Neurophysiol.* 2010;103(1):479–489.
145. Opazo P, Choquet D. A three-step model for the synaptic recruitment of AMPA receptors. *Mol Cell Neurosci.* 2011;46(1):1–8.
146. Zhu JJ, Qin Y, Zhao M, Van Aelst L, Malinow R. Ras and Rap control AMPA receptor trafficking during synaptic plasticity. *Cell.* 2002;110(4):443–455.
147. Snyder EM, Philpot BD, Huber KM, Dong X, Fallon JR, Bear MF. Internalization of ionotropic glutamate receptors in response to mGluR activation. *Nat Neurosci.* 2001;4(11):1079–1085.
148. Mockett BG, Guevremont D, Wutte M, Hulme SR, Williams JM, Abraham WC. Calcium/calmodulin-dependent protein kinase II mediates group I metabotropic glutamate receptor-dependent protein synthesis and long-term depression in rat hippocampus. *J Neurosci.* 2011;31(20):7380–7391.
149. Hansel C, de Jeu M, Belmeguenai A, et al. alphaCaMKII Is essential for cerebellar LTD and motor learning. *Neuron.* 2006;51(6):835–843.
150. Kawaguchi SY, Hirano T. Gating of long-term depression by Ca2+/calmodulin-dependent protein kinase II through enhanced cGMP signalling in cerebellar Purkinje cells. *J Physiol.* 2013;591(Pt 7):1707–1730.
151. van Woerden GM, Hoebeek FE, Gao Z, et al. betaCaMKII controls the direction of plasticity at parallel fiber-Purkinje cell synapses. *Nat Neurosci.* 2009;12(7):823–825.
152. Sheng M, Hoogenraad CC. The postsynaptic architecture of excitatory synapses: a more quantitative view. *Annu Rev Biochem.* 2007;76:823–847.

CHAPTER FOUR

The Transcription Factor Zif268/ Egr1, Brain Plasticity, and Memory

Alexandra Veyrac[*,†], Antoine Besnard[‡], Jocelyne Caboche[§,¶,||], Sabrina Davis[*,†], Serge Laroche[*,†]

[*]CNRS, Centre de Neurosciences Paris-Sud, UMR 8195, Orsay, France
[†]Centre de Neurosciences Paris-Sud, Univ Paris-Sud, UMR 8195, Orsay, France
[‡]Harvard Stem Cell Institute, Harvard Medical School, Center for Regenerative Medicine, Massachusetts General Hospital, Boston, Massachusetts, USA
[§]INSERM, UMRS 952, Physiopathologie des Maladies du Système Nerveux Central, Paris, France
[¶]CNRS, UMR7224, Physiopathologie des Maladies du Système Nerveux Central, Paris, France
[||]UPMC University Paris 6, Paris, France

Contents

Progress in Molecular Biology and Translational Science, Volume 122
ISSN 1877-1173
http://dx.doi.org/10.1016/B978-0-12-420170-5.00004-0

Abstract

The capacity to remember our past experiences and organize our future draws on a number of cognitive processes that allow our brain to form and store neural representations that can be recalled and updated at will. In the brain, these processes require mechanisms of neural plasticity in the activated circuits, brought about by cellular and molecular changes within the neurons activated during learning. At the cellular level, a wealth of experimental data accumulated in recent years provides evidence that signaling from synapses to nucleus and the rapid regulation of the expression of immediate early genes encoding inducible, regulatory transcription factors is a key step in the mechanisms underlying synaptic plasticity and the modification of neural networks required for the laying down of memories. In the activated neurons, these transcriptional events are thought to mediate the activation of selective gene programs and subsequent synthesis of proteins, leading to stable functional and structural remodeling of the activated networks, so that the memory can later be reactivated upon recall. Over the past few decades, novel insights have been gained in identifying key transcriptional regulators that can control the genomic response of synaptically activated neurons. Here, as an example of this approach, we focus on one such activity-dependent transcription factor, Zif268, known to be implicated in neuronal plasticity and memory formation. We summarize current knowledge about the regulation and function of Zif268 in different types of brain plasticity and memory processes.

1. INTRODUCTION

A leading theory of how the brain encodes, stores, and represents information postulates that memories for events are represented as specific spatio-temporal patterns of synchronized firing activity in dynamically associated assemblies of neurons forming transiently active networks, and that changes take place at the cellular level within the activated networks to store these representations. This is based theoretically on the notion that the mechanisms subserving the formation and storage of memory representations reside in activity-driven modifications of synaptic strength and remodeling of neural networks so that the specific distribution of activity can later form a readout pattern during recall. Extensive work over the last decades has begun to shed light into the cellular and molecular events in neurons that subserve the mechanisms of brain plasticity required to store and maintain memories.

Among the many important advances that have been made in discovering some of these molecular mechanisms, one of the most intriguing is the realization that a cardinal step in the mechanisms underlying brain plasticity and memory storage is the regulation of selective gene programs leading to the synthesis of specific proteins in the activated neurons. As in many aspects of cell function involving regulated gene transcription, a critical step in triggering the genomic response of neurons is the expression of a class of immediate early genes (IEGs) encoding inducible transcription factors which interact with promoter regulatory elements of a host of downstream effector genes, to regulate their expression. This early genomic response generally occurs within minutes after cell-surface receptor activation and is mediated via the activation of multiple kinase cascades converging onto constitutively expressed transcriptional regulators. These, when activated by posttranslational modification such as phosphorylation in the nucleus, or after translocation to the nucleus, bind to specific response elements in the promoter region of IEGs and induce their expression. This constitutes an early genomic response required for triggering the mechanisms underlying persistent cell modification. In essence, it suggests a molecular process implemented in neurons that, when the neurons are sufficiently activated, is capable of changing their functional properties and, as a consequence, the properties of the circuits within which they are embedded, based on activation of transcriptional events initiating the regulated transcription of effector genes. To illustrate this molecular process and its potential role in long-term memory formation and synaptic plasticity, this review will focus on one such activity-dependent IEG encoding a transcriptional regulator, the gene *Zif268*.

2. ZIF268 AND THE EGR FAMILY OF TRANSCRIPTION FACTORS

2.1. The discovery of Zif268

Zif268/Egr1 (also known as Krox-24, NGF1-A, Egr1, TIS8, and Zenk) belongs to the early growth response (Egr) family of proteins, simultaneously described as an inducible transcription factor that controls the expression of genes and as such constitutes a master regulator of cell development and function, and as an immediate early or primary response gene, whose activation is rapid and transient and requires no *de novo* protein synthesis.[1–4] It was discovered in the late 1980s shortly after the discovery of the proto-oncogenes, c-Myc, c-Fos, and c-Jun. This class of genes became rapidly incorporated into a conceptual framework that they served a mechanism of "third messenger" linking neurotransmitter action with altered gene expression that underpinned rapid, activity-dependent modification to the cell upon membrane excitation.[5] The gene is evolutionary conserved[6]; it is present in the mammalian genome as a single-copy gene on chromosome 18 in the mouse. It encodes an 82 and an 88 kDa protein, depending on the initiation sequence,[7] and its predominant distinguishing feature, shared by the other members of the Egr family, is the three-zinc-finger sequence in the DNA-binding domain that allows the protein to regulate the transcription of other genes.[8,9] There is a high homology in the three-zinc-finger sequence in all members of the Egr family, suggesting that the proteins may bind to the same sites of at least a subset of the same target genes. The protein binds to cognate GC-rich sequences in the DNA containing the consensus 9-base-pair binding site GCG(G/T)GGGCG, the EGR response element (ERE).

2.2. Zif268 brain expression in developing and adult brain

Both gene and protein are relatively ubiquitously distributed both in the central nervous system and peripheral organs in the adult. As expected by the inducible nature of the gene, the mRNA and protein show very low basal expression in peripheral organs. In contrast basal expression in the brain is high by comparison, although varying according to brain structures; the highest levels being in several areas of the neocortex, primary olfactory structures, hippocampus, amygdala, striatum, thalamus, and cerebellar cortex. In the neocortex, Zif268 is expressed at its highest levels in layers IV and VI, and in the hippocampus, it is most highly expressed in CA1, to a lesser extent in CA2–CA4, with almost negligible expression in the dentate gyrus.[10] Early research showed that basal expression of both *Zif268* gene and protein in the brain appears to be tightly linked with activity of the NMDA receptors,[11] but are also responsive to AMPA receptor[12] and L-type voltage-sensitive calcium channel[13] activity. This led to the suggestion that high basal levels of Zif268 are maintained by synaptic activity in response to physiological stimuli[10] and may reflect ongoing physiological regulation by incoming stimuli.

During development, Zif268/Egr1 mRNA expression is completely absent in the embryonic nervous system but is expressed at low levels postnatally, where, over a period of about 2 weeks, expression levels increase to that observed in the adult, with different brain regions peaking at different times within this 2-week period. For example, hippocampal expression peaks at PND 14 and by PND 17 remain stable in CA1 but decline in DG to negligible levels throughout adulthood.

2.3. Mechanisms of Zif268 regulation

Zif268/Egr1 is regulated by classical mechanisms through transcription of the gene, and posttranslational modification of the protein. The promoter region of the *Zif268* gene contains six adjacent serum response element (SRE) sites, which constitute the dominant transcriptional regulatory sites. The transcription factor Elk-1 is associated with two molecules of serum response factor at the SRE site on the Zif268 promoter.[14,15] Its phosphorylation, which occurs in response to activation of MAPK/ERK, JNK, and p38MAPK signaling pathways, allows the recruitment of coactivators, including CBP (CREB binding protein) and p300, then leading to the activation of their intrinsic histone acetyltransferases activity. But Elk-1 has a complex function since it is also capable of recruiting histone deacetylases to repress activation of target genes.[16,17] The promoter also contains several specificity protein 1 consensus sequences, a putative AP1 binding site, and at least one CRE site. Zif268 can also regulate its own transcription by binding to its own Egr binding sites, where it suppresses its activity, a mechanism also seen with c-Fos.[18] More recently, other sites have been identified, allowing the binding of NF-κB-p65 to the promoter region of *Zif268* to mediate UV-irradiating cell death[19] and binding of the tumor suppressor p53 at two functional nonconsensus binding sites to sustain Zif268 expression and trigger apoptosis in tumor cells.[20] Finally, two conserved ATF5 response elements (AREs) have been identified upstream of the SRE sites in the Zif268 promoter. Interacting with, and being acetylated by p300, this ATF5/p300 complex stabilizes to allow it to bind to the ARE sites and promote cell proliferation and survival in cancer cells.[21]

The Zif268 protein comprises functionally independent domains with a threonine-serine rich N-terminal domain and a threonine-serine-proline rich C-terminal domain. A highly conserved DNA-binding domain between αα331 and αα419 contains three Cys2-His2-type zinc finger proteins that bind to prototypic CG-rich promoter sequences on downstream target genes. The protein also contains a bipartite nuclear localization signal at the N-terminal side of the first zinc finger and between zinc fingers 2 and 3.[22] A strong activator domain is followed by two repressor domains that can inhibit *Zif268* transcriptional activity by direct interaction with NAB1 and NAB2.[23,24] NAB2 can in fact be activated by the same stimuli that activate Zif268 expression, but its regulation is delayed compared with that of Zif268.[25,26] Importantly, Zif268 induces expression of *NAB2* gene via 11 ERE sites on its promoter region[27]; thus, Zif268 can restrict its own regulation of downstream targets. Together, this suggests that mechanisms are strategically poised on the *Zif268* gene and protein to very rapidly shut down its activity and ensure the transient nature of its activity.

As with most proteins, Zif268 proteins can be regulated by a number of different posttranslational modifications. It contains a number of O- and N-linked glycosylation sites but little is known of their function except that an N-linked glycosylation site is located on the second zinc finger that may be associated with its binding activity.[28] This may constitute a mechanism that allows the second and third zinc fingers to be nonspecifically bound to DNA during the search mode where Zif268 identifies its correct target sequence.[29] Zif268 also undergoes phosphorylation, with identified sites for at least casein kinase II, tyrosine kinases, and PKC, and this results in an increase in the half-life of the protein and its DNA-binding activity[18,30,31] and more recently, it has been shown that Akt can phosphorylate Zif268, at serine 350 and threonine 309.[32] An acetylation site close to the third zinc finger forming the end part of the basic nuclear localization signal[22] can be regulated by the P300/CBP complex, which prevents further transcriptional activity of Zif268.[33] Yu and colleagues[32] showed that Akt phosphorylation of Zif268 is necessary for its sumoylation that is mediated by Zif268 interacting with an ARF/sumoylation ligase, E2-conjugating enzyme complex that results in activation of the protein PTEN. Finally, two studies have shown Zif268 to be ubiquitinated, one by an indirect passive mechanism, whereby upon its physical interaction with the fusion protein PAX3-FOX01 that becomes ubiquitinated and subsequently degraded, Zif268 is also ubiquitinated and degraded.[34] The other has shown that Zif268 ubiquitination and degradation is triggered by its prior sumoylation[35] with the net outcome of reducing Zif268 protein levels. Apart from its physical interaction with proteins during sumoylation and ubiquitination, Zif268 can also bind to other transcription factors to direct expression of other genes. For example, Zif268 can form dimeric complexes with the transcription factors NFAT,[36] members of the NF-κB family,[37,38] c-Fos,[39] and c-Jun,[40] indicating that Zif268 can participate in the regulation of distinct gene programs.

2.4. Signaling via Zif268

As Zif268 is a transcription factor that regulates genomic programs potentially leading to enduring functional modification of the cell, a key question this begs is what are the downstream target genes activated by Zif268.

To our knowledge there have been no studies that have clearly identified direct downstream target genes of Zif268 that are activated in association with synaptic plasticity or memory processing. Much of our understanding of the target genes of Zif268 comes from research in pathology, notably cancer and analyses in artificial cell systems, and as such the number of potential target genes has been suggested to be in the high hundreds. Identification of target genes with ChIP-on-chip, gene-chip, or analysis of specific potential targets of Zif268 have shown a major class of Zif268 target genes associated with cell proliferation such as platelet-derived growth factor A and B (PDGF-A, PDGF-B) and insulin growth factor II,[41–43] and cell survival and differentiation via BCL-2, fibronectin, and nuclear factor-B.[44,45] Conversely, other target genes include those that induce apoptosis via p53, p73, PTEN and TNFα, FasL, MAX, and RRAS2[46,47] or inhibition of cell growth via TGF-β1 and p57/KIP2, which it does in a cell-type-specific manner.[48,49] These results suggest that Zif268 can act to either promote pathological and tumorigenic progression or to arrest it. On the whole, Zif268 regulation of growth factors has been observed in prostate cell cancer and fibrosis, whereas it initiates growth arrest in other cancers including lung cancer and glioblastomas.[48,50] The general consensus is that these opposing effects are cell-type specific, which would logically preclude these types of mechanisms being associated with plasticity and memory in neurons.

To a certain extent, this does appear to be the case as shown in a study using large-scale screen of hippocampal cells subjected to stimulation by NMDA.[51] In this study, overexpression of Zif268 led to regulation of 153 genes, the majority being categorized in functions including signaling, synapse formation, the immune response, the proteasome, and a number of transcription factors including *Zif268* itself. More importantly, whereas target genes in pathological conditions were upregulated with overexpression of *Zif268*, overexpression of NMDA-stimulated Zif268 led to downregulation in all but three genes, suggesting that Zif268 acts as a functional suppressor, a notion that is somewhat counterintuitive to the traditional view of a role for Zif268 in plasticity and memory. Focusing on downregulation of proteins associated with the proteasome, James and colleagues suggest reducing protein degradation via proteasomal mechanisms may constitute an important means for mediating postsynaptic mechanisms necessary for plasticity.[51] In other conditions, however, overexpression of Zif268 can result in upregulation of the target gene encoding the glucocorticoid receptor in 7-day-old pups born to mothers with a high level of maternal care via epigenetic mechanisms.[52] In addition, a number of genes that have been associated with pathological states have been identified as having Zif268 binding sites on their promoter regions that include p300 and CPB,[53] ARF,[32] neuroglobin,[54] and the T-type calcium channel Cav3.2.[55] Finally, Zif268 can mediate epigenetic programming via DNA methylation and histone posttranslational modifications to influence downstream gene regulation.[52,56]

3. THE ROLE OF ZIF268 IN SYNAPTIC PLASTICITY

3.1. Long-term potentiation

The idea that Zif268 could be an important transcriptional regulator involved in synaptic plasticity can be traced back to the discovery in the early 1990s that the induction of LTP in the dentate gyrus of the hippocampus is associated with rapid and robust transcription of Zif268 in the activated granule cells.[57,58] The induction of Zif268 mRNA occurs between 10 min and 2 h following LTP-inducing stimulation, is dependent on activation of the NMDA receptor and appears associated with the persistence of the protein synthesis-dependent late phase of LTP, rather than with the magnitude of LTP induction.[59–63] In the dentate gyrus, *Zif268* gene transcription and protein translation is associated with increased functional binding of the protein to its cognate ERE DNA sequence, suggesting activity-dependent transcription has a functional readout for the activation of target genes.[64]

The evidence that Zif268 plays a crucial role in LTP was established in 2001 by examination of LTP in *Zif268* mutant mice.[65] In awake mutant mice, the absence of Zif268 did not affect induction and early phases of dentate gyrus LTP, but LTP maintenance could not be maintained over 24 h. Interestingly, a similar rapid decay of LTP was observed in heterozygous mice, thus indicating that half the complement of Zif268 is not sufficient for LTP persistence in this brain area. Conversely, using a gain-of-function strategy, it was recently shown in transgenic mice that forebrain overexpression of *Zif268* can enhance LTP in the dentate gyrus.[66] Characterization of the cell-signaling mechanisms implicated in LTP-dependent transcription of Zif268 in dentate granule cells (DGCs) revealed the requirement of the MAPK/ERK signaling cascade, a molecular cascade known to be essential for LTP[67] and for IEG expression in cell systems.[68] Induction of LTP led to coordinate activation of ERK1/2, Elk-1 and CREB, two ERK downstream transcription factors within minutes and in the presence of an inhibitor of the upstream kinase, MEK, LTP-induced hyperphosphorylation of ERK1/2, Elk-1 and CREB, and the overexpression of Zif268 mRNA were blocked, resulting in rapidly declining LTP.[69] These findings support a model in which ERK-dependent phosphorylation of CREB and Elk-1, binding to CRE and SRE response elements on the *Zif268* promoter, contributes critically to the maintenance of LTP.

Zif268 is also upregulated following the induction of LTP in other brain structures such as area CA1 of the hippocampus in slices[70,71] and *in vivo*,[39] in the insular cortex following stimulation of the basolateral amygdala[72] and in the visual cortex following tetanic stimulation of the thalamocortical tract.[73] It is yet to be established, however, that Zif268 is required for the expression of LTP in these structures.

Although the key role of Zif268 expression for the maintenance of LTP is well established, the contribution of other members of the Egr family is less clear. Like Zif268, Egr2 and Egr3 are upregulated after LTP in the dentate gyrus.[74,75] A recent study, however, showed that while the mRNA for Egr2 and Egr3 are upregulated following LTP in area CA1, this is not associated with an apparent increase in protein levels.[39] To date, there is no report on whether Egr2 plays a role in LTP and that of Egr3 is unclear. In Egr3 mutant mice, LTP has been reported as either normal[76] or altered at the induction phase,[77] suggesting lack of Egr3 may affect synaptic proteins essential for induction of LTP rather than acting via activity-dependent regulated transcription.

3.2. Other forms of synaptic plasticity

Numerous studies have shown Zif268 to be regulated by seizure activity, electroconvulsive shock, kindling, and pathological stimuli or brain injury, which includes spreading depression, hypoxia-ischemia, nerve transection, neurodegeneration, apoptosis, drug tolerance and dependence, physiological stress, and drug-induced stress. The reader is referred to several reviews[10,78,79] for more in-depth discussion. Zif268 can also be activated by transcranial direct current stimulation that increases Schaffer collateral to CA1 pyramidal cell LTP[80] and by transcranial magnetic stimulation, another technique of brain stimulation capable of inducing plastic changes.[81] Studies in the spinal cord have also reported a form of LTP-like changes in excitability at dorsal horn neurons that is associated with Zif268 overexpression and sensitive to Zif268 knockdown.[82]

Few studies, however, examined the expression of Zif268 in association with long-term depression (LTD) of synaptic efficacy. An early attempt using chemical LTD induced by group I metabotropic glutamate receptor agonist DHPG in organotypic slices of the hippocampus was shown to induce the expression of several IEGs, including Zif268.[83] However, electrophysiological experiments *in vivo* reported no evidence for modulation of Zif268 expression following LTD in area CA1.[84] Three other forms of plasticity associated with Zif268 expression have been examined in *Zif268* mutant mice, but no clear phenotype was identified in the mutant mice. For example, despite a strong induction of Zif268 in the suprachiasmatic nucleus by photic stimuli capable of inducing a phase shift in the circadian rhythm, entrainment of phase shifting is not affected in *Zif268* mutant mice.[85] Similarly, the development of kindling by amygdala stimulation, which is associated with overexpression of Zif268 and axonal sprouting of mossy fibers, is not affected in mice carrying a null mutation of *Zif268*, suggesting that neither constitutive nor seizure-induced Zif268 expression is required for kindling development and seizure-associated synaptic reorganization.[86] Finally, Zif268 expression increases in the visual cortex with eye opening and during the critical period (P24–28), as well as following photo-stimulation after a period of dark adaptation in the adult, suggesting activity-dependent regulation of Zif268 in visual cortex.[87] However, the authors showed ocular dominance plasticity was not affected in *Zif268* mutant mice, suggesting that Zif268 is not required in developmental plasticity in this system. Thus, several forms of plasticity associated with Zif268 expression are not impaired in *Zif268* mutant mice, and the only clear phenotype in terms of neural plasticity at this point is a complete abolition of the late phases of hippocampal LTP.

4. THE ROLE OF ZIF268 IN LEARNING AND MEMORY

4.1. Learning and memory consolidation

As soon as the evidence that IEGs and, in particular, Zif268 can be regulated in an activity-dependent manner by LTP *in vivo* came about, and driven by the increasing knowledge about the role of LTP mechanisms in learning and memory, several laboratories examined whether Zif268 could be regulated in neurons in relation to naturally occurring environmental experience. Over the years, a wealth of experimental data has provided extensive evidence that the expression of *Zif268* is sensitive to natural environmental stimuli[88,89] or in a learning context (see Refs. 90–92 for reviews). Learning-related increases in Zif268 expression have been reported in many learning and memory paradigms and in different brain structures in relation to the task at hand. To cite a few examples, Zif268, but not c-Fos or JunD, is activated in the infero-temporal cortex of macaque monkeys trained in a form of explicit memory task, visual paired-associate learning, but not in a simple visual discrimination task,[93] and in the hippocampus after active avoidance learning[94] or brightness discrimination.[95] In songbirds, regulation of the Zif268 homolog Zenk (an acronym for Zif268, Egr1, NGFI-A, and Krox-24), occurs after presentation of con-specific songs[96] or associative learning,[97] and this response in the neostriatum and ventral hyperstriatum correlates with the strength of tutor song learning.[98] Learning-induced regulation of *Zif268* has also been examined in the hippocampus and amygdala in relation to contextual fear conditioning. In one study, an increase in *Zif268* expression, but not *c-Fos* or other members of the Egr family, has been observed in the lateral nucleus of the amygdala, but not in the hippocampus, after contextual conditioning.[99] Zif268 is also overexpressed in the lateral amygdala following cued-fear conditioning.[100] Other studies reported an increase in expression in both the lateral nucleus of the amygdala and area CA1 of the hippocampus,[88] although suggesting this might be merely due to novelty rather than learning *per se*. A recent study, however, confirmed activation of Zif268 in relation to the formation of contextual fear memory, highlighting specific dynamics of coordinated ERK1/2 phosphorylation and Zif268 activation, along with other IEGs, in the dentate gyrus after contextual conditioning.[101] Further analyses reported an increase in Zif268 binding activity to the ERE sequence after the establishment of a fear memory in an inhibitory avoidance learning task, indicating a functional readout of the increase in protein expression.[39] Conversely, downregulation of Zif268 expression occurs in the prefrontal cortex during extinction of fear memory.[102] In this region, extinction of cued-fear memory was shown to increase expression of nuclear p300/CBP-associated factor PCAF, which recruits the repressive transcription factor ATF4/CREB2 to the Zif268 promoter, resulting in decreased expression of Zif268; an epigenetic mechanism that may constrain post-recall reconsolidation of the fear memory and promote fear extinction.[102] Finally, *Zif268* and other IEGs are also overexpressed in the hippocampus after training in spatial or cued navigation tasks,[103] and in the dentate gyrus after spatial exploration of objects,[104] a task associated with synaptic potentiation in the hippocampus.[105]

As in LTP, the induction of Zif268 following learning, although with different dynamics of activation in different tasks, generally occurs rapidly, resulting in protein expression within 60–90 min after training, and is transient, suggesting a role in the initiation of mechanisms underlying consolidation of memory. In many studies, examination of Zif268 expressing neurons following learning is used as a marker for the mapping of behaviorally activated circuits across brain structures. This approach is particularly powerful for identifying neuronal ensembles activated in association with specific learning or memory processes, as exemplified in a recent study revealing Zif268 activation in distinct ensembles of DGCs in behavioral conditions requiring disambiguation of events that differ in memory demands.[106] However, the tight relationship between the expression of Zif268 and synaptic plasticity on one hand and the nature of the gene as a transcriptional regulator on the other is suggestive of the recruitment of molecular mechanisms controlling expression of specific gene programs that may serve memory stabilization.

The first demonstration of a functional role of Zif268 in memory formation came from the study of Jones and collaborators in *Zif268* knockout mice. Typically, *Zif268* mutant mice displayed intact short-term memory in several types of tasks; however, long-term memory was drastically impaired in tasks such as social transmission of food preference task, taste aversion memory, spatial memory, object recognition memory, and object-place recognition memory.[65,107,108] *Zif268* knockout mice were also recently shown to be impaired in consolidation of contextual fear memory.[109] However, previous studies using antisense oligonucleotides to partially knockdown Zif268 in specific structures showed that knockdown of Zif268 in the hippocampus does not impair contextual fear conditioning,[110] while Zif268 knockdown in the amygdala impairs both contextual[111] and cued-fear memory formation.[100]

The role of Zif268 in memory consolidation is also supported by recent studies investigating epigenetic regulation at the Zif268 promoter. Accumulating evidence suggests that transcriptional changes underlying long-term memory are supported by epigenetic modifications across diverse brain regions, including DNA methylation and posttranslational modifications of histone tails such as phosphorylation, acetylation, and methylation.[112] Several findings suggest an important role of Zif268 in experience-dependent epigenetic mechanisms that underlie memory. For example, regulation of histone methylation at the *Zif268* promoter can facilitate contextual fear memory,[113] histone H4 acetylation at the *Zif268* promoter correlates with Zif268 expression in the hippocampus,[114] and histone acetylation critically modulates object recognition memory consolidation.[115] Furthermore, exploration of objects was recently shown to trigger rapid phosphorylation, acetylation, and methylation of histones at the *Zif268* promoter in the hippocampus and the prefrontal cortex,[116] and blocking these epigenetic marks at the *Zif268* promoter impairs recognition memory, while their enhancement by intensive training or by transgenic intervention favors recognition memory.[116]

While the role of Zif268 in memory is well established, that of other members of the Egr family is less well known. The behavioral phenotype of *Egr3* mutant mice has also been characterized, and major differences with the neural and behavioral phenotype of *Zif268* mutant mice have been found.[76,77] *Egr3* mutant mice were found to have motor abnormalities due to defects in muscle spindles morphogenesis,[117] heightened reactivity to stressful stimuli, abnormalities in their adaptation to novel and stressful stimuli, and abnormalities in social interactions.[76] In addition, clear short-term memory deficits were identified in contextual fear conditioning and object recognition, with a consequential defect in long-term memory performance, together with compromised early phase of LTP in area CA1 of the hippocampus.[77] Deficits in NR2B-containing NMDA receptor function and in CA1 hippocampal LTD have also been reported in these mice.[76] More recently, we examined conditional, forebrain-specific *Egr2* mutant mice.[118] *Egr2* transgenic mice showed no sign of anxiety disturbance, nor was their motor activity and coordination impaired. Both conditional homozygous and heterozygous transgenic mice demonstrated normal learning and memory performance in spatial navigation, normal associative memory in a conditioned taste aversion paradigm, and normal fear learning and memory in a trace-fear conditioning paradigm. Surprisingly, however, the mice had superior implicit motor skill-learning abilities in an accelerating rotarod task and enhanced long-term object recognition memory.[118] In all, these studies suggest that Egr family members have different, and in certain circumstances antagonistic, functions in the adult brain. While there is evidence that Zif268 is required for the consolidation of long-term memory, Egr3 is primarily essential for short-term memory and Egr2, as opposed to Egr1 and Egr3, is dispensable for several forms of learning and memory and, on the contrary, can act as an inhibitory constraint for certain cognitive functions.[119]

The large variety of types of memories that are affected by impaired Zif268 expression suggests that Zif268 function in neurons may subserve a generalized mechanism implicated in memory consolidation, which, as expected, operates differentially in distinct brain structures and circuits depending on the contribution these structures and circuits play in the different types of memory. In addition to structure-specificity in relation to the type of memory, a second level of specificity is the dosage of *Zif268* gene expression. This is highlighted when comparing performance in different tasks of *Zif268* homozygous and heterozygous mice. Heterozygous mice, which carry half the complement of the protein, are able to learn as well as the wild-type mice in certain tasks such as object recognition, falls midway between the performance level of the wild-type and homozygous mice in tasks such as conditioned taste aversion and social transmission of food preference, and are as much impaired as the homozygous mice in other tasks such as spatial learning and memory in the water maze and memory of the spatial location of objects.[65,120] The data are consistent with a key role of Zif268 in hippocampal functions with an increasing sensitivity to Zif268 deficiency as the complexity and the explicit spatial demand of the task increase. Consistent with this, electrophysiological recordings of CA1 hippocampal pyramidal cells in *Zif268* knockout mice showed that while basic firing properties of place cells or the characteristics of place fields such as coherence and information content in a familiar environment are not degraded, Zif268 deficiency renders established spatial representations more sensitive to interference and impairs the formation of stable hippocampal place cell representations of novel environments.[121] The lack of long-term stability of newly formed neural representations and the destabilization of more remotely established cellular representations in *Zif268* knockout mice both suggest that *Zif268*-dependent mechanisms underlying long-term synaptic plasticity are essential for stabilizing CA1 place cell representations and the formation of long-term spatial memory.

The influence of *Zif268* gene expression dosage is also illustrated by two recent studies using a gain-of-function strategy. In transgenic forebrain-specific *Zif268* overexpressing mice, enhanced neuronal expression of Zif268 was shown to result in resistance to extinction in a conditioned taste aversion memory task[122] and to enhance the capacity to form a long-term memory of the spatial location of objects, without otherwise affecting non-spatial object recognition.[66] In all, the evidence to date indicate that transcriptional mechanisms engaging Zif268 contribute to determining the strength of newly encoded memories and that this operates in an intricate relationship between Zif268 expression dosage, the type of memory process involved, and the brain structures that are processing the ongoing information.

4.2. Systems consolidation

Contemporary literature provides strong evidence that the hippocampus is implicated in the initial storage of facts and events, however, the temporal gradients of retrograde amnesia observed in patients with hippocampal/temporal lobe damage have long suggested that recent memories are more severely affected in amnesic patients than remote memories. This has led to the suggestion that certain items in explicit memory are only temporarily stored in the hippocampus, while with time cortical areas become progressively capable of supporting retrieval independently, a process called systems consolidation that implies that the final repository of remote explicit memory resides in the neocortex.[123,124] In line with this notion, lesion and imaging studies in humans and animals suggest that the hippocampus and related structures are necessary for early storage of explicit memories, while as memories age, the hippocampus is disengaged and specific cortical sites mediate access to remote memories (see Refs. 125,126 for reviews). In rodents, lesion or inactivation of specific areas of the prefrontal cortex, including the anterior cingulate cortex and prelimbic area, affects specifically recall of remote spatial and contextual memories initially processed by the hippocampus.[127,128] A highly debated issue, however, arising from human studies, is whether or not certain items of information, in particular, details of episodic memory, do undergo complete systems consolidation with time or rely on a continuous hippocampo-cortical dialogue.[129] This view opposes a "standard" model of consolidation to a "multitrace" model which states that the cortex becomes increasingly, rather than exclusively engaged in recall, at least for certain types of information (see Refs. 126,130 for reviews).

Although this issue is not as yet fully resolved, a related question is what molecular/cellular mechanisms underlie the reorganization of memory during systems consolidation. To date, there is evidence that glutamatergic activity in the anterior cingulate cortex is crucial for processing remote memory[131] and that NMDA receptor activation in the prefrontal cortex is required during a critical window of 1–2 weeks following learning.[132] Further, mice with a heterozygous CaMKII mutation that affects cortical but not hippocampal synaptic plasticity impedes cortical reorganization necessary for consolidation of remote memory.[133] A recent study also reported that increases in spine density occur in the hippocampus shortly after learning, while occurring in the anterior cingulate cortex at a remote delay.[134] These results suggest that mechanisms of synaptic plasticity in prefrontal cortex areas may mediate the maturation of cortical networks in the process of systems consolidation. Does systems consolidation involves transcriptional events in hippocampo-cortical networks and does Zif268 play a role in this process? Several studies have shown that retrieval of recently formed (1-day) spatial of contextual fear memories are associated with Zif268 and c-Fos overexpression in the hippocampus, but not in the prefrontal cortex, whereas in contrast, recall of remote (28–36 days) memories result in robust activation of the genes in the prefrontal cortex, but not in the hippocampus.[127,128,135] Moreover, hippocampal inactivation by lidocaine infusion was shown to abolish post-recall Zif268 expression in CA1 and impair retrieval of the recent, but not remote, memory,[128] and there is evidence to suggest that certain forms of cortical-dependent remote memories require ERK activation and histone H3 acetylation as a mark of early tagging of cortical networks necessary for the formation of remote memory.[136] In certain situations such as when new hippocampal-dependent associative information can be incorporated into a preexisting cortical schema, systems consolidation can occur quickly.[137] In this case, the assimilation of new information into an existing schema was also shown to be associated with immediate Zif268 expression in cortical regions involved in remote memory.[138] Altogether, these observations provide a hint as to a potential role of Zif268-dependent transcriptional regulation in systems consolidation. Proof of causality is, however, still lacking, as there is as yet no evidence that Zif268 activation in prefrontal cortex areas is required for sustaining remote memory or for the rapid incorporation of new information into a schema.

4.3. Memory reconsolidation

For a long time, it was widely held that memories, once consolidated, are stable, permanently stored in hardwired circuits resilient to disruption. This view, however, has been considerably challenged over the past decade or so, reinforcing the notion that memories are fundamentally dynamic and reconstructive processes. While systems consolidation is a mark of a slow process of memory remodeling over time, compelling evidence nowadays also suggest that upon recall, memories can reenter a state of transient instability after which they can be strengthened, modified, or lost (see Refs. 139,140 for reviews). This notion was revived by the demonstration that a previously consolidated auditory fear memory can be impaired by microinjection of the protein synthesis inhibitor anisomycin into the basolateral nucleus of the amygdala immediately after reexposure to the tone.[141] Thus, during recall, reactivation of a previously consolidated memory can destabilize that memory and initiate a further consolidation process, generally called reconsolidation.

This landmark report stimulated a rapidly growing body of literature that extended this observation to several species ranging from nematodes to humans and provided evidence that post-recall reconsolidation can affect several types of mammalian memories (see Refs. 142,143 for review). A great deal of effort was placed in the determination of the cellular and molecular mechanisms underlying the reconsolidation of memory, exploring whether or not reconsolidation is a simple recapitulation of the initial consolidation mechanisms (see Refs. 139,142–144 for reviews). Although accumulating evidence supports the idea that the mechanisms underlying reconsolidation are not a mere repetition of those underlying consolidation, multiple reports highlighted Zif268 expression as a landmark of memory reconsolidation in distinct brain regions and learning and memory tasks. Early on the rediscovery of the memory reconsolidation process, fear memory retrieval was reported to trigger Zif268 mRNA expression in structures, such as the hippocampus and amygdala,[135] nucleus accumbens, and prefrontal cortex.[145] The transcriptional regulation of Zif268 in response to fear memory recall was recently confirmed by the demonstration that *Zif268* promoter undergoes epigenetic remodeling upon contextual fear memory recall.[146] Chromatin immunoprecipitation assay combined with quantitative real-time PCR unveiled the increase in histone H3 phosphoacetylation and acetylation levels at the *Zif268* promoter upon contextual fear memory recall.[146] Zif268 mRNA expression was further shown to precede the synthesis of Zif268 protein in the hippocampus[39,101,110] and amygdala[100,147] upon fear memory recall.

The critical role of Zif268 for memory reconsolidation was further demonstrated via different approaches in different learning and memory paradigms. Using a targeted deletion of the *Zif268* gene in mice, Zif268 was shown to be required for the reconsolidation of novel-object recognition[107] and object-place recognition memories.[108] Knockdown of Zif268 using antisense oligodeoxynucleotides infused in delimited brain regions before the learning or retrieval of several types of memories demonstrated a double dissociation between BDNF and Zif268 expression in contextual fear memory consolidation and reconsolidation, respectively.[110] A similar experiment followed by microarray-based exploration of gene regulation reported nonoverlapping lists of genes regulated in area CA1 during contextual fear memory consolidation and reconsolidation, particularly those associated with proinflammatory cytokine signaling.[148] The involvement of Zif268 in reconsolidation is not restricted to hippocampal function since Zif268 expression in the amygdala was also shown to be involved in the reconsolidation of cued-fear memory.[100,149]

If the critical role of Zif268 expression in memory reconsolidation is relatively well established, the upstream molecular mechanisms gating Zif268 expression upon recall remain poorly understood. Recent studies highlighted the involvement of IKKα/NF-κB signaling in the regulation of chromatin remodeling at the *Zif268* promoter[146] and in Zif268 expression[150] during reconsolidation of contextual fear memory. We recently suggested, however, that Zif268 expression triggered by memory recall does not fully recapitulate its regulation during the initial learning experience. In area CA1 of the hippocampus, for example, recruitment of Zif268 to its DNA consensus sequence occurs in a more limited time window following retrieval of a fear memory than following initial learning.[39] In addition, in the dentate gyrus the magnitude of Zif268 expression is lower following contextual fear memory reconsolidation than following its initial consolidation.[101] This could be in part linked to the nature of the stimuli presented during contextual fear memory training (footshock associated with the context) and retrieval (reexposure to the context in the absence of the footshock). However, we recently showed that *Zif268* knockout mice display a striking impairment of memory consolidation, whereas *Zif268* heterozygous mice show a selective impairment of reconsolidation.[109] These data reinforce the idea that *Zif268* gene expression dosage is an important feature of the different phases of memory processing and highlight the higher vulnerability of reconsolidation to Zif268 expression levels, compared to consolidation, echoing previous reports suggesting that reconsolidation is more sensitive to interfering treatments than consolidation (see Refs. 139,143 for reviews).

Finally, the doubly dissociable involvement of hippocampal BDNF and Zif268 expression in memory consolidation and reconsolidation[110] has recently been extended to explore the issue of the role of reconsolidation in the updating of an original memory. In this context, Zif268 downregulation was shown to prevent the strengthening of a contextual fear memory when applied before the second, but not the initial conditioning session,[151] and to prevent the updating of the hippocampal memory content in a context preexposure facilitation effect paradigm.[152]

4.4. Drug memory

Drugs of abuse have long been suspected to dysregulate the neural circuitry underlying learning and memory of natural rewards.[153] Psychostimulants,[154–156] opioids,[157] and Delta9-tetrahydrocannabinol (THC)[158,159] have been shown to trigger Zif268 expression in the brain reward circuit. Within this neural pathway, the striatum is a key brain region involved in the selection and execution of action, mostly composed of GABAergic medium-sized spiny neurons integrating glutamatergic and dopaminergic inputs from multiple brain regions that convey sensorimotor, emotional, and cognitive information.[153] Striatal Zif268 induction in response to drugs of abuse was consistently reported to depend on dopaminergic activity.[154,158,160–163] Acute cocaine treatment triggers Zif268 expression in medium-sized spiny neurons expressing D1, and to a lower extent D2, dopaminergic receptors.[164] These results are in accordance with the early report that striatal D2 coupling to Zif268 expression switches from negative at early stages of ontogeny to slightly positive in adults.[165]

The molecular mechanisms underlying Zif268 expression in response to cocaine was proposed to continuously develop during adolescence.[166,167] This phenomenon could account for the differential sensitivity to drugs of abuse observed in adolescence. In adults, the activity of the MAP kinase ERK1/2 pathway was proposed to participate in the induction of Zif268 in response to cocaine.[168] The downstream molecular events linking ERK1/2 activation to Zif268 expression involved the activity of the transcription factor Elk-1,[169] but not the nuclear kinase MSK-1.[170] In addition, the expression of the glucocorticoid receptor in medium-sized spiny neurons expressing D1 receptors is also critical for the expression of Zif268 in response to cocaine, but not morphine.[171]

The rapid increase in Zif268 mRNA expression in response to cocaine is followed by a prolonged period of marked reduction of Zif268 transcription in the striatum.[172] This effect could persist for several days after multiple cocaine injections[172] and likely depends on the pharmacological effect of the drug since chronic morphine exposure leads to a progressive increase in Zif268 transcription.[173] Therefore, repeated noncontingent exposure (passive administration) to cocaine,[174] amphetamine,[175,176] or THC[177] dampens the transcriptional activation of Zif268 in the striatum. In the same line, repeated cocaine contingent exposure (active administration) dampens Zif268 mRNA expression in the striatum 2 weeks following a single self-administration session.[178] It should be noted that in the latter paradigm, the contingency does not contribute to Zif268 mRNA expression but rather reflects the pharmacological effects of the drug.[179]

A hallmark of drug addiction is the high rate of relapse occurring even after an extended period of withdrawal.[180] Distinct factors have been implicated in relapse to drug taking, such as exposure to cues explicitly associated with drug use, stress, and reexposure to the drug itself, namely, drug priming.[180] In this framework, Zif268 mRNA expression was shown to increase in the striatum upon the presentation of a discrete cue associated to cocaine or heroin administrated contingently or noncontingently.[181] In addition, cue or drug priming-induced cocaine or heroin-seeking reinstatement elicited Zif268 mRNA expression throughout key brain regions of the reward pathway such as the hippocampus, amygdala, striatum, and prefrontal cortex, even after several weeks of withdrawal.[182–184] Social stress was also shown to trigger long-lasting adaptations resulting in an increased sensitivity to cocaine and altered levels of Zif268 expression within the prefrontal cortex and amygdala.[185] In the same vein, the reinstatement to cocaine conditioned-place preference (CPP) induced by an acute stressor was also shown to enhance Zif268 expression in the striatum.[186]

Different approaches were then used to explore the causal relationship between Zif268 expression and drug memories. Mice baring a targeted deletion of the *Zif268* gene display altered establishment of psychomotor sensitization and CPP induced by cocaine.[168] In light of the critical role of Zif268 in the reconsolidation of memory, Zif268 antisense oligodeoxynucleotides were extensively used to examine whether Zif268 downregulation could prevent the reconsolidation of drug memories. In a first study, Zif268 downregulation into the basolateral amygdala prior to the reactivation of the memory of a cocaine-associated cue was shown to suppress the ability of this cocaine-associated cue to elicit the acquisition of drug seeking under a second-order schedule of reinforcement.[149] Zif268 downregulation in the basolateral amygdala mediating the disruption of drug-associated memory reconsolidation was also shown to prevent cue-induced relapse to cocaine seeking[187] and cue-induced heroin withdrawal.[188] Conversely, infusion of D-cycloserine, a partial NMDA receptor agonist, into the amygdala enhances Zif268 expression and memory reconsolidation of a cocaine-associated cue, resulting in an increase in cue-induced relapse to cocaine seeking.[189] Interestingly, the contingence of cocaine administration seems to influence the role of Zif268 on the reconsolidation of drug-associated memories. Whereas Zif268 expression in the nucleus accumbens and basolateral amygdala is critical for the reconsolidation of drug-associated memory in a CPP paradigm, it is dispensable in the nucleus accumbens in a self-administration paradigm,[190] but critical in the basolateral amygdala.[149]

The influence of *Zif268* gene expression dosage is also an important element in the case of drug-related memories. For instance, cocaine-induced psychomotor sensitization induced by repeated cocaine administrations was similarly altered in *Zif268* heterozygous and homozygous mice.[168] However, cocaine-induced CPP which consisted in the association between the appetitive effects of cocaine and environmental clues in a three-pairing schedule was normal in *Zif268* heterozygous but completely prevented in *Zif268* homozygous mice.[168] In accordance with these data, we recently reported that the TAT-DEF-Elk-1 peptide that decreases Elk-1 activation and subsequently Zif268 expression in response to cocaine administration similarly interfered with cocaine-induced psychomotor sensitization, but not CPP with the notable exception of a CPP paradigm using a mild cocaine-pairing schedule.[169] Thus, we cannot rule out the possibility that *Zif268* heterozygous mice display a similar deficit in situation of mild cocaine pairing in the CPP paradigm which involves the hippocampus.[191]

5. ADULT NEUROGENESIS: A NEW ROLE FOR ZIF268

5.1. Mechanisms and function of adult neurogenesis

New neurons are continuously generated in the brain of adult mammals. Neurogenesis throughout life has been clearly demonstrated at two locations under normal conditions: the subventricular zone (SVZ) of the lateral ventricles and the subgranular zone (SGZ) of the dentate gyrus of the hippocampus. In both areas, proliferating glial precursor cells give rise to transient amplifying progenitors, which in turn generate neuroblasts.[192,193] The SVZ neuroblasts proceed toward the olfactory bulb (OB) along a tangential path of migration. Once reaching the core of the OB, immature neurons migrate radially and mature into inhibitory GABAergic interneurons. Newborn OB granule cells and periglomerular cells, which lack axons, form dendro-dendritic synapses with mitral and tufted cells, the two relay cells of the OB. Within a month, almost half of the newcomers are eliminated, while the other half is durably integrated into the OB. The synaptic integration of adult-born neurons is a rapid process. They receive GABAergic and glutamatergic inputs as soon as they reach the OB. In 1–2 weeks, dendrites and spines start to form and the synaptic contacts continue to grow over several months. However, the existence of a critical period for the development of functional outputs, the activity-dependant development of glutamatergic inputs onto newborn cells and their ability to sustain LTP has been reported between 2 and 8 weeks after birth. Such properties may allow integrated adult-born neurons to make a unique contribution to olfactory functions during this period.[194,195] Using paradigms of olfactory deprivation, behavioral enrichment or learning experience, several studies reported a strong correlation between circuit activity and the maturation/survival of newborn OB neurons, suggesting experience-dependent selection of new neurons and their participation to olfactory information processing. The exact function of adult neurogenesis in the OB remains largely unresolved. However, an increasing number of studies explored at cellular and network levels the effect of regulating and ablating neurogenesis and reported major consequences for olfactory behavior. In addition to their role in the maintenance of OB circuitry by replacing dying interneurons, newborn neurons may tune information processing by providing inhibitory activity necessary for sharp pattern segregation, and have an impact on olfactory detection, perceptual learning, short and long-term associative memory, and relevant social memories such as that of the mating partner or the progeny.[196–199]

In the adult SGZ, neuroblasts migrate into the inner granule cell layer and differentiate into glutamatergic DGCs in the dentate gyrus of the hippocampus. The majority of DGCs follows a 2-month rate of development to become fully mature and integrated into existing circuits. From the second week after birth, axons of DGCs elongate and project toward CA3 area, dendrites extend in the molecular layer and spines start to appear. Newborn neurons are initially activated by ambient depolarizing GABA and then hyperpolarized after conversion of chloride cotransporters expression, along with expression of glutamate receptors and their progressive responsiveness to glutamatergic inputs. Between 3 and 6 weeks after birth, newborn neurons exhibit hyperexcitability, they display properties of enhanced synaptic plasticity and are prone to integrate and shape the existing hippocampal network (see Ref. 200 for a review). As their functional maturation progresses, however, newborn DGCs compete to survive, leaving the majority eliminated by cell death. As in the OB, several behavioral and activity-dependent manipulations can regulate their maturation and survival. Accumulating evidence from studies using perturbation or ablation of hippocampal neurogenesis have revealed significant contribution of adult newborn neurons to several types of learning and memory, such as spatial learning and long-term memory, spatial and contextual pattern discrimination, trace and contextual fear conditioning, clearance of hippocampal memory traces, reorganization of memory to extrahippocampal substrates and to certain antidepressant-induced behavioral responses.[193,196,200–203]

In the OB as in the dentate gyrus, intrinsic and extrinsic mechanisms regulate different aspects of neurogenesis. Several molecules of different canonical signaling pathways have been identified as important players in the regulation of neurogenesis, including niche factors/receptors, cytoplasmic factors, neurotransmitters, hormones, neurotrophins growth factors, transcription factors, and epigenetic regulators.[192,193] Nevertheless, the molecular mechanisms that govern maturation, selection, and experience-dependent integration of new neurons are only starting to be delineated.[204,205] IEGs encoding inducible transcription factors such as Zif268 are interesting candidates because they are transiently and sometimes preferentially induced in newborn neurons in response to synaptic activity and behavioral experience.

5.2. Expression and regulation of Zif268 in newborn neurons

Few studies have explored the exact conditions of expression and regulation of Zif268 in adult newly generated neurons, the majority using the induction of Zif268 (or other IEGs) as an index of newborn neurons activation/integration in response to behavioral manipulations. In the OB, Zif268 is mainly expressed in preexisting and adult-born interneurons. The exact kinetic of Zif268 expression during maturation of newborn neurons remains uncertain, the available data thus far concerning mostly 3–8-week-old neurons. Without any specific olfactory stimulation, 15–20% of 4-week-old newborn granule cells express Zif268.[206,207] The proportion of Zif268 expressing newborn neurons can be modified by extended olfactory stimulation, but depends on the nature of odor presented, the hormonal status, and olfactory experience of the animals. After passive olfactory exposure to pure odors[208,209] or social odors,[207] the proportion of newborn neurons that express Zif268 reaches 30–45%. However, this strong expression in response to male urinary odors requires both postnatal and adult normal exposure of females to estradiol, suggesting a complementary role of Zif268 and hormones in the responsiveness of newborn neurons to socially relevant odors.[207] Interestingly, after odor stimulation, the proportion of Zif268 expressing newborn neurons can be reinforced by familiarized odors,[208] repeated daily exposure to single novel odorants,[206] olfactory perceptual learning of similar odorants,[209] and paternal recognition of adult offspring odors.[210]

These data suggest a potential relationship between the induced-expression of Zif268 in newborn neurons and their contribution to olfactory learning and memory. Several studies used Zif268 expression as an index of functional recruitment of adult-born neurons during olfactory learning and memory recall. If the two first published studies did not directly examine Zif268 expression in newborn neurons, both reported a correlation between the regionalized expression of Zif268 in the OB and the proportion of newborn neurons selected to survive during initial olfactory discrimination learning.[211,212] More recently, a set of data demonstrated higher proportions (40–55%) of Zif268 expressing newborn neurons in various conditions of associative olfactory learning that correlates with olfactory memory performance.[213–216] Unfortunately, none of these studies compared the expression of Zif268 in newborn neurons of different ages with that of preexisting older neurons. It is therefore difficult to know whether Zif268 is or is not preferentially induced in newborn neurons of the OB during their recruitment by olfactory learning and memory consolidation.

Zif268 expression in newborn neurons of the dentate gyrus is more precisely characterized, both in terms of kinetic of maturation and in relation to specific behavioral contexts. In basal conditions, a small proportion of newborn DGCs express Zif268, a proportion that, however, increases over time (from 1–2% to ~25% between 1 and 6 weeks after birth) and becomes higher than in preexisting mature neurons (1–2%),[217–220] indicating that Zif268 might be preferentially induced in newborn DGCs during their critical period of integration.[221] As in the OB, the proportion of Zif268 expressing newborn neurons can be modified by various manipulations, such as induction of seizures, high-frequency stimulation of hippocampal afferents, exposure to novel/enriched environments, exercise, or spatial learning and memory recall. Kainic acid, pentylenetetrazol seizures, or high-frequency stimulation are the most efficient methods to induce Zif268 in newborn DGCs. Similarly to what is reported in basal conditions, 2–4-week-old newborn neurons are more prone to respond to high levels of neuronal activity.[222–226] Due to a more rapid functional maturation in rats than in mice, this proportion is higher for rats in 2–4-week-old neurons.[223–226] The responsiveness of newborn neurons in terms of Zif268 expression after physiological and behavioral manipulations is generally lower. Exposure to a novel/enriched environment was shown to induce Zif268 in 4–6% of 5–6-week-old newborn DGCs, a proportion similar to that observed in preexisting neurons, indicating no preferential activation of Zif268 in newborn neurons in these conditions.[217] Only running and swimming sessions seem to induce Zif268 in a preferential manner at this maturation stage.[218] A low proportion of Zif268-expressing newly generated neurons have also been reported after spatial training in the water maze.[217,218] In rats, again, Zif268 activation is more pronounced during the critical period of functional integration of newborn DGCs (10% at 2 weeks, 25% at 3 weeks, and 5% at 10 weeks).[225,227] Older neurons, however, do not preferentially induce Zif268 in a learning-specific manner.[217,218] In contrast, one study reported a greater proportion of Zif68 expressing newborn neurons (1–3 week old) after spatial training that increased when animals were previously trained in multiple contexts.[219] Notably, induction of Zif268 expression in 6–7-week-old neurons (11%) was reported either when neurons are early tagged by massed spatial learning (9 days after birth) and later recruited to update a memory trace,[228] or when newborn neurons are recruited during the critical period of functional integration (18 days after birth) (35%).[220] These findings suggest that Zif268 expression in newborn neurons could play an essential role at an age when they are more prone to be incorporated into functional networks during spatial memory formation and long-term consolidation.[200,203]

5.3. Functional role of Zif268 in newborn neurons integration

At present only one study explored the functional role of Zif268 in the neurogenic processes and the recruitment of newborn DGCs in relation to learning and memory.[220] Using the birth-dating marker 5-bromo-2′-deoxyuridine (BrdU) to follow the fate of newly generated cells in *Zif268* knockout mice, we first demonstrated that *Zif268* actively controls the survival of newborn DGCs during their critical period of maturation in the first 2–3 weeks of their birth. At this age, the absence of Zif268 resulted in an accelerated death of newborn neurons both in the OB and dentate gyrus, with a similar accelerated death in *Zif268* heterozygous mice, indicating a strong dependence of Zif268 in the neurogenic process. To assess the functional consequences of the absence of Zif268 during this critical period, we examined the recruitment of 9- and 18-day-old newborn DGCs by massed spatial learning, their functional incorporation, and subsequent long-term survival. In wild-type mice, recall of spatial memory 30 days after training resulted in preferential activation of Zif268 in newborn DGCs aged 18 days, but not 9 days, at the time of initial learning, and promoted their survival. In contrast, in the absence of Zif268, the occurrence of spatial learning during the critical period (at 18 days) failed to recruit newborn DGCs, to promote their survival and prevented their activation upon recall in association with long-term memory impairment. In all, these findings indicate that Zif268 is required for the functional incorporation of newborn DGCs into memory networks. Mechanistically, we found that the absence of Zif268 critically affects the neurochemical and morphological maturation of the newborn neurons, as indicated by lower GluR1 expression, reduced NKCC1/KCC2b chloride cotransporter ratio, altered dendritic development, and marked spines growth defect, a cellular phenotype likely related to their inability to be recruited during learning and participate in establishing long-term memory.

6. ZIF268 IN PATHOLOGY

Although understanding the role of Zif268 in pathology is a far cry from understanding its role in plasticity and memory, it is reminiscent of when the NMDA receptor was discovered as being involved in memory and LTP in the hippocampus and was also known to be involved in seizure activity and cell death. As the NMDA receptor subsequently became regarded as a coincidence detector, zif268 is considered to be a gateway into regulating gene programs to induce long-lasting changes where the type of stimulation and the state of cellular viability would determine the genetic programs it initiates. Zif268 has similarly been shown to be associated with many pathologies including cancers such as breast, lung, colon, and prostate cancers; ischemic injury, arthrosclerosis, vascular injury, oxidative stress, seizure activity, wound healing; and neurodegenerative diseases such as Alzheimer's disease. To date, we can glean little on the mechanisms whereby it can play a role in brain pathologies; the most well-documented understanding comes from peripheral cancers and is worthy of note if only to understand the wide ranging possible mechanisms by which Zif268 can be regulated that has been relatively neglected in research into its role in learning and memory.

In general, Zif268 can bestow protective or detrimental effects in pathology depending on the downstream target genes it binds to and activates. It can promote the progression of various forms of cancer via activation of genes encoding growth factors, or it can attempt to arrest progression by activating genes encoding apoptotic proteins. In prostate cancer cell where Zif268 promotes the carcinogenic progression by regulation of genes encoding growth factors, this occurs in conjunction with a high level of overexpression of the gene that correlates with the degree of malignancy and loss of expression of its corepressor, NAB2.[229,230] In addition, some of the target genes activated by Zif268 encode proteins that induce upregulation of *Zif268* through MAP kinase signaling, thereby bestowing a positive feedback loop designed to maintain continuous mitogenic signaling to promote progression of the tumorigenic process.

Contrasting this mechanism, in other forms of cancer, Zif268 activates proapoptotic genes such as *PTEN* and growth inhibitors such as p53. Where the tumorigenic process continues, it could be considered due to a failure of

Zif268 to activate apoptotic genes, and evidence suggests that Zif268 in these cases shows much lower levels of expression or is mutated.[46,231] As the consensus suggests these opposing effects are cell-type specific, regulation in neurons might also be different, and this is in fact supported by micro-array analyses of NMDA-mediated overexpression of Zif268 in neurons.[51] However, in a rather more complex situation, PTEN, a proapoptotic protein that inhibits Akt signaling, is silent in many cancers, including prostate cancer, where Zif268 promotes progression of cell growth. In at least 50% of prostate cancer cases, the promoter region of PTEN is hypermethylated and this may well preclude Zif268's ability to bind to and activate it.[46] Rather than the effect of Zif268 being dependent on cell type, it suggests that it may be dependent on the viability of its downstream target genes.

In addition to positive feedback loops between Zif268 downstream target genes and proteins that stimulate Zif268, a negative feedback loops has also been shown, where Zif268 binding to the promoter regions of p300 and CPB acetylates Zif268 and promotes its binding to them both, a complex that then becomes inhibitory for the transcription of p300/CPB and *Zif268* itself.[33] Whether Zif268 leads to activation of genes encoding growth factors or apoptototic proteins largely depends on the level of activation of *Zif268*. High activation of *Zif268* can induce casein kinase-mediated phosphorylation of Zif268, reduce transactivation of p300/CBP and growth-related genes, and the activation of proapoptotic genes such as *p53*, *p73*, and *PTEN*.[232,233] Under low levels of stimulation of Zif268 by serum, or low levels of growth factors, Zif268 is acetylated, preventing transactivation of p300/CPB, and *Zif268* itself is silenced while proliferation and survival genes are activated.

In other pathologies, much less is understood about the functional role of Zif268 and the mechanisms of its regulation. Downregulation of Zif268 has been observed in aging or certain diseases leading to cognitive deficits.[234–236] One study reported that Zif268 binding to and transactivation of the pore-forming calcium channel subunit CaV3.2 is counteracted by the binding of the repressor element 1-silencing transcription factor to modulate calcium homeostasis[55] and this may be dysfunctional during epileptic seizures and cardiac arrhythmias. In Alzheimer's disease, Zif268 mRNA has been shown to be upregulated in human brain tissue in correlation with the progression of the disease[237,238] and can indirectly lead to hyperphosphorylation of tau at pathological sites to promote formation of neurofibrillary tangles via Cdk5.[239] In pathological states, regulation of Zif268 therefore is highly complex and dependent on a number of external factors that control its regulation.

7. CONCLUSIONS AND PERSPECTIVES

As this survey shows, the transcription factor Zif268 plays an essential role in at least two major aspects of brain plasticity, synaptic plasticity as exemplified by its role in the maintenance of LTP, and adult neurogenesis as shown by its role in the selection, maturation, and functional integration of dentate gyrus newborn neurons. These two forms of brain plasticity are themselves essential for many cognitive functions. As a consequence, and as the results of the past decades of research has firmly established, Zif268 plays a critical role in many different forms of explicit memories, operating in distinct brain structures in relation to the type of memory processes that these structures subserve. A most notable conclusion derived from research in recent years is that Zif268 seems to play a key role not only in different forms of memory but also in different processes operating on memory, from post-learning memory consolidation and systems consolidation to reconsolidation, updating, and extinction. Recent findings also highlight the complex tripartite relationship between Zif268 expression dosage, its function in different forms of memory (and therefore presumably in different brain structures), and distinct processes of memory. This, together with the approaches investigating the consequences of increasing neuronal Zif268 expression, suggests the hypothesis that one main function of Zif268 in brain plasticity and cognitive functions is a control of the strength and duration of memory, and of the underlying plastic changes. Key challenges in the field remain, however, and new types of experiments are clearly needed to attain a richer understanding of the role of Zif268 in a broader context of the molecular mechanisms of memory function and dysfunction. On the one side, what type of neural activity and what combination of active molecular cascades trigger Zif268 expression in specific neurons in relation to distinct memory processes remains relatively uncertain. Posttranslational modification of Zif268 proteins and the complex interactions between Zif268, its coactivators and repressors, and the transcriptional machinery are also poorly understood *in vivo*, as are the molecular events controlling its epigenetic regulation. On the other side, the numerous targets of Zif268 that have been documented need to be identified as genuine effectors of the plasticity mechanisms and cognitive processes in which Zif268 is engaged. Further, Zif268 is not the only transcription factor involved in neuronal plasticity and memory; several others are activated with similar or different kinetics, and how specific ensembles of transcription factors cooperate in initiating and instructing downstream gene programs remains unexplored. Finally, although a great deal of effort has been made in uncovering the role that Zif268 may have in various forms of pathological cell processes, such as cancer, alterations of its function in brain pathologies associated with cognitive deficiencies remains poorly investigated. Whether or not manipulating Zif268 dosage in certain neuron types or brain structures may represent a promising approach for alleviating certain cognitive dysfunctions thus remains an unsolved issue.

ACKNOWLEDGMENTS

Supported by Centre National de la Recherche Scientifique (CNRS), University Paris-Sud, University Paris 6, and by a grant from Agence Nationale pour la Recherche (ANR-2010-BLAN-1413-01) to S. L. A. V. was supported by the ANR grant and A. B. by the Bettencourt Schueller Foundation and the Philippe Foundation.

REFERENCES

1. Christy BA, Lau LF, Nathans D. A gene activated in mouse 3T3 cells by serum growth factors encodes a protein with "zinc finger" sequences. *Proc Natl Acad Sci U S A.* 1988;85(21):7857–7861.
2. Lemaire P, Revelant O, Bravo R, Charnay P. Two mouse genes encoding potential transcription factors with identical DNA-binding domains are activated by growth factors in cultured cells. *Proc Natl Acad Sci U S A.* 1988;85(13):4691–4695.
3. Milbrandt J. A nerve growth factor-induced gene encodes a possible transcriptional regulatory factor. *Science.* 1987;238(4828):797–799.
4. Sukhatme VP, Cao XM, Chang LC, et al. A zinc finger-encoding gene coregulated with c-fos during growth and differentiation, and after cellular depolarization. *Cell.* 1988;53(1):37–43.
5. Curran T, Rauscher 3rd FJ, Cohen DR, Franza Jr BR. Beyond the second messenger: oncogenes and transcription factors. *Cold Spring Harb Symp Quant Biol.* 1988;53: 769–777.
6. Burmeister SS, Fernald RD. Evolutionary conservation of the egr-1 immediate-early gene response in a teleost. *J Comp Neurol.* 2005;481(2):220–232.
7. Lemaire P, Vesque C, Schmitt J, Stunnenberg H, Frank R, Charnay P. The serum-inducible mouse gene Krox-24 encodes a sequence-specific transcriptional activator. *Mol Cell Biol.* 1990;10(7):3456–3467.
8. Christy B, Nathans D. DNA binding site of the growth factor-inducible protein Zif268. *Proc Natl Acad Sci U S A.* 1989;86(22):8737–8741.
9. Swirnoff AH, Milbrandt J. DNA-binding specificity of NGFI-A and related zinc finger transcription factors. *Mol Cell Biol.* 1995;15(4):2275–2287.
10. Beckmann AM, Wilce PA. Egr transcription factors in the nervous system. *Neurochem Int.* 1997;31(4):477–510, discussion 517–476.
11. Worley PF, Christy BA, Nakabeppu Y, Bhat RV, Cole AJ, Baraban JM. Constitutive expression of zif268 in neocortex is regulated by synaptic activity. *Proc Natl Acad Sci U S A.* 1991;88(12):5106–5110.

12. Wang JQ, Daunais JB, McGinty JF. Role of kainate/AMPA receptors in induction of striatal zif/268 and preprodynorphin mRNA by a single injection of amphetamine. *Brain Res Mol Brain Res.* 1994;27(1):118–126.
13. Murphy TH, Worley PF, Baraban JM. L-type voltage-sensitive calcium channels mediate synaptic activation of immediate early genes. *Neuron.* 1991;7(4):625–635.
14. Karin M, Hunter T. Transcriptional control by protein phosphorylation: signal transmission from the cell surface to the nucleus. *Curr Biol.* 1995;5(7):747–757.
15. Silverman ES, Collins T. Pathways of Egr-1-mediated gene transcription in vascular biology. *Am J Pathol.* 1999;154(3):665–670.
16. Yang SH, Vickers E, Brehm A, Kouzarides T, Sharrocks AD. Temporal recruitment of the mSin3A-histone deacetylase corepressor complex to the ETS domain transcription factor Elk-1. *Mol Cell Biol.* 2001;21(8):2802–2814.
17. Besnard A, Galan-Rodriguez B, Vanhoutte P, Caboche J. Elk-1 a transcription factor with multiple facets in the brain. *Front Neurosci.* 2011;5:35.
18. Cao X, Mahendran R, Guy GR, Tan YH. Detection and characterization of cellular EGR-1 binding to its recognition site. *J Biol Chem.* 1993;268(23):16949–16957.
19. Thyss R, Virolle V, Imbert V, Peyron JF, Aberdam D, Virolle T. NF-kappaB/Egr-1/Gadd45 are sequentially activated upon UVB irradiation to mediate epidermal cell death. *EMBO J.* 2005;24(1):128–137.
20. Yu J, Baron V, Mercola D, Mustelin T, Adamson ED. A network of p73, p53 and Egr1 is required for efficient apoptosis in tumor cells. *Cell Death Differ.* 2007;14(3):436–446.
21. Li G, Li W, Angelastro JM, Greene LA, Liu DX. Identification of a novel DNA binding site and a transcriptional target for activating transcription factor 5 in c6 glioma and mcf-7 breast cancer cells. *Mol Cancer Res.* 2009;7(6):933–943.
22. Matheny C, Day ML, Milbrandt J. The nuclear localization signal of NGFI-A is located within the zinc finger DNA binding domain. *J Biol Chem.* 1994;269(11):8176–8181.
23. Russo MW, Sevetson BR, Milbrandt J. Identification of NAB1, a repressor of NGFI-A- and Krox20-mediated transcription. *Proc Natl Acad Sci USA.* 1995;92(15):6873–6877.
24. O'Donovan KJ, Tourtellotte WG, Millbrandt J, Baraban JM. The EGR family of transcription-regulatory factors: progress at the interface of molecular and systems neuroscience. *Trends Neurosci.* 1999;22(4):167–173.
25. Svaren J, Sevetson BR, Apel ED, Zimonjic DB, Popescu NC, Milbrandt J. NAB2, a corepressor of NGFI-A (Egr-1) and Krox20, is induced by proliferative and differentiative stimuli. *Mol Cell Biol.* 1996;16(7):3545–3553.
26. Ehrengruber MU, Muhlebach SG, Sohrman S, Leutenegger CM, Lester HA, Davidson N. Modulation of early growth response (EGR) transcription factor-dependent gene expression by using recombinant adenovirus. *Gene.* 2000;258(1–2):63–69.
27. Kumbrink J, Gerlinger M, Johnson JP. Egr-1 induces the expression of its corepressor nab2 by activation of the nab2 promoter thereby establishing a negative feedback loop. *J Biol Chem.* 2005;280(52):42785–42793.
28. Müller HJ, Skerka C, Bialonski A, Zipfel PF. AClone pAT 133 identifies a gene that encodes another human member of a class of growth factor-induced genes with almost identical zinc-finger domains. *Proc Natl Acad Sci U S A.* 1991;88(22):10079–10083.
29. Zandarashvili L, Vuzman D, Esadze A, et al. Asymmetrical roles of zinc fingers in dynamic DNA-scanning process by the inducible transcription factor Egr-1. *Proc Natl Acad Sci U S A.* 2012;109(26):E1724–E1732.
30. Cao X, Mahendran R, Guy GR, Tan YH. Protein phosphatase inhibitors induce the sustained expression of the Egr-1 gene and the hyperphosphorylation of its gene product. *J Biol Chem.* 1992;267(18):12991–12997.
31. Huang RP, Fan Y, deBelle I, Ni Z, Matheny W, Adamson ED. Egr-1 inhibits apoptosis during the UV response: correlation of cell survival with Egr-1 phosphorylation. *Cell Death Differ.* 1998;5(1):96–106.

32. Yu J, Zhang SS, Saito K, et al. PTEN regulation by Akt-EGR1-ARF-PTEN axis. *EMBO J*. 2009;28(1):21–33.
33. Yu J, de Belle I, Liang H, Adamson ED. Coactivating factors p300 and CBP are transcriptionally crossregulated by Egr1 in prostate cells, leading to divergent responses. *Mol Cell*. 2004;15(1):83–94.
34. Roeb W, Boyer A, Cavenee WK, Arden KC. Guilt by association: PAX3-FOXO1 regulates gene expression through selective destabilization of the EGR1 transcription factor. *Cell Cycle*. 2008;7(7):837–841.
35. Manente AG, Pinton G, Tavian D, Lopez-Rodas G, Brunelli E, Moro L. Coordinated sumoylation and ubiquitination modulate EGF induced EGR1 expression and stability. *PLoS One*. 2011;6(10):e25676.
36. Decker EL, Nehmann N, Kampen E, Eibel H, Zipfel PF, Skerka C. Early growth response proteins (EGR) and nuclear factors of activated T cells (NFAT) form heterodimers and regulate proinflammatory cytokine gene expression. *Nucleic Acids Res*. 2003;31(3):911–921.
37. Cogswell PC, Mayo MW, Baldwin Jr AS. Involvement of Egr-1/RelA synergy in distinguishing T cell activation from tumor necrosis factor-alpha-induced NF-kappa B1 transcription. *J Exp Med*. 1997;185(3):491–497.
38. Wieland GD, Nehmann N, Muller D, et al. Early growth response proteins EGR-4 and EGR-3 interact with immune inflammatory mediators NF-kappaB p50 and p65. *J Cell Sci*. 2005;118(Pt. 14):3203–3212.
39. Cheval H, Chagneau C, Levasseur G, et al. Distinctive features of Egr transcription factor regulation and DNA binding activity in CA1 of the hippocampus in synaptic plasticity and consolidation and reconsolidation of fear memory. *Hippocampus*. 2012;22(3):631–642.
40. Levkovitz Y, Baraban JM. A dominant negative Egr inhibitor blocks nerve growth factor-induced neurite outgrowth by suppressing c-Jun activation: role of an Egr/c-Jun complex. *J Neurosci*. 2002;22(10):3845–3854.
41. Khachigian LM, Williams AJ, Collins T. Interplay of Sp1 and Egr-1 in the proximal platelet-derived growth factor A-chain promoter in cultured vascular endothelial cells. *J Biol Chem*. 1995;270(46):27679–27686.
42. Khachigian LM, Lindner V, Williams AJ, Collins T. Egr-1-induced endothelial gene expression: a common theme in vascular injury. *Science*. 1996;271(5254):1427–1431.
43. Svaren J, Ehrig T, Abdulkadir SA, Ehrengruber MU, Watson MA, Milbrandt J. EGR1 target genes in prostate carcinoma cells identified by microarray analysis. *J Biol Chem*. 2000;275(49):38524–38531.
44. Townsend KJ, Zhou P, Qian L, et al. Regulation of MCL1 through a serum response factor/Elk-1-mediated mechanism links expression of a viability-promoting member of the BCL2 family to the induction of hematopoietic cell differentiation. *J Biol Chem*. 1999;274(3):1801–1813.
45. Tang KF, Song GB, Shi YS, Yuan L, Li YH. Dicer knockdown induces fibronectin-1 expression in HEK293T cells via induction of Egr1. *Biochim Biophys Acta*. 2010;1800(3):380–384.
46. Baron V, Adamson ED, Calogero A, Ragona G, Mercola D. The transcription factor Egr1 is a direct regulator of multiple tumor suppressors including TGFbeta1, PTEN, p53, and fibronectin. *Cancer Gene Ther*. 2006;13(2):115–124.
47. Arora S, Wang Y, Jia Z, et al. Egr1 regulates the coordinated expression of numerous EGF receptor target genes as identified by ChIP-on-chip. *Genome Biol*. 2008;9(11):R166.
48. Adamson ED, Mercola D. Egr1 transcription factor: multiple roles in prostate tumor cell growth and survival. *Tumour Biol*. 2002;23(2):93–102.
49. Gitenay D, Baron VT. Is EGR1 a potential target for prostate cancer therapy? *Future Oncol*. 2009;5(7):993–1003.

50. Rong Y, Hu F, Huang R, et al. Early growth response gene-1 regulates hypoxia-induced expression of tissue factor in glioblastoma multiforme through hypoxia-inducible factor-1-independent mechanisms. *Cancer Res.* 2006;66(14):7067–7074.
51. James AB, Conway AM, Morris BJ. Genomic profiling of the neuronal target genes of the plasticity-related transcription factor – Zif268. *J Neurochem.* 2005;95(3):796–810.
52. Weaver IC, D'Alessio AC, Brown SE, et al. The transcription factor nerve growth factor-inducible protein a mediates epigenetic programming: altering epigenetic marks by immediate-early genes. *J Neurosci.* 2007;27(7):1756–1768.
53. Adamson ED, Yu J, Mustelin T. Co-factors p300 and CBP catch Egr1 in their network. *Prostate.* 2005;63(4):407–410.
54. Liu N, Yu Z, Xiang S, et al. Transcriptional regulation mechanisms of hypoxia-induced neuroglobin gene expression. *Biochem J.* 2012;443(1):153–164.
55. van Loo KM, Schaub C, Pernhorst K, et al. Transcriptional regulation of T-type calcium channel CaV3.2: bi-directionality by early growth response 1 (Egr1) and repressor element 1 (RE-1) protein-silencing transcription factor (REST). *J Biol Chem.* 2012;287(19):15489–15501.
56. Xiong F, Xiao D, Zhang L. Norepinephrine causes epigenetic repression of PKCepsilon gene in rodent hearts by activating Nox1-dependent reactive oxygen species production. *FASEB J.* 2012;26(7):2753–2763.
57. Cole AJ, Saffen DW, Baraban JM, Worley PF. Rapid increase of an immediate early gene messenger RNA in hippocampal neurons by synaptic NMDA receptor activation. *Nature.* 1989;340(6233):474–476.
58. Wisden W, Errington ML, Williams S, et al. Differential expression of immediate early genes in the hippocampus and spinal cord. *Neuron.* 1990;4(4):603–614.
59. Abraham WC, Dragunow M, Tate WP. The role of immediate early genes in the stabilization of long-term potentiation. *Mol Neurobiol.* 1991;5(2–4):297–314.
60. Abraham WC, Mason SE, Demmer J, et al. Correlations between immediate early gene induction and the persistence of long-term potentiation. *Neuroscience.* 1993;56(3):717–727.
61. Richardson CL, Tate WP, Mason SE, Lawlor PA, Dragunow M, Abraham WC. Correlation between the induction of an immediate early gene, zif/268, and long-term potentiation in the dentate gyrus. *Brain Res.* 1992;580(1–2):147–154.
62. Worley PF, Bhat RV, Baraban JM, Erickson CA, McNaughton BL, Barnes CA. Thresholds for synaptic activation of transcription factors in hippocampus: correlation with long-term enhancement. *J Neurosci.* 1993;13(11):4776–4786.
63. Dragunow M. A role for immediate-early transcription factors in learning and memory. *Behav Genet.* 1996;26(3):293–299.
64. Williams JM, Beckmann AM, Mason-Parker SE, Abraham WC, Wilce PA, Tate WP. Sequential increase in Egr-1 and AP-1 DNA binding activity in the dentate gyrus following the induction of long-term potentiation. *Brain Res Mol Brain Res.* 2000;77(2):258–266.
65. Jones MW, Errington ML, French PJ, et al. A requirement for the immediate early gene Zif268 in the expression of late LTP and long-term memories. *Nat Neurosci.* 2001;4(3):289–296.
66. Penke Z, Morice E, Veyrac A, et al. Zif268/Egr1 gain of function facilitates hippocampal synaptic plasticity and long-term spatial recognition memory. *Phil Trans R Soc Lond Biol Sci.* 2014;269(1633):20130159.
67. English JD, Sweatt JD. A requirement for the mitogen-activated protein kinase cascade in hippocampal long term potentiation. *J Biol Chem.* 1997;272(31):19103–19106.
68. Treisman R. Regulation of transcription by MAP kinase cascades. *Curr Opin Cell Biol.* 1996;8(2):205–215.
69. Davis S, Vanhoutte P, Pages C, Caboche J, Laroche S. The MAPK/ERK cascade targets both Elk-1 and cAMP response element-binding protein to control long-term

potentiation-dependent gene expression in the dentate gyrus in vivo. *J Neurosci.* 2000;20(12):4563–4572.
70. Mackler SA, Brooks BP, Eberwine JH. Stimulus-induced coordinate changes in mRNA abundance in single postsynaptic hippocampal CA1 neurons. *Neuron.* 1992;9(3):539–548.
71. Roberts LA, Higgins MJ, O'Shaughnessy CT, Stone TW, Morris BJ. Changes in hippocampal gene expression associated with the induction of long-term potentiation. *Brain Res Mol Brain Res.* 1996;42(1):123–127.
72. Jones MW, French PJ, Bliss TV, Rosenblum K. Molecular mechanisms of long-term potentiation in the insular cortex in vivo. *J Neurosci.* 1999;19(21):RC36.
73. Heynen AJ, Bear MF. Long-term potentiation of thalamocortical transmission in the adult visual cortex in vivo. *J Neurosci.* 2001;21(24):9801–9813.
74. Yamagata K, Kaufmann WE, Lanahan A, et al. Egr3/Pilot, a zinc finger transcription factor, is rapidly regulated by activity in brain neurons and colocalizes with Egr1/zif268. *Learn Mem.* 1994;1(2):140–152.
75. Williams J, Dragunow M, Lawlor P, et al. Krox20 may play a key role in the stabilization of long-term potentiation. *Brain Res Mol Brain Res.* 1995;28(1):87–93.
76. Gallitano-Mendel A, Izumi Y, Tokuda K, et al. The immediate early gene early growth response gene 3 mediates adaptation to stress and novelty. *Neuroscience.* 2007;148(3): 633–643.
77. Li L, Yun SH, Keblesh J, et al. Egr3, a synaptic activity regulated transcription factor that is essential for learning and memory. *Mol Cell Neurosci.* 2007;35(1):76–88.
78. Hughes P, Dragunow M. Induction of immediate-early genes and the control of neurotransmitter-regulated gene expression within the nervous system. *Pharmacol Rev.* 1995;47(1):133–178.
79. Herdegen T, Leah JD. Inducible and constitutive transcription factors in the mammalian nervous system: control of gene expression by Jun, Fos and Krox, and CREB/ATF proteins. *Brain Res Brain Res Rev.* 1998;28(3):370–490.
80. Ranieri F, Podda MV, Riccardi E, et al. Modulation of LTP at rat hippocampal CA3-CA1 synapses by direct current stimulation. *J Neurophysiol.* 2012;107(7):1868–1880.
81. Aydin-Abidin S, Trippe J, Funke K, Eysel UT, Benali A. High- and low-frequency repetitive transcranial magnetic stimulation differentially activates c-Fos and zif268 protein expression in the rat brain. *Exp Brain Res.* 2008;188(2):249–261.
82. Rygh LJ, Suzuki R, Rahman W, et al. Local and descending circuits regulate long-term potentiation and zif268 expression in spinal neurons. *Eur J Neurosci.* 2006;24(3):761–772.
83. Lindecke A, Korte M, Zagrebelsky M, et al. Long-term depression activates transcription of immediate early transcription factor genes: involvement of serum response factor/Elk-1. *Eur J Neurosci.* 2006;24(2):555–563.
84. Yilmaz-Rastoder E, Miyamae T, Braun AE, Thiels E. LTP- and LTD-inducing stimulations cause opposite changes in arc/arg3.1 mRNA level in hippocampal area CA1 in vivo. *Hippocampus.* 2011;21(12):1290–1301.
85. Kilduff TS, Vugrinic C, Lee SL, et al. Characterization of the circadian system of NGFI-A and NGFI-A/NGFI-B deficient mice. *J Biol Rhythms.* 1998;13(4):347–357.
86. Zheng D, Butler LS, McNamara JO. Kindling and associated mossy fibre sprouting are not affected in mice deficient of NGFI-A/NGFI-B genes. *Neuroscience.* 1998;83(1):251–258.
87. Mataga N, Fujishima S, Condie BG, Hensch TK. Experience-dependent plasticity of mouse visual cortex in the absence of the neuronal activity-dependent marker egr1/zif268. *J Neurosci.* 2001;21(24):9724–9732.
88. Hall J, Thomas KL, Everitt BJ. Rapid and selective induction of BDNF expression in the hippocampus during contextual learning. *Nat Neurosci.* 2000;3(6):533–535.

89. Brennan PA, Schellinck HM, Keverne EB. Patterns of expression of the immediate-early gene egr-1 in the accessory olfactory bulb of female mice exposed to pheromonal constituents of male urine. *Neuroscience*. 1999;90(4):1463–1470.
90. Tischmeyer W, Grimm R. Activation of immediate early genes and memory formation. *Cell Mol Life Sci*. 1999;55(4):564–574.
91. Davis S, Bozon B, Laroche S. How necessary is the activation of the immediate early gene zif268 in synaptic plasticity and learning? *Behav Brain Res*. 2003;142(1–2):17–30.
92. Knapska E, Kaczmarek L. A gene for neuronal plasticity in the mammalian brain: Zif268/Egr-1/NGFI-A/Krox-24/TIS8/ZENK? *Prog Neurobiol*. 2004;74(4):183–211.
93. Okuno H, Miyashita Y. Expression of the transcription factor Zif268 in the temporal cortex of monkeys during visual paired associate learning. *Eur J Neurosci*. 1996;8(10):2118–2128.
94. Nikolaev E, Kaminska B, Tischmeyer W, Matthies H, Kaczmarek L. Induction of expression of genes encoding transcription factors in the rat brain elicited by behavioral training. *Brain Res Bull*. 1992;28(3):479–484.
95. Grimm R, Tischmeyer W. Complex patterns of immediate early gene induction in rat brain following brightness discrimination training and pseudotraining. *Behav Brain Res*. 1997;84(1–2):109–116.
96. Mello CV, Vicario DS, Clayton DF. Song presentation induces gene expression in the songbird forebrain. *Proc Natl Acad Sci U S A*. 1992;89(15):6818–6822.
97. Jarvis ED, Mello CV, Nottebohm F. Associative learning and stimulus novelty influence the song-induced expression of an immediate early gene in the canary forebrain. *Learn Mem*. 1995;2(2):62–80.
98. Bolhuis JJ, Zijlstra GG, den Boer-Visser AM, Van Der Zee EA. Localized neuronal activation in the zebra finch brain is related to the strength of song learning. *Proc Natl Acad Sci U S A*. 2000;97(5):2282–2285.
99. Malkani S, Rosen JB. Specific induction of early growth response gene 1 in the lateral nucleus of the amygdala following contextual fear conditioning in rats. *Neuroscience*. 2000;97(4):693–702.
100. Maddox SA, Monsey MS, Schafe GE. Early growth response gene 1 (Egr-1) is required for new and reactivated fear memories in the lateral amygdala. *Learn Mem*. 2011;18(1):24–38.
101. Besnard A, Laroche S, Caboche J. Comparative dynamics of MAPK/ERK signalling components and immediate early genes in the hippocampus and amygdala following contextual fear conditioning and retrieval. *Brain Struct Funct*. 2013; http://dx.doi.org/10.1007/s00429-013-0505-y.
102. Wei W, Coelho CM, Li X, et al. p300/CBP-associated factor selectively regulates the extinction of conditioned fear. *J Neurosci*. 2012;32(35):11930–11941.
103. Guzowski JF, Setlow B, Wagner EK, McGaugh JL. Experience-dependent gene expression in the rat hippocampus after spatial learning: a comparison of the immediate-early genes Arc, c-fos, and zif268. *J Neurosci*. 2001;21(14):5089–5098.
104. Soule J, Penke Z, Kanhema T, Alme MN, Laroche S, Bramham CR. Object-place recognition learning triggers rapid induction of plasticity-related immediate early genes and synaptic proteins in the rat dentate gyrus. *Neural Plast*. 2008;2008:269097.
105. Clarke JR, Cammarota M, Gruart A, Izquierdo I, Delgado-Garcia JM. Plastic modifications induced by object recognition memory processing. *Proc Natl Acad Sci U S A*. 2010;107(6):2652–2657.
106. Satvat E, Schmidt B, Argraves M, Marrone DF, Markus EJ. Changes in task demands alter the pattern of zif268 expression in the dentate gyrus. *J Neurosci*. 2011;31(19):7163–7167.

107. Bozon B, Davis S, Laroche S. A requirement for the immediate early gene zif268 in reconsolidation of recognition memory after retrieval. *Neuron*. 2003;40(4):695–701.
108. Davis S, Renaudineau S, Poirier R, Poucet B, Save E, Laroche S. The formation and stability of recognition memory: what happens upon recall? *Front Behav Neurosci*. 2010;4:177.
109. Besnard A, Caboche J, Laroche S. Recall and reconsolidation of contextual fear memory: differential control by ERK and Zif268 expression dosage. *PLoS One*. 2013;8: e72006. http://dx.doi.org/10.1371/journal.pone.0072006.
110. Lee JL, Everitt BJ, Thomas KL. Independent cellular processes for hippocampal memory consolidation and reconsolidation. *Science*. 2004;304(5672):839–843.
111. Malkani S, Wallace KJ, Donley MP, Rosen JB. An egr-1 (zif268) antisense oligodeoxynucleotide infused into the amygdala disrupts fear conditioning. *Learn Mem*. 2004;11(5):617–624.
112. Zovkic IB, Guzman-Karlsson MC, Sweatt JD. Epigenetic regulation of memory formation and maintenance. *Learn Mem*. 2013;20(2):61–74.
113. Gupta S, Kim SY, Artis S, et al. Histone methylation regulates memory formation. *J Neurosci*. 2010;30(10):3589–3599.
114. Xie L, Korkmaz KS, Braun K, Bock J. Early life stress-induced histone acetylations correlate with activation of the synaptic plasticity genes Arc and Egr1 in the mouse hippocampus. *J Neurochem*. 2013;125(3):457–464.
115. Zhao Z, Fan L, Fortress AM, Boulware MI, Frick KM. Hippocampal histone acetylation regulates object recognition and the estradiol-induced enhancement of object recognition. *J Neurosci*. 2012;32(7):2344–2351.
116. Graff J, Woldemichael BT, Berchtold D, Dewarrat G, Mansuy IM. Dynamic histone marks in the hippocampus and cortex facilitate memory consolidation. *Nat Commun*. 2012;3:991.
117. Tourtellotte WG, Milbrandt J. Sensory ataxia and muscle spindle agenesis in mice lacking the transcription factor Egr3. *Nat Genet*. 1998;20(1):87–91.
118. Poirier R, Cheval H, Mailhes C, Charnay P, Davis S, Laroche S. Paradoxical role of an Egr transcription factor family member, Egr2/Krox20, in learning and memory. *Front Behav Neurosci*. 2007;1:6.
119. Poirier R, Cheval H, Mailhes C, et al. Distinct functions of egr gene family members in cognitive processes. *Front Neurosci*. 2008;2(1):47–55.
120. Bozon B, Davis S, Laroche S. Regulated transcription of the immediate-early gene Zif268: mechanisms and gene dosage-dependent function in synaptic plasticity and memory formation. *Hippocampus*. 2002;12(5):570–577.
121. Renaudineau S, Poucet B, Laroche S, Davis S, Save E. Impaired long-term stability of CA1 place cell representation in mice lacking the transcription factor zif268/egr1. *Proc Natl Acad Sci U S A*. 2009;106(28):11771–11775.
122. Baumgartel K, Genoux D, Welzl H, et al. Control of the establishment of aversive memory by calcineurin and Zif268. *Nat Neurosci*. 2008;11(5):572–578.
123. Scoville WB, Milner B. Loss of recent memory after bilateral hippocampal lesions. *J Neurol Neurosurg Psychiatry*. 1957;20(1):11–21.
124. Squire LR, Alvarez P. Retrograde amnesia and memory consolidation: a neurobiological perspective. *Curr Opin Neurobiol*. 1995;5(2):169–177.
125. Bontempi B, Laurent-Demir C, Destrade C, Jaffard R. Time-dependent reorganization of brain circuitry underlying long-term memory storage. *Nature*. 1999;400(6745):671–675.
126. Frankland PW, Bontempi B. The organization of recent and remote memories. *Nat Rev Neurosci*. 2005;6(2):119–130.
127. Frankland PW, Bontempi B, Talton LE, Kaczmarek L, Silva AJ. The involvement of the anterior cingulate cortex in remote contextual fear memory. *Science*. 2004;304(5672):881–883.

128. Maviel T, Durkin TP, Menzaghi F, Bontempi B. Sites of neocortical reorganization critical for remote spatial memory. *Science*. 2004;305(5680):96–99.
129. Nadel L, Moscovitch M. Memory consolidation, retrograde amnesia and the hippocampal complex. *Curr Opin Neurobiol*. 1997;7(2):217–227.
130. Wiltgen BJ, Brown RA, Talton LE, Silva AJ. New circuits for old memories: the role of the neocortex in consolidation. *Neuron*. 2004;44(1):101–108.
131. Teixeira CM, Pomedli SR, Maei HR, Kee N, Frankland PW. Involvement of the anterior cingulate cortex in the expression of remote spatial memory. *J Neurosci*. 2006;26(29):7555–7564.
132. Takehara-Nishiuchi K, Nakao K, Kawahara S, Matsuki N, Kirino Y. Systems consolidation requires postlearning activation of NMDA receptors in the medial prefrontal cortex in trace eyeblink conditioning. *J Neurosci*. 2006;26(19):5049–5058.
133. Frankland PW, O'Brien C, Ohno M, Kirkwood A, Silva AJ. Alpha-CaMKII-dependent plasticity in the cortex is required for permanent memory. *Nature*. 2001;411(6835):309–313.
134. Restivo L, Vetere G, Bontempi B, Ammassari-Teule M. The formation of recent and remote memory is associated with time-dependent formation of dendritic spines in the hippocampus and anterior cingulate cortex. *J Neurosci*. 2009;29(25):8206–8214.
135. Hall J, Thomas KL, Everitt BJ. Cellular imaging of zif268 expression in the hippocampus and amygdala during contextual and cued fear memory retrieval: selective activation of hippocampal CA1 neurons during the recall of contextual memories. *J Neurosci*. 2001;21(6):2186–2193.
136. Lesburgueres E, Gobbo OL, Alaux-Cantin S, Hambucken A, Trifilieff P, Bontempi B. Early tagging of cortical networks is required for the formation of enduring associative memory. *Science*. 2011;331(6019):924–928.
137. Tse D, Langston RF, Kakeyama M, et al. Schemas and memory consolidation. *Science*. 2007;316(5821):76–82.
138. Tse D, Takeuchi T, Kakeyama M, et al. Schema-dependent gene activation and memory encoding in neocortex. *Science*. 2011;333(6044):891–895.
139. Nader K, Hardt O. A single standard for memory: the case for reconsolidation. *Nat Rev Neurosci*. 2009;10(3):224–234.
140. Lee JL. Reconsolidation: maintaining memory relevance. *Trends Neurosci*. 2009;32(8): 413–420.
141. Nader K, Schafe GE, Le Doux JE. Fear memories require protein synthesis in the amygdala for reconsolidation after retrieval. *Nature*. 2000;406(6797):722–726.
142. Finnie PS, Nader K. The role of metaplasticity mechanisms in regulating memory destabilization and reconsolidation. *Neurosci Biobehav Rev*. 2012;36(7):1667–1707.
143. Besnard A, Caboche J, Laroche S. Reconsolidation of memory: a decade of debate. *Prog Neurobiol*. 2012;99(1):61–80.
144. Tronson NC, Taylor JR. Molecular mechanisms of memory reconsolidation. *Nat Rev Neurosci*. 2007;8(4):262–275.
145. Thomas KL, Hall J, Everitt BJ. Cellular imaging with zif268 expression in the rat nucleus accumbens and frontal cortex further dissociates the neural pathways activated following the retrieval of contextual and cued fear memory. *Eur J Neurosci*. 2002;16(9):1789–1796.
146. Lubin FD, Sweatt JD. The IkappaB kinase regulates chromatin structure during reconsolidation of conditioned fear memories. *Neuron*. 2007;55(6):942–957.
147. Kwon JT, Jhang J, Kim HS, Lee S, Han JH. Brain region-specific activity patterns after recent or remote memory retrieval of auditory conditioned fear. *Learn Mem*. 2012;19(10):487–494.
148. Barnes P, Kirtley A, Thomas KL. Quantitatively and qualitatively different cellular processes are engaged in CA1 during the consolidation and reconsolidation of contextual fear memory. *Hippocampus*. 2012;22(2):149–171.

149. Lee JL, Di Ciano P, Thomas KL, Everitt BJ. Disrupting reconsolidation of drug memories reduces cocaine-seeking behavior. *Neuron.* 2005;47(6):795–801.
150. Lee JL, Hynds RE. Divergent cellular pathways of hippocampal memory consolidation and reconsolidation. *Hippocampus.* 2013;23(3):233–244.
151. Lee JL. Memory reconsolidation mediates the strengthening of memories by additional learning. *Nat Neurosci.* 2008;11(11):1264–1266.
152. Lee JL. Memory reconsolidation mediates the updating of hippocampal memory content. *Front Behav Neurosci.* 2010;4:168.
153. Kelley AE. Memory and addiction: shared neural circuitry and molecular mechanisms. *Neuron.* 2004;44(1):161–179.
154. Moratalla R, Robertson HA, Graybiel AM. Dynamic regulation of NGFI-A (zif268, egr1) gene expression in the striatum. *J Neurosci.* 1992;12(7):2609–2622.
155. Nguyen TV, Kosofsky BE, Birnbaum R, Cohen BM, Hyman SE. Differential expression of c-fos and zif268 in rat striatum after haloperidol, clozapine, and amphetamine. *Proc Natl Acad Sci U S A.* 1992;89(10):4270–4274.
156. Hirata H, Asanuma M, Cadet JL. Superoxide radicals are mediators of the effects of methamphetamine on Zif268 (Egr-1, NGFI-A) in the brain: evidence from using CuZn superoxide dismutase transgenic mice. *Brain Res Mol Brain Res.* 1998;58(1–2):209–216.
157. Bhat RV, Worley PF, Cole AJ, Baraban JM. Activation of the zinc finger encoding gene krox-20 in adult rat brain: comparison with zif268. *Brain Res Mol Brain Res.* 1992;13(3):263–266.
158. Valjent E, Pages C, Rogard M, Besson MJ, Maldonado R, Caboche J. Delta 9-tetrahydrocannabinol-induced MAPK/ERK and Elk-1 activation in vivo depends on dopaminergic transmission. *Eur J Neurosci.* 2001;14(2):342–352.
159. Derkinderen P, Valjent E, Toutant M, et al. Regulation of extracellular signal-regulated kinase by cannabinoids in hippocampus. *J Neurosci.* 2003;23(6):2371–2382.
160. Cole AJ, Bhat RV, Patt C, Worley PF, Baraban JM. D1 dopamine receptor activation of multiple transcription factor genes in rat striatum. *J Neurochem.* 1992;58(4):1420–1426.
161. Bhat RV, Cole AJ, Baraban JM. Role of monoamine systems in activation of zif268 by cocaine. *J Psychiatry Neurosci.* 1992;17(3):94–102.
162. Alonso R, Gnanadicom H, Frechin N, Fournier M, Le Fur G, Soubrie P. Blockade of neurotensin receptors suppresses the dopamine D1/D2 synergism on immediate early gene expression in the rat brain. *Eur J Neurosci.* 1999;11(3):967–974.
163. Tien LT, Ho IK, Ma T. Methamphetamine-induced expression of zif268 mRNA is prevented by haloperidol in mice lacking mu-opioid receptor. *Neurotoxicology.* 2010;31(3):326–330.
164. Bertran-Gonzalez J, Bosch C, Maroteaux M, et al. Opposing patterns of signaling activation in dopamine D1 and D2 receptor-expressing striatal neurons in response to cocaine and haloperidol. *J Neurosci.* 2008;28(22):5671–5685.
165. Jung AB, Bennett Jr JP. Development of striatal dopaminergic function. II: Dopaminergic regulation of transcription of the immediate early gene zif268 and of D1 (D1a) and D2 (D2a) receptors during pre- and postnatal development. *Brain Res Dev Brain Res.* 1996;94(2):121–132.
166. Caster JM, Kuhn CM. Maturation of coordinated immediate early gene expression by cocaine during adolescence. *Neuroscience.* 2009;160(1):13–31.
167. Hollis F, Gaval-Cruz M, Carrier N, Dietz DM, Kabbaj M. Juvenile and adult rats differ in cocaine reward and expression of zif268 in the forebrain. *Neuroscience.* 2012;200:91–98.
168. Valjent E, Aubier B, Corbille AG, et al. Plasticity-associated gene Krox24/Zif268 is required for long-lasting behavioral effects of cocaine. *J Neurosci.* 2006;26(18):4956–4960.

169. Besnard A, Bouveyron N, Kappes V, et al. Alterations of molecular and behavioral responses to cocaine by selective inhibition of Elk-1 phosphorylation. *J Neurosci.* 2011;31(40):14296–14307.
170. Brami-Cherrier K, Valjent E, Herve D, et al. Parsing molecular and behavioral effects of cocaine in mitogen- and stress-activated protein kinase-1-deficient mice. *J Neurosci.* 2005;25(49):11444–11454.
171. Barik J, Parnaudeau S, Saint Amaux AL, et al. Glucocorticoid receptors in dopaminoceptive neurons, key for cocaine, are dispensable for molecular and behavioral morphine responses. *Biol Psychiatry.* 2010;68(3):231–239.
172. Ennulat DJ, Babb S, Cohen BM. Persistent reduction of immediate early gene mRNA in rat forebrain following single or multiple doses of cocaine. *Brain Res Mol Brain Res.* 1994;26(1–2):106–112.
173. Ziolkowska B, Korostynski M, Piechota M, Kubik J, Przewlocki R. Effects of morphine on immediate-early gene expression in the striatum of C57BL/6J and DBA/2J mice. *Pharmacol Rep.* 2012;64(5):1091–1104.
174. Hope B, Kosofsky B, Hyman SE, Nestler EJ. Regulation of immediate early gene expression and AP-1 binding in the rat nucleus accumbens by chronic cocaine. *Proc Natl Acad Sci U S A.* 1992;89(13):5764–5768.
175. Persico AM, Schindler CW, O'Hara BF, Brannock MT, Uhl GR. Brain transcription factor expression: effects of acute and chronic amphetamine and injection stress. *Brain Res Mol Brain Res.* 1993;20(1–2):91–100.
176. Shilling PD, Kelsoe JR, Kuczenski R, Segal DS. Differential regional zif268 messenger RNA expression in an escalating dose/binge model of amphetamine-induced psychosis. *Neuroscience.* 2000;96(1):83–90.
177. Boucher AA, Vivier L, Metna-Laurent M, et al. Chronic treatment with Delta(9)-tetrahydrocannabinol impairs spatial memory and reduces zif268 expression in the mouse forebrain. *Behav Pharmacol.* 2009;20(1):45–55.
178. Mutschler NH, Miczek KA, Hammer Jr RP. Reduction of zif268 messenger RNA expression during prolonged withdrawal following "binge" cocaine self-administration in rats. *Neuroscience.* 2000;100(3):531–538.
179. Fumagalli F, Franchi C, Caffino L, Racagni G, Riva MA, Cervo L. Single session of cocaine intravenous self-administration shapes goal-oriented behaviours and up-regulates Arc mRNA levels in rat medial prefrontal cortex. *Int J Neuropsychopharmacol.* 2009;12(3):423–429.
180. Robinson TE, Berridge KC. The neural basis of drug craving: an incentive-sensitization theory of addiction. *Brain Res Brain Res Rev.* 1993;18(3):247–291.
181. Thomas KL, Arroyo M, Everitt BJ. Induction of the learning and plasticity-associated gene Zif268 following exposure to a discrete cocaine-associated stimulus. *Eur J Neurosci.* 2003;17(9):1964–1972.
182. Schmidt ED, Voorn P, Binnekade R, Schoffelmeer AN, De Vries TJ. Differential involvement of the prelimbic cortex and striatum in conditioned heroin and sucrose seeking following long-term extinction. *Eur J Neurosci.* 2005;22(9):2347–2356.
183. Hearing MC, Schochet TL, See RE, McGinty JF. Context-driven cocaine-seeking in abstinent rats increases activity-regulated gene expression in the basolateral amygdala and dorsal hippocampus differentially following short and long periods of abstinence. *Neuroscience.* 2010;170(2):570–579.
184. Ziolkowska B, Kielbinski M, Gieryk A, Soria G, Maldonado R, Przewlocki R. Regulation of the immediate-early genes arc and zif268 in a mouse operant model of cocaine seeking reinstatement. *J Neural Transm.* 2011;118(6):877–887.
185. Covington 3rd HE, Kikusui T, Goodhue J, Nikulina EM, Hammer Jr RP, Miczek KA. Brief social defeat stress: long lasting effects on cocaine taking during a binge and zif268

mRNA expression in the amygdala and prefrontal cortex. *Neuropsychopharmacology*. 2005;30(2):310–321.
186. Briand LA, Blendy JA. Not all stress is equal: CREB is not necessary for restraint stress reinstatement of cocaine-conditioned reward. *Behav Brain Res*. 2013;246:63–68.
187. Lee JL, Milton AL, Everitt BJ. Cue-induced cocaine seeking and relapse are reduced by disruption of drug memory reconsolidation. *J Neurosci*. 2006;26(22):5881–5887.
188. Hellemans KG, Everitt BJ, Lee JL. Disrupting reconsolidation of conditioned withdrawal memories in the basolateral amygdala reduces suppression of heroin seeking in rats. *J Neurosci*. 2006;26(49):12694–12699.
189. Lee JL, Gardner RJ, Butler VJ, Everitt BJ. D-cycloserine potentiates the reconsolidation of cocaine-associated memories. *Learn Mem*. 2009;16(1):82–85.
190. Theberge FR, Milton AL, Belin D, Lee JL, Everitt BJ. The basolateral amygdala and nucleus accumbens core mediate dissociable aspects of drug memory reconsolidation. *Learn Mem*. 2010;17(9):444–453.
191. Meyers RA, Zavala AR, Neisewander JL. Dorsal, but not ventral, hippocampal lesions disrupt cocaine place conditioning. *Neuroreport*. 2003;14(16):2127–2131.
192. Zhao C, Deng W, Gage FH. Mechanisms and functional implications of adult neurogenesis. *Cell*. 2008;132(4):645–660.
193. Ming GL, Song H. Adult neurogenesis in the mammalian brain: significant answers and significant questions. *Neuron*. 2011;70(4):687–702.
194. Lledo PM, Alonso M, Grubb MS. Adult neurogenesis and functional plasticity in neuronal circuits. *Nat Rev Neurosci*. 2006;7(3):179–193.
195. Nissant A, Pallotto M. Integration and maturation of newborn neurons in the adult olfactory bulb–from synapses to function. *Eur J Neurosci*. 2011;33(6):1069–1077.
196. Sahay A, Wilson DA, Hen R. Pattern separation: a common function for new neurons in hippocampus and olfactory bulb. *Neuron*. 2011;70(4):582–588.
197. Lazarini F, Lledo PM. Is adult neurogenesis essential for olfaction? *Trends Neurosci*. 2011;34(1):20–30.
198. Feierstein CE. Linking adult olfactory neurogenesis to social behavior. *Front Neurosci*. 2012;6:173.
199. Alonso M, Lepousez G, Sebastien W, et al. Activation of adult-born neurons facilitates learning and memory. *Nat Neurosci*. 2012;15(6):897–904.
200. Piatti VC, Ewell LA, Leutgeb JK. Neurogenesis in the dentate gyrus: carrying the message or dictating the tone. *Front Neurosci*. 2013;7:50.
201. Deng W, Aimone JB, Gage FH. New neurons and new memories: how does adult hippocampal neurogenesis affect learning and memory? *Nat Rev Neurosci*. 2010;11(5):339–350.
202. Aimone JB, Deng W, Gage FH. Resolving new memories: a critical look at the dentate gyrus, adult neurogenesis, and pattern separation. *Neuron*. 2011;70(4):589–596.
203. Gu Y, Arruda-Carvalho M, Wang J, et al. Optical controlling reveals time-dependent roles for adult-born dentate granule cells. *Nat Neurosci*. 2012;15(12):1700–1706.
204. Sun J, Ming GL, Song H. Epigenetic regulation of neurogenesis in the adult mammalian brain. *Eur J Neurosci*. 2011;33(6):1087–1093.
205. Mu Y, Lee SW, Gage FH. Signaling in adult neurogenesis. *Curr Opin Neurobiol*. 2010;20(4):416–423.
206. Veyrac A, Sacquet J, Nguyen V, Marien M, Jourdan F, Didier A. Novelty determines the effects of olfactory enrichment on memory and neurogenesis through noradrenergic mechanisms. *Neuropsychopharmacology*. 2009;34(3):786–795.
207. Veyrac A, Bakker J. Postnatal and adult exposure to estradiol differentially influences adult neurogenesis in the main and accessory olfactory bulb of female mice. *FASEB J*. 2011;25(3):1048–1057.

208. Magavi SS, Mitchell BD, Szentirmai O, Carter BS, Macklis JD. Adult-born and pre-existing olfactory granule neurons undergo distinct experience-dependent modifications of their olfactory responses in vivo. *J Neurosci*. 2005;25(46):10729–10739.
209. Moreno MM, Linster C, Escanilla O, Sacquet J, Didier A, Mandairon N. Olfactory perceptual learning requires adult neurogenesis. *Proc Natl Acad Sci U S A*. 2009;106(42):17980–17985.
210. Mak GK, Weiss S. Paternal recognition of adult offspring mediated by newly generated CNS neurons. *Nat Neurosci*. 2010;13(6):753–758.
211. Mandairon N, Sacquet J, Garcia S, Ravel N, Jourdan F, Didier A. Neurogenic correlates of an olfactory discrimination task in the adult olfactory bulb. *Eur J Neurosci*. 2006;24(12):3578–3588.
212. Alonso M, Viollet C, Gabellec MM, Meas-Yedid V, Olivo-Marin JC, Lledo PM. Olfactory discrimination learning increases the survival of adult-born neurons in the olfactory bulb. *J Neurosci*. 2006;26(41):10508–10513.
213. Sultan S, Mandairon N, Kermen F, Garcia S, Sacquet J, Didier A. Learning-dependent neurogenesis in the olfactory bulb determines long-term olfactory memory. *FASEB J*. 2010;24(7):2355–2363.
214. Sultan S, Rey N, Sacquet J, Mandairon N, Didier A. Newborn neurons in the olfactory bulb selected for long-term survival through olfactory learning are prematurely suppressed when the olfactory memory is erased. *J Neurosci*. 2011;31(42):14893–14898.
215. Mandairon N, Sultan S, Nouvian M, Sacquet J, Didier A. Involvement of newborn neurons in olfactory associative learning? The operant or non-operant component of the task makes all the difference. *J Neurosci*. 2011;31(35):12455–12460.
216. Moreno MM, Bath K, Kuczewski N, Sacquet J, Didier A, Mandairon N. Action of the noradrenergic system on adult-born cells is required for olfactory learning in mice. *J Neurosci*. 2012;32(11):3748–3758.
217. Tashiro A, Makino H, Gage FH. Experience-specific functional modification of the dentate gyrus through adult neurogenesis: a critical period during an immature stage. *J Neurosci*. 2007;27(12):3252–3259.
218. Clark PJ, Bhattacharya TK, Miller DS, Kohman RA, DeYoung EK, Rhodes JS. New neurons generated from running are broadly recruited into neuronal activation associated with three different hippocampus-involved tasks. *Hippocampus*. 2012;22(9):1860–1867.
219. Snyder JS, Clifford MA, Jeurling SI, Cameron HA. Complementary activation of hippocampal-cortical subregions and immature neurons following chronic training in single and multiple context versions of the water maze. *Behav Brain Res*. 2012;227(2):330–339.
220. Veyrac A, Gros A, Bruel-Jungerman E, et al. Zif268/egr1 gene controls the selection, maturation and functional integration of adult hippocampal newborn neurons by learning. *Proc Natl Acad Sci U S A*. 2013;110(17):7062–7067.
221. Aasebo IE, Blankvoort S, Tashiro A. Critical maturational period of new neurons in adult dentate gyrus for their involvement in memory formation. *Eur J Neurosci*. 2011;33(6):1094–1100.
222. Jessberger S, Kempermann G. Adult-born hippocampal neurons mature into activity-dependent responsiveness. *Eur J Neurosci*. 2003;18(10):2707–2712.
223. Bruel-Jungerman E, Davis S, Rampon C, Laroche S. Long-term potentiation enhances neurogenesis in the adult dentate gyrus. *J Neurosci*. 2006;26(22):5888–5893.
224. Kuipers SD, Tiron A, Soule J, Messaoudi E, Trentani A, Bramham CR. Selective survival and maturation of adult-born dentate granule cells expressing the immediate early gene Arc/Arg3.1. *PLoS One*. 2009;4(3):e4885.
225. Snyder JS, Choe JS, Clifford MA, et al. Adult-born hippocampal neurons are more numerous, faster maturing, and more involved in behavior in rats than in mice. *J Neurosci*. 2009;29(46):14484–14495.

226. Jungenitz T, Radic T, Jedlicka P, Schwarzacher SW. High-frequency stimulation induces gradual immediate early gene expression in maturing adult-generated hippocampal granule cells. *Cereb Cortex*. 2013; http://dx.doi.org/10.1093/cercor/bht035.
227. Epp JR, Spritzer MD, Galea LA. Hippocampus-dependent learning promotes survival of new neurons in the dentate gyrus at a specific time during cell maturation. *Neuroscience*. 2007;149(2):273–285.
228. Trouche S, Bontempi B, Roullet P, Rampon C. Recruitment of adult-generated neurons into functional hippocampal networks contributes to updating and strengthening of spatial memory. *Proc Natl Acad Sci U S A*. 2009;106(14):5919–5924.
229. Eid MA, Kumar MV, Iczkowski KA, Bostwick DG, Tindall DJ. Expression of early growth response genes in human prostate cancer. *Cancer Res*. 1998;58(11):2461–2468.
230. Abdulkadir SA, Carbone JM, Naughton CK, Humphrey PA, Catalona WJ, Milbrandt J. Frequent and early loss of the EGR1 corepressor NAB2 in human prostate carcinoma. *Hum Pathol*. 2001;32(9):935–939.
231. Thiel G, Cibelli G. Regulation of life and death by the zinc finger transcription factor Egr-1. *J Cell Physiol*. 2002;193(3):287–292.
232. Jain N, Mahendran R, Philp R, Guy GR, Tan YH, Cao X. Casein kinase II associates with Egr-1 and acts as a negative modulator of its DNA binding and transcription activities in NIH 3T3 cells. *J Biol Chem*. 1996;271(23):13530–13536.
233. Virolle T, Adamson ED, Baron V, et al. The Egr-1 transcription factor directly activates PTEN during irradiation-induced signalling. *Nat Cell Biol*. 2001;3(12):1124–1128.
234. Blalock EM, Chen KC, Sharrow K, et al. Gene microarrays in hippocampal aging: statistical profiling identifies novel processes correlated with cognitive impairment. *J Neurosci*. 2003;23(9):3807–3819.
235. Dickey CA, Loring JF, Montgomery J, Gordon MN, Eastman PS, Morgan D. Selectively reduced expression of synaptic plasticity-related genes in amyloid precursor protein + presenilin-1 transgenic mice. *J Neurosci*. 2003;23(12):5219–5226.
236. Gersten M, Alirezaei M, Marcondes MC, et al. An integrated systems analysis implicates EGR1 downregulation in simian immunodeficiency virus encephalitis-induced neural dysfunction. *J Neurosci*. 2009;29(40):12467–12476.
237. MacGibbon GA, Lawlor PA, Walton M, et al. Expression of Fos, Jun, and Krox family proteins in Alzheimer's disease. *Exp Neurol*. 1997;147(2):316–332.
238. Gomez Ravetti M, Rosso OA, Berretta R, Moscato P. Uncovering molecular biomarkers that correlate cognitive decline with the changes of hippocampus' gene expression profiles in Alzheimer's disease. *PLoS One*. 2010;5(4):e10153.
239. Lu Y, Li T, Qureshi HY, Han D, Paudel HK. Early growth response 1 (Egr-1) regulates phosphorylation of microtubule-associated protein tau in mammalian brain. *J Biol Chem*. 2011;286(23):20569–20581.

CHAPTER FIVE

Mechanisms of Translation Control Underlying Long-Lasting Synaptic Plasticity and the Consolidation of Long-Term Memory

Emanuela Santini, Thu N. Huynh, Eric Klann
Center for Neural Science, New York University, New York, USA

Contents

Abstract

The complexity of memory formation and its persistence is a phenomenon that has been studied intensely for centuries. Memory exists in many forms and is stored in various brain regions. Generally speaking, memories are reorganized into broadly distributed cortical networks over time through systems level consolidation. At the cellular level, storage of information is believed to initially occur via altered synaptic strength by processes such as long-term potentiation. New protein synthesis is required for long-lasting synaptic

Progress in Molecular Biology and Translational Science, Volume 122
ISSN 1877-1173
http://dx.doi.org/10.1016/B978-0-12-420170-5.00005-2

plasticity as well as for the formation of long-term memory. The mammalian target of rapamycin complex 1 (mTORC1) is a critical regulator of cap-dependent protein synthesis and is required for numerous forms of long-lasting synaptic plasticity and long-term memory. As such, the study of mTORC1 and protein factors that control translation initiation and elongation has enhanced our understanding of how the process of protein synthesis is regulated during memory formation. Herein we discuss the molecular mechanisms that regulate protein synthesis as well as pharmacological and genetic manipulations that demonstrate the requirement for proper translational control in long-lasting synaptic plasticity and long-term memory formation.

1. THE STUDY AND CHARACTERIZATION OF MEMORY SYSTEMS

1.1. Historical overview

The term consolidation refers to the progressive postacquisition stabilization of long-term memory and to the phase(s) of the memory stabilization process.[1,2] During the consolidation process, newly acquired information is gradually transformed into long-term memory. Historically, the term consolidation is attributed to Müller and Pilzecker, who first studied the acquisition and retrieval of syllable pairs in human subjects.[1,2] These studies demonstrated that the correct recall of syllable pairs required some time before reaching a certain success criterion, and that the presentation of new material before the consolidation of syllable pairs damaged its recall. Thus, new memories require time to be stabilized or "consolidated." Furthermore, these studies indicated that there is a post-training time window during which it is possible to interfere with newly learned information. As a result, new information is susceptible to destruction before being consolidated into a memory. Overall, these studies pointed out two important properties of the consolidation process: time dependency and distinct stages of memory. Indeed, during the consolidation process, memory is converted from a labile to stable form in a time-dependent manner that eventually becomes resilient to interference.

The idea that memory is gradually transformed was hypothesized first over a hundred years ago. In 1882, the French psychologist, Théodule-Armand Ribot, proposed that memories are gradually reorganized with the passage of time.[3] Moreover, he made the observation that following brain injury the degree of memory loss was correlated with the age of the memory. Specifically, there was a greater loss of newer memories, compared to older memories. This time-dependent component of memory reorganization is known as Ribot's Law and provides the basis for research performed on temporally graded retrograde amnesia. Today, we know of this memory transformation process as systems consolidation.

One of the most well-described cases of a patient that exhibited retrograde amnesia was patient Henry Gustav Molaison (known for many years as patient H.M. until after his death in 2008, when his name was finally revealed), first described by Milner and Scoville.[4,5] H.M. was affected by severe temporal lobe epilepsy, which was treated by the surgical transection of his hippocampus, parahippocampal gyrus, and amygdala. Following this procedure, H.M. suffered from moderate retrograde amnesia; however, the amnesia was not complete. Indeed, only his most recent memories were lost, whereas memories from his childhood were retained.

H.M. also suffered from persistent anterograde amnesia, which is the inability to form new memories, specifically restricted to declarative memory, which is defined as memory that can be recollected consciously. His impairment included the inability to form new episodic (memory of events) and semantic (memory of general knowledge or facts) information. However, other forms of non-declarative memory, such as procedural memory and perceptual learning, were largely spared. Overall, the studies of H.M.'s memory deficits indicated that memory is not a unitary phenomenon but rather involves multiple memory systems that exist simultaneously in different brain regions. In fact, lesions in the hippocampus critically impaired declarative memory but left procedural memory intact, suggesting that the latter is different from the former and is not reliant on the hippocampus. Moreover, these lesions only impaired recent memories, indicating that hippocampus is the anatomical locus where newly acquired memories reside, whereas older memories are stored permanently elsewhere.

It is clear that the general term "consolidation" includes at least two different types of processes characterized by distinct temporal dynamics (Fig. 5.1). Heretofore we refer to fast consolidation processes as synaptic consolidation, which is completed in hours after training and involves the stabilization of changes in synaptic connectivity within localized circuits. By contrast, systems consolidation is a more prolonged process and requires the gradual reorganization of brain regions that support memory. Hereafter we focus on the molecular mechanisms underlying synaptic consolidation.

1.2. Cellular mechanisms of memory consolidation

Memory is encoded and stored in the brain via carefully regulated interactions between neuronal networks. Synapses are the essential unit of memory

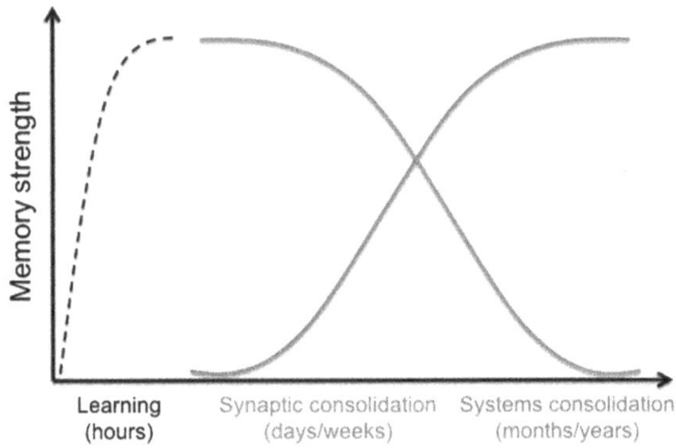

Figure 5.1 The reorganization of memory occurs with the passage of time. Learning occurs within hours and is initially stored in local synaptic circuits. This "synaptic consolidation" persists for days and up to weeks. Synaptic consolidation is gradually converted to systems consolidation, where the memory can persist for years, in broadly distributed cortical networks. (See the color plate.)

and are the site of electrochemical communication between neurons. These connections are "plastic," in that the physiological responsiveness (i.e., the "strength" of the synaptic connection) is modifiable. Importantly, changes in synaptic strength (synaptic plasticity) also are defined temporally, with some alterations lasting only seconds and others persisting over the lifetime of the organism.[6,7] The most well-established model for persistent activity-dependent strengthening of individual synapses is long-term potentiation (LTP).

The first full description of LTP by Bliss and Lomo[8] reported that trains of high-frequency stimulation (HFS) to the perforant path caused a sustained increase in efficiency of synaptic transmission in the granule cells of the dentate gyrus in the hippocampus of the rabbit. These findings suggested that the synaptic changes underlying certain forms of learning and memory might be similar to the mechanisms responsible for LTP. Indeed, activity-dependent synaptic plasticity at appropriate synapses during memory formation is necessary and sufficient for storage of information. Some of the characteristics of LTP, such as cooperativity, associativity, input specificity,[9] and durability,[10] have been put forth as solid arguments in support of the hypothesis that LTP is a biological substrate for at least some forms of memory.

Other similarities have been described between memory consolidation and synaptic plasticity. For instance, there are common molecular changes that occur after the induction of LTP and during memory acquisition, which can be targeted and disrupted using the same pharmacological agents. For example, memory is divided into two different temporal phases: short-term memory, which lasts for a few hours, and long-term memory, which persists for much longer. At the cellular level, the storage of long-term memory is associated with changes in gene expression, *de novo* protein synthesis, and the formation of new synaptic connections.[11–13] Consistent with these observations, protein synthesis inhibitors block long-term memory but leave short-term memory unaffected, suggesting that stable long-lasting memories rely on new gene expression triggered at the time of the experience.[14–18] Similarly, LTP consists of distinct phases involving different molecular mechanisms. Early phase LTP (E-LTP), which lasts 2–3 h, is independent of protein synthesis, whereas more long-lasting late phase LTP (L-LTP), which persists several hours *in vitro* and either weeks or months *in vivo*, requires the synthesis of new proteins.[9] Thus, memory and synaptic plasticity share similar properties and require similar molecular events for their persistence.

Hippocampal circuits have been a valuable system for studying synaptic plasticity in the context of information storage mechanisms in the brain. Typically, synaptic plasticity is measured using electrophysiological techniques. Moreover, the hippocampus is one of the principal brain regions studied because of the presence of the trisynaptic circuit (see below). The laminar anatomical arrangement of the hippocampus allows for the preservation of this circuit in transverse slices, which facilitates experimentation. Additionally, the neuronal pathways within the hippocampus display remarkably pronounced plasticity.[19]

The trisynaptic circuit refers to the major hippocampal synapses and is comprised of three main groups of excitatory connections: (1) the perforant pathway generated by axons of pyramidal neurons in the entorhinal cortex that synapse on the granule neurons in the dentate gyrus, (2) the mossy fiber pathway generated by axons of granule neurons of the dentate gyrus that synapse onto CA3 pyramidal neurons, and (3) the Schaffer collateral pathway consisting of axonal projections of the CA3 pyramidal neurons that synapse onto CA1 pyramidal neurons. Most of the electrophysiological experiments described in this chapter were performed on the synaptic connections between CA3 and CA1 pyramidal neurons.[20]

The standard recording arrangement used to study synaptic plasticity between CA3 and CA1 pyramidal neurons is achieved by placing a stimulating electrode proximal to the CA3 Schaffer collateral axons. The recording electrode is placed into the stratum radiatum of the CA1, where it will record electrical activity induced by the stimulation of axonal inputs originating in the CA3 (Fig. 5.2A). LTP can be induced using a wide variety of induction protocols at these synapses. The standard protocols include either a single (for E-LTP) or repeated (usually 2–4 trains for L-LTP) delivery of 1-s 100 Hz stimulation (HFS), where repeated trains of 100 Hz are separated by 20 s or greater (Fig. 5.2B). A variation of this protocol employed to induce LTP is the use of theta burst stimulation (TBS). In this protocol, 3 trains of 100 Hz are delivered with and inter-train intervals of 20 s.[19]

The dependence of protein synthesis for long-lasting memory correlates with the necessity of *de novo* protein synthesis for synaptic plasticity.[11] The classical view of how translation impacts L-LTP is that proteins are synthesized in the cell body and delivered to the potentiated synapses. However, it is now clear that proteins can be synthesized locally in dendrites in response to changes in synaptic plasticity. For instance, local dendritic protein synthesis has been demonstrated with biochemical fractionation techniques, which

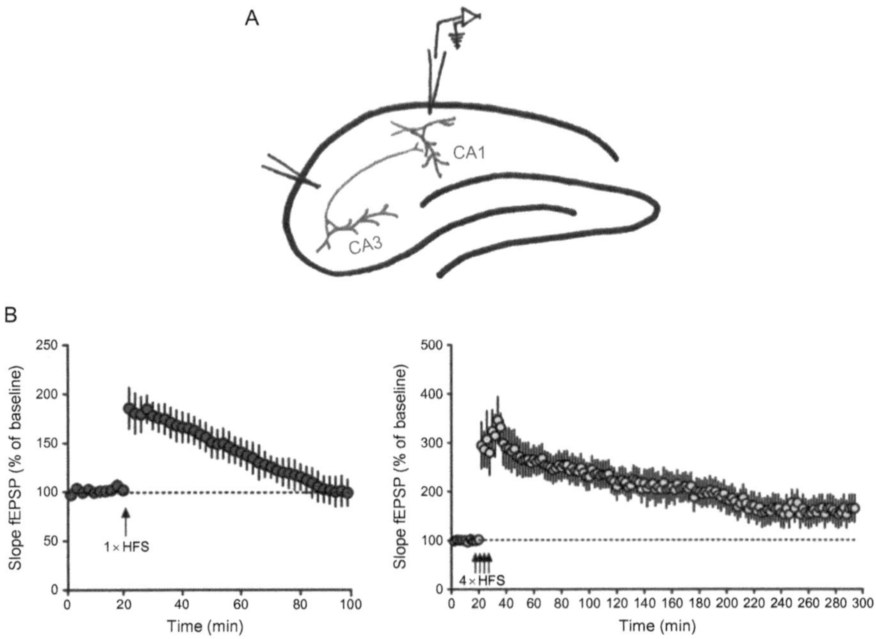

Figure 5.2 LTP: A cellular substrate for memory formation. (A) *Hippocampal LTP can be induced by high-frequency stimulation (HFS) of the Schaffer Collateral pathway.* Shown is a diagram of a stimulating electrode, which is placed near the axonal tract originating from CA3 pyramidal neurons, and a recording electrode, which is placed in the dendritic tree of CA1 pyramidal neurons. (B) *E-LTP and L-LTP induction is achieved using HFS.* Early phase LTP (E-LTP) is induced by 1 train of HFS and long-lasting, late phase LTP (L-LTP) is induced using 4 spaced trains of HFS. E-LTP persists approximately 1 h, whereas L-LTP persists for several hours. (See the color plate.)

permits the isolation of dendritic fragments (e.g., synaptoneurosome preparations). In these studies, different kinds of stimulation, such as membrane depolarization with high K^+,[21] brain-derived neurotrophic factor (BDNF),[22] and group I metabotropic glutamate receptor (mGluR) agonists,[23] were shown to increase protein synthesis. Several proteins were shown to be synthesized, including the α subunit of calcium/calmodulin-dependent protein kinase II (α-CaMKII), Arc, and fragile X mental retardation protein, respectively. Overall, these experiments are consistent with the idea that *de novo* protein synthesis occurs in active synapses, which supports the notion that local protein synthesis is involved in synaptic plasticity.

The demonstration that local protein synthesis occurs in synaptoneurosome preparations was confirmed with studies in neuronal cultures and brain slices. For example, it was shown that in hippocampal slices, BDNF-induced synaptic plasticity is blocked by protein synthesis inhibitors.[24] In the same study, Schaffer collateral-CA1 synapses isolated from their pre- and postsynaptic cell bodies still exhibited the protein synthesis-dependent plasticity.[24] A similar dependence on *de novo* protein synthesis has also been observed for a different form of synaptic plasticity: metabotropic glutamate receptor-dependent long-term depression (mGluR-LTD) at the same synaptic location.[25,26] Moreover, in cultured hippocampal neurons, the presence of local protein synthesis was demonstrated first with a protein synthesis reporter.[27] In this study, the reporter contained the coding sequence of green fluorescent protein (GFP), which was flanked by the 5′- and 3′-untranslated regions (UTRs) of α-CaMKII, which confers mRNA localization and translational regulation. After stimulation with BDNF in isolated dendrites, the increase in GFP fluorescence was followed with time-lapse imaging.[27] These experiments strongly suggest that BDNF-induced synaptic plasticity requires local translation machinery and, thus, dendritic protein synthesis.

1.3. Behavioral characterization of the memory trace

The study of molecular and synaptic neuronal modifications triggered by learning, which enables long-term storage of memories, is accompanied by separate studies that have been performed to investigate the functional role of these modifications in behavioral paradigms. These studies have contributed to the characterization of molecular mechanisms in memory consolidation.

As discussed earlier, a distinct feature of long-term memories is that it is more stable than short-term memory and thus more resistant to interference.

Experimentally, this window of susceptibility depends on the behavioral task and the type of interference or blockers used, and can range from minutes (e.g., electroconvulsive shock, ECS) to hours (e.g., distractor tasks, protein synthesis drug inhibitors).

Initially, ECS was one of the first blockers used to interfere with the formation of long-term memory and was shown to mimic retrograde amnesia in animal models.[11] In these studies, rodents showed impaired memory retention only when ECS was given shortly after training, but had no effect when administered after a longer time period following training.[28,29] Although the ability of ECS to disrupt memory is restricted to minutes after memory formation, certain pharmacological agents are capable of interfering with memory more robustly. Furthermore, these pharmacological treatments can result in effects that are more persistent, such as the case with protein synthesis inhibitors. The first demonstration that protein synthesis is required for the formation of long-lasting memories was provided by the work of Louis and Josefa Flexner. They demonstrated that temporal lobe injections of the protein synthesis inhibitor puromycin after learning were effective in blocking long-term memory in mice.[30]

Later, a series of pharmacological experiments in learning and memory paradigms confirmed that protein synthesis occurs after training and is required for the formation of long-term memory. Importantly, blocking protein synthesis does not impair the acquisition of short-term memory (extensively reviewed in Ref. 11). Commonly used protein synthesis inhibitors in memory studies include puromycin, anisomycin, acetoxycycloheximide, and cycloheximide. In a typical pharmacological study, an animal is treated with the protein synthesis inhibitor shortly after training. Following this, acquisition of short-term memory and long-term memory formation is tested.[31] The expected outcomes from these studies are that the acquisition of short-term memory will be intact, whereas the formation of long-term memory will be impaired. Thus, consolidation of short-term memories into long-term memories is dependent on protein synthesis.[32,33]

2. MECHANISMS OF mTOR-DEPENDENT TRANSLATIONAL CONTROL

The mammalian target of rapamycin (mTOR) is an evolutionarily conserved kinase that has emerged as one of the critical checkpoints of protein synthesis and plays a major role in synaptic plasticity and memory. mTOR activity is modulated by a wide variety of signals, ranging from neurotransmitters to changes in energy metabolism status, which result in the tight control of cap-dependent protein synthesis through multiple downstream effectors.

2.1. Upstream regulation of mTOR

mTOR is activated by multiple neurotransmitters and neurotrophic factors via neuronal surface receptors and channels. For instance, a-amino-3-hydroxy-5-methyl-4-isoxazolepropionic acid receptors (AMPA-R), *N*-methyl-D-aspartate receptors (NMDA-R), dopamine receptors, mGluRs, and BDNF receptors (i.e., TrkB-R) activate mTOR and are vital for the induction and expression of synaptic plasticity and memory.[34] mTOR acts as a node of convergence downstream of these receptors as well as several signaling pathways that include phosphoinositide-dependent kinase 1, phosphatidylinositol 3-kinase (PI3K), PKB/Akt, and tuberous sclerosis complex proteins 1 and 2 (TSC1/2).[34–36] The extracellular signal-regulated kinase (ERK) signaling pathway also activates mTOR, although it is unclear whether this results from phosphorylation of TSC1/2 by ERK or via its downstream effector p70 S6 kinase 1(S6K1).[37,38]

In addition to activation via neurotransmitters, mTOR signaling is also regulated by amino acid and cellular energy status. Amino acid starvation typically inhibits mTOR signaling via a mechanism that involves the PI3K-related protein Vps34 and is independent of TSC1/2-Rheb. On the other hand, reduction of adenosine triphosphate levels inhibits mTOR signaling via AMP-activated protein kinase (AMPK), which is a sensor of cellular energy status, and leads to the subsequent phosphorylation of TSC1/2.[39,40] All of these signals are ultimately funneled and integrated in the control of cap-dependent protein synthesis through direct or indirect regulation of mTOR downstream effectors.

mTOR is typically the catalytic component in one of two multiprotein signaling complexes, termed mTOR complex 1 and 2 (mTORC1 and mTORC2, respectively)[41] (Fig. 5.3). The other proteins that are components of the mTORCs control subcellular localization, substrate specificity, and signaling. Because functional mTOR is always observed in one of the two complexes, hereafter we will refer to only mTORCs.

mTOR is in mTORC1 when it is associated to the regulatory scaffolding protein Raptor. The function of Raptor is to bind proteins containing a short amino acid sequence called the mTOR signaling (TOS) motif and shuttle them to the mTOR catalytic subunit. The best-characterized

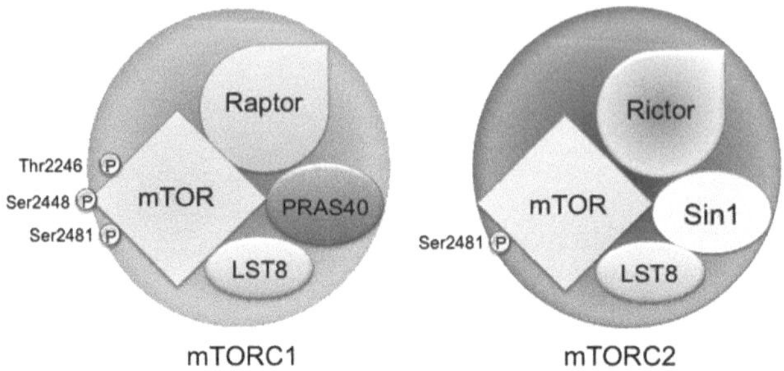

Figure 5.3 mTOR exists in two complexes. mTORC1 is comprised of mTOR, raptor, LST8, and PRAS40. mTORC2 is comprised of mTOR, rictor, LST8, and sin1. (See the color plate.)

TOS motif-containing proteins are those that are involved in the regulation of protein synthesis, S6K1, and eIF4E-binding protein (4E-BP).[34,42] mTORC1 is sensitive to inhibition by the antifungal rapamycin because of the competition between Raptor and the rapamycin–FKBP12 complex for binding to the FKBP12–rapamycin-binding (FRB) domain.[43,44] In addition to Raptor, mTORC1 contains other proteins that fall into two categories: common constituents of the mTORCs, such as the G-beta-like protein family member LST8, and exclusive constituents of mTORC1. For instance, proline-rich Akt/PKB substrate of 40 KDa (PRAS40) is only present in mTORC1. PRAS40 is important for the regulation of mTOR–Raptor interactions and mTOR signaling.[45,46] The best-characterized function of mTORC1 during synaptic plasticity and memory is the regulation of protein synthesis, achieved by controlling two critical downstream effectors proteins: S6K1 and 4E-BP, which are involved in the initiation and elongation phases of translation (see below).

The other mTOR complex, mTORC2, is formed when mTOR is associated with the rapamycin-insensitive companion of mTOR, Rictor. Although insensitive to acute rapamycin exposure, prolonged chronic treatment also disrupts mTORC2 function.[44,47] mTORC2 contains its own unique component, the SAPK-interacting protein1 (Sin1), which is believed to be required for the proper formation of mTORC2, as well as for its activity.[48–50] Based on its unique composition, mTORC2 acts on a different set of targets and regulates different intracellular processes. There is less known about the role of mTORC2 in synaptic plasticity and memory, but recent studies with mice that lack Rictor have demonstrated that similar to mTORC1, mTORC2 is required for L-LTP and long-term memory.[51]

The process of protein synthesis results in the sequential translation of mRNA into amino acids that are added to a growing polypeptide chain to make a protein. The biochemical signaling mechanisms that regulate protein synthesis can be divided in three consecutive phases: initiation, elongation, and termination. These phases are facilitated and regulated by proteins termed eukaryotic initiation factors (eIFs), elongation factors (eEFs), and release factors based on the particular phase in which they are active. mTORC1 controls a number of components involved in the initiation and elongation phases of translation.

It is important to note that mTORC1 signaling likely contributes to both the short-term (minutes) activation of translation and long-term (hours) increases in the translational capacity of neurons by regulating the synthesis of ribosomal proteins and other translational components. Both forms of translational regulation are important for synaptic plasticity and memory formation.

2.2. Translation initiation

The initiation phase of protein synthesis refers to the assembly of the small ribosome subunit (40S) to mRNA (mediated by eIF4 group of eIFs) and the recruitment of the initiator methionyl-tRNA (Met-$tRNA_i^{Met}$) to the first AUG start codon. The recognition of the first AUG codon at the beginning of the coding region of the mRNA occurs after a process termed "scanning," where the 5′-UTR of the mRNA is inspected for a suitable start codon.[34,35,52,53]

A wide variety of signals are transduced and integrated to regulate the initiation phase of protein synthesis. This phase is controlled by two main mechanisms: phosphorylation of eIFs, which modulates their activity and the association of regulatory proteins to specific sequences of mRNAs, usually in either the 5′- or 3′-UTRs, which either promotes or represses their translation. In the next sections, we mostly focus on the molecular mechanisms modulating the initiation phase of protein synthesis. We also review the regulatory mechanisms acting at the 5′ end of the mRNAs.[52–54] Additional proteins also have been described that regulate translation by acting at the 3′ end of mRNAs, such as cytoplasmic polyadenylation element-binding proteins and polyA-binding proteins (PABPs), which will not be discussed in this chapter.[55–59]

The recruitment of the Met-$tRNA_i^{Met}$ to the 40S ribosomal subunit starts with the binding of the translation factor eIF2 to the initiator Met-$tRNA_i^{Met}$ in a GTP-dependent manner. This complex associates to the small (40S) ribosomal subunit together with other initiation factors, such as eIF1, eIF1A, eIF3, and eIF5, to form a 43S preinitiation complex. The association of the 43S preinitiation complex to the mRNA is facilitated by the eIF4 group of eIFs and structures present on the mRNA. The structures important for translation initiation are the 7-methyl-GTP cap at the 5′ end and the poly(A) tail at the 3′ end of the mRNA. The 5′ and 3′ structures are bound by specific proteins, such as the cap-binding protein eIF4E and PABPs, respectively.[34,35,52,53,60,61]

The eIF4 group of eIFs involved in the initiation of translation forms the cap-binding complex, eIF4F, which consists of eIF4E, the DEAD-box RNA helicase eIF4A, and the scaffolding protein eIF4G. eIF4G binds eIF3 of the 43S complex and circularizes the mRNA by interacting simultaneously with eIF4E and PABPs (Fig. 5.4). The circularized mRNA enhances the association of the cap-binding complex eIF4F and, consequently, the 48S complex formation. The eIF4A helicase unwinds the secondary structure at the 5′ end of the mRNA, thus facilitating ribosomal binding and scanning. The helicase activity of eIF4A is enhanced by eIF4B when it is phosphorylated by S6K1, which in turn is controlled by mTORC1 (see below).[34,35,44,52–54,60,61]

The formation of the cap-binding complex eIF4F is thought to promote the binding of the 43S complex to the mRNA, which forms the 48S preinitiation complex. This complex scans the mRNA from the 5′ end and stops in proximity of the first AUG start codon. Recognition of the start codon triggers the hydrolysis of GTP by eIF2 and promotes the release of the eIFs from the 48S complex. This allows the complex eIF5B, another GTP-bound protein, and the large ribosomal subunit (60S) to join the complex and form the 80S ribosome. Then, eIF2-GDP is released from the 48S complex and the guanine exchange factor (GEF) eIF2B catalyzes the exchange of GDP for GTP on eIF2, thereby recycling eIF2 for another round of translation.[34,35,52–54]

2.3. Translation elongation

The elongation phase of protein synthesis refers to the translocation of the ribosome along the mRNA, the decoding of the mRNA codons into amino acids, and the formation of new peptide bonds that extend the newly synthesized polypeptide chain. This phase of protein synthesis requires two factors, eEF1 and eEF2, and represents the second checkpoint of the regulation of translation.[62,63]

eEF2 is a GTP-binding protein that promotes the translocation step of elongation. During this phase, the ribosome moves along the mRNA and the peptidyl transfer RNA (tRNA) translocates from the A site into the P site on the ribosome, following formation of the new peptide bond.

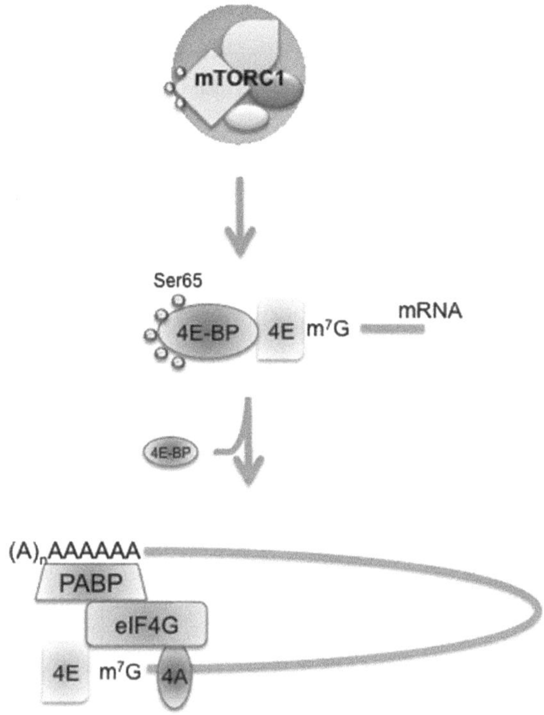

Figure 5.4 The phosphorylation of 4E-BP by mTORC1 leads to the formation of eIF4F, the cap-dependent translation initiation complex. The eIF4 group of eIFs form the cap-binding complex eIF4F, which consists of the cap-binding protein eIF4E, the DEAD-box RNA helicase eIF4A, and the scaffolding protein eIF4G. eIF4G circularizes the mRNA by interacting simultaneously with eIF4E and PABPs. (See the color plate.)

eEF2 undergoes phosphorylation at Thr56 in its GTP-binding domain which inhibits the binding of eEF2 to ribosomes, thereby reducing its activity. The phosphorylation of eEF2 is catalyzed by a highly specific Ca^{2+}/calmodulin-dependent kinase, eEF2 kinase.[63–65]

Stimuli which induce protein synthesis leads to the rapid dephosphorylation of eEF2, which results in increased activity and accelerated elongation.[66] The dephosphorylation of eEF2 appears to be caused by inhibition of eEF2 kinase. Both of these events are blocked by rapamycin,[66] suggesting that mTORC1/S6K1 is involved in the control of eEF2 function. The activity of eEF2 kinase is modulated via phosphorylation at several sites. The residue Ser366 is phosphorylated by S6K1 and RSK, leading to the inhibition of eEF2 kinase activity, at least at submaximal Ca^{2+} ion concentrations.[67] Phosphorylation at Ser78 is controlled by mTORC1 and results in the inhibition of the association calmodulin/eEF2 kinase, thus blocking its activity.[68] Also, phosphorylation at Ser359 is regulated by mTORC1 and inhibits eEF2 kinase activity.[69] Thus, mTORC1 activates translation elongation either directly or via S6K1. The mTORC1-dependent regulation of translation at the elongation phase is an important mechanism that, in conjuction with the regulation at the initiation phase, tightly controls the set of synthesized proteins under certain stimuli.[68,69]

3. TRANSLATIONAL CONTROL MOLECULES INVOLVED IN SYNAPTIC PLASTICITY AND MEMORY

Genetic engineering of mice is an approach that has been utilized to selectively and subtly manipulate discreet molecular effectors that regulate protein synthesis. The study of a number of mutant mice that either lack or overexpress translational control molecules has resulted in the accumulation of a large body of evidence supporting a role of protein synthesis in memory formation. The role several of the aforementioned translational control molecules in memory formation have been validated using a variety of commonly used behavioral tests, which are described in Box 5.1.

Box 5.1 Behavioral assessment of learning and memory

Morris Water Maze

The MWM is a commonly used task that measures hippocampus-dependent memory by training rodents to acquire the location of a hidden platform within a circular pool of water. Memory is measured in several ways, the first being the determination of the latency to locate the hidden platform. After a given amount of training, rodents will acquire the platform position and the latency to find the platform will decrease. Memory also is measured using a probe test. In this phase of the task, the platform is removed and memory is measured both by the number of times the rodent crosses the platform position and by the amount of times the rodent spends in the quadrant in which the platform was formerly located, also known as the target quadrant. One of the most critical aspects of this test is the presence of visual cues surrounding the water maze such that the rodents can use these cues to guide them in learning the platform position.

Fear Conditioning

Auditory fear conditioning (FC) is behavioral test that measures a type of associative memory. In this task a conditioned stimulus (CS), typically a tone, is paired with an unconditioned stimulus (US), typically a footshock. Two types of memory can be tested following this CS–US pairing: contextual fear memory and cued fear memory. To measure contextual fear memory, rodents are placed in the context in which they received the CS–US pairing and memory is measured by the time spent freezing in this context. Cued fear memory, on the other hand, is measured by the amount of time spent freezing during the CS presentation in a novel context. Importantly, contextual fear memory is hippocampus-dependent, whereas cued fear memory is not.

Condition Taste Aversion

In the CTA task, rodents are introduced to a novel taste and immediately injected with a malaise-inducing agent such as LiCl. Memory of this association is measured using a taste aversion index, where a higher value represents a greater memory of the association. Latent inhibition of CTA is a similar task, but in this task rodents are pre-exposed to the novel taste, which reduces the aversion index in the memory phase of the task.

Novel Object Recognition

The novel object recognition task is a measure of the innate tendency of rodents to explore novel objects. Although the details vary from laboratory to laboratory, the basic framework of this task is the same. In the first phase of the task, the rodent is habituated to the testing arena, which does not contain any objects. This is designed to reduce any possible anxiety that may be associated with being placed in an unfamiliar environment that could confound behavioral outcomes. Following a set intertrial interval, the rodent is once again placed in the testing arena where the presence of two identical objects has been added, which they are allowed to explore. The last phase of the task occurs after another set intertrial interval and the rodents are placed back into the same environment; however, one of the familiar objects is replaced with a novel object. Because rodents display a tendency to explore novelty, they will spend a greater amount of time interacting with the novel object compared to the familiar object. Typically, this value is represented as a ratio of time spent interacting with the novel object, over the total time spend interacting with either object. A preference index of 0.5 would represent an equal amount of time spend with both objects; thus, a preference index >0.5 reflects a preference for the novel object.

3.1. mTOR

3.1.1 mTOR structure and function

mTOR is a large (approximately 250 KDa) multidomain serine/threonine kinase and member of the family of phosphoinositide (PI) 3-kinase-related kinases. mTOR is highly conserved from yeast to humans.[34,70] In mTORC1, it is a target of rapamycin, which is an immunosuppressant drug (see below). Structurally, mTOR is characterized by more than 20 HEAT (*H*untingtin, *E*longation Factor 3, *A* subunit of PP2A, *T*OR1) repeats located in at its N-terminal. The HEAT repeats fold into a helical structure that functionally interacts with regulatory proteins such as Raptor and Rictor.[34,41,71,72]

The C-terminal portion contains FRAP and FATC domains required for the catalytic functions of mTOR. Adjacent to these domains is the FRB domain that serves as binding site for the immunophilin FKBP12, which appears to occur only in the presence of rapamycin. The binding of the FKBP12–rapamycin complex to FRB disrupts the interactions of mTOR with Raptor, thereby inhibiting mTORC1. Other proteins interacting with mTOR, such as the Ras homolog enriched in the brain (Rheb), also binds the FRB domain.[34,41,42,71–73]

The kinase catalytic domain (KIN) of mTOR is comprised of a region termed negative regulatory domain (NRD). NRD contains three amino acid residues (Thr2246, Ser2448, and Ser2481) that upon phosphorylation regulate mTOR kinase activity. Thr2246 is phosphorylated by AMPK and S6K1, Ser2448 is the target of Akt and S6K1, and finally Ser2481 is an autocatalytic phosphorylation site. Phosphorylation of these residues, in particular, Ser2448, has been shown to be correlated with high mTOR kinase activity. Overall, the phosphorylation state of mTOR at these multiple sites represents an important mechanism of feedback regulation as some of them are substrates of downstream effectors of mTOR (i.e., S6K1).[44,68,72,74–76]

Deletion of the *mTOR* gene results in a lethal phenotype, suggesting that *mTOR* is an essential gene. Indeed, mice lacking *mTOR* die *in utero* shortly after implantation and embryonic development appears to be arrested at E5.5 with multiple developmental aberrations.[77,78] Consequently, the only method of studying the function of mTORC1 has been via its pharmacological inhibition. Rapamycin has been used widely to study mTORC1 signaling in long-lasting synaptic plasticity and memory formation across multiple experimental preparations and species.

3.1.2 The role of mTOR in synaptic plasticity and memory consolidation

Rapamycin (or sirolimus) is a macrolide produced by the bacterium *Streptomyces hygroscopicus*. It was named after the island of Rapa Nui where it was discovered first in a soil sample.[79] It has immunosuppressive and antiproliferative properties and is an FDA-approved immunosuppressant drug used clinically to prevent rejection after organ transplant.

Rapamycin is a potent inhibitor of mTORC1, which is defined as the complex of mTOR and the adaptor protein Raptor (see above). Rapamycin acts by binding FKBP12, the intracellular receptor of rapamycin, and disrupts the interaction between Raptor and mTOR, thereby blocking mTORC1 signaling.[34,42,73,80]

The first demonstration of the involvement of mTORC1 in synaptic plasticity comes from experiments performed in rodent hippocampal slices where pretreatment with rapamycin abolished the late phase of L-LTP.[36,81] Moreover, L-LTP was sensitive to rapamycin only when the drug was bath applied during induction and not after the establishment of this form of synaptic plasticity.[82] These experiments suggest that activation of mTORC1 is required for the expression of L-LTP, and that this molecular event is tightly regulated within a specific temporal window.

mTOR, its effector protein 4E-BP, and eIF4E are colocalized with postsynaptic markers, suggesting that rapamycin blocks L-LTP by inhibiting the postsynaptic translational machinery.[36,81] Subsequently, it was shown that L-LTP leads to the activation of mTORC1 signaling resulting in the translation of mRNAs, such as the one encoding αCaMKII[83,84] and eEF1A.[85] The eEF1A mRNA contains a 5′-UTR oligopyrimidine tract (TOP) sequence, which is one of the structures that operate as a regulatory element for protein synthesis. TOP mRNAs generally encode for ribosomal proteins and translation factors, and their translation is specifically regulated by mTORC1. Thus, mTORC1 controls the synthesis of proteins important for synaptic plasticity (i.e., αCaMKII) and amplifies translation by increasing the availability of factors involved in the protein synthesis machinery.

Consistent with the involvement of mTORC1 in synaptic plasticity, there is a great deal of experimental evidence that demonstrates the requirement of mTORC1-dependent protein synthesis in memory formation *in vivo*. For example, hippocampus-dependent long-term spatial memory has been shown to be mTORC1-dependent. To study the role of mTORC1 in the Morris water maze (MWM), rats were given post-training infusions of rapamycin in the dorsal hippocampus. Inhibition of mTORC1 with rapamycin blocked long-term memory in a retention test 48 h after training.[86] Finally, consolidation of fear memory, an amygdala-dependent and fear-motivated learning task, also has been shown to be an mTORC1-dependent process. Rats trained in a cued FC paradigm exhibited increased phosphorylation of S6K1 and post-training infusions of rapamycin into the amygdala prevented the formation of cued fear memory.[87]

Overall, pharmacological studies using rapamycin to inhibit mTORC1 signaling have shown that blocking this molecular signaling impairs the formation of long-lasting memories. However, limitations of this kind of approach reside in the nature of pharmacological tools, which are not specific enough to parse out the fine molecular mechanisms of translational control that contribute to memory formation. Fortunately, there have been numerous studies in which investigators have utilized genetic models targeting proteins within the mTORC1 signaling pathway; these studies have provided critical information regarding the specific roles of these translational control molecules in synaptic plasticity and memory.

3.2. FKBP12

3.2.1 FKBP12 structure and function

FK506-binding proteins (FKBPs) are enzymes with peptidyl-prolyl *cis-trans* isomerase (PPiase) activity. To date, 15 mammalian FKBPs have been found, most of which are expressed in the brain. FKBPs have multiple functions in the nervous system, the best known being a receptor for immunosuppressant drugs. These agents act as "prodrug" and are thus active only when complexed to immunophilins, as is the case with both FK506 and rapamycin, which bind FKBP12.[88]

FKBP12 is the smallest member of the FKBP family, with a molecular weight of 12 kDa, and has a minimal peptide sequence that harbors the catalytic domain and a drug-binding pocket. In the absence of a macrolide immunosuppressant, it appears that FKBP12 does not bind to mTOR but instead is associated with either receptors or Ca^{2+} channels, such as the inositol-(1,4,5)-triphosphate receptor and ryanodine receptors, respectively. When the FKBP12–rapamycin complex is formed, it associates with the FRB domain of mTOR, thereby inhibiting its function (see above).[88,89]

The *FKBP12* gene was deleted in mice to study the function of the protein; however, the resulting deletion led to an embryonic lethal phenotype due to alterations in cardiac development.[90] Consequently, conditional knockout (cKO) mice were generated by crossing a mouse line with floxed alleles of *FKBP12* gene with another line expressing forebrain-specific *Cre recombinase*. In these mice, the expression of Cre recombinase mediates a forebrain-specific deletion of the *FKBP12* gene, which occurs approximately 21 days after the birth.[80] The forebrain-specific FKBP12 cKO mice became an invaluable tool to study the influence of FKBP12 on mTORC1, hippocampal synaptic plasticity, and memory consolidation.[80]

3.2.2 The role of FKPB12 in synaptic plasticity and memory consolidation

Surprisingly, the genetic deletion of FKBP12 resulted in increased mTORC1 formation under basal conditions in the absence of rapamycin, suggesting that one of the normal functions of FKBP12 is to suppress mTORC1. In fact, FKBP12 cKO mice exhibited increased phosphorylation at Ser2448 of mTOR, a phosphorylation site that is correlated with activation of mTOR. Furthermore, phosphorylation of S6K1 was enhanced specifically at Thr389, suggesting that activity of the mTORC1 was elevated in the absence of FKBP12. Accordingly, the mTOR–Raptor association also was increased as indicated by enhanced interactions of these proteins in immunoprecipitation experiments in brain tissue from FKBP12 cKO mice.[80]

Studies examining synaptic plasticity at Schaffer collateral-CA1 synapses of the hippocampus demonstrated that FKBP12 cKO mice displayed enhanced L-LTP. However, basal synaptic transmission, paired-pulse facilitation, and E-LTP were normal in the absence of FKBP12. These results indicate that FKBP12 usually suppresses basal mTORC1 activity and genetic removal of this molecular break enhances L-LTP. Moreover, mTORC1 levels were increased following L-LTP induction in slices from wild-type (WT) mice, but not in slices from FKBP12 cKO mice. The elevated levels of mTORC1 eventually returned to baseline in WT slices; however, mTORC1 levels remained elevated in slices from FKBP12 cKO mice, indicating that there may be an impairment in the downregulation of the mTOR/Raptor interactions in these mice. As expected, the enhanced L-LTP observed in the FKBP12 cKO mice was resistant to rapamycin treatment; however, it was still sensitive to anisomycin. It is interesting to note the presence of an enhancement in the initial phase of E-LTP induction, a stage of LTP that occurs independently of protein synthesis. This observation suggests that FKBP12 also mediates processes that are protein synthesis-independent.[80]

In agreement with the enhanced mTORC1 signaling and the increased L-LTP, FKBP12 cKO mice exhibited enhanced formation of contextual fear memory compared to their WT littermates. Additionally, the formation of contextual FC in WT mice was accompanied by increased mTOR/Raptor interactions. Taken together, these findings suggested that long-term memory is more robust in FKBP12 KO mice. However, in the novel object recognition (NOR) test, FKBP12 cKO mice exhibited a preference for the familiar, but not novel object. Moreover, in the MWM, spatial learning and memory were normal, but when the platform was moved to a different location, the FKBP12 cKO mice exhibited impaired reversal learning. This type of perseverative behavior was confirmed in a water-based Y-maze.[80,91] These findings suggest that elevated mTORC1 signaling enhances certain types of associative memory but impairs reversal learning, a more demanding cognitive task.

3.3. 4E-BPs

3.3.1 4E-BP structure and function

4E-BPs modulate the formation of the eIF4G initiation complex. 4E-BP normally binds to and sequesters eIF4E, thereby preventing its association with eIF4G and blocking eIF4F formation. When not bound to 4E-BP, eIF4E is free to associate with eIF4G, a molecular event that promotes the formation of eIF4F and cap-dependent translation initiation.[61,92,93]

The binding of 4E-BP to eIF4E is regulated by phosphorylation.[94,95] 4E-BP has multiple phosphorylation sites that are regulated by the activity of several kinases, such as ERK, PI3K, and mTORC1. Phosphorylation of these sites occurs in an ordered, hierarchical fashion: residues Thr37 and Thr46 are phosphorylated first, which primes the phosphorylation of Thr70 and Ser65 is phosphorylated last.[94,95] Phosphorylation at these sites is sensitive to inhibitors of ERK, PI3K, and mTORC1, but the identity of the kinase that phosphorylates each individual site has not been established. Nonetheless, it was shown that phosphorylation of Ser65 occurs last and is rapamycin-dependent, suggesting that mTORC1 activity is necessary to inhibit the binding of 4E-BP to eIF4E, thereby promoting cap-dependent protein synthesis.[34] Numerous studies have shown that proper regulation of ERK, PI3K, and mTORC1 is required for protein synthesis-dependent synaptic plasticity, memory formation, and storage.[61,92,94,95] Consistent with this idea, the expression of 4E-BP has been shown to be necessary for normal synaptic plasticity and memory.

3.3.2 The role of 4E-BP in synaptic plasticity and memory consolidation

There are multiple isoforms of 4E-BPs, but 4E-BP2 is the isoform with the most robust expression in the brain. Therefore, 4E-BP2 KO mice were generated to assess the role of 4E-BP in synaptic plasticity and memory.[96–98]

Various forms of synaptic plasticity were studied in Schaffer collateral-CA1 synapses in 4E-BP2 KO mice. A stimulation protocol that usually induces E-LTP in WT hippocampal slices (1 train of HFS) produced an enhanced and sustained L-LTP in 4E-BP2 KO mice. On the other hand, a protocol that typically elicits L-LTP in WT slices (4 trains of HFS) was unable to elicit L-LTP in 4E-BP2 KO slices. Indeed, the L-LTP that was induced in the 4E-BP2 KO slices decayed to baseline after 2 h, compared to the sustained L-LTP in WT slices that persisted for over 4 h. This suggests that there is a reduced threshold to induce L-LTP in 4E-BP2 KO mice. Consistent with this, 4E-BP2 KO mice displayed an increase in eIF4E–eIF4G interactions, which is an index of increased cap-dependent protein synthesis and presumably underlies this enhanced plasticity. Indeed, the sustained plasticity observed in 4E-BP2 KO slices was blocked by anisomycin. Importantly, abalation of 4E-BP2 did not alter basal synaptic transmission. Paired pulse facilitation, as well as PTP, a form of short-term plasticity, also was unaffected in slices from the 4E-BP2 KO mice.[96,98]

Hippocampus-dependent memory was assessed in the 4E-BP2 KO mice using the MWM and FC paradigms. 4E-BP2 KO mice exhibited longer escape latencies during MWM training and did not exhibit a preference for the target quadrant in the probe trials compared to WT littermates. 4E-BP2 KO mice also exhibited impaired contextual FC.[96–98] Overall, these results suggest that the relationship between protein synthesis, synaptic plasticity, and memory is very tightly regulated. Specifically, these findings suggest that there is an optimal level of protein synthesis that underlies the coordination between synaptic plasticity and memory, and that exaggerated levels of protein synthesis are detrimental to cognitive function.

3.4. eIF4E

3.4.1 eIF4E structure and function

eIF4E, the cap-binding protein promotes protein synthesis by binding to eIF4G. Beside its role in translation regulation, eIF4E is also important for mRNA stability and the proper subcellular localization of mRNA. eIF4E is regulated by phosphorylation via the mitogen-activate protein kinase-interacting serine/threonine kinase 1 and 2 (Mnk1/2). Mnk1/2 is activated by the ERK and p38 mitogen-activated protein kinase pathways, and subsequently phosphorylates eIF4E on serine 209. In fact, mice that lack Mnk1/2 did not exhibit phosphorylation of eIF4E at Ser209.[99–105]

It has been proposed that Mnk1/2 binds to the scaffolding protein, eIF4G. This molecular event, concomitant with the binding of eIF4E to eIF4G, is required for the subsequent phosphorylation of eIF4E.[63,105] Moreover, it has been shown that phosphorylation of eIF4E reduces its cap-binding affinity.[99,101,106] These data contribute to the formulation of a model in which dephosphorylated eIF4E binds to the mRNA cap structure, promotes eIF4F formation, followed by the recruitment of ribosomes to the mRNA.

Subsequently, the phosphorylation of eIF4E reduces its affinity for the cap structure on the mRNA and allows the ribosome to scan for the AUG start codon.[63,107,108] However, evidence regarding the functional effect of eIF4E phosphorylation is contradictory. It has been reported that mice lacking Mnk1/2 are viable and apparently normal, although their synaptic plasticity and memory have not yet been analyzed.[109] On the other hand, it has been shown that pharmacological treatments, such as BDNF and NMDA-R agonists, that induce protein synthesis-dependent synaptic plasticity increase the phosphorylation of eIF4E in an ERK-dependent manner.[110,111]

3.4.2 The role of eIF4E in synaptic plasticity and memory consolidation

The involvement of eIF4E in synaptic plasticity and memory formation has been addressed with 4EGI-1, a compound that specifically inhibits the interaction of eIF4E with eIF4G.[112] Blockade of eIF4E/eIF4G interactions with bath application of 4EGI-1 in hippocampal slices interferes with L-LTP without affecting basal synaptic transmission, PPF and E-LTP. Moreover, the inhibition of L-LTP by 4EGI-1 was accompanied by a reduction in HFS-induced protein synthesis in the slices.[113] These experiments suggest that the activity of eIF4E is necessary for the maintenance of protein synthesis-dependent L-LTP. Consistent with the LTP studies, infusions of 4EGI-1 in the amygdala of rats trained in a cued FC paradigm blocked the consolidation of long-term memory.[114] However, the same infusions administered after memory retrieval did not affect the reconsolidation of cued fear memory. Moreover, it also was shown that the association of eIF4E with eIF4G was increased during consolidation but not reconsolidation of cued fear memory.[114] These data suggest a different requirement for the formation of the initiation complex eIF4F, and consequently cap-dependent translation, in the consolidation and reconsolidation of cued fear memory.

3.5. S6K1

3.5.1 S6K1 structure and function

The *S6Ks* are a family of genes that, broadly speaking, are involved in many facets of cellular homeostasis. The dysregulation of S6Ks has been implicated in a variety of pathological states including, but not limited to obesity, cancer, aging, and aberrant metabolism.[76,115–117] The *S6K* genes were discovered during the investigation of the family of kinases that phosphorylate the 40S ribosomal protein S6. One of these genes was originally described as the 70 kDa ribosomal S6K and is now known as p70 S6K1, or simply S6K1.[118] The linking of the S6Ks to the mTORC1 signaling pathway was not discovered until later.

In 1995, a novel phosphorylation site was discovered on S6K1 that was sensitive to rapamycin treatment.[119] Shortly thereafter, *in vitro* kinase assays identified S6K1 to be one of the substrates of mTORC1, along with 4E-BP1.[120] This phosphorylation of S6K1 by mTORC1 then was found to be linked to the phosphorylation of S6.

S6K1 and S6K2 are part of the AGC kinase family, so named after the protein kinase A, G, and C families, and kinase domains with a two-lobe fold structure. Activation of these kinases occurs near the C-terminus C-lobe loop by T-loop kinases. Both S6K1 and S6K2 have two isoforms, known as S6K1S and S6K1L, and S6K2S and S6K2L. S6K1 is the isoform that has been studied most intensely and whose function is best understood, especially in the nervous system. Similar to many proteins, S6K1 contains multiple phosphorylation sites. However, the most well-studied activators of S6K1 are in the insulin/IGF pathway. As mentioned earlier, this pathway activates mTORC1 via PI3K, which is now known to result in the phosphorylation of S6K1 on Thr389.[76,115,116] S6K1 can also be activated in a PI3K-independent manner through Ras/ERK signaling, which results in phosphorylation of Ser421/Ser424.[121]

The consequences of S6K1 phosphorylation are many; however, we will focus on the effects that relate to the regulation of protein synthesis. As mentioned, S6K1 phosphorylates S6, eIF4B, as well as eEF2 kinase, suggesting that S6K1 is involved with both the initiation and elongation phases of protein synthesis.[95] It was originally thought that the phosphorylation of S6 by S6K1 was responsible for the translation of TOP mRNAs. However, although multiple studies have questioned the validity of this hypothesis,[122,123] the involvement of S6 phosphorylation in the translation of TOP mRNAs remains a possibility.[124] The phosphorylation of eIF4B functions to enhance the helicase activity of eIF4A, a component of eIF4F, which helps unwind complex 5′-UTR structures. S6K1 also phosphorylates an eIF4A inhibitor, PDCD4, which enhances the helicase activity of eIF4A.[95] Finally, S6K1 phosphorylates and inhibits eEF2 kinase, which is responsible for the phosphorylation of eEF2. In this manner, S6K1 indirectly regulates eEF2 phosphorylation and ribosomal translocation in peptide chain elongation.[67] Thus, S6K1 also plays a role in the elongation phase of protein synthesis.

3.5.2 The role of S6K1 in synaptic plasticity and memory consolidation

Genetic deletion of both S6K1 and S6K2 is embryonic lethal; however single mutants are born at traditional Mendelian ratios.[122] S6K1 and S6K2 KO mice were utilized to study synaptic plasticity and memory.[125] S6K1 and S6K2 KO did not exhibit any major impairments in either basal synaptic transmission or synaptic plasticity. The S6K1 and S6K2 KO mice exhibited a mild impairment in the acquisition phase of the FC.[125] For contextual fear memory, S6K1 KO mice exhibited deficient short- and long-term memory; however, S6K2 KO mice did not exhibit impairments in memory until 7 days after training.[125] These deficits also were observed in the S6K1 and S6K2 heterozygous KO mice[125], suggesting that full expression levels of these proteins are required for proper contextual fear memory.

S6K1 and S6K2 KO mice were tested for conditioned taste aversion (CTA), another associative learning task, and latent inhibition of CTA.[125] In the CTA task, the S6K1 KO mice displayed a lower aversion index than their WT littermates, and the extinction of this memory indicated that the S6K1 KO mice rapidly extinguished the association between the saccharin and LiCl injection.[125] Together, these results suggest that CTA memory is weak in S6K1 KO mice. In contrast, S6K2 KO mice did not exhibit impaired CTA; however, the latent inhibition of CTA was significantly higher than their WT littermates, suggesting that S6K2 is required for novel taste experiences.

The MWM was used to assess the spatial learning and memory abilities of the S6K1 and S6K2 KO mice. During a probe trial on day 3, the S6K1 KO mice exhibited fewer platform crossings and spent less time in the target quadrant; this effect was not observed in the day 7 probe trial. The S6K2 KO mice did not exhibit impairments in the MWM.[125]

In summary, S6K mutant mice did not display learning and memory impairments, and to date, a major role for these kinases in LTP and memory has not been demonstrated.

3.6. eIF2 and eIF2α kinases

3.6.1 eIF2 structure and function

eIF2 is composed of three subunits, α, β, and γ, and phosphorylation of the α subunit at serine 51 (Ser51) converts eIF2 into an inhibitor of eIF2B, the GEF that catalyzes the conversion of inactive eIF2-GDP into active eIF2-GTP. Thus, phosphorylation of Ser51 on eIF2α impairs eIF2 recycling by inhibiting eIF2B such that it remains blocked in an inactive eIF2-GDP state. Even if the relative abundance of eIF2/eIF2B has not been reported in the nervous system, in other tissues it has been shown that eIF2 is more highly expressed than eIF2B, and phosphorylation of a small fraction of eIF2α is sufficient to inhibit eIF2B and blocks general protein synthesis.[35,52]

Phosphorylation of eIF2α at Ser51 is a highly dynamic and regulated event that is mediated by four different kinases and two phosphatases. The kinases that phosphorylate eIF2α are protein kinase-RNA regulated (PKR), heme-regulated initiation (HRI) factor 2α kinase, eIF2α kinase 3 (PERK), and general control nonderepressible 2 (GCN2). All of these kinases are present in the brain and are activated by different cellular stressors: PKR is activated by double-stranded RNA (dsRNA), HRI is induced by low heme levels, PERK is activated by endoplasmic reticulum (ER) stress as well as unfolded proteins in the ER, and GCN2 is induced by amino acid starvation.[52] Although all of the eIF2α kinases are present in the brain, GCN2 appears to be the most abundant, consistent with the indication that GCN2 mRNA is strongly expressed in the mouse brain.[126,127]

The phosphatase complexes that dephosphorylate eIF2α are the complex of the eIF2α-specific regulatory subunit constitutive repressor of eIF2α phosphorylation (CReP) and protein phosphatase I catalytic subunit (PPIc), and the complex of growth arrest and DNA damage-inducible gene 34 (GADD34) and PPIc. The regulatory subunits CReP and GADD34 provide specificity for the complex toward eIF2α.[128]

eIF2α kinases historically have been described as stress-responsive regulators of general protein synthesis. Thus, phosphorylation of eIF2α on Ser51 suppresses general protein synthesis but at the same time stimulates translation of a subset of mRNAs that contain an open reading frame in the 5′-UTR.[129,130] This type of gene-specific translational control has been reported for the transcriptional activator GCN4 in yeast[131,132] and for activating transcription factor 4 (ATF4, also known as CREB2) in mammalian cells.[133,134] The increase in ATF4 that results from increased eIF2α phosphorylation is negatively correlated with CREB-dependent transcription, synaptic plasticity, and memory (see below).

Several studies have been performed to indicate an important role for eIF2α phosphorylation in memory formation. We describe the studies that have performed with GCN2 KO mice, PERK cKO mice, PKR KO mice, and eIF2α^{+}/S51A mutant mice.

3.6.2 GCN2

The first studies dissecting the role of eIF2α phosphorylation in synaptic plasticity and memory were performed in mice with a constitutive deletion of GCN2, the eIF2α kinase that is most robustly expressed in the brain.[135]

Various forms of hippocampal synaptic plasticity were studied in GCN2 KO mice. A stimulation protocol that usually induces E-LTP in

WT hippocampal slices (1 train of HFS) produced an enhanced and sustained L-LTP in GCN2 KO mice. Induction of E-LTP in GCN2 KO mice resulted in long-lasting LTP, which was dependent on transcription, translation, as well as cAMP-dependent protein kinase activity. Moreover, anisomycin and actinomycin D, inhibitors of translation and transcription, respectively, also decreased the potentiation during the early phase of L-LTP. Overall, these results suggest that there is a reduced threshold for LTP induction in the GCN2 KO mice that is sensitive to inhibitors of protein synthesis and transcription. In addition, L-LTP was studied in the GCN2 KO mice by using electrical (4 trains of HFS) and chemical stimulation (bath application of forskolin) protocols. In these conditions, L-LTP decayed to baseline after 3 h in GCN2 KO slices, whereas in WT slices, L-LTP persisted over 3 h. Despite the aberrant synaptic plasticity observed in various forms of LTP in GCN2 KO mice, the mice did not exhibit any differences in basal synaptic transmission compared to their WT littermates.[135,136]

Biochemical analysis in WT slices following induction of L-LTP showed a decrease in GCN2 and eIF2α phosphorylation levels. However, these biochemical effects were not observed following the induction of E-LTP. Thus, the decrease in GCN2 activity and eIF2α phosphorylation is elicited following strong stimulation, but not weak stimulation, consistent with a role for protein synthesis in L-LTP, but not E-LTP. In agreement with these results, basal levels of phosphorylated eIF2α and ATF4 were reduced in GCN2 KO mice. Interestingly, anisomycin and actinomycin D both block the long-lasting LTP induced in the GCN2 KO mice with a weak stimulation protocol, suggesting that there may be an enhancement in the translation of pre-existing transcripts, as well as transcription of mRNA, in the GCN2 KO mice.[135,136]

In behavioral studies of the GCN2 KO mice, a deficit in contextual FC, but not cued FC, was observed. In the MWM, WT mice were able to learn and remember the platform position using the standard training protocol (3 training sessions/day), but the GCN2 KO mice were slower to learn and remember the position of the platform. Based on the observation that weak electrical stimulation led to L-LTP, the GCN2 KO mice were trained in the MWM using a weak training protocol (1 session per day), which revealed shorter escape latencies in the GCN2 KO mice, as well as increased time spent in the target quadrant on day 5 of training. Thus, the spatial memory of the GCN2 KO mice was enhanced following weak training.[135] Overall, these behavioral studies indicate that the role of GCN2, both in synaptic plasticity and memory, is tightly regulated, and most likely linked to the regulation of optimal levels of protein synthesis during these processes.

3.6.3 eIF2α S51A

The studies with the GCN2 KO mice suggested that changes in eIF2α phosphorylation play a significant role in synaptic plasticity and memory consolidation. To more directly study the role of eIF2α phosphorylation in synaptic plasticity and memory, a mutant mouse was generated with a point mutation on the eIF2α phosphorylation site. The mice were heterozygous mutants where Ser51 was mutated to an alanine (S51A), and exhibited an approximate 50% reduction in eIF2α phosphorylation.[137]

Similar to the previous studies performed in GCN2 KO mice, these eIF2α S51A mutant mice are expected to exhibit L-LTP using a weak stimulation protocol. Consistent with this expectation, 1 train of HFS resulted in a sustained L-LTP in eIF2α mutant mice. This long-lasting LTP was blocked by anisomycin, as well as by actinomycin D, demonstrating the requirement of both translation and transcription. The eIF2α S51A mutant mice did not exhibit any differences in basal synaptic transmission. Interestingly, the eIF2α mutants exhibited stable L-LTP using both electrical and chemical stimulation protocols, which was not observed in GCN2 KO mice (see above).[137]

Because of the previous studies that indicated that GCN2 KO mice were able to acquire knowledge of the platform position in the MWM using a weak training protocol, the eIF2α S51A mutant mice were tested on the MWM using a similar weak training protocol. eIF2α S51A mutant mice were trained once a day for 6 days on the MWM and by day 6, the mutant mice exhibited a shorter escape latency and spent a greater amount of time in the target quadrant compared to WT littermates. However, using the standard MWM training protocol, eIF2α S51A mutant mice also demonstrated shorter escape latency during training compared to their WT littermates. In the probe test, eIF2α mutants also spent more time in the target quadrant compared to their WT littermates. Thus, although GCN2 KO mice showed impairments in the MWM using standard training protocols, these effects were not observed in the eIF2α S51A mutant mice.[137] These results indicate that decreased eIF2α phosphorylation leads to enhanced hippocampus-dependent learning and memory, regardless of the strength of the training protocol.

Consistent with results of the MWM studies, the eIF2α S51A mutant mice also were tested on contextual and auditory FC and showed enhanced memory despite the use of differing footshock intensities. In addition, in the CTA task, eIF2α S51A mutant mice did not exhibit a difference in the aversion index to sucrose compared to WT mice; however, latent inhibition of CTA in the mutant mice was decreased compared to their WT littermates.[137]

These differing outcomes may be a result of ceiling effects in the CTA task that are unmasked in the latent inhibition of CTA.

The results of the behavioral studies of the eIF2α S51A mutant mice suggested that inhibiting the dephosphorylation of eIF2α would inhibit L-LTP and memory formation. Thus, Sal003, an inhibitor of eIF2α phosphatases, was used to assess the influence of inhibiting the dephosphorylation of eIF2α on synaptic plasticity and memory. As expected, inhibiting the dephosphorylation of eIF2α in WT hippocampal slices led to an impairment in L-LTP. Sal003 did not have this effect in slices from ATF4KO mice, suggesting that the Sal003-mediated impairment in L-LTP is ATF4-dependent. The effect of Sal003 in blocking L-LTP was specific to LTP induction, as the application of this compound 45 min after LTP induction had no effect on potentiation.[137]

Contextual FC experiments in WT mice demonstrated that eIF2α phosphorylation was reduced 30 min after training. Thus, blocking this dephosphorylation would be expected to interfere with contextual fear memory. Indeed, Sal003 infusion into the dorsal hippocampus blocked the formation of contextual fear memory. Moreover, using a standard training protocol in the MWM, Sal003 infusion also blocked the formation of long-term memory as evidenced by an increase in escape latency and reduced target quadrant occupancy.[137]

The results of the experiments performed in the eIF2α S51A mutant mice confirmed observations made in the GCN2 KO mice. Specifically, that the phosphorylation and dephosphorylation of eIF2α are strictly regulated during various forms of LTP and various types of memory tasks.

3.6.4 PERK

PERK is an eIF2α kinase that is activated during ER stress to reduce protein synthesis. Similar to GCN2, the phosphorylation of eIF2α by PERK leads to the translation of ATF4. Forebrain-specific PERK cKO mice were generated and tested on several different behavioral paradigms to measure learning and memory.[138] The PERK cKO mice exhibited an enhanced preference for the novel object in the NOR task. PERK cKO mice also exhibited an impairment in the extinction of cued FC. In the MWM, PERK cKO mice did not exhibit any differences in learning the position of the platform, and performed similar to WT during probe tests. However, the PERK cKO were impaired in learning the location of the platform when the position of the platform was switched to the opposite quadrant.[138] These results suggested that the PERK cKO mice exhibit perseverative behavior. The perseverative behavior of the PERK cKO mice was confirmed by examining reversal learning a water-based Y-maze task, an additional task used to measure cognitive inflexibility. Again, the PERK cKO mice were unable to acquire the location of the new platform position.[138] WT mice were sacrificed to measure levels of eIF2α phosphorylation during reversal learning and a decrease in eIF2α phosphorylation was observed,[138] suggesting that the dephosphorylation of eIF2α is required for reversal learning. GCN2 KO mice also were tested for reversal learning in the Y-maze, but they did not exhibit an impairment in this task. The results of these studies suggest that although eIF2α phosphorylation is decreased during reversal learning, the pool of eIF2α that is dephosphorylated is normally phosphorylated by PERK and not other eIF2α kinases.

3.6.5 PKR

PKR is an eIF2α kinase that is activated by dsRNA species during viral infection.[139] PKR is another regulator of protein synthesis via its ability to phosphorylate eIF2α; however, another function of PKR is to regulate protein levels via degradative mechanisms, which is a process that is critically involved in synaptic plasticity and memory.[140] LTP and memory function were studied in a PKR cKO in which the levels of PKR were increased in the CA1 region of the hippocampus via chemical induction.[141] Hippocampal synaptic plasticity was also studied in the PKR KO mice, and E-LTP was found to be similar to their WT littermates. However, L-LTP induced with TBS was impaired in the PKR KO mice, suggesting that PKR is required for this form of synaptic plasticity.[141] Moreover, PKR KO mice exhibited impairments in contextual FC, but not auditory FC, suggesting that the fear memory deficit was hippocampus-dependent.[141] Importantly, the PKR KO mice do not exhibit any reduction in general translation,[141] indicating that the L-LTP and contextual fear memory impairments are due to reduced eIF2α phosphorylation causing gene-specific changes in translation in these mice.

Collectively, the vast array of studies performed in eIF2α kinase (GCN2, PERK, and PKR) mutant mice and in the eIF2α S51A mutant mice clearly demonstrate that the regulation of eIF2α phosphorylation is involved in long-lasting synaptic plasticity and several types of long-term memory. Specifically, there is an optimal level of eIF2α phosphorylation that must be achieved for these different forms of synaptic plasticity and memory, and that the mechanisms involved in the control of eIF2α phosphorylation are not conserved across various experimental paradigms.

4. CONCLUDING REMARKS

Today we have a substantial understanding of the molecular mechanisms, including the regulation of translation, underlying long-term memory formation, and the importance of the synaptic changes that are required for it. However, many of the molecular details of the regulation of protein synthesis are still missing and we are just now beginning to understand how translation is locally regulated.

Although, synaptic plasticity is a key step in memory storage, it is becoming more apparent that a simple enhancement in the efficacy of a synapse is not sufficient to form and store a complex memory. It is believed, instead, that changes in synaptic strength must occur within a subset of neurons to produce a specific alteration in information flow through a neural circuit. How translation impacts synaptic changes that ultimately affect the neuronal circuitry is currently under investigation.

Finally, there is compelling evidence that defects in memory storage result from pathological changes in the fundamental mechanisms that control molecular signaling, including translational control, and synaptic plasticity. A future challenge that remains is to understand how the molecular mechanisms that control translation during long-lasting synaptic plasticity and memory formation are impaired with age or disease, such as in Alzheimer's disease, neuropsychiatric disorders, and neurodevelopmental disorders. Thus, it is of critical importance to understand in greater detail, the molecular mechanisms that are required for memory formation as well as the abnormal changes that take place in neuropathological conditions in order to design specific therapeutic compounds that could be used to restore cognitive function.

REFERENCES

1. McGaugh JL. Memory—a century of consolidation. *Science*. 2000;287(5451):248–251.
2. Dudai Y. The neurobiology of consolidations, or, how stable is the engram? *Annu Rev Psychol*. 2004;55:51–86.
3. Ribot T. *Diseases of Memory*. London: Kegan Paul Trench & co.; 1977.
4. Sciville WB, Milner B. Loss of recent memory after bilateral hippocampal lesions. *J Neurol Neurosurg Psychiatry*. 1957;20(1):11–21.
5. Penfield W, Milner B. Memory deficit produced by bilateral lesions in the hippocampal zone. *AMA Arch Neurol Psychiatry*. 1958;79(5):475–497.
6. Dudai Y, Eisenberg M. Rites of passage of the engram: reconsolidation and the lingering consolidation hypothesis. *Neuron*. 2004;44(1):93–100.
7. Kandel E. The molecular biology of memory storage: a dialogue between genes and synapses. *Science*. 2001;294(5544):1030–1038.

8. Bliss TV, Lomo T. Long-lasting potentiation of synaptic transmission in the dentate area of the anaesthetized rabbit following stimulation of the perforant path. *J Physiol Lond.* 1973;232(2):331–356.
9. Bliss TV, Collingridge GL. A synaptic model of memory: long-term potentiation in the hippocampus. *Nature.* 1993;361(6407):31–39.
10. Abraham WC, Mason-Parker SE, Williams J, Dragunow M. Analysis of the decremental nature of LTP in the dentate gyrus. *Brain Res Mol Brain Res.* 1995;30(2):367–372.
11. Davis HP, Squire LR. Protein synthesis and memory: a review. *Psychol Bull.* 1984;96(3):518–559.
12. Bailey CH, Bartsch D, Kandel ER. Toward a molecular definition of long-term memory storage. *Proc Natl Acad Sci U S A.* 1996;93(24):13445–13452.
13. Steward O, Schuman EM. Protein synthesis at synaptic sites on dendrites. *Annu Rev Neurosci.* 2001;24:299–325.
14. Otani S, Abraham WC. Inhibition of protein synthesis in the dentate gyrus, but not the entorhinal cortex, blocks maintenance of long-term potentiation in rats. *Neurosci Lett.* 1989;106(1–2):175–180.
15. Otani S, Marshall CJ, Tate WP, Goddard GV, Abraham WC. Maintenance of long-term potentiation in rat dentate gyrus requires protein synthesis but not messenger RNA synthesis immediately post-tetanization. *Neuroscience.* 1989;28(3):519–526.
16. Nguyen PV, Kandel ER. A macromolecular synthesis-dependent late phase of long-term potentiation requiring cAMP in the medial perforant pathway of rat hippocampal slices. *J Neurosci.* 1996;16(10):3189–3198.
17. Frey U, Krug M, Reymann KG, Matthies H. Anisomycin, an inhibitor of protein synthesis, blocks late phases of LTP phenomena in the hippocampal CA1 region in vitro. *Brain Res.* 1988;452(1–2):57–65.
18. Stanton PK, Sarvey JM. Blockade of long-term potentiation in rat hippocampal CA1 region by inhibitors of protein synthesis. *J Neurosci.* 1984;4(12):3080–3088.
19. Neves G, Cooke SF, Bliss TVP. Synaptic plasticity, memory and the hippocampus: a neural network approach to causality. *Nat Rev Neurosci.* 2008;9(1):65–75. Available at, http://www.nature.com/nrn/journal/v9/n1/pdf/nrn2303.pdf.
20. Amaral D, Lavenex P. Hippocampal neuroanatomy. In: Andersen P, Morris RGM, Amaral DG, Bliss TV, O'Keefe J, eds. *The Hippocampus Book.* Oxford: Oxford University Press; 2007:37–114.
21. Bagni C, Mannucci L, Dotti CG, Amaldi F. Chemical stimulation of synaptosomes modulates alpha -Ca2+/calmodulin-dependent protein kinase II mRNA association to polysomes. *J Neurosci.* 2000;20(10):RC76.
22. Yin Y, Edelman GM, Vanderklish PW. The brain-derived neurotrophic factor enhances synthesis of Arc in synaptoneurosomes. *Proc Natl Acad Sci U S A.* 2002;99(4):2368–2373.
23. Weiler IJ, Irwin SA, Klintsova AY, et al. Fragile X mental retardation protein is translated near synapses in response to neurotransmitter activation. *Proc Natl Acad Sci U S A.* 1997;94(10):5395–5400.
24. Kang H, Schuman EM. A requirement for local protein synthesis in neurotrophin-induced hippocampal synaptic plasticity. *Science.* 1996;273(5280):1402–1406.
25. Huber KM, Kayser MS, Bear MF. Role for rapid dendritic protein synthesis in hippocampal mGluR-dependent long-term depression. *Science.* 2000;288(5469):1254–1257.
26. Huber KM, Roder JC, Bear MF. Chemical induction of mGluR5- and protein synthesis—dependent long-term depression in hippocampal area CA1. *J Neurophysiol.* 2001;86(1):321–325.
27. Aakalu G, Smith WB, Nguyen N, Jiang C, Schuman EM. Dynamic visualization of local protein synthesis in hippocampal neurons. *Neuron.* 2001;30(2):489–502.
28. Gerard RW. Physiology and psychiatry. *Am J Psychiatry.* 1949;106(3):161–173.

29. Duncan CP. The retroactive effect of electroshock on learning. *J Comp Physiol Psychol.* 1949;42(1):32–44.
30. Flexner JB, Flexner LB, Stellar E. Memory in mice as affected by intracerebral puromycin. *Science.* 1963;141:57–59.
31. Barraco RA, Stettner LJ. Antibiotics and memory. *Psychol Bull.* 1976;83(2):242–302.
32. Barondes SH. Cerebral protein synthesis inhibitors block long-term memory. *Int Rev Neurobiol.* 1970;12:177–205.
33. Squire LR, Barondes SH. Variable decay of memory and its recovery in cycloheximide-treated mice. *Proc Natl Acad Sci U S A.* 1972;69(6):1416–1420.
34. Hay N, Sonenberg N. Upstream and downstream of mTOR. *Genes Dev.* 2004;18(16):1926–1945.
35. Costa-Mattioli M, Sossin WS, Klann E, Sonenberg N. Translational control of long-lasting synaptic plasticity and memory. *Neuron.* 2009;61(1):10–26.
36. Tang SJ, Schuman EM. Protein synthesis in the dendrite. *Philos Trans R Soc Lond B Biol Sci.* 2002;357(1420):521–529.
37. Ma L, Chen Z, Erdjument-Bromage H, Tempst P, Pandolfi PP. Phosphorylation and functional inactivation of TSC2 by Erk implications for tuberous sclerosis and cancer pathogenesis. *Cell.* 2005;121(2):179–193.
38. Roux PP, Ballif BA, Anjum R, Gygi SP, Blenis J. Tumor-promoting phorbol esters and activated Ras inactivate the tuberous sclerosis tumor suppressor complex via p90 ribosomal S6 kinase. *Proc Natl Acad Sci U S A.* 2004;101(37):13489–13494.
39. Inoki K, Kim J, Guan K-L. AMPK and mTOR in cellular energy homeostasis and drug targets. *Annu Rev Pharmacol Toxicol.* 2012;52:381–400.
40. Inoki K, Zhu T, Guan K-L. TSC2 mediates cellular energy response to control cell growth and survival. *Cell.* 2003;115(5):577–590.
41. Jacinto E, Hall MN. Tor signalling in bugs, brain and brawn. *Nat Rev Mol Cell Biol.* 2003;4(2):117–126.
42. Kim D-H, Sarbassov DD, Ali SM, et al. mTOR interacts with raptor to form a nutrient-sensitive complex that signals to the cell growth machinery. *Cell.* 2002;110(2):163–175.
43. Sarbassov DD, Guertin DA, Ali SM, Sabatini DM. Phosphorylation and regulation of Akt/PKB by the rictor-mTOR complex. *Science.* 2005;307(5712):1098–1101.
44. Gingras AC, Raught B, Sonenberg N. mTOR signaling to translation. *Curr Top Microbiol Immunol.* 2004;279:169–197.
45. Wang L, Harris TE, Roth RA, Lawrence JC. PRAS40 regulates mTORC1 kinase activity by functioning as a direct inhibitor of substrate binding. *J Biol Chem.* 2007;282(27):20036–20044.
46. Vander Haar E, Lee S-I, Bandhakavi S, Griffin TJ, Kim D-H. Insulin signalling to mTOR mediated by the Akt/PKB substrate PRAS40. *Nat Cell Biol.* 2007;9(3):316–323.
47. Sarbassov DD, Ali SM, Sengupta S, et al. Prolonged rapamycin treatment inhibits mTORC2 assembly and Akt/PKB. *Mol Cell.* 2006;22(2):159–168.
48. Jacinto E, Facchinetti V, Liu D, et al. SIN1/MIP1 maintains rictor-mTOR complex integrity and regulates Akt phosphorylation and substrate specificity. *Cell.* 2006;127(1):125–137.
49. Yang Q, Inoki K, Ikenoue T, Guan K-L. Identification of Sin1 as an essential TORC2 component required for complex formation and kinase activity. *Genes Dev.* 2006;20(20):2820–2832.
50. Frias MA, Thoreen CC, Jaffe JD, et al. mSin1 is necessary for Akt/PKB phosphorylation, and its isoforms define three distinct mTORC2s. *Curr Biol.* 2006;16(18):1865–1870.

51. Huang W, Zhu PJ, Zhang S, et al. mTORC2 controls actin polymerization required for consolidation of long-term memory. *Nat Neurosci.* 2013;16(4):441–448.
52. Klann E, Dever TE. Biochemical mechanisms for translational regulation in synaptic plasticity. *Nat Rev Neurosci.* 2004;5(12):931–942.
53. Costa-Mattioli M, Sonenberg N, Richter JD. Translational regulatory mechanisms in synaptic plasticity and memory storage. *Prog Mol Biol Transl Sci.* 2009;90:293–311.
54. Gingras AC, Raught B, Sonenberg N. eIF4 initiation factors: effectors of mRNA recruitment to ribosomes and regulators of translation. *Annu Rev Biochem.* 1999;68:913–963.
55. Richter JD. CPEB: a life in translation. *Trends Biochem Sci.* 2007;32(6):279–285.
56. Alarcon JM, Hodgman R, Theis M, Huang Y-S, Kandel ER, Richter JD. Selective modulation of some forms of schaffer collateral-CA1 synaptic plasticity in mice with a disruption of the CPEB-1 gene. *Learn Mem.* 2004;11(3):318–327.
57. Udagawa T, Swanger SA, Takeuchi K, et al. Bidirectional control of mRNA translation and synaptic plasticity by the cytoplasmic polyadenylation complex. *Mol Cell.* 2012;47(2):253–266.
58. Derry MC, Yanagiya A, Martineau Y, Sonenberg N. Regulation of poly(A)-binding protein through PABP-interacting proteins. *Cold Spring Harb Symp Quant Biol.* 2006;71:537–543.
59. Khoutorsky A, Yanagiya A, Gkogkas CG, et al. Control of synaptic plasticity and memory via suppression of poly(A)-binding protein. *Neuron.* 2013;78(2):298–311.
60. Sossin WS, Lacaille J-C. Mechanisms of translational regulation in synaptic plasticity. *Curr Opin Neurobiol.* 2010;20(4):450–456.
61. Topisirovic I, Svitkin YV, Sonenberg N, Shatkin AJ. Cap and cap-binding proteins in the control of gene expression. *Wiley Interdiscip Rev RNA.* 2011;2(2):277–298.
62. Proud CG. Signalling to translation: how signal transduction pathways control the protein synthetic machinery. *Biochem J.* 2007;403:217–234.
63. Sonenberg N, Hershey JWB, Mathews MB. *Translational Control of Gene Expression.* Cold Spring Harbor, NY: CSHL Press; 2001.
64. Carlberg U, Nilsson A, Nygård O. Functional properties of phosphorylated elongation factor 2. *Eur J Biochem.* 1990;191(3):639–645.
65. Ryazanov AG, Ward MD, Mendola CE, et al. Identification of a new class of protein kinases represented by eukaryotic elongation factor-2 kinase. *Proc Natl Acad Sci U S A.* 1997;94(10):4884–4889.
66. Redpath NT, Foulstone EJ, Proud CG. Regulation of translation elongation factor-2 by insulin via a rapamycin-sensitive signalling pathway. *EMBO J.* 1996;15(9):2291–2297.
67. Wang X, Li W, Williams M, Terada N, Alessi DR, Proud CG. Regulation of elongation factor 2 kinase by p90(RSK1) and p70 S6 kinase. *EMBO J.* 2001;20(16):4370–4379.
68. Browne GJ, Proud CG. A novel mTOR-regulated phosphorylation site in elongation factor 2 kinase modulates the activity of the kinase and its binding to calmodulin. *Mol Cell Biol.* 2004;24(7):2986–2997.
69. Browne GJ, Proud CG. Regulation of peptide-chain elongation in mammalian cells. *Eur J Biochem.* 2002;269(22):5360–5368.
70. Gingras AC, Raught B, Sonenberg N. Regulation of translation initiation by FRAP/mTOR. *Genes Dev.* 2001;15(7):807–826.
71. Guertin DA, Sabatini DM. The pharmacology of mTOR inhibition. *Sci Signal.* 2009;2(67):e24.
72. Ikenoue T, Hong S, Inoki K. Monitoring mammalian target of rapamycin (mTOR) acitivty. *Methods Enzymol.* 2009;452:165–180.

73. Beretta L, Gingras AC, Svitkin YV, Hall MN, Sonenberg N. Rapamycin blocks the phosphorylation of 4E-BP1 and inhibits cap-dependent initiation of translation. *EMBO J*. 1996;15(3):658–664.
74. Holz MK, Ballif BA, Gygi SP, Blenis J. mTOR and S6K1 mediate assembly of the translation preinitiation complex through dynamic protein interchange and ordered phosphorylation events. *Cell*. 2005;123(4):569–580.
75. Proud CG. mTOR-mediated regulation of translation factors by amino acids. *Biochem Biophys Res Commun*. 2004;313(2):429–436.
76. Thoreen CC, Chantranupong L, Keys HR, Wang T, Gray NS, Sabatini DM. A unifying model for mTORC1-mediated regulation of mRNA translation. *Nature*. 2012;485(7396):109–113.
77. Gangloff Y-G, Mueller M, Dann SG, et al. Disruption of the mouse mTOR gene leads to early postimplantation lethality and prohibits embryonic stem cell development. *Mol Cell Biol*. 2004;24(21):9508–9516.
78. Murakami M, Ichisaka T, Maeda M, et al. mTOR is essential for growth and proliferation in early mouse embryos and embryonic stem cells. *Mol Cell Biol*. 2004;24(15):6710–6718.
79. Dobashi Y, Watanabe Y, Miwa C, Suzuki S, Koyama S. Mammalian target of rapamycin: a central node of complex signaling cascades. *Int J Clin Exp Pathol*. 2011;4(5):476–495.
80. Hoeffer CA, Tang W, Wong H, et al. Removal of FKBP12 enhances mTOR-raptor interactions, LTP, memory, and perseverative/repetitive behavior. *Neuron*. 2008;60(5):832–845.
81. Tang SJ, Reis G, Kang H, Gingras A-C, Sonenberg N, Schuman EM. A rapamycin-sensitive signaling pathway contributes to long-term synaptic plasticity in the hippocampus. *Proc Natl Acad Sci U S A*. 2002;99(1):467–472.
82. Cammalleri M, Lütjens R, Berton F, et al. Time-restricted role for dendritic activation of the mTOR-p70S6K pathway in the induction of late-phase long-term potentiation in the CA1. *Proc Natl Acad Sci U S A*. 2003;100(24):14368–14373.
83. Ouyang Y, Rosenstein A, Kreiman G, Schuman EM, Kennedy MB. Tetanic stimulation leads to increased accumulation of Ca(2+)/calmodulin-dependent protein kinase II via dendritic protein synthesis in hippocampal neurons. *J Neurosci*. 1999;19(18):7823–7833.
84. Gong R, Park CS, Abbassi NR, Tang S-J. Roles of glutamate receptors and the mammalian target of rapamycin (mTOR) signaling pathway in activity-dependent dendritic protein synthesis in hippocampal neurons. *J Biol Chem*. 2006;281(27):18802–18815.
85. Tsokas P, Grace EA, Chan P, et al. Local protein synthesis mediates a rapid increase in dendritic elongation factor 1A after induction of late long-term potentiation. *J Neurosci*. 2005;25(24):5833–5843.
86. Dash PK, Orsi SA, Moore AN. Spatial memory formation and memory-enhancing effect of glucose involves activation of the tuberous sclerosis complex-Mammalian target of rapamycin pathway. *J Neurosci*. 2006;26(31):8048–8056.
87. Parsons RG, Gafford GM, Helmstetter FJ. Translational control via the mammalian target of rapamycin pathway is critical for the formation and stability of long-term fear memory in amygdala neurons. *J Neurosci*. 2006;26(50):12977–12983.
88. Sabatini DM, Lai MM, Snyder SH. Neural roles of immunophilins and their ligands. *Mol Neurobiol*. 1997;15(2):223–239.
89. Harrar Y, Bellini C, Faure JD. FKBPs: at the crossroads of folding and transduction. *Trends Plant Sci*. 2001;6(9):426–431.
90. Shou W, Aghdasi B, Armstrong DL, et al. Cardiac defects and altered ryanodine receptor function in mice lacking FKBP12. *Nature*. 1998;391(6666):489–492.

91. Hoeffer CA, Klann E. mTOR signaling: at the crossroads of plasticity, memory and disease. *Trends Neurosci.* 2010;33(2):67–75.
92. Sonenberg N, Hinnebusch AG. Regulation of translation initiation in eukaryotes: mechanisms and biological targets. *Cell.* 2009;136(4):731–745.
93. Richter JD, Sonenberg N. Regulation of cap-dependent translation by eIF4E inhibitory proteins. *Nature.* 2005;433(7025):477–480.
94. Gingras AC, Raught B, Gygi SP, et al. Hierarchical phosphorylation of the translation inhibitor 4E-BP1. *Genes Dev.* 2001;15(21):2852–2864.
95. Raught B, Peiretti F, Gingras A-C, et al. Phosphorylation of eucaryotic translation initiation factor 4B Ser422 is modulated by S6 kinases. *EMBO J.* 2004;23(8):1761–1769.
96. Banko JL, Merhav M, Stern E, Sonenberg N, Rosenblum K, Klann E. Behavioral alterations in mice lacking the translation repressor 4E-BP2. *Neurobiol Learn Mem.* 2007;87(2):248–256.
97. Banko JL, Klann E. Cap-dependent translation initiation and memory. *Prog Brain Res.* 2008;169:59–80.
98. Banko JL, Poulin F, Hou L, DeMaria CT, Sonenberg N, Klann E. The translation repressor 4E-BP2 is critical for eIF4F complex formation, synaptic plasticity, and memory in the hippocampus. *J Neurosci.* 2005;25(42):9581–9590.
99. Scheper GC, van Kollenburg B, Hu J, Luo Y, Goss DJ, Proud CG. Phosphorylation of eukaryotic initiation factor 4E markedly reduces its affinity for capped mRNA. *J Biol Chem.* 2002;277(5):3303–3309.
100. Kleijn M, Scheper GC, Wilson ML, Tee AR, Proud CG. Localisation and regulation of the eIF4E-binding protein 4E-BP3. *FEBS Lett.* 2002;532(3):319–323.
101. Scheper GC, Morrice NA, Kleijn M, Proud CG. The mitogen-activated protein kinase signal-integrating kinase Mnk2 is a eukaryotic initiation factor 4E kinase with high levels of basal activity in mammalian cells. *Mol Cell Biol.* 2001;21(3):743–754.
102. Scheper GC, Proud CG. Does phosphorylation of the cap-binding protein eIF4E play a role in translation initiation? *Eur J Biochem.* 2002;269(22):5350–5359.
103. Waskiewicz AJ, Flynn A, Proud CG, Cooper JA. Mitogen-activated protein kinases activate the serine/threonine kinases Mnk1 and Mnk2. *EMBO J.* 1997;16(8):1909–1920.
104. Waskiewicz AJ, Johnson JC, Penn B, Mahalingam M, Kimball SR, Cooper JA. Phosphorylation of the cap-binding protein eukaryotic translation initiation factor 4E by protein kinase Mnk1 in vivo. *Mol Cell Biol.* 1999;19(3):1871–1880.
105. Pyronnet S, Imataka H, Gingras AC, Fukunaga R, Hunter T, Sonenberg N. Human eukaryotic translation initiation factor 4G (eIF4G) recruits mnk1 to phosphorylate eIF4E. *EMBO J.* 1998;18(1):270–279.
106. Volpon L, Osborne MJ, Topisirovic I, Siddiqui N, Borden KLB. Cap-free structure of eIF4E suggests a basis for conformational regulation by its ligands. *EMBO J.* 2006;25(21):5138–5149.
107. Sonenberg N. eIF4E, the mRNA cap-binding protein: from basic discovery to translational researchThis paper is one of a selection of papers published in this special issue, entitled CSBMCB—Systems and Chemical Biology, and has undergone the journal's usual peer review process. *Biochem Cell Biol.* 2008;86(2):178–183.
108. Marcotrigiano J, Gingras AC, Sonenberg N, Burley SK. Cap-dependent translation initiation in eukaryotes is regulated by a molecular mimic of eIF4G. *Mol Cell.* 1999;3(6):707–716.
109. Ueda T, Watanabe-Fukunaga R, Fukuyama H, Nagata S, Fukunaga R. Mnk2 and Mnk1 are essential for constitutive and inducible phosphorylation of eukaryotic initiation factor 4E but not for cell growth or development. *Mol Cell Biol.* 2004;24(15):6539–6549.

110. Kelleher RJ, Govindarajan A, Jung H-Y, Kang H, Tonegawa S. Translational control by MAPK signaling in long-term synaptic plasticity and memory. *Cell.* 2004;116(3):467–479.
111. Banko JL, Hou L, Klann E. NMDA receptor activation results in PKA- and ERK-dependent Mnk1 activation and increased eIF4E phosphorylation in hippocampal area CA1. *J Neurochem.* 2004;91(2):462–470.
112. Moerke NJ, Aktas H, Chen H, et al. Small-molecule inhibition of the interaction between the translation initiation factors eIF4E and eIF4G. *Cell.* 2007;128(2):257–267.
113. Hoeffer CA, Santini E, Ma T, et al. Multiple components of eIF4F are required for protein synthesis-dependent hippocampal long-term potentiation. *J Neurophysiol.* 2013;109(1):68–76.
114. Hoeffer CA, Cowansage KK, Arnold EC, et al. Inhibition of the interactions between eukaryotic initiation factors 4E and 4G impairs long-term associative memory consolidation but not reconsolidation. *Proc Natl Acad Sci U S A.* 2011;108(8):3383–3388.
115. Zoncu R, Efeyan A, Sabatini DM. mTOR: from growth signal integration to cancer, diabetes and ageing. *Nat Rev Mol Cell Biol.* 2011;12(1):21–35.
116. Zoncu R, Bar-Peled L, Efeyan A, Wang S, Sancak Y, Sabatini DM. mTORC1 senses lysosomal amino acids through an inside-out mechanism that requires the vacuolar H(+)-ATPase. *Science.* 2011;334(6056):678–683.
117. Dann SG, Selvaraj A, Thomas G. mTOR Complex1-S6K1 signaling: at the crossroads of obesity, diabetes and cancer. *Trends Mol Med.* 2007;13(6):252–259.
118. Banerjee P, Ahmad MF, Grove JR, Kozlosky C, Price DJ, Avruch J. Molecular structure of a major insulin/mitogen-activated 70-kDa S6 protein kinase. *Proc Natl Acad Sci U S A.* 1990;87(21):8550–8554.
119. Pearson G, Robinson F, Beers Gibson T, et al. Mitogen-activated protein (MAP) kinase pathways: regulation and physiological functions. *Endocr Rev.* 2001;22(2):153–183.
120. Burnett PE, Barrow RK, Cohen NA, Snyder SH, Sabatini DM. RAFT1 phosphorylation of the translational regulators p70 S6 kinase and 4E-BP1. *Proc Natl Acad Sci U S A.* 1998;95(4):1432–1437.
121. Ferrari S. A rapid purification protocol for the mitogen-activated p70 S6 kinase. *Protein Expr Purif.* 1998;13(2):170–176.
122. Shima H, Pende M, Chen Y, Fumagalli S, Thomas G, Kozma SC. Disruption of the p70(s6k)/p85(s6k) gene reveals a small mouse phenotype and a new functional S6 kinase. *EMBO J.* 1998;17(22):6649–6659.
123. Tang H, Hornstein E, Stolovich M, et al. Amino acid-induced translation of TOP mRNAs is fully dependent on phosphatidylinositol 3-kinase-mediated signaling, is partially inhibited by rapamycin, and is independent of S6K1 and rpS6 phosphorylation. *Mol Cell Biol.* 2001;21(24):8671–8683.
124. Pende M, Um SH, Mieulet V, et al. S6K1(−/−)/S6K2(−/−) mice exhibit perinatal lethality and rapamycin-sensitive 5'-terminal oligopyrimidine mRNA translation and reveal a mitogen-activated protein kinase-dependent S6 kinase pathway. *Mol Cell Biol.* 2004;24(8):3112–3124.
125. Antion MD, Merhav M, Hoeffer CA, et al. Removal of S6K1 and S6K2 leads to divergent alterations in learning, memory, and synaptic plasticity. *Learn Mem.* 2008;15(1):29–38.
126. Berlanga JJ, Santoyo J, De Haro C. Characterization of a mammalian homolog of the GCN2 eukaryotic initiation factor 2alpha kinase. *Eur J Biochem.* 1999;265(2):754–762.
127. Sood R, Porter AC, Olsen DA, Cavener DR, Wek RC. A mammalian homologue of GCN2 protein kinase important for translational control by phosphorylation of eukaryotic initiation factor-2alpha. *Genetics.* 2000;154(2):787–801.

128. Jiang H-Y, Wek SA, Mcgrath BC, et al. Activating transcription factor 3 is integral to the eukaryotic initiation factor 2 kinase stress response. *Mol Cell Biol.* 2004;24(3):1365–1377.
129. Hinnebusch AG, Asano K, Olsen DS, Phan L, Nielsen KH, Valášek L. Study of translational control of eukaryotic gene expression using yeast. *Ann N Y Acad Sci.* 2004;1038:60–74.
130. Ron D, Walter P. Signal integration in the endoplasmic reticulum unfolded protein response. *Nat Rev Mol Cell Biol.* 2007;8(7):519–529.
131. Hinnebusch AG. eIF2alpha kinases provide a new solution to the puzzle of substrate specificity. *Nat Struct Mol Biol.* 2005;12:835–838.
132. Hinnebusch AG. Translational regulation of GCN4 and the general amino acid control of yeast. *Annu Rev Microbiol.* 2005;59:407–450.
133. Harding HP, Novoa I, Zhang Y, et al. Regulated translation initiation controls stress-induced gene expression in mammalian cells. *Mol Cell.* 2000;6(5):1099–1108.
134. Vattem KM, Wek RC. Reinitiation involving upstream ORFs regulates ATF4 mRNA translation in mammalian cells. *Proc Natl Acad Sci U S A.* 2004;101(31):11269–11274.
135. Costa-Mattioli M, Gobert D, Harding H, et al. Translational control of hippocampal synaptic plasticity and memory by the eIF2alpha kinase GCN2. *Nature.* 2005;436(7054):1166–1173.
136. Gkogkas C, Sonenberg N, Costa-Mattioli M. Translational control mechanisms in long-lasting synaptic plasticity and memory. *J Biol Chem.* 2010;285(42):31913–31917.
137. Costa-Mattioli M, Gobert D, Stern E, et al. eIF2alpha phosphorylation bidirectionally regulates the switch from short- to long-term synaptic plasticity and memory. *Cell.* 2007;129(1):195–206.
138. Trinh MA, Kaphzan H, Wek RC, Pierre P, Cavener DR, Klann E. Brain-specific disruption of the eIF2α kinase PERK decreases ATF4 expression and impairs behavioral flexibility. *Cell Rep.* 2012;1(6):676–688.
139. Dever TE, Sripriya R, McLachlin JR, et al. Disruption of cellular translational control by a viral truncated eukaryotic translation initiation factor 2alpha kinase homolog. *Proc Natl Acad Sci U S A.* 1998;95(8):4164–4169.
140. Bingol B, Sheng M. Deconstruction for reconstruction: the role of proteolysis in neural plasticity and disease. *Neuron.* 2011;69(1):22–32.
141. Jiang Z, Belforte JE, Lu Y, et al. eIF2alpha Phosphorylation-dependent translation in CA1 pyramidal cells impairs hippocampal memory consolidation without affecting general translation. *J Neurosci.* 2010;30(7):2582–2594.

CHAPTER SIX

BDNF–TrkB Receptor Regulation of Distributed Adult Neural Plasticity, Memory Formation, and Psychiatric Disorders

Raül Andero[*,†], **Dennis C. Choi**[*,†], **Kerry J. Ressler**[*,†,‡]
[*]Department of Psychiatry and Behavioral Sciences, Emory University School of Medicine, Atlanta, Georgia, USA
[†]Yerkes National Primate Research Center, Atlanta, Georgia, USA
[‡]Howard Hughes Medical Institute, Chevy Chase, Maryland, USA

Contents

Abstract

Brain-derived neurotrophic factor (BDNF) and its single transmembrane receptor, tropomysin-related kinase B (TrkB), are essential for adult synaptic plasticity and the formation of memories. However, there are regional and task-dependent differences underlying differential mechanisms of BDNF–TrkB function in the formation of these memories. Additionally, the BDNF pathway has been implicated in several psychiatric disorders including posttraumatic stress disorder, phobia, and panic disorder. Gaining a better understanding of this pathway and the neurobiology of memory through fundamental research may be helpful to identify effective prevention and treatment

Progress in Molecular Biology and Translational Science, Volume 122
ISSN 1877-1173
http://dx.doi.org/10.1016/B978-0-12-420170-5.00006-4

approaches both for diseases of memory deficit as well as in cases of enhanced aversive memory, such as in anxiety disorders.

1. BRAIN-DERIVED NEUROTROPHIC FACTOR

Neurotrophins are classically known for their effects on neuronal survival. More specifically, they are well known to modulate axonal and dendritic growth and branching, synaptic structure, neurotransmission, and synaptic plasticity.[1–3] The most abundant neurotrophin in the central nervous system related to these processes is brain-derived neurotrophic factor (BDNF).

BDNF is a 119-amino acid (aa) residue nonglycosylated polypeptide.[4] Initially synthesized as a 247-aa residue prepropeptide, the BDNF molecule is divided into a 18-aa residue signal sequence, a 110-aa residue prosequence (ProBDNF), and a 119-aa residue mature segment.[5] It is also the best-characterized neurotrophin, and it is highly present in brain areas such as the hypothalamus, amygdala, hippocampus, and neocortex. Moreover, its expression and release are dependent on neural activity. With development and learning processes, BDNF elicits changes in the structure of dendrites and spines in mammalian neurons.[2] Cells known to express BDNF in the central nervous system include fibroblasts,[6] astrocytes,[7] and neurons of varying phenotype and location.[7–9]

The activity of BDNF acts through two distinct receptor systems, its high-affinity tropomysin-related kinase B (TrkB) receptor and the low-affinity receptor p75 which is a common nonspecific receptor for all neurotrophins.[10–12] BDNF binding modulates dimerization of its receptors, and activates several signaling pathways through the intracellular kinase-binding domain[1,13] (see Fig. 6.1).

2. BDNF, LONG-TERM POTENTIATION, AND SYNAPTIC PLASTICITY

BDNF is essential for synaptic plasticity processes which are required for long-term learning and memory.[14] Synaptic plasticity is the process by which connections between two neurons, or synapses, change in strength. These changes involve structural and functional alterations of synapses which underlie long-term potentiation (LTP).[14,15]

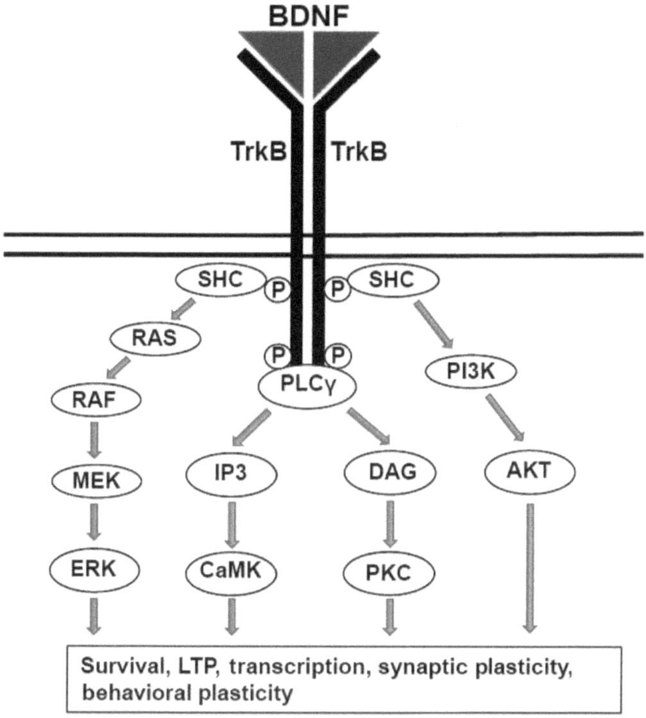

Figure 6.1 Brain-derived neurotrophic factor (BDNF) and TrkB signaling pathways. BDNF binds to the extracellular domain of TrkB forming homodimers to activate downstream intracellular signaling cascades, including Ras/MAPK, phospholipase C (PLC)-γ, and phospatidyl inositol-3 kinase (PI3K)/AKT pathways. Note that TrkB can also heterodimerize with other single transmembrane growth factor receptors.

Transgenic mice lacking *BDNF* are known to not survive postnatally, while *BDNF* heterozygous mice are viable but are found to exhibit profound deficits (see Ref. 1). Most notably, decreased BDNF leads deficits in LTP in the hippocampus.[16,17] BDNF increases plasticity at the neuromuscular synapse[18] and enhances synaptic transmission via the TrkB receptor in hippocampal neurons.[19,20] Moreover, the deficits in hippocampal LTP in *BDNF* mutant mice can be rescued by exogenous BDNF or viral *BDNF* gene transfer into the CA1 of the hippocampus.[17,21] Conditional mutations of TrkB also result in deficits in hippocampal-dependent learning and memory tests.[22] This suggests that BDNF likely modulates LTP through TrkB downstream signaling.[23] Additionally, p75 mutant mice also show deficits in learning tasks,[24,25] although this may be mediated via a number of different growth factors. However, it remains possible that BDNF may affect learning and plasticity through both p75 and TrkB receptors and their downstream signaling pathways (Fig. 6.1).

2.1. BDNF- and hippocampal-dependent learning

Memory processes characterized by conscious recall are called declarative memory, whereas memories that are recalled unconsciously, are considered implicit or associative memories. The hippocampus is critical for a variety of explicit, declarative, and spatial memory functions. Functional distinctions have also been attributed to the dorsal hippocampus—cognition, spatial, declarative, and emotional memory and ventral hippocampus—control of behavioral inhibition, stress response, and emotional memory.[26] It is generally believed that the hippocampus plays a time-limited role in consolidating labile new short-term memory into more stable long-term memory. At the completion of hippocampus-dependent consolidation, these memories are eventually stored via a distributed circuit within cortex without major hippocampal contribution.[27] In brief, it is believed that BDNF–TrkB activation is required for the consolidation of hippocampal-dependent learning followed by a delayed stabilization phase which is specifically required for the maintenance of that memory.[28,29]

2.2. Declarative memory

BDNF rapidly potentiates synaptic transmission in hippocampal neurons,[20] an effect mediated by the MAP kinase pathway (MAPK, also known as Ras–Raf–Mek–Erk pathway).[30] Concordantly, specific *BDNF* deletion in the dorsal hippocampus impairs declarative memory evaluated by the novel object recognition and water maze.[31] This is consistent with the impairment of hippocampal LTP and spatial learning in heterozygous *BDNF* knock-out mice (+/−) and the rescue of these phenotypes by recombinant BDNF.[17] Additionally, unilateral infusion of recombinant BDNF in the hippocampus is sufficient to enhance declarative memory in wild-type mice evaluated with the water maze.[32] Also, genetic manipulation of the *TrkB* gene shows that mice overexpressing full-length, functional TrkB have increased activation of the TrkB–PLC-γ pathway and enhanced hippocampal-dependent memory.[33] In contrast, conditional forebrain restriction of TrkB expression in adult mice revealed reduced CA1-LTP and impaired hippocampal-dependent memory.[14] These findings support the role of BDNF–TrkB as necessary for the consolidation of declarative memory.

2.3. Contextual fear learning

Although the amygdala and medial prefrontal cortex (mPFC) also modulate the storage of contextual fear memory, the most critical area in the contextual memory component of this process is the hippocampus. *BDNF* mRNA is rapidly upregulated in the hippocampus,[34] whereas heterozygous *BDNF* knock-out mice (+/−) present impairment of contextual fear learning which is partially rescued by recombinant BDNF in the hippocampus.[35] This BDNF upregulation may be associated with increased signaling via inhibitory gamma-aminobutyric acid (GABA)ergic synapses.[36] Concordantly, mice overexpressing full-length TrkB present enhanced contextual fear learning and enhanced LTP in the hippocampus.[33] Consistent with this, *BDNF* deletion in the dorsal hippocampus impairs fear extinction.[31] Moreover, epigenetic processes, such as methylation within the *BDNF* gene in the hippocampus, have been found upregulated after contextual fear conditioning.[37] A separate epigenetic study found that contextual fear conditioning induced differential regulation of exon-specific *BDNF* mRNAs in the hippocampus. They showed that DNA methylation across the *BDNF* gene is a dynamic process sufficient to alter differential *BDNF* transcriptional regulation.[38] Moreover, contextual fear conditioning increases histone H3 trimethylation around the *BDNF* gene promoter, and inhibition of histone deacetylase (HDAC) activity alters histone H3 trimethylation after contextual fear conditioning.[39] Together, these data are consistent with a model in which dynamic regulation of the *BDNF* gene within the hippocampus, via genetic, epigenetic, and active regulatory processes, is key for the consolidation of contextual fear learning.

2.4. BDNF- and amygdala-dependent learning

The amygdala is a key structure required for the consolidation process of associative fear memory, although it is modulated by other structures such as the hippocampus and different cortical areas (see the review by Johansen and collaborators[40]). Rattiner and colleagues first reported the association of BDNF and TrkB in amygdala-dependent fear conditioning.[41] Specifically, they found that cued-fear conditioning by pairing a CS (tone) and a US (electric foot shock) elicited an increase in *BDNF* mRNA expression and TrkB phosphorylation levels in the basolateral amygdala (BLA). Additionally, pharmacological and genetic *TrkB* inhibition blocked fear acquisition, which suggested that TrkB activation is necessary for the consolidation of fear memories. Further studies revealed that *BDNF* is transcriptionally upregulated after the same fear conditioning paradigm.[42] There was also a positive correlation between expression of conditioned fear and amygdala–BDNF protein levels.[43] Moreover, heterozygous *BDNF* (+/−) knock-out mice have around ~50% reduction in BDNF protein levels in the hippocampus, amygdala, and mPFC, accompanied by a deficit in the extinction of cued-fear learning.[44] Concordantly, BDNF or 7,8-dihydroxyflavone (7,8-DHF, a TrkB agonist[45]) reduced the threshold for LTP induction, using *in vitro* whole-cell patch clamp recordings from BLA principal neurons.[46] In regard to the amygdala signaling pathway required for the consolidation of the fear memories, it has been described that fear conditioning elicits TrkB activation by BDNF, which results in activation of MAPK and PI3 kinase (PI3K).[47] Additionally, MAPK signaling cascade is required for the consolidation of auditory fear memory and synaptic plasticity.[48]

Additionally, BDNF signaling is also necessary for cued-fear extinction. To examine this, Chhatwal and colleagues trained rats with fear conditioning, and 24 h later, the BLA was infected with a lentivirus encoding a dominant-negative *TrkB* (*TrkB*.t1) to antagonize TrkB signaling. Two weeks later, *TrkB*.t1-infected rats had impaired consolidation of fear extinction, but normal within-session extinction, compared to control virus infected rats.[49] Concordantly, a single systemic dose of the TrkB agonist, 7,8-DHF enhances the acquisition and extinction of cued-fear conditioning[50] (see Table 6.1). Interestingly, 7,8-DHF given before fear extinction also enhances within-session extinction.[50] This apparent contradiction with the previous study suggests that BDNF signaling is not required for within-session extinction but an unusually high level of TrkB activation may be sufficient to enhance this process. A recent study suggests that a chronic dosage of 7,8-DHF enhances amygdala synaptic plasticity and prevents age-related declines in fear learning in rats.[58]

BDNF seems to interact with other neurotransmitters such as GABA and dopamine during the consolidation of amygdala-dependent memories. For example, GABA receptors are rapidly downregulated during fear conditioning in the amygdala, which is predicted to elicit a transient membrane hyperexcitability, contributing to cellular mechanisms of memory formation.[60] Moreover, amygdala cell cultures show that BDNF-dependent TrkB activation may underlie the rapid internalization of $GABA_AR\alpha1$, and that this appears to be PKC-dependent.[61] Another example is provided by the reduction of LTP threshold induced by GBR12783 (a dopamine reuptake inhibitor). This dopamine-dependent reduced LTP threshold appears to be

Table 6.1 Recent adult memory studies using the TrkB agonist, 7,8-dihydroxyflavone

Reference	Route	Dose	Animal/cells	Actions
Choi *et al.*[51]	Intraperitoneal	5 mg/kg	Male neocortical BDNF knock-out mouse	Rescues a fear-expression deficit
Andero *et al.*[50]	Intraperitoneal	5 mg/kg	Male C57 mouse	Enhances both the acquisition of fear and its extinction and rescues an extinction deficit in a PTSD-like model
Zeng *et al.*[52]	Perfusion	10 μM	Male Fischer 344 rat	Rescues synaptic plasticity in aging
Andero *et al.*[53]	Intraperitoneal	5 mg/kg	Male Sprague-Dawley rat	Blocks stress-induced long-term memory impairment in a PTSD-like model
Chen *et al.*[54]	Incubation	20 μM	HT-22 cells	Neuroprotection against glutamate-induced toxicity
Devi and Ohno[55]	Intraperitoneal	5 mg/kg	Male 5XFAD mouse	Amelioration of Alzheimer disease-associated memory deficits
Choi *et al.*[56]	Intraperitoneal	5 mg/kg	Male TrkB(F616A) mouse	Rescues both apetitive and aversive deficits in BDNF deletion within the prelimbic cortex
Zeng *et al.*[57]	Intraperitoneal	5 mg/kg	Male Sprague-Dawley rat	Reversal of age-dependent memory impairment in cognitive aging
Zeng *et al.*[58]	Intraperitoneal	5 mg/kg	Male Sprague-Dawley rat	Prevention of age-related declines in fear learning and memory
Baker-Andresen *et al.*[59]	Intraperitoneal	25 mg/kg	Female C57 mouse	Enhances fear extinction

blocked by prior application of a TrkB-immunoglobulin-G fusion protein (TrkB-Fc, which scavenges BDNF and prevents TrkB receptor activation).[46] Overall, these studies suggest that BDNF–TrkB signaling, accompanied by a variety of neurotransmitter modulation, is necessary, and at times sufficient, for the consolidation of amygdala-dependent fear memories.

2.5. BDNF- and prefrontal cortical-dependent learning

The mPFC is positioned to support higher order cognitive functions including modulating emotional learning and memory. Notably, the prefrontal cortex is also ideally positioned to regulate the consolidation and expression of both appetitive and aversive memories, by directly influencing the nucleus accumbens and amygdala (Fig. 6.2; Refs. 62,63). The prelimbic

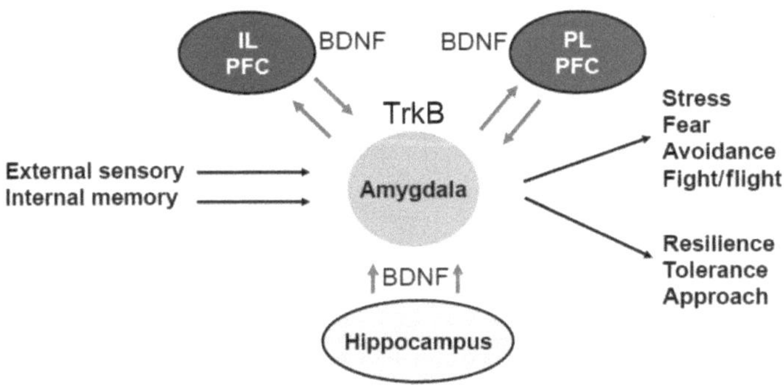

Figure 6.2 The PFC–amygdala–hippocampal circuit regulation of fear. (A) Schematic diagram is shown illustrating that BDNF–TrkB activation is critical for consolidation of fear learning (amygdala), fear expression (prelimbic prefrontal cortex, PL PFC), and extinction of fear (infralimbic prefrontal cortex, IL PFC). The hippocampus utilizes the BDNF–TrkB pathway for both contextual fear memory as well as the contextualization of extinction memory. *BDNF* mRNA expression and TrkB activation are correlated in the basolateral amygdala (BLA) with consolidation of both fear learning and extinction of fear. Moreover, blockade of TrkB receptor function with a lentiviral-mediated dominant-negative *TrkB* in BLA prevents consolidation of both conditioned fear and extinction of conditioned fear. Thus, TrkB activation, within BLA is required for fear-related synaptic plasticity and learning. Additionally, BDNF activation of TrkB is required in amygdala for synaptic plasticity encoding the consolidation of fear and extinction whereas *BDNF* mRNA expression in PL and amygdala is required for consolidation of learned fear. Also, BDNF infused into IL enhances inhibition and extinction of learned fear. This is consistent with the transient enhancement of BDNF functioning, using a TrkB agonist, during and after extinction which augments the reduction of fear. Green arrows indicate excitatory activity whereas red arrows represent inhibitory activity. (See the color plate.)

cortex (PL) region of the mPFC is known to drive the expression and consolidation of fear and cocaine seeking, while the more ventral, infralimbic cortex (IL) of the mPFC is known to mediate the extinction of fear as well as drug seeking (see Ref. 64).

Considerable evidence indicates that BDNF and TrkB signaling in the prefrontal cortex play a role in regulating both appetitive and aversive learning (see Refs. 65,66). Contextual fear extinction elicits an increase in *BDNF* mRNA in the mPFC, although in this study PL and IL were not differentiated.[67] Another group reported that after repeated cocaine exposure in rats, a gradual increase in the expression BDNF in the mPFC facilitates activity-induced LTP of excitatory synapses.[68] This enhanced synaptic plasticity could be attributed to BDNF-induced suppression of GABAergic inhibition in the mPFC by reducing the surface expression of $GABA_AR$. The BDNF effect was found to be mediated by TrkB signaling pathway. Therefore, this study provides exciting evidence of how prefrontal BDNF expression sensitizes the excitatory synapses in the mPFC to undergo activity-dependent plasticity that may contribute to cue-induced drug craving and drug-seeking behavior.

Conditioned taste aversion (CTA) also has been found to induce increased BDNF levels in the ventromedial prefrontal cortex (vmPFC). Interestingly, it was found that the enhanced TrkB phosphorylation occurred at a time point before the increased BDNF expression, suggesting rapid induction of activity-dependent BDNF secretion by CTA learning, or possibly compensatory replacement of *BDNF* mRNA after it is translated to replace released BDNF peptide.[69]

Recent evidence suggests that HDAC inhibitors facilitate extinction of rewarding drug memories, as well as extinction of aversive memories. It is possible that the HDAC epigenetic regulation effects are principally mediated through altering *BDNF* expression. A recent study examined this using conditioned place aversion (CPA). CPA extinction training induced an increase in histone H3 acetylation at the *BDNF* gene and increased *BDNF* mRNA and protein expression in the vmPFC of acute morphine-dependent rats. They determined that such epigenetic regulation of *BDNF* gene transcription could be facilitated or diminished by intra-vmPFC infusion of the HDAC inhibitor trichostatin A before extinction training. Correspondingly, disruption of regulation of *BDNF* gene transcription with a MAPK inhibitor, U0126, or suppression of BDNF signaling with Trk receptor antagonist, K252a, or a BDNF scavenger (TrkB-Fc), blocked extinction of CPA behavior. These data suggest that extinction of the aversive memory of withdrawal requires epigenetic regulation of *BDNF* gene transcription in the vmPFC through downstream TrkB signaling.[70]

Recent work also found that there are sex differences in fear extinction, where female mice are resistant to fear extinction. This extinction resistance is accompanied by increased DNA methylation of the *BDNF* gene and decreased *BDNF* mRNA expression within the mPFC.[59] Furthermore, activation of BDNF signaling by the TrkB agonist 7,8-DHF blocked the return of fear in female mice after extinction training.

More recently, work from our group demonstrated that selective neocortical *BDNF* knock-out mice had deficits in consolidation of fear memories.[51] Additionally, viral Cre-mediated *BDNF* deletion in the PL in floxed *BDNF* transgenic mice[71] also replicated this fear deficit.[51] We further discovered that PL *BDNF* gene deletion also leads to similar blunted cocaine preference following cocaine conditioned place preference (CPP).[56] This is consistent with another report demonstrating that PL *BDNF* knockdown (but not orbitofrontal cortex), also accelerates the extinction of instrumental responding for food reward.[72] These findings suggest that BDNF in PL is ideally positioned to regulate both appetitive and aversive learning. However, it remains to be determined if PL BDNF specifically targets TrkB receptors within the prefrontal cortex or at downstream target regions including the amygdala.

Notably, the deficits in learned fear and CPP in neocortical BDNF knock-down mice or PL BDNF knockdown are both rescued by systemic injections of the TrkB agonist, 7,8-DHF (Table 6.1; Refs. 51,56). Infusions of BDNF into the dorsal medial prefrontal cortex (anterior cingulate cortex) have been found to suppress cocaine seeking in rats, suggesting that TrkB in the cingulate is critical for the suppression of appetitive behavior,[73] while nonspecific Trk inhibitor, K252a, blocks the suppressive effect of BDNF in the anterior cingulate.[74] Another report found that following fear conditioning, infusions of BDNF into the IL-enhanced cue-dependent extinction in rats, even without extinction training.[75] This BDNF effect is dependent on *N*-methyl D-aspartate (NMDA) receptors. This also provides evidence that TrkB signaling in the IL is critical for extinction learning. Thus, these data support the widely accepted idea explained above that there is a clear dissociation in the role of different mPFC areas in fear learning and memory processes, with IL involved in fear extinction and PL in fear acquisition. *TrkB*^{F616A} mutant mice[76] allow for pharmacogenetic studies with exogenous delivery of specific compounds (NM-PP1) that have high affinity binding to the mutant *TrkB*^{F616A} kinase domain, to potently inhibit TrkB signaling. Using these *TrkB*^{F616A} mice, we found that blockade of TrkB signaling immediately following fear conditioning can also lead to deficits in fear consolidation.[56] This data provide convincing evidence that PL TrkB downstream signaling is also necessary for modulating consolidation of fear memories.

There is a growing line of work focused on the role of BDNF in regulating stress-related learning and memory (see Ref. 77). One recent report demonstrated that a moderate stressor enhanced late reversal learning in a mouse choice task.[78] Most interestingly, stress facilitation of reversal learning was prevented by BDNF infusion into the vmPFC. Another report found that 3 weeks of exposure to high dose corticosterone in rats lead to impaired fear extinction to contextual freezing.[79] Moreover, cortical *BDNF* mRNA was decreased in the orbitofrontal cortex but not in mPFC, although this subarea has been extensively related to fear processes. A possible explanation of this lack of effect on the *BDNF* mRNA levels in the mPFC after chronic glucocorticoid exposure might be that stress differentially regulates PL and IL BDNF levels. Then, when analyzing BDNF across the entire mPFC, this differential effect may be masked. Additional specific studies are needed for a more complete understanding of the HPA regulation of BDNF and fear extinction processes, a cornerstone of posttraumatic stress disorder (PTSD).

3. THE ROLE OF BDNF IN HUMAN MEMORY AND TRANSLATIONAL STUDIES

The first evidence of a relationship between *BDNF* and cognitive processes in humans came from the seminal study by Egan and colleagues.[80] A single nucleotide polymorphism (SNP) at nucleotide 196 (G/A) producing an amino acid substitution (valine to methionine) at codon 66 (Val66Met SNP, rs6265) was shown to be related to poorer episodic memory and abnormal hippocampal activation measured with functional magnetic resonance imaging. Additionally, the authors showed in hippocampus cell culture that a methionine *BDNF* viral construct elicited reduced activity-dependent BDNF secretion, when compared to the valine *BDNF* construct, without alterations in the constitutive release. Concordantly, later studies found that Met carriers present impaired fear extinction,[81] declarative memory,[82] spatial learning,[83] working memory,[84] decreased emotional responses to rewarding stimuli,[85] poorer performance on the cognitive test of verbal recall,[86] and reduced hippocampal volume,[87,88] which is suggested to be independent of psychiatric status.[88] Moreover, mice engineered to carry a knock-in allele of the human Val66Met allele and cell cultures show reduced: neuronal BDNF availability, neuronal survival, hippocampal dendritic arborization, hippocampal volume, and LTP, which may result in the declarative memory and fear extinction deficits found in these mice.[81,89–96] However, there are discordant studies which suggest no effect of the Val66Met SNP in hippocampal volume,[97] acquisition and generalization of fear conditioning,[98] episodic memory,[99] and different memory neuropsychological tests.[97] These inconsistencies may be due to a number of variables including study design, age, gender, or ethnicity (see Dincheva and colleagues for a review[100]). In brief, it appears that there are more studies suggesting that nonclinical subjects carrying the Met allele present impaired memory functions, altered hippocampal volume and functioning (suggesting decreased synaptic plasticity), rather than no relationship. This is in agreement with the myriad of animal studies described above showing that BDNF is necessary for synaptic plasticity and memory processes.

Thus, BDNF signaling is required for memory function, and it is hypothesized that alterations in this system may be a susceptibility factor for developing and maintaining poor recovery during treatment across psychiatric disorders. On the contrary, optimal functioning of the BDNF–TrkB system may be an important protective factor against developing psychiatric disorders. Additionally, environmental factors, which are discussed below, such as physical exercise may also influence BDNF and TrkB signaling.

3.1. BDNF and the formation of memories in clinical psychiatry and animal models

BDNF has been associated with different psychiatric disorders including depression and schizophrenia.[101] Among learning and memory processes, fear memories are key in some anxiety disorders such as panic disorder (PD), phobia, and PTSD. Specifically relevant for these disorders is the fear memory consolidation phase following emotional learning, since it is required to stabilize the memory trace. Thus, understanding the molecular mechanisms of the formation of fear memory, environmental risk factors and resilience markers may help to prevent disease, develop new biological markers, and improve the treatment of these disorders.

PD is one of the most common anxiety disorders, and it is diagnosed when frequent and unexpected panic attacks occur. The typical attack involves intense fear or discomfort and is accompanied by symptoms such as palpitation, sweating, trembling or shacking, fear of losing control, and fear of dying. It has been postulated that the beginning and/or maintenance of PD is could partially be explained by a memory bias for threatening information.[102] Actually, patients with PD may present enhanced memory for words related to anxiety and PD (e.g., collapse).[102] The *BDNF* Val66Met SNP has been related to PD in one study[103] whereas other studies failed to find a relationship.[104–106] Again, demographics might highly influence these apparent contradictory results. Interestingly, a recent study suggested that a SNP (rs6502892) in the microRNA mIR-22 is associated with PD as well as reduced activity of the *BDNF* gene.[107] Of note, another study found increased serum BDNF levels in patients with PD that have significant improvement after cognitive behavioral therapy.[108] In brief, more studies are needed to further understand the molecular mechanisms of the altered fear memory process in PD where BDNF may play a role.

Phobia is a cue-specific activation of many of the same panic responses as PD. There are few studies addressing the relationship between phobias memory and BDNF. One study suggests that the *BDNF* polymorphism rs10835210 (but not any other *BDNF* SNP) is associated with phobia[109] and children carrying the met allele Val66Met SNP with social phobia present decreased gray volume matter in the amygdala and hippocampus.[110] Replication of these findings and new studies are needed to elucidate the putative relationship between BDNF and memory in phobia.

PTSD occurs in some individuals after experiencing or witnessing extreme traumatic events and presents a prevalence as high as 20–30% in highly traumatized individuals.[111] The symptoms of PTSD include reexperiencing memories of these traumatic events through intrusive thoughts, flashbacks, and nightmares. PTSD is also generally accompanied by hyperarousal symptoms. Moreover, persistent highly aversive memories related to the trauma, potentially over-consolidated memories, and the inability of these memories to be extinguished, are frequent characteristics of this disorder. Studies of genetic variants such as the *BDNF* Val66Met SNP may help to understand the mechanisms of individual differences in fear memory formation, contributing to an explanation of why some individuals develop PTSD whereas other remain resilient. A current hypothesis is that PTSD Met carriers (those with the less functional *BDNF* allele) may present higher risk of developing the disorder, worse symptomatology, and lower rates of recovery. However, there is only one study that has clearly shown to date an association of the Val66Met SNP with PTSD whereas two others find no relationship.[112–114] A possible reason of this discrepancy may be the limited power of the studies due to the low number of homozygous Met carriers in the population and demographic factors.[93,115] Another study shows that Met carriers of the Val66Met SNP in the *BDNF* gene present worse response to exposure treatment of PTSD.[116]

However, new studies with more participants would be desirable since there are other reports suggesting that BDNF is related to PTSD. For example, blood BDNF levels have been reported to be increased in serum shortly after trauma and decreasing over time.[117] Additional studies have found decreased in plasma of PTSD patients,[118] which may reflect a time-course effect. However, the interpretation of blood BDNF levels remains unclear because the relationship between BDNF in blood versus brain is unclear, although they may be correlated.[119]

Animal models of PTSD-like symptoms, which elicit enhanced anxiety and alterations in fear learning, also provide evidence of a role for BDNF. For example, rats previously submitted to single prolonged stress (SPS, exposure to 2 h restraint followed by 20 min of forced swimming and then anesthesia with isoflurane) present enhanced contextual freezing[120] and deficits in contextual and cued-fear extinction retention.[121] One of the molecular mechanisms underlying this phenotype may be the enhanced *BDNF* mRNA and protein levels in the hippocampus after contextual fear conditioning, followed by increased acetylation of histone H3 and H4.[122] Additionally, this PTSD-like paradigm elicits increased calmodulin (CaM) and decreased BDNF-activated CaMKII in the BLA.[123] Thus, traumatic stress exposure might be causing epigenetic changes in the BDNF gene, enhanced BDNF protein levels, and increased TrkB activation and downstream signaling, which may contribute to increased consolidation of new fear memories. Interestingly, there is another PTSD-like model, immobilization stress, which elicits PTSD-like symptoms including impaired cued-fear extinction, resembling to what occurs in PTSD.[50,53,124] This deficit in fear extinction is rescued by a single dose of 7,8-DHF, suggesting that TrkB activation during extinction learning may provide a successful way to enhance psychotherapies based on fear extinction.[50]

4. POTENTIAL THERAPEUTIC USES OF BDNF

Among the studies on BDNF–TrkB signaling and addiction, most of them focus on the regulation of this pathway in reward, which is beyond the scope of this chapter. However, specific activation of BDNF–TrkB, accompanied by other therapeutic strategies such as psychotherapy and pharmacology, may help to treat comorbidities such as addictions. Additionally, enhancing the BDNF–TrkB pathway may be attractive approach to improve the treatment of fear extinction disorders.[125]

Recombinant BDNF has been used extensively in rodents studies, but negative results have been reported in clinical trials,[126] probably due to the poor pharmacokinetic properties of BDNF and/or the fact that BDNF also binds to p75, which might decrease availability of BDNF to activate TrkB.[127,128] Additionally, there are no TrkB agonists approved by the food and drug administration, so it is not yet possible to use these compounds in humans. Besides 7,8-DHF (Table 6.1), there are other TrkB agonists, although to date, we are not aware if they have been tested yet in animal models of anxiety disorders with altered fear learning. 4′-dimethylamino-7,8-dihydroxyflavone appears as a new synthetic derivative of 7,8-DHF and displays more potent TrkB stimulatory effect than 7,8-DHF when given both by parenteral administration.[129] Moreover, both compounds present antidepressant actions.[129] Also a flavone-derivative, 7,8,3′-trihydroxyflavone, appears to act as a TrkB agonist and rescues degeneration of spiral ganglion neurons *in vitro* and *in vivo*.[130] LM22A-4 is another potential TrkB agonist, which when chronically administered intranasally in rodents crosses the blood–brain barrier and activates its downstream signaling.[131] Thus, more research is needed to study the potential beneficial use of these and other putative TrkB agonists along with establishment of their safety profiles, especially when chronically administered. However, there may be indirect ways of enhancing BDNF levels, with potential beneficial effects for synaptic plasticity and memory, as described below.

4.1. BDNF and the treatment of psychiatric disorders

Although we lack compounds that are approved for clinical use that elicit an effective direct augmentation of BDNF–TrkB signaling, there are indirect and potentially therapeutic ways to accomplish this. A study by Karpova and colleagues showed that the SSRI fluoxetine, given chronically, enhances BDNF and synaptic plasticity in the amygdala, which elicits augmentation of fear extinction in wild-type mice.[132] According to this study, it would be reasonable to hypothesize that antidepressants may enhance fear extinction in phobias, PD, and PTSD. Notably, antidepressants alone are the most common treatment for these fear disorders. Unfortunately, there is no clear evidence that antidepressants alone or in combination with psychotherapy are more effective than placebo or no treatment.[133] A possible explanation and mechanism of this lack of effect comes from a study where the knock-in humanized mice carrying the *BDNF* Val66Met SNP were given fluoxetine chronically.[134] Above it was explained that the *BDNF* homozygous Met/Met mice have reduced BDNF, synaptic plasticity, and impaired memory functions. Thus, in this study,[134] fluoxetine treatment increased LTP and BDNF protein in the hippocampus of *BDNF* (Val/Val) mice, in accordance with the study by Karpova and colleagues,[132] whereas these changes did not occur in *BDNF* (Met/Met) mice. Moreover, there is a very recent paper which suggests that antidepressants can impair fear extinction.[135] Thus, it is reasonable to hypothesize that anxiety disorders with altered fear learning are accompanied by BDNF signaling alterations, which cannot be rescued with antidepressants. Besides antidepressants, other drugs have shown drug-induced increases in BDNF, which may provide therapeutic effects. For example, the NMDA receptor partial agonist D-cycloserine has been shown to effectively enhance fear learning, in both clinical populations and animal models, possibly in part through BDNF modulation.[133] Also, it has been suggested that endocannabinoids released as retrograde messengers, induce postsynaptic TrkB activation, which may modulate the consolidation of extinction memories.[133,136]

Physical exercise in rats has been shown to enhance proBDNF, BDNF, and tPA (a serine proteinase which facilitates proBDNF cleavage into BDNF) in the hippocampus.[137] Moreover, tPa blockade reduced TrkB activation and its downstream signaling.[137] Also, *BDNF* mRNA levels are increased in the dentate gyrus, a region within the hippocampus, which is accompanied by enhanced contextual freezing and context discrimination.[138] Thus, rats that had physical exercise were more able to discriminate between a safe environment versus the environment were they received electric shocks.

These findings may have a potential translational interest since PTSD may be caused, in part, by a lack of an inability to inhibit learned fear during conditions of safety. So, it is possible that enhancing endogenous BDNF levels may increase the ability of these patients to inhibit fear in safety conditions which may improve their symptoms.

5. CONCLUSIONS

Overall, there has been great progress in furthering the molecular understanding of BDNF–TrkB signaling, as well as their roles in differentially modulating plasticity across neural circuits. However, much remains to be answered, and the elucidation of the molecular-, cellular-, and circuit-based processes underlying synaptic plasticity, with BDNF–TrkB playing a central role, is critical for appreciation of the brain mechanisms of psychiatric disorders and comorbidities.

ACKNOWLEDGMENTS AND DISCLOSURES

This work was supported through NIH (MH071537, MH096764) and from the NIH/National Center for Research Resources base Grant P51RR000165 to Yerkes National Primate Research Center. K. J. R. is a founding member of Extinction Pharmaceuticals/Therapade Technologies, which have provided no equity or income, but which exist to develop D-cycloserine for use to augment the effectiveness of psychotherapy.

REFERENCES

1. Chao MV. Neurotrophins and their receptors: a convergence point for many signalling pathways. *Nat Rev Neurosci.* 2003;4(4):299–309.
2. McAllister AK, Katz LC, Lo DC. Neurotrophins and synaptic plasticity. *Annu Rev Neurosci.* 1999;22:295–318.
3. Poo MM. Neurotrophins as synaptic modulators. *Nat Rev Neurosci.* 2001;2(1):24–32.
4. Barbacid M. Structural and functional properties of the TRK family of neurotrophin receptors. *Ann N Y Acad Sci.* 1995;766:442–458.
5. Maisonpierre PC, Le Beau MM, Espinosa 3rd R, et al. Human and rat brain-derived neurotrophic factor and neurotrophin-3: gene structures, distributions, and chromosomal localizations. *Genomics.* 1991;10(3):558–568.
6. Cartwright M, Mikheev AM, Heinrich G. Expression of neurotrophin genes in human fibroblasts: differential regulation of the brain-derived neurotrophic factor gene. *Int J Dev Neurosci.* 1994;12(8):685–693.
7. Moretto G, Xu RY, Walker DG, Kim SU. Co-expression of mRNA for neurotrophic factors in human neurons and glial cells in culture. *J Neuropathol Exp Neurol.* 1994;53(1):78–85.
8. Wetmore C, Olson L. Neuronal and nonneuronal expression of neurotrophins and their receptors in sensory and sympathetic ganglia suggest new intercellular trophic interactions. *J Comp Neurol.* 1995;353(1):143–159.
9. Barakat-Walter I. Brain-derived neurotrophic factor-like immunoreactivity is localized mainly in small sensory neurons of rat dorsal root ganglia. *J Neurosci Methods.* 1996;68(2):281–288.
10. Chao MV, Hempstead BL. p75 and Trk: a two-receptor system. *Trends Neurosci.* 1995;18(7):321–326.
11. Dechant G, Barde YA. The neurotrophin receptor p75(NTR): novel functions and implications for diseases of the nervous system. *Nat Neurosci.* 2002;5(11):1131–1136.
12. Reichardt LF. Neurotrophin-regulated signalling pathways. *Philos Trans R Soc Lond B Biol Sci.* 2006;361(1473):1545–1564.
13. Kaplan DR, Miller FD. Neurotrophin signal transduction in the nervous system. *Curr Opin Neurobiol.* 2000;10(3):381–391.
14. Minichiello L. TrkB signalling pathways in LTP and learning. *Nat Rev Neurosci.* 2009;10(12):850–860.
15. Xu B, Gottschalk W, Chow A, et al. The role of brain-derived neurotrophic factor receptors in the mature hippocampus: modulation of long-term potentiation through a presynaptic mechanism involving TrkB. *J Neurosci.* 2000;20(18):6888–6897.
16. Korte M, Carroll P, Wolf E, Brem G, Thoenen H, Bonhoeffer T. Hippocampal long-term potentiation is impaired in mice lacking brain-derived neurotrophic factor. *Proc Natl Acad Sci U S A.* 1995;92(19):8856–8860.

17. Patterson SL, Abel T, Deuel TA, Martin KC, Rose JC, Kandel ER. Recombinant BDNF rescues deficits in basal synaptic transmission and hippocampal LTP in BDNF knockout mice. *Neuron*. 1996;16(6):1137–1145.
18. Lohof AM, Ip NY, Poo MM. Potentiation of developing neuromuscular synapses by the neurotrophins NT-3 and BDNF. *Nature*. 1993;363(6427):350–353.
19. Kang H, Schuman EM. Long-lasting neurotrophin-induced enhancement of synaptic transmission in the adult hippocampus. *Science*. 1995;267(5204):1658–1662.
20. Levine ES, Dreyfus CF, Black IB, Plummer MR. Brain-derived neurotrophic factor rapidly enhances synaptic transmission in hippocampal neurons via postsynaptic tyrosine kinase receptors. *Proc Natl Acad Sci U S A*. 1995;92(17):8074–8077.
21. Korte M, Griesbeck O, Gravel C, et al. Virus-mediated gene transfer into hippocampal CA1 region restores long-term potentiation in brain-derived neurotrophic factor mutant mice. *Proc Natl Acad Sci U S A*. 1996;93(22):12547–12552.
22. Minichiello L, Korte M, Wolfer D, et al. Essential role for TrkB receptors in hippocampus-mediated learning. *Neuron*. 1999;24(2):401–414.
23. Minichiello L, Calella AM, Medina DL, Bonhoeffer T, Klein R, Korte M. Mechanism of TrkB-mediated hippocampal long-term potentiation. *Neuron*. 2002;36(1):121–137.
24. Peterson DA, Dickinson-Anson HA, Leppert JT, Lee KF, Gage FH. Central neuronal loss and behavioral impairment in mice lacking neurotrophin receptor p75. *J Comp Neurol*. 1999;404(1):1–20.
25. von Schack D, Casademunt E, Schweigreiter R, Meyer M, Bibel M, Dechant G. Complete ablation of the neurotrophin receptor p75NTR causes defects both in the nervous and the vascular system. *Nat Neurosci*. 2001;4(10):977–978.
26. Segal M, Richter-Levin G, Maggio N. Stress-induced dynamic routing of hippocampal connectivity: a hypothesis. *Hippocampus*. 2010;20(12):1332–1338.
27. Zola-Morgan SM, Squire LR. The primate hippocampal formation: evidence for a time-limited role in memory storage. *Science*. 1990;250(4978):288–290.
28. Lee JL, Everitt BJ, Thomas KL. Independent cellular processes for hippocampal memory consolidation and reconsolidation. *Science*. 2004;304(5672):839–843.
29. Bekinschtein P, Cammarota M, Igaz LM, Bevilaqua LR, Izquierdo I, Medina JH. Persistence of long-term memory storage requires a late protein synthesis- and BDNF-dependent phase in the hippocampus. *Neuron*. 2007;53(2):261–277.
30. Tyler WJ, Alonso M, Bramham CR, Pozzo-Miller LD. From acquisition to consolidation: on the role of brain-derived neurotrophic factor signaling in hippocampal-dependent learning. *Learn Mem*. 2002;9(5):224–237.
31. Heldt SA, Stanek L, Chhatwal JP, Ressler KJ. Hippocampus-specific deletion of BDNF in adult mice impairs spatial memory and extinction of aversive memories. *Mol Psychiatry*. 2007;12(7):656–670.
32. Cirulli F, Berry A, Chiarotti F, Alleva E. Intrahippocampal administration of BDNF in adult rats affects short-term behavioral plasticity in the Morris water maze and performance in the elevated plus-maze. *Hippocampus*. 2004;14(7):802–807.
33. Koponen E, Voikar V, Riekki R, et al. Transgenic mice overexpressing the full-length neurotrophin receptor trkB exhibit increased activation of the trkB-PLCgamma pathway, reduced anxiety, and facilitated learning. *Mol Cell Neurosci*. 2004;26(1):166–181.
34. Hall J, Thomas KL, Everitt BJ. Rapid and selective induction of BDNF expression in the hippocampus during contextual learning. *Nat Neurosci*. 2000;3(6):533–535.
35. Liu IY, Lyons WE, Mamounas LA, Thompson RF. Brain-derived neurotrophic factor plays a critical role in contextual fear conditioning. *J Neurosci*. 2004;24(36):7958–7963.
36. Chen J, Kitanishi T, Ikeda T, Matsuki N, Yamada MK. Contextual learning induces an increase in the number of hippocampal CA1 neurons expressing high levels of BDNF. *Neurobiol Learn Mem*. 2007;88(4):409–415.

37. Mizuno K, Dempster E, Mill J, Giese KP. Long-lasting regulation of hippocampal Bdnf gene transcription after contextual fear conditioning. *Genes Brain Behav*. 2012;11(6):651–659.
38. Lubin FD, Roth TL, Sweatt JD. Epigenetic regulation of BDNF gene transcription in the consolidation of fear memory. *J Neurosci*. 2008;28(42):10576–10586.
39. Gupta S, Kim SY, Artis S, et al. Histone methylation regulates memory formation. *J Neurosci*. 2010;30(10):3589–3599.
40. Johansen JP, Cain CK, Ostroff LE, LeDoux JE. Molecular mechanisms of fear learning and memory. *Cell*. 2011;147(3):509–524.
41. Rattiner LM, Davis M, French CT, Ressler KJ. Brain-derived neurotrophic factor and tyrosine kinase receptor B involvement in amygdala-dependent fear conditioning. *J Neurosci*. 2004;24(20):4796–4806.
42. Rattiner LM, Davis M, Ressler KJ. Differential regulation of brain-derived neurotrophic factor transcripts during the consolidation of fear learning. *Learn Mem*. 2004;11(6):727–731.
43. Yee BK, Zhu SW, Mohammed AH, Feldon J. Levels of neurotrophic factors in the hippocampus and amygdala correlate with anxiety- and fear-related behaviour in C57BL6 mice. *J Neural Transm*. 2007;114(4):431–444.
44. Psotta L, Lessmann V, Endres T. Impaired fear extinction learning in adult heterozygous BDNF knock-out mice. *Neurobiol Learn Mem*. 2013;103:34–38.
45. Jang SW, Liu X, Chan CB, et al. Deoxygedunin, a natural product with potent neurotrophic activity in mice. *PLoS One*. 2010;5(7):e11528.
46. Li C, Dabrowska J, Hazra R, Rainnie DG. Synergistic activation of dopamine D1 and TrkB receptors mediate gain control of synaptic plasticity in the basolateral amygdala. *PLoS One*. 2011;6(10):e26065.
47. Ou LC, Gean PW. Regulation of amygdala-dependent learning by brain-derived neurotrophic factor is mediated by extracellular signal-regulated kinase and phosphatidylinositol-3-kinase. *Neuropsychopharmacology*. 2006;31(2):287–296.
48. Schafe GE, Atkins CM, Swank MW, Bauer EP, Sweatt JD, LeDoux JE. Activation of ERK/MAP kinase in the amygdala is required for memory consolidation of pavlovian fear conditioning. *J Neurosci*. 2000;20(21):8177–8187.
49. Chhatwal JP, Stanek-Rattiner L, Davis M, Ressler KJ. Amygdala BDNF signaling is required for consolidation but not encoding of extinction. *Nat Neurosci*. 2006;9(7):870–872.
50. Andero R, Heldt SA, Ye K, Liu X, Armario A, Ressler KJ. Effect of 7,8-dihydroxyflavone, a small-molecule TrkB agonist, on emotional learning. *Am J Psychiatry*. 2011;168(2):163–172.
51. Choi DC, Maguschak KA, Ye K, Jang SW, Myers KM, Ressler KJ. Prelimbic cortical BDNF is required for memory of learned fear but not extinction or innate fear. *Proc Natl Acad Sci U S A*. 2010;107(6):2675–2680.
52. Zeng Y, Tan M, Kohyama J, et al. Epigenetic enhancement of BDNF signaling rescues synaptic plasticity in aging. *J Neurosci*. 2011;31(49):17800–17810.
53. Andero R, Daviu N, Escorihuela RM, Nadal R, Armario A. 7,8-Dihydroxyflavone, a TrkB receptor agonist, blocks long-term spatial memory impairment caused by immobilization stress in rats. *Hippocampus*. 2012;22(3):399–408.
54. Chen J, Chua KW, Chua CC, et al. Antioxidant activity of 7,8-dihydroxyflavone provides neuroprotection against glutamate-induced toxicity. *Neurosci Lett*. 2011;499(3):181–185.
55. Devi L, Ohno M. 7,8-Dihydroxyflavone, a small-molecule TrkB agonist, reverses memory deficits and BACE1 elevation in a mouse model of Alzheimer's disease. *Neuropsychopharmacology*. 2012;37(2):434–444.

56. Choi DC, Gourley SL, Ressler KJ. Prelimbic BDNF and TrkB signaling regulates consolidation of both appetitive and aversive emotional learning. *Transl Psychiatry*. 2012;2: e205.
57. Zeng Y, Lv F, Li L, Yu H, Dong M, Fu Q. 7,8-dihydroxyflavone rescues spatial memory and synaptic plasticity in cognitively impaired aged rats. *J Neurochem*. 2012;122(4):800–811.
58. Zeng Y, Liu Y, Wu M, Liu J, Hu Q. Activation of TrkB by 7,8-dihydroxyflavone prevents fear memory defects and facilitates amygdalar synaptic plasticity in aging. *J Alzheimers Dis*. 2012;31(4):765–778.
59. Baker-Andresen D, Flavell CR, Li X, Bredy TW. Activation of BDNF signaling prevents the return of fear in female mice. *Learn Mem*. 2013;20(5):237–240.
60. Heldt SA, Ressler KJ. Training-induced changes in the expression of GABAA-associated genes in the amygdala after the acquisition and extinction of Pavlovian fear. *Eur J Neurosci*. 2007;26(12):3631–3644.
61. Mou L, Heldt SA, Ressler KJ. Rapid brain-derived neurotrophic factor-dependent sequestration of amygdala and hippocampal GABA(A) receptors via different tyrosine receptor kinase B-mediated phosphorylation pathways. *Neuroscience*. 2011;176:72–85.
62. Vertes RP. Interactions among the medial prefrontal cortex, hippocampus and midline thalamus in emotional and cognitive processing in the rat. *Neuroscience*. 2006;142(1):1–20.
63. Vertes RP. Differential projections of the infralimbic and prelimbic cortex in the rat. *Synapse*. 2004;51(1):32–58.
64. Peters J, Kalivas PW, Quirk GJ. Extinction circuits for fear and addiction overlap in prefrontal cortex. *Learn Mem*. 2009;16(5):279–288.
65. McGinty JF, Whitfield Jr TW, Berglind WJ. Brain-derived neurotrophic factor and cocaine addiction. *Brain Res*. 2010;1314:183–193.
66. Musumeci G, Minichiello L. BDNF-TrkB signalling in fear learning: from genetics to neural networks. *Rev Neurosci*. 2011;22(3):303–315.
67. Bredy TW, Wu H, Crego C, Zellhoefer J, Sun YE, Barad M. Histone modifications around individual BDNF gene promoters in prefrontal cortex are associated with extinction of conditioned fear. *Learn Mem*. 2007;14(4):268–276.
68. Lu H, Cheng PL, Lim BK, Khoshnevisrad N, Poo MM. Elevated BDNF after cocaine withdrawal facilitates LTP in medial prefrontal cortex by suppressing GABA inhibition. *Neuron*. 2010;67(5):821–833.
69. Ma L, Wang DD, Zhang TY, et al. Region-specific involvement of BDNF secretion and synthesis in conditioned taste aversion memory formation. *J Neurosci*. 2011;31(6):2079–2090.
70. Wang WS, Kang S, Liu WT, et al. Extinction of aversive memories associated with morphine withdrawal requires ERK-mediated epigenetic regulation of brain-derived neurotrophic factor transcription in the rat ventromedial prefrontal cortex. *J Neurosci*. 2012;32(40):13763–13775.
71. Rios M, Fan G, Fekete C, et al. Conditional deletion of brain-derived neurotrophic factor in the postnatal brain leads to obesity and hyperactivity. *Mol Endocrinol*. 2001;15(10):1748–1757.
72. Gourley SL, Howell JL, Rios M, DiLeone RJ, Taylor JR. Prelimbic cortex bdnf knock-down reduces instrumental responding in extinction. *Learn Mem*. 2009;16(12):756–760.
73. Berglind WJ, See RE, Fuchs RA, et al. A BDNF infusion into the medial prefrontal cortex suppresses cocaine seeking in rats. *Eur J Neurosci*. 2007;26(3):757–766.
74. Whitfield Jr TW, Shi X, Sun WL, McGinty JF. The suppressive effect of an intra-prefrontal cortical infusion of BDNF on cocaine-seeking is Trk receptor and

extracellular signal-regulated protein kinase mitogen-activated protein kinase dependent. *J Neurosci.* 2011;31(3):834–842.
75. Peters J, Dieppa-Perea LM, Melendez LM, Quirk GJ. Induction of fear extinction with hippocampal-infralimbic BDNF. *Science.* 2010;328(5983):1288–1290.
76. Chen X, Ye H, Kuruvilla R, et al. A chemical-genetic approach to studying neurotrophin signaling. *Neuron.* 2005;46(1):13–21.
77. Gray JD, Milner TA, McEwen BS. Dynamic plasticity: the role of glucocorticoids, brain-derived neurotrophic factor and other trophic factors. *Neuroscience.* 2013;239:214–227.
78. Graybeal C, Feyder M, Schulman E, et al. Paradoxical reversal learning enhancement by stress or prefrontal cortical damage: rescue with BDNF. *Nat Neurosci.* 2011;14(12):1507–1509.
79. Gourley SL, Kedves AT, Olausson P, Taylor JR. A history of corticosterone exposure regulates fear extinction and cortical NR2B, GluR2/3, and BDNF. *Neuropsychopharmacology.* 2009;34(3):707–716.
80. Egan MF, Kojima M, Callicott JH, et al. The BDNF val66met polymorphism affects activity-dependent secretion of BDNF and human memory and hippocampal function. *Cell.* 2003;112(2):257–269.
81. Soliman F, Glatt CE, Bath KG, et al. A genetic variant BDNF polymorphism alters extinction learning in both mouse and human. *Science.* 2010;327(5967):863–866.
82. Hariri AR, Goldberg TE, Mattay VS, et al. Brain-derived neurotrophic factor val66met polymorphism affects human memory-related hippocampal activity and predicts memory performance. *J Neurosci.* 2003;23(17):6690–6694.
83. Banner H, Bhat V, Etchamendy N, Joober R, Bohbot VD. The brain-derived neurotrophic factor Val66Met polymorphism is associated with reduced functional magnetic resonance imaging activity in the hippocampus and increased use of caudate nucleus-dependent strategies in a human virtual navigation task. *Eur J Neurosci.* 2011;33(5):968–977.
84. Cerasa A, Tongiorgi E, Fera F, et al. The effects of BDNF Val66Met polymorphism on brain function in controls and patients with multiple sclerosis: an imaging genetic study. *Behav Brain Res.* 2010;207(2):377–386.
85. Gasic GP, Smoller JW, Perlis RH, et al. BDNF, relative preference, and reward circuitry responses to emotional communication. *Am J Med Genet B Neuropsychiatr Genet.* 2009;150B(6):762–781.
86. Schofield PR, Williams LM, Paul RH, et al. Disturbances in selective information processing associated with the BDNF Val66Met polymorphism: evidence from cognition, the P300 and fronto-hippocampal systems. *Biol Psychol.* 2009;80(2):176–188.
87. Bueller JA, Aftab M, Sen S, Gomez-Hassan D, Burmeister M, Zubieta JK. BDNF Val66Met allele is associated with reduced hippocampal volume in healthy subjects. *Biol Psychiatry.* 2006;59(9):812–815.
88. Molendijk ML, van Tol MJ, Penninx BW, et al. BDNF val66met affects hippocampal volume and emotion-related hippocampal memory activity. *Transl Psychiatry.* 2012;2:e74.
89. Chen ZY, Patel PD, Sant G, et al. Variant brain-derived neurotrophic factor (BDNF) (Met66) alters the intracellular trafficking and activity-dependent secretion of wild-type BDNF in neurosecretory cells and cortical neurons. *J Neurosci.* 2004;24(18):4401–4411.
90. Chen ZY, Ieraci A, Teng H, et al. Sortilin controls intracellular sorting of brain-derived neurotrophic factor to the regulated secretory pathway. *J Neurosci.* 2005;25(26):6156–6166.
91. Chen ZY, Jing D, Bath KG, et al. Genetic variant BDNF (Val66Met) polymorphism alters anxiety-related behavior. *Science.* 2006;314(5796):140–143.

92. Cao L, Dhilla A, Mukai J, et al. Genetic modulation of BDNF signaling affects the outcome of axonal competition in vivo. *Curr Biol.* 2007;17(11):911–921.
93. Frielingsdorf H, Bath KG, Soliman F, Difede J, Casey BJ, Lee FS. Variant brain-derived neurotrophic factor Val66Met endophenotypes: implications for posttraumatic stress disorder. *Ann N Y Acad Sci.* 2010;1208:150–157.
94. Ninan I, Bath KG, Dagar K, et al. The BDNF Val66Met polymorphism impairs NMDA receptor-dependent synaptic plasticity in the hippocampus. *J Neurosci.* 2010;30(26):8866–8870.
95. Spencer JL, Waters EM, Milner TA, Lee FS, McEwen BS. BDNF variant Val66Met interacts with estrous cycle in the control of hippocampal function. *Proc Natl Acad Sci U S A.* 2010;107(9):4395–4400.
96. Yu H, Wang Y, Pattwell S, et al. Variant BDNF Val66Met polymorphism affects extinction of conditioned aversive memory. *J Neurosci.* 2009;29(13):4056–4064.
97. Karnik MS, Wang L, Barch DM, Morris JC, Csernansky JG. BDNF polymorphism rs6265 and hippocampal structure and memory performance in healthy control subjects. *Psychiatry Res.* 2010;178(2):425–429.
98. Torrents-Rodas D, Fullana MA, Arias B, et al. Acquisition and generalization of fear conditioning are not modulated by the BDNF-val66met polymorphism in humans. *Psychophysiology.* 2012;49(5):713–719.
99. Dennis NA, Cabeza R, Need AC, Waters-Metenier S, Goldstein DB, LaBar KS. Brain-derived neurotrophic factor val66met polymorphism and hippocampal activation during episodic encoding and retrieval tasks. *Hippocampus.* 2011;21(9): 980–989.
100. Dincheva I, Glatt CE, Lee FS. Impact of the BDNF Val66Met polymorphism on cognition: implications for behavioral genetics. *Neuroscientist.* 2012;18(5):439–451.
101. Autry AE, Monteggia LM. Brain-derived neurotrophic factor and neuropsychiatric disorders. *Pharmacol Rev.* 2012;64(2):238–258.
102. McNally RJ. Memory and anxiety disorders. *Philos Trans R Soc Lond B Biol Sci.* 1997;352(1362):1755–1759.
103. Ishii T, Akiyoshi J, Hanada H, et al. Association between the obestatin and BDNF gene polymorphism and panic disorder, and depressive disorder. *Psychiatr Genet.* 2009;19(3):159.
104. Lam P, Cheng CY, Hong CJ, Tsai SJ. Association study of a brain-derived neurotrophic factor (Val66Met) genetic polymorphism and panic disorder. *Neuropsychobiology.* 2004;49(4):178–181.
105. Shimizu E, Hashimoto K, Koizumi H, et al. No association of the brain-derived neurotrophic factor (BDNF) gene polymorphisms with panic disorder. *Prog Neuropsychopharmacol Biol Psychiatry.* 2005;29(5):708–712.
106. Otowa T, Shimada T, Kawamura Y, et al. No association between the brain-derived neurotrophic factor gene and panic disorder in Japanese population. *J Hum Genet.* 2009;54(8):437–439.
107. Muinos-Gimeno M, Espinosa-Parrilla Y, Guidi M, et al. Human microRNAs miR-22, miR-138-2, miR-148a, and miR-488 are associated with panic disorder and regulate several anxiety candidate genes and related pathways. *Biol Psychiatry.* 2011;69(6): 526–533.
108. Kobayashi K, Shimizu E, Hashimoto K, et al. Serum brain-derived neurotrophic factor (BDNF) levels in patients with panic disorder: as a biological predictor of response to group cognitive behavioral therapy. *Prog Neuropsychopharmacol Biol Psychiatry.* 2005;29(5):658–663.
109. Xie B, Wang B, Suo P, et al. Genetic association between BDNF gene polymorphisms and phobic disorders: a case–control study among mainland Han Chinese. *J Affect Disord.* 2011;132(1–2):239–242.

110. Mueller SC, Aouidad A, Gorodetsky E, Goldman D, Pine DS, Ernst M. Gray matter volume in adolescent anxiety: an impact of the brain-derived neurotrophic factor Val(66)Met polymorphism? *J Am Acad Child Adolesc Psychiatry*. 2013;52(2):184–195.
111. Breslau N, Kessler RC, Chilcoat HD, Schultz LR, Davis GC, Andreski P. Trauma and posttraumatic stress disorder in the community: the 1996 Detroit Area Survey of Trauma. *Arch Gen Psychiatry*. 1998;55(7):626–632.
112. Valente NL, Vallada H, Cordeiro Q, et al. Candidate-gene approach in posttraumatic stress disorder after urban violence: association analysis of the genes encoding serotonin transporter, dopamine transporter, and BDNF. *J Mol Neurosci*. 2011;44(1):59–67.
113. Zhang H, Ozbay F, Lappalainen J, et al. Brain derived neurotrophic factor (BDNF) gene variants and Alzheimer's disease, affective disorders, posttraumatic stress disorder, schizophrenia, and substance dependence. *Am J Med Genet B Neuropsychiatr Genet*. 2006;141B(4):387–393.
114. Zhang L, Benedek DM, Fullerton CS, et al. PTSD risk is associated with BDNF Val66Met and BDNF overexpression. *Mol Psychiatry*. 15 January 2013 in press; Epub ahead of Print.
115. Rakofsky JJ, Ressler KJ, Dunlop BW. BDNF function as a potential mediator of bipolar disorder and post-traumatic stress disorder comorbidity. *Mol Psychiatry*. 2012;17(1):22–35.
116. Felmingham KL, Dobson-Stone C, Schofield PR, Quirk GJ, Bryant RA. The brain-derived neurotrophic factor Val66Met polymorphism predicts response to exposure therapy in posttraumatic stress disorder. *Biol Psychiatry*. 2013;73(11):1059–1063.
117. Hauck S, Kapczinski F, Roesler R, et al. Serum brain-derived neurotrophic factor in patients with trauma psychopathology. *Prog Neuropsychopharmacol Biol Psychiatry*. 2010;34(3):459–462.
118. Dell'osso L, Carmassi C, Del Debbio A, et al. Brain-derived neurotrophic factor plasma levels in patients suffering from post-traumatic stress disorder. *Prog Neuropsychopharmacol Biol Psychiatry*. 2009;33(5):899–902.
119. Karege F, Schwald M, Cisse M. Postnatal developmental profile of brain-derived neurotrophic factor in rat brain and platelets. *Neurosci Lett*. 2002;328(3):261–264.
120. Takahashi T, Morinobu S, Iwamoto Y, Yamawaki S. Effect of paroxetine on enhanced contextual fear induced by single prolonged stress in rats. *Psychopharmacology (Berl)*. 2006;189(2):165–173.
121. Knox D, George SA, Fitzpatrick CJ, Rabinak CA, Maren S, Liberzon I. Single prolonged stress disrupts retention of extinguished fear in rats. *Learn Mem*. 2012;19(2):43–49.
122. Takei S, Morinobu S, Yamamoto S, Fuchikami M, Matsumoto T, Yamawaki S. Enhanced hippocampal BDNF/TrkB signaling in response to fear conditioning in an animal model of posttraumatic stress disorder. *J Psychiatr Res*. 2011;45(4):460–468.
123. Xiao B, Han F, Shi YX. Dysfunction of Ca2+/CaM kinase IIalpha cascades in the amygdala in post-traumatic stress disorder. *Int J Mol Med*. 2009;24(6):795–799.
124. Armario A, Escorihuela RM, Nadal R. Long-term neuroendocrine and behavioural effects of a single exposure to stress in adult animals. *Neurosci Biobehav Rev*. 2008;32(6):1121–1135.
125. Monteggia LM. Toward neurotrophin-based therapeutics. *Am J Psychiatry*. 2011;168(2):114–116.
126. Wu D. Neuroprotection in experimental stroke with targeted neurotrophins. *NeuroRx*. 2005;2(1):120–128.
127. Webster NJ, Pirrung MC. Small molecule activators of the Trk receptors for neuroprotection. *BMC Neurosci*. 2008;9(Suppl. 2):S1.
128. Price RD, Milne SA, Sharkey J, Matsuoka N. Advances in small molecules promoting neurotrophic function. *Pharmacol Ther*. 2007;115(2):292–306.

129. Liu X, Chan CB, Jang SW, et al. A synthetic 7,8-dihydroxyflavone derivative promotes neurogenesis and exhibits potent antidepressant effect. *J Med Chem*. 2010; 53(23): 8274–8286.
130. Yu Q, Chang Q, Liu X, Gong S, Ye K, Lin X. 7,8,3′-Trihydroxyflavone, a potent small molecule TrkB receptor agonist, protects spiral ganglion neurons from degeneration both in vitro and in vivo. *Biochem Biophys Res Commun*. 2012;422(3):387–392.
131. Massa SM, Yang T, Xie Y, et al. Small molecule BDNF mimetics activate TrkB signaling and prevent neuronal degeneration in rodents. *J Clin Invest*. 2010;120(5): 1774–1785.
132. Karpova NN, Pickenhagen A, Lindholm J, et al. Fear erasure in mice requires synergy between antidepressant drugs and extinction training. *Science*. 2011;334(6063): 1731–1734.
133. Andero R, Ressler KJ. Fear extinction and BDNF: translating animal models of PTSD to the clinic. *Genes Brain Behav*. 2012;11(5):503–512.
134. Bath KG, Jing DQ, Dincheva I, et al. BDNF Val66Met impairs fluoxetine-induced enhancement of adult hippocampus plasticity. *Neuropsychopharmacology*. 2012;37(5):1297–1304.
135. Burghardt NS, Sigurdsson T, Gorman JM, McEwen BS, Ledoux JE. Chronic antidepressant treatment impairs the acquisition of fear extinction. *Biol Psychiatry*. 2013;73(11):1078–1086.
136. Lemtiri-Chlieh F, Levine ES. BDNF evokes release of endogenous cannabinoids at layer 2/3 inhibitory synapses in the neocortex. *J Neurophysiol*. 2010;104(4):1923–1932.
137. Ding Q, Ying Z, Gomez-Pinilla F. Exercise influences hippocampal plasticity by modulating brain-derived neurotrophic factor processing. *Neuroscience*. 2011;192:773–780.
138. Greenwood BN, Strong PV, Foley TE, Fleshner M. A behavioral analysis of the impact of voluntary physical activity on hippocampus-dependent contextual conditioning. *Hippocampus*. 2009;19(10):988–1001.

CHAPTER SEVEN

Mechanisms of Synaptic Plasticity and Recognition Memory in the Perirhinal Cortex

P.J. Banks, E.C. Warburton, M.W. Brown, Z.I. Bashir
School of Physiology and Pharmacology, University of Bristol, Bristol, United Kingdom

Contents

Abstract

Learning is widely believed to involve synaptic plasticity, employing mechanisms such as those used in long-term potentiation (LTP) and long-term depression (LTD). In this chapter, we will review work on mechanisms of synaptic plasticity in perirhinal cortex *in vitro* and relate these findings to studies underlying recognition memory *in vivo*. We describe how antagonism of different glutamate and acetylcholine receptors, inhibition of nitric oxide synthase, inhibition of CREB phosphorylation, and interfering with glutamate AMPA receptor internalization can produce deficits in synaptic plasticity *in vitro*. Inhibition of each of these different mechanisms *in vivo* also results in recognition memory deficits. Therefore, we provide strong evidence that synaptic plastic mechanisms are necessary for the information processing and storage that underlies object recognition memory.

Progress in Molecular Biology and Translational Science, Volume 122
ISSN 1877-1173
http://dx.doi.org/10.1016/B978-0-12-420170-5.00007-6

ABBREVIATIONS

AMPA α amino 3 hydroxy 5 methyl 4 isoxazolepropionic acid receptor
CNQX 6-cyano-7-nitroquinoxaline-2,3-dione
LTD long-term depression
LTP long-term potentiation
NBQX 2,3-dihydroxy-6-nitro-7-sulfamoyl-benzo[f]quinoxaline-2,3-dione
NMDA *N*-methyl-D-aspartate
OR object recognition

1. INTRODUCTION

Visual recognition memory is the ability to discriminate whether a visual stimulus is novel or familiar. Such memory is essential for normal everyday life and decline of such memory, as occurs in dementias and in Alzheimer's disease, can be devastating. The perirhinal cortex is essential for recognition memory.[1–9] Perirhinal cortex is juxtallocortex, transitional between the hippocampal formation and the neocortex of the temporal lobe.[10] While learning is widely believed to depend on synaptic plasticity,[11–13] the precise relationships between learning and synaptic plasticity are still not well understood. Mechanisms of synaptic plasticity, notably long-term potentiation (LTP) and long-term depression (LTD), are chiefly studied *in vitro*.[14,15] Accordingly, it is important to establish links between learning and such plasticity. This review will consider what is known of recognition memory mechanisms, what is known of perirhinal synaptic plasticity mechanisms, and therefore what this reveals about the synaptic and cellular changes that most likely underlie perirhinal cortex-dependent recognition memory.

2. PERIRHINAL CORTEX AND RECOGNITION MEMORY

2.1. Lesion studies

Ablation studies have demonstrated the crucial role of perirhinal cortex for object recognition memory: lesions of perirhinal cortex in rats and primates produce impairments in recognition memory tasks.[2,9,16–19] In rats, these impairments are seen in tests of recognition memory that rely upon spontaneous preference for novel objects and for the purposes of this review, we will focus upon studies of novel object recognition memory.

2.2. Recording studies

Neuronal recording studies have provided important evidence concerning the likely means by which perirhinal cortex neurons encode visual recognition memory. While most data about perirhinal neuronal responses come from monkey studies,[1,20–22] there is work in the rodent that is supportive of the monkey findings.[23,24] In monkeys, performing recognition memory tasks up to 25% of recorded responses in perirhinal, lateral, entorhinal, and adjacent anterior inferior medial temporal cortex (visual association area TE) are reduced when an initially new visual stimulus is seen again.[20] These response reductions are maintained for a long time (for $\geq$24 h in many cases) and remain selective for previously seen stimuli, even after many 100s of stimuli have been seen. Therefore, these response reductions potentially convey information that can be used to encode or solve long-term recognition memory tasks.[20]

One can hypothesize from the decreased responsiveness to familiar stimuli that visual recognition memory relies on a decrease in synaptic efficacy within perirhinal cortex.[20] Computational models that include a component that weakens synapses perform familiarity discrimination with very high efficiency and very high storage capacity.[5,25–27] Thus, both experimental and theoretical evidence suggest that a process that reduces synaptic efficacy will be important for recognition memory in the perirhinal cortex.

2.3. Pharmacological infusion studies

Pharmacological investigations using localized infusions via cannulae targeted at perirhinal cortex have transiently inactivated perirhinal cortex by locally infusing CNQX (an AMPA and kainate glutamate receptor antagonist) to block synaptic transmission. Such studies have shown that perirhinal cortex has to be active at both acquisition and retrieval stages of the object recognition task.[28] The impairment produced by CNQX acting during acquisition does not differentiate between a role for perirhinal cortex in transmission or in storage. However, the effect at retrieval indicates that impairment can be produced by blocking transmission even when storage has already occurred.

In conclusion, there is very strong evidence that perirhinal cortex is essential for recognition memory, that blockade of activity within perirhinal cortex prevents recognition memory, and that such memory relies on lasting decrements in perirhinal neuronal activity.

3. MECHANISMS OF SYNAPTIC PLASTICITY AND VISUAL RECOGNITION MEMORY

In the next section, we will discuss experiments that address the question of the neurotransmitter receptors and subsequent biochemical pathways that are important in synaptic plasticity and recognition memory.[9,29] The synaptic plasticity studies rely on electrophysiological recording from slices of perirhinal cortex and interfering with transmitter receptors or downstream cascades of synaptic plasticity. The *in vivo* studies have relied largely on localized infusions of pharmacological ligands via cannulae targeted at perirhinal cortex. Many of these studies target transmitter receptors and biochemical mechanisms that are thought to be important in synaptic plasticity. Importantly, such infusions can be given so that they selectively affect different stages of memory processing: acquisition, consolidation, storage, and retrieval.

NMDARs in synaptic plasticity: One of the transmitter receptors commonly involved in induction of synaptic plasticity is the NMDA subtype of glutamate receptor (NMDAR). The first reports of synaptic plasticity in perirhinal cortex showed that LTP could be induced by high-frequency stimulation and that the induction of LTP relied on the activation of NMDARs.[30,31] Furthermore, activity-dependent reversal of LTP (depotentiation) by low-frequency (1 Hz) stimulation also required NMDAR activation. However, LTD was not readily induced in perirhinal slices from adult animals[31] using standard extracellular recording methods, but could be induced using whole cell recording by pairing depolarization with 1 Hz stimulation; this form of LTD was at least partially NMDAR-dependent.[32,33] Therefore, under appropriate conditions, the activity-dependent induction of the major forms of long-term synaptic plasticity can be induced in perirhinal cortex and can rely on NMDAR activation.

NMDARs in recognition memory: Given the above-described role of NMDARs in synaptic plasticity in perirhinal cortex, an obvious experiment to examine whether synaptic plasticity plays a role in recognition memory is to block NMDARs *in vivo* during behavioral tasks that assess visual recognition memory. Thus, AP5 (an antagonist of NMDARs) infused into perirhinal cortex to be active during acquisition impairs object recognition memory measured at long memory delays (1–24 h).[28,34,35] In contrast, AP5 does not produce a retrieval impairment[34,36] though see Ref. 37. The role of NMDARs in recognition memory provides a potentially strong link between perirhinal plasticity mechanisms and recognition memory formation. However, AP5 infused into perirhinal cortex does not impair recognition memory when the memory delay is only 20 min[34] though see Ref. 37. One potential explanation for this failure of impairment is that there is a region outside perirhinal cortex that can independently support recognition memory at such short delays. A second potential explanation is that there are other mechanisms within perirhinal cortex, which are able to maintain memory across shorter retention delays that are not dependent on NMDA receptor transmission.

Because NMDAR antagonism prevented all forms of long-term plasticity in perirhinal cortex *in vitro*, elucidating whether visual recognition memory processes could be related to one specific form of plasticity (i.e., LTP or LTD) cannot be achieved easily by blocking NMDARs. However, subsequent investigations of the roles of NMDAR subunits, of other glutamate receptors, and other transmitter systems revealed that different forms of plasticity could be induced by activation of different receptors and this has allowed conclusions to be drawn about the plasticity mechanisms that are most likely to be important in recognition memory. A description of different glutamate receptors and other transmitters involved in perirhinal synaptic plasticity will now be considered.

Subtypes of NMDARs in synaptic plasticity: NMDARs are heteromers[38,39] composed of two essential NR1 subunits[40] and two or three NR2 subunits.[41,42] There are four NR2 subunits (A–D), but the NR2A and NR2B subunits predominate in the forebrain. Each of the NR2 subunits imparts different characteristics on functional NMDARs[43,44]; for example, NR2A-containing NMDARs have rapid kinetics compared with those of NR2B-containing NMDARs.[45] In perirhinal cortex, NMDARs with different subunit compositions can preferentially induce LTP or LTD.[46] Thus, NR2A-containing NMDARs are preferentially involved in induction of LTP and depotentiation, and NR2B-containing NMDARs are preferentially involved in induction of LTD. Similar findings were also reported in hippocampus.[47–49]

Subtypes of NMDARs in recognition memory: The above findings concerning the selective role of NR2A or NR2B subunits in LTP or LTD mechanisms allow experiments to be conducted that can ask whether mechanisms specifically related to LTP or LTD are critical for visual recognition memory. *In vivo*, when antagonism is both of receptors of NR2A (using NPV-AAM077) and of NR2B (using Ro256981)—containing subunits, a similar pattern of amnesia is produced to that by produced AP5 (impairment at long memory delays).[34] However, when either antagonist is given on its own, there is no impairment.[34] This result suggests that when only one of the two types of NMDA receptor is inactive, activity in the other can compensate as far as recognition memory formation is concerned. This result may suggest that both LTP- and LTD-like mechanisms are required for visual recognition memory. An alternative possibility that has not been examined is that under *in vivo* conditions, selective blockade of either NR2A or NR2B does not produce a selective block of LTP or LTD, respectively, as under *in vitro* conditions.[46]

3.1. Kainate receptors

Antagonism of kainate glutamate receptors (GluK1) within perirhinal cortex produces amnesia after a 20-min memory delay but with remembrance after a 24-h delay.[34] Hence, infusing a GluK1 antagonist into perirhinal cortex establishes that kainate receptors in perirhinal cortex are necessary for recognition memory after a 20-min delay. In such circumstances, no other brain region is able to support recognition memory at this delay. Since interfering with perirhinal cortex function can produce impairment at a 20-min delay and no other region can compensate, it would seem most likely that the lack of AP5-induced memory impairment at 20 min is the result of an alternative kainate-dependent memory mechanism within perirhinal cortex. However, it is not yet possible to make a link between such memory mechanisms and kainate-dependent plasticity since there have not yet been any reports of plasticity mechanisms that rely on kainate receptors changing AMPA receptor transmission.

3.2. Metabotropic glutamate receptors in synaptic plasticity

That there was a potential role for mGluRs in LTD in perirhinal cortex came from studies showing that application of the group I/II mGluR agonist ACPD or the group I mGluR agonist DHPG produced a lasting depression of synaptic transmission[50] in a manner similar to that seen in the hippocampus.[51] Initial studies on the role of metabotropic glutamate receptor (mGluR) involvement in activity-dependent synaptic plasticity in perirhinal cortex showed that the group I/II mGluR antagonist MCPG did not block activity-dependent LTP or depotentiation.[31] However, subsequent studies demonstrated a role of mGluRs in perirhinal LTD.[32] In experiments using whole cell recording, it was shown that at resting membrane potential (~ -70 mV) the induction of LTD required activation of NMDARs, group

I and group II mGluRs. However, at depolarized potentials (~ -40 mV), the requirement for group II mGluRs was removed.[32] These results established that mGluRs were involved in activity-dependent LTD in perirhinal cortex.

3.3. mGluRs in recognition memory

The NMDAR antagonist AP5 prevents the induction of both LTP and LTD in perirhinal cortex *in vitro* (see below). Therefore, NMDAR blockade cannot determine whether recognition memory might rely on an LTP- or an LTD-like mechanism. Antagonism of group I and group II mGluRs (with MCPG) in perirhinal cortex *in vitro* impairs the induction of LTD but not LTP.[31,32] Therefore, the use of appropriate metabotropic glutamatergic antagonists may provide clues as to whether LTP or LTD relates more closely to the plastic process underlying visual recognition memory. *In vivo* antagonism of group I and group II metabotropic receptors (with MPEP and EGLU) results in object recognition memory impairment.[52] As for NMDA receptor antagonism, mGluR antagonism produced impairments after long memory delays (24 h) but not after a delay of 20 min.[52] That mGluRs are involved in LTD but not LTP induction indicates LTD-like mechanisms may support recognition memory. Furthermore, this suggests that LTP-like plasticity mechanisms are insufficient to support recognition memory when LTD-like mechanisms are impaired.

3.4. Muscarinic cholinergic receptors in synaptic plasticity

Application of the cholinergic agonist carbachol to perirhinal slices results in the induction of long-lasting depression of synaptic transmission.[53] This form of plasticity relied on a release of calcium from intracellular stores and on protein synthesis.[53] Interestingly, paralleling the results for mGluRs, the broad spectrum muscarinic antagonist scopolamine prevented the induction of activity-dependent LTD but had no effect on induction of LTP in slices of perirhinal cortex.[54] This again provides a means of examining the role of LTD-like mechanisms in visual recognition memory.

Cholinergic receptors and recognition memory: *In vivo* antagonism of perirhinal muscarinic cholinergic receptors produces recognition memory impairment at shorter (≤ 3 h) but not longer (≥ 6 h) intervals,[54–56] though see Ref. 57. Further studies have demonstrated that this memory impairment is produced by antagonism of muscarinic M1 receptors.[56] Conversely, infusion of the $\alpha 7$ nicotinic antagonist methyllycaconitine produces recognition memory deficits at longer (24 h) but not shorter (20 min) delays,[56] although the role of α7 receptors in synaptic plasticity is unknown.

Together, these findings further support the hypothesis that LTD-like mechanisms are critical for recognition memory.

3.5. L-type voltage-gated calcium channels in synaptic plasticity and recognition memory

There is a great deal of evidence that the activation of voltage-gated calcium channels (VGCCs) is important in the induction of synaptic plasticity in different brain regions and under different conditions.[58,59] In perirhinal cortex, the VGCC inhibitor verapamil prevented the induction of both LTD and depotentiation while having no effect on LTP.[60] VGCC inhibition within perirhinal cortex impaired recognition memory,[60] so providing a further link between the plasticity mechanisms involved in LTD and those underlying recognition memory.

Endocannabinoid receptors and synaptic plasticity: Retrograde signaling is critical in synaptic plasticity, coordinating pre- and postsynaptic changes following induction of LTP or LTD. Endocannabinoids are synthesized following postsynaptic stimulation of Gq-coupled receptors by different neurotransmitters. In the CNS, eCBs decrease transmitter release through activation of presynaptic cannabinoid receptor 1 (CB1). Furthermore, eCBs have been implicated in mechanisms of LTD in the striatum, cortex, and hippocampus.[61–64] Recently, it has been demonstrated that LTP in perirhinal cortex is prevented by AM251, an antagonist of endocannabinoid receptors; however, LTD was unaffected by AM251.[65] These results provide an additional means to test the hypothesis that LTD but not LTP is crucial for visual recognition memory.

Endocannabinoid receptors and recognition memory: There is evidence that activation of endocannabinoid receptors is important in hippocampal and amygdala-dependent associative learning and memory.[66,67] Importantly, CB1 knockout mice have been described to have improved memory performance in a 24-h delay object recognition task.[68,69] In contrast, however, in our recent study, localized infusion of the CB1 receptor antagonist AM251 into perirhinal cortex had no effect on short- (20 min) or long-term (24 h) visual recognition memory.[65] Several issues may explain these differences. First, the results in the study by Reibaud *et al.*[69] were based on a global CB1 knockout; therefore, the behavioral effects observed may be due to effects outside perirhinal cortex. Second, there are procedural differences in the assessment of recognition memory between the two studies. In the study by Reibaud *et al.*,[69] only one object was presented in the sample phase and two objects were presented in the test phase. Thus, a spatial memory component that does not involve perirhinal cortex may have been introduced into the design of that experiment. We would argue that our results showing a block of LTP (but not LTD) by CB1 antagonism and no effect of AM251 on recognition memory provide support for the hypothesis that LTP-like mechanisms are less likely to be critical for recognition memory.

Nitric oxide and synaptic plasticity: In physiological conditions, NO is synthesized postsynaptically in neurons and in blood vessels by constitutive isoforms of nitric oxide synthase (neuronal, nNOS; endothelial, eNOS) that are activated by Ca^{2+}–calmodulin (Garthwaite, 2008[70,71]). Nitric oxide can play a role in retrograde signaling in LTD in the cerebellum, hippocampus, and prefrontal cortex[72–74] and in LTP in the hippocampus and visual cortex.[75–79] We have recently demonstrated that activity-dependent LTD and muscarinic agonist-induced LTD in perirhinal cortex *in vitro* are both prevented by nitric oxide synthase (NOS) inhibitors (LNAME or NPA). In contrast, the induction of LTP was unaffected by inhibition of NOS.[65]

Nitric oxide and recognition memory: Previous work has suggested that NO is implicated in learning and memory, including spatial[80] and motor learning[81,82] and recognition memory.[83] We have recently described that the inhibitor of neuronal nitric oxide synthase, NPA, infused locally into perirhinal cortex produces deficits in recognition memory at 24-h delay but not at 20-min delay.[65] Together, the results with inhibition of NOS again suggest that LTD-type mechanisms rather than LTP-like mechanisms play a role in visual recognition memory.

Intracellular signaling mechanisms in synaptic plasticity and recognition memory: Induction of LTP or LTD leads to a cascade of biochemical processes that have been well studied in the hippocampus and other brain areas,[84–87] but rather less extensively studied in perirhinal cortex. Nevertheless, several parallels have been established between processes necessary for the persistence of the synaptic changes induced during LTP or LTD, and impairment of recognition memory produced by interference with these processes during memory consolidation in perirhinal cortex. Once such example has been established by interfering with the phosphorylation of cyclic AMP response element-binding (CREB) protein.[88–90] CREB is activated (phosphorylated to form pCREB) by MAPK and CamKs[91] that are activated following activation of a variety of transmitter receptors (including glutamate and acetylcholine receptors). Preventing pCREB from activating nuclear transcription and hence preventing the generation of new protein in perirhinal cortex also impairs recognition memory.[92] In this study, LTP was also shown to be impaired in transfected perirhinal slices[92]; effects on LTD were not investigated. Therefore, this result leaves open the possibility that while most findings point to LTD-like mechanisms being critical for visual recognition, LTP-like processes may also have some role to play in recognition memory processing.

3.6. Expression mechanisms of synaptic plasticity

As described above, it has been possible to show that under different conditions, certain transmitter receptors couple preferentially to the induction of LTD or LTP. This has been valuable in trying to dissect whether either of these plasticity mechanisms is more important in recognition memory. However, in addition to any effects specifically on plasticity mechanisms, blocking a particular transmitter receptor is likely to have effects more generally on cellular excitability and intracellular biochemical cascades. It is not possible to rule out that these more general effects of receptor antagonism, rather than effects specifically on plasticity, may be responsible for any observed deficits in learning and memory. One approach to circumvent such concerns is to understand the expression mechanisms of synaptic plasticity. For LTD, there is now a wealth of evidence that a common final step that underlies the decrease in synaptic transmission is a removal of AMPA receptors from the synaptic membrane. The steps that lead to the activity-dependent endocytosis of AMPA receptors, resulting in LTD, are now well defined.[14,93,94] In perirhinal cortex an interaction between the clathrin adaptor protein (AP2) and the GluR2 subunit of AMPA receptors is required for activity-dependent LTD,[33] as for hippocampus.[95] These results were established by introducing into single neurons a peptide (DA849-Q853) that prevented the normal interaction between GluR2 and AP2. This intervention prevented the induction of activity-dependent LTD, importantly without affecting basal AMPA receptor synaptic transmission or NMDA receptor-mediated synaptic currents.[33]

Expression mechanisms of recognition memory: As described above the removal of synaptic AMPA receptors can be blocked (and hence LTD can be blocked) by preventing the binding of intracellular clathrin AP2 to the c-terminal of the GluR2 subunit of AMPA receptors by the peptide pepDA849-Q853. In *in vitro* conditions this can be achieved relatively easily by including the peptide within the whole cell filling solution. However, under *in vivo* conditions delivery of peptides into neurons is more problematic. Nevertheless this can be achieved by viral transduction of specific peptides. Thus a block of AP2-GluR2 interaction was achieved *in vivo* using stereotaxic injections to transduce perirhinal cortex with lentiviral vectors expressing pepDA849-Q853. Blocking the AP2-GluR2 interaction in perirhinal cortex *in vivo* impairs object recognition memory at both short (5 min) and long memory (24 h) delays.[33] Therefore, these data provide strong evidence for the necessity to recognition memory of LTD-like plasticity occurring within perirhinal cortex.

4. LEARNING-RELATED CHANGES IN SYNAPTIC TRANSMISSION

One method that has been used extensively for investigating the relationship between learning and plasticity has been to monitor whether learning produces changes in synaptic transmission—either directly *in vivo* (e.g., Ref. 96) or indirectly *in vitro*.[97] Presentations of many novel visual stimuli to one eye or many familiar stimuli to the other eye at the same time under the same conditions allowed the subsequent assessment of whether these visual stimuli resulted in any changes of synaptic transmission or of synaptic plasticity in perirhinal cortex *in vitro*.[98] The results of these studies showed that LTP *in vitro* was unaffected by either form of stimulation; however, LTD and depotentiation were both absent *in vitro* following the previous *in vivo* experience of the repeated presentation of familiar stimuli. Critically, there was no effect on LTD and depotentiation if the viewing of stimuli had been carried out after systemic administration of the muscarinic antagonist scopolamine. These results demonstrate that it is possible to detect learning-related changes in synaptic plasticity *in vitro* and establish the importance of learning-related changes in plasticity gain. Unexpectedly, plasticity mechanisms were disrupted by viewing familiar rather than novel stimuli. It was suggested[98] that the lack of effect of viewing novel stimuli was because the potential capacity of perirhinal cortex for registering new visual stimuli was so high[26] that the hundreds of pictures viewed were nonetheless too few to have a measurable effect on overall levels of perirhinal plasticity. In contrast, the frequently repeated exposure to a limited number of familiar stimuli produced a bias, a change in the perirhinal plasticity gain.[98] This change in plasticity gain might relate to reconsolidation mechanisms[99] evoked when the previously seen stimuli are re-encountered or to the network changes induced when stimuli are repeatedly viewed.[100]

5. CONCLUSIONS

The considerable body of evidence that has been obtained from pharmacological infusion studies in recognition memory and of perirhinal synaptic plastic processes provides strong evidence that particular plasticity processes are likely used during memory formation. A key type of plasticity for such recognition memory is synaptic weakening within perirhinal cortex. However, while there is now strong evidence pointing to the importance of LTD-like mechanisms underlying object recognition memory there are also possible links between LTP mechanisms and recognition memory. It is not known whether there is a necessity for LTP in the initial formation of recognition memory or a necessary requirement for optimal rebalancing of circuit excitability following LTD. All of the synaptic plasticity experiments referred to rely on *in vitro* induction of LTP and LTD. It must be borne in mind that if the pathway being studied is in a brain slice there is uncertainty concerning the exact match of *in vitro* and *in vivo* conditions—this can relate to the balance of excitation and inhibition within the local microcircuitry, the importance of internal and external influences on this circuitry by cholinergic, monoaminergic and other modulatory systems, and the relationship between induction mechanisms of plasticity and learning *in vivo*. Thus although LTP and LTD are the foremost models for studying plasticity, there are reasons for being cautious concerning the details of the parallels with what may happen during learning itself.

REFERENCES

1. Brown MW, Warburton EC, Aggleton JP. Recognition memory: material, processes, and substrates. *Hippocampus*. 2010;20(11):1228–1244.
2. Eichenbaum H, Yonelinas AP, Ranganath C. The medial temporal lobe and recognition memory. *Annu Rev Neurosci*. 2007;30(1):123–152.
3. Montaldi D, Mayes AR. The role of recollection and familiarity in the functional differentiation of the medial temporal lobes. *Hippocampus*. 2010;20(11):1291–1314.
4. Murray E, Bussey T, Saksida L. Visual perception and memory: a new view of medial temporal lobe function in primates and rodents. *Annu Rev Neurosci*. 2007;30:99–122.
5. Norman KA. How hippocampus and cortex contribute to recognition memory: revisiting the complementary learning systems model. *Hippocampus*. 2010;20(11):1217–1227.
6. Squire L, Wixted J. The cognitive J Neurosci of human memory since H.M. *Annu Rev Neurosci*. 2011;34:259–288.
7. Vann SD, Tsivilis D, Denby CE, et al. Impaired recollection but spared familiarity in patients with extended hippocampal system damage revealed by 3 convergent methods. *Proc Natl Acad Sci U S A*. 2009;106(13):5442–5447.

8. Vann SD, Albasser MM. Hippocampus and neocortex: recognition and spatial memory. *Curr Opin Neurobiol.* 2011;21(3):440–445.
9. Winters BD, Saksida LM, Bussey TJ. Object recognition memory: neurobiological mechanisms of encoding, consolidation and retrieval. *Neurosci Biobehav Rev.* 2008;32(5):1055–1070.
10. Burwell R. Borders and cytoarchitecture of the perirhinal and postrhinal cortices in the rat. *J Comp Neurol.* 2001;437(1):17–41.
11. Abraham WC, Robins A. Memory retention—the synaptic stability versus plasticity dilemma. *Trends Neurosci.* 2005;28:73–78.
12. Bliss TV, Collingridge GL. A synaptic model of memory: long-term potentiation in the hippocampus. *Nature.* 1993;361(6407):31–39.
13. Martin SJ, Morris RGM. New life in an old idea: the synaptic plasticity and memory hypothesis revisited. *Hippocampus.* 2002;12:609–636.
14. Kemp N, Bashir ZI. Long-term depression: a cascade of induction and expression mechanisms. *Prog Neurobiol.* 2001;65(4):339–365.
15. Malenka RC, Bear MF. LTP and LTD: an embarrassment of riches. *Neuron.* 2004;44:5–21.
16. Brown MW, Aggleton JP. Recognition memory: what are the roles of the perirhinal cortex and hippocampus? *Nat Rev Neurosci.* 2001;2(1):51–61.
17. Clark RE, Squire LR. An animal model of recognition memory and medial temporal lobe amnesia: history and current issues. *Neuropsychologia.* 2010;48(8):2234–2244.
18. Guderian S, Brigham D, Mishkin M. Two processes support visual recognition memory in rhesus monkeys. *Proc Natl Acad Sci.* 2011;108(48):19425–19430.
19. Warburton EC, Brown MW. Findings from animals concerning when interactions between perirhinal cortex, hippocampus and medial prefrontal cortex are necessary for recognition memory. *Neuropsychologia.* 2010;48(8):2262–2272.
20. Brown MW, Xiang JZ. Recognition memory: neuronal substrates of the judgement of prior occurrence. *Prog Neurobiol.* 1998;55(2):149–189.
21. Miller EK, Erickson CA, Desimone R. Neural mechanisms of visual working memory in prefrontal cortex of the macaque. *J Neurosci.* 1996;16(16):5154–5167.
22. Ringo JL. Stimulus specific adaptation in inferior temporal and medial temporal cortex of the monkey. *Behav Brain Res.* 1996;76(1–2):191–197.
23. Zhu XO, Brown MW. Changes in neuronal activity related to the repetition and relative familiarity of visual stimuli in rhinal and adjacent cortex of the anaesthetised rat. *Brain Res.* 1995;689(1):101–110.
24. Zhu XO, Brown MW, Aggleton JP. Neuronal signalling of information important to visual recognition memory in rat rhinal and neighbouring cortices. *Eur J Neurosci.* 1995;7(4):753–765.
25. Androulidakis Z, Lulham A, Bogacz R, Brown M. Computational models can replicate the capacity of human recognition memory. *Network.* 2008;19(3):161–182.
26. Bogacz R, Brown MW. Comparison of computational models of familiarity discrimination in the perirhinal cortex. *Hippocampus.* 2003;13(4):494–524.
27. Lulham A, Bogacz R, Vogt S, Brown M. An Infomax algorithm can perform both familiarity discrimination and feature extraction in a single network. *Neural Comput.* 2011;23(4):909–926.
28. Winters BD, Bussey TJ. Glutamate receptors in perirhinal cortex mediate encoding, retrieval, and consolidation of object recognition memory. *J Neurosci.* 2005;25(17):4243–4251.
29. Brown MW, Barker GRI, Aggleton JP, Warburton EC. What pharmacological interventions indicate concerning the role of the perirhinal cortex in recognition memory. *Neuropsychologia.* 2012;50(13):3122–3140.
30. Bilkey DK. Long-term potentiation in the in vitro perirhinal cortex displays associative properties. *Brain Res.* 1996;733(2):297–300.

31. Ziakopoulos Z, Tillett CW, Brown MW, Bashir ZI. Input-and layer-dependent synaptic plasticity in the rat perirhinal cortex in vitro. *J Neurosci.* 1999; 92(2):459–472.
32. Cho K, Kemp N, Noel J, Aggleton JP, Brown MW, Bashir ZI. A new form of long-term depression in the perirhinal cortex. *Nat Neurosci.* 2000;3(2):150–156.
33. Griffiths S, Scott H, Glover C, et al. Expression of long-term depression underlies visual recognition memory. *Neuron.* 2008;58(2):186–194.
34. Barker GRI, Warburton EC, Koder T, et al. The different effects on recognition memory of perirhinal kainate and NMDA glutamate receptor antagonism: implications for underlying plasticity mechanisms. *J Neurosci.* 2006;26(13):3561–3566.
35. Barker GRI, Warburton EC. NMDA receptor plasticity in the perirhinal and prefrontal cortices is crucial for the acquisition of long-term object-in-place associative memory. *J Neurosci.* 2008;28(11):2837–2844.
36. Winters BD, Bussey TJ. Transient inactivation of perirhinal cortex disrupts encoding, retrieval, and consolidation of object recognition memory. *J Neurosci.* 2005;25(1): 52–61.
37. Abe H, Ishida Y, Iwasaki T. Perirhinal N-methyl-D-aspartate and muscarinic systems participate in object recognition in rats. *Neurosci Lett.* 2004;356(3):191–194.
38. Cull-Candy SG, Brickley SG. *NMDA Receptors. Encyclopedia of Life Sciences.* Chichester: John Wiley & Sons, Ltd; 2001.
39. Dingledine R, Borges K, Bowie D, Traynelis SF. The glutamate receptor ion channels. *Pharmacol Rev.* 1999;51(1):7–61.
40. Behe P, Stern P, Wyllie DJ, Nassar M, Schoepfer R, Colquhoun D. Determination of NMDA NR1 subunit copy number in recombinant NMDA receptors. *Proc Biol Sci.* 1995;262(1364):205–213.
41. Laube B, Kuhse J, Betz H. Evidence for a tetrameric structure of recombinant NMDA receptors. *J Neurosci.* 1998;18(8):2954–2961.
42. Premkumar LS, Auerbach A. Stoichiometry of recombinant N-methyl-D-aspartate receptor channels inferred from single-channel current patterns. *J Gen Physiol.* 1997;110(5):485–502.
43. Kutsuwada T, Kashiwabuchi N, Mori H, et al. Molecular diversity of the NMDA receptor channel. *Nature.* 1992;358(6381):36–41.
44. Loftis JM, Janowsky A. The N-methyl-D-aspartate receptor subunit NR2B: localization, functional properties, regulation, and clinical implications. *Pharmacol Ther.* 2003;97(1):55–85.
45. Monyer H, Burnashev N, Laurie DJ, Sakmann B, Seeburg PH. Developmental and regional expression in the rat brain and functional properties of four NMDA receptors. *Neuron.* 1994;12(3):529–540.
46. Massey PV, Johnson BE, Moult PR, et al. Differential roles of NR2A and NR2B-containing NMDA receptors in cortical long-term potentiation and long-term depression. *J Neurosci.* 2004;24(36):7821–7828.
47. Liu L, Wong TP, Pozza MF, et al. Role of NMDA receptor subtypes in governing the direction of hippocampal synaptic plasticity. *Science.* 2004;304(5673):1021–1024.
48. Bartlett TE, Bannister NJ, Collett VJ, et al. Differential roles of NR2A and NR2B-containing NMDA receptors in LTP and LTD in the CA1 region of two-week old rat hippocampus. *Neuropharmacology.* 2007;52(1):60–70.
49. Morishita W, Lu W, Smith GB, Nicoll RA, Bear MF, Malenka RC. Activation of NR2B-containing NMDA receptors is not required for NMDA receptor-dependent long-term depression. *Neuropharmacology.* 2007;52(1):71–76.
50. McCaffery B, Cho K, Bortolotto ZA, et al. Synaptic depression induced by pharmacological activation of metabotropic glutamate receptors in the perirhinal cortex in vitro. *J Neurosci.* 1999;93(3):977–984.

51. Fitzjohn SM, Palmer MJ, May JE, Neeson A, Morris SA, Collingridge GL. A characterisation of long-term depression induced by metabotropic glutamate receptor activation in the rat hippocampus in vitro. *J Physiol.* 2001;537(Pt. 2):421–430.
52. Barker GRI, Bashir ZI, Brown MW, Warburton EC. A temporally distinct role for group I and group II metabotropic glutamate receptors in object recognition memory. *Learn Mem.* 2006;13(2):178–186.
53. Massey PV, Bhabra G, Cho K, Brown MW, Bashir ZI. Activation of muscarinic receptors induces protein synthesis-dependent long-lasting depression in the perirhinal cortex. *Eur J Neurosci.* 2001;14(1):145–152.
54. Warburton EC, Koder T, Cho K, et al. Cholinergic neurotransmission is essential for perirhinal cortical plasticity and recognition memory. *Neuron.* 2003;38(6):987–996.
55. Abe H, Iwasaki T. NMDA and muscarinic blockade in the perirhinal cortex impairs object discrimination in rats. *Neuroreport.* 2001;12(15):3375–3379.
56. Tinsley CJ, Fontaine-Palmer NS, Vincent M, et al. Differing time dependencies of object recognition memory impairments produced by nicotinic and muscarinic cholinergic antagonism in perirhinal cortex. *Learn Mem.* 2011;18(7):484–492.
57. Winters BD, Saksida LM, Bussey TJ. Paradoxical facilitation of object recognition memory after infusion of scopolamine into perirhinal cortex: implications for cholinergic system function. *J Neurosci.* 2006;26(37):9520–9529.
58. Bolshakov VY, Siegelbaum SA. Postsynaptic induction and presynaptic expression of hippocampal long-term depression. *Science.* 1994;264(5162):1148–1152.
59. Grover LM, Teyler TJ. Two components of long-term potentiation induced by different patterns of afferent activation. *Nature.* 1990;347(6292):477–479.
60. Seoane A, Massey PV, Keen H, Bashir ZI, Brown MW. L-type voltage-dependent calcium channel antagonists impair perirhinal long-term recognition memory and plasticity processes. *J Neurosci.* 2009;29(30):9534–9544.
61. Lafourcade M, Elezgarai I, Mato S, Bakiri Y, Grandes P, Manzoni OJ. Molecular components and functions of the endocannabinoid system in mouse prefrontal cortex. *PLoS ONE.* 2007;2(8):e709.
62. Robbe D, Kopf M, Remaury A, Bockaert J, Manzoni OJ. Endogenous cannabinoids mediate long-term synaptic depression in the nucleus accumbens. *Proc Natl Acad Sci U S A.* 2002;99:8384–8388.
63. Sergeeva OA, Doreulee N, Chepkova AN, Kazmierczak T, Haas HL. Long-term depression of cortico-striatal synaptic transmission by DHPG depends on endocannabinoid release and nitric oxide synthesis. *Eur J Neurosci.* 2007;26:1889–1894.
64. Yasuda H, Huang Y, Tsumoto T. Regulation of excitability and plasticity by endocannabinoids and PKA in developing hippocampus. *Proc Natl Acad Sci U S A.* 2008;105:3106–3111.
65. Tamagnini F, Barker G, Warburton EC, Burattini C, Aicardi G, Bashir ZI. Nitric oxide-dependent long-term depression but not endocannabinoid-mediated long term potentiation is crucial for visual recognition memory. *J Physiol.* 2013; 591(Pt 16):3963–3979.
66. Marsicano G, Wotjak CT, Azad SC, et al. The endogenous cannabinoid system controls extinction of aversive memories. *Nature.* 2002;418:530–534.
67. Varvel SA, Wise LE, Niyuhire F, Cravatt BF, Lichtman AH. Inhibition of fatty-acid amide hydrolase accelerates acquisition and extinction rates in a spatial memory task. *Neuropsychopharmacology.* 2007;32:1032–1041.
68. Lutz B. The endocannabinoid system and extinction learning. *Mol Neurobiol.* 2007;36:92–101.
69. Reibaud M, Obinu MC, Ledent C, Parmentier M, Bohme GA, Imperato A. Enhancement of memory in cannabinoid CB1 receptor knock-out mice. *Eur J Pharmacol.* 1999;379:R1–R2.

70. Garthwaite J. Concepts of neural nitric oxide-mediated transmission. *Eur J Neurosci.* 2008;27:2783–2802.
71. Steinert JR, Chernova T, Forsythe ID. Nitric oxide signaling in brain function, dysfunction, and dementia. *Neuroscientist.* 2010;16:435–452.
72. Huang CC, Hsu KS. Activation of muscarinic acetylcholine receptors induces a nitric oxide-dependent long-term depression in rat medial prefrontal cortex. *Cereb Cortex.* 2010;20(4):982–996.
73. Reyes-Harde M, Potter BV, Galione A, Stanton PK. Induction of hippocampal LTD requires nitric-oxide-stimulated PKG activity and Ca2+ release from cyclic ADP-ribose-sensitive stores. *J Neurophysiol.* 1999;82:1569–1576.
74. Shin JH, Linden DJ. An NMDA receptor/nitric oxide cascade is involved in cerebellar LTD but is not localized to the parallel fiber terminal. *J Neurophysiol.* 2005;94: 4281–4289.
75. Arancio O, Antonova I, Gambaryan S, et al. Presynaptic role of cGMP-dependent protein kinase during long-lasting potentiation. *J Neurosci.* 2001;21:143–149.
76. Arancio O, Kandel ER, Hawkins RD. Activity-dependent long-term enhancement of transmitter release by presynaptic 3′,5′-cyclic GMP in cultured hippocampal neurons. *Nature.* 1995;376:74–80.
77. Arancio O, Kiebler M, Lee CJ, et al. Nitric oxide acts directly in the presynaptic neuron to produce long-term potentiation in cultured hippocampal neurons. *Cell.* 1996;87:1025–1035.
78. Haghikia A, Mergia E, Friebe A, Eysel UT, Koesling D, Mittmann T. Long-term potentiation in the visual cortex requires both nitric oxide receptor guanylyl cyclases. *J Neurosci.* 2007;27:818–823.
79. Wang HG, Lu FM, Jin I, et al. Presynaptic and postsynaptic roles of NO, cGK, and RhoA in long-lasting potentiation and aggregation of synaptic proteins. *Neuron.* 2005;45:389–403.
80. Bohme GA, Bon C, Lemaire M, et al. Altered synaptic plasticity and memory formation in nitric oxide synthase inhibitor-treated rats. *Proc Natl Acad Sci U S A.* 1993;90:9191–9194.
81. Allen MT, Steinmetz JE. A nitric oxide synthase inhibitor delays the formation of learning-related neural activity in the cerebellar interpositus nucleus during rabbit eyelid conditioning. *Pharmacol Biochem Behav.* 1996;53:147–153.
82. Nagao S, Kitazawa H, Osanai R, Hiramatsu T. Acute effects of tetrahydrobiopterin on the dynamic characteristics and adaptability of vestibulo-ocular reflex in normal and flocculus lesioned rabbits. *Neurosci Lett.* 1997;231:41–44.
83. Pitsikas N, Rigamonti AE, Bonomo SM, Cella SG, Muller EE. Molsidomine antagonizes L-NAME-induced acquisition deficits in a recognition memory task in the rat. *Pharmacol Res.* 2003;47(4):311–315.
84. Bading H, Ginty D, Greenberg M. Regulation of gene expression in hippocampal neurons by distinct calcium signaling pathways. *Science.* 1993;260(5105):181–186.
85. Citri A, Malenka R. Synaptic plasticity: multiple forms, functions, and mechanisms. *Neuropsychopharmacology.* 2008;33(1):18–41.
86. Kotaleski J, Blackwell K. Modelling the molecular mechanisms of synaptic plasticity using systems biology approaches. *Nat Rev Neurosci.* 2010;11(4):239–251.
87. MacDonald JF, Jackson MF, Beazely MA. Hippocampal long-term synaptic plasticity and signal amplification of NMDA receptors. *Crit Rev Neurobiol.* 2006;18(1–2):71–84.
88. Ahn S, Ginty DD, Linden DJ. A late phase of cerebellar long-term depression requires activation of CaMKIV and CREB. *Neuron.* 1999;23(3):559–568.
89. Deisseroth K, Bito H, Tsien RW. Signaling from synapse to nucleus: postsynaptic CREB phosphorylation during multiple forms of hippocampal synaptic plasticity. *Neuron.* 1996;16(1):89–101.

90. Silva AJ, Kogan JH, Frankland PW, Kida S. CREB and memory. *Annu Rev Neurosci.* 1998;21:127–148.
91. Sun P, Enslen H, Myung PS, Maurer RA. Differential activation of CREB by Ca2+/calmodulin-dependent protein kinases type II and type IV involves phosphorylation of a site that negatively regulates activity. *Genes Dev.* 1994;8(21):2527–2539.
92. Warburton EC, Glover CP, Massey PV, et al. cAMP responsive element-binding protein phosphorylation is necessary for perirhinal long-term potentiation and recognition memory. *J Neurosci.* 2005;25(27):6296–6303.
93. Collingridge GL, Peineau S, Howland JG, Wang YT. Long-term depression in the CNS. *Nat Rev Neurosci.* 2010;11(7):459–473.
94. Massey PV, Bashir ZI. Long-term depression: multiple forms and implications for brain function. *Trends Neurosci.* 2007;30(4):176–184.
95. Lee SH, Liu L, Wang YT, Sheng M. Clathrin adaptor AP2 and NSF interact with overlapping sites of GluR2 and play distinct roles in AMPA receptor trafficking and hippocampal LTD. *Neuron.* 2002;36(4):661–674.
96. Whitlock JR, Heynen AJ, Shuler MG, Bear MF. Learning induces long-term potentiation in the hippocampus. *Science.* 2006;313(5790):1093–1097.
97. Rioult-Pedotti MS, Friedman D, Hess G, Donoghue JP. Strengthening of horizontal cortical connections following skill learning. *Nat Neurosci.* 1998;1(3):230–234.
98. Massey PV, Phythian D, Narduzzo K, Warburton EC, Brown MW, Bashir ZI. Learning-specific changes in long-term depression in adult perirhinal cortex. *J Neurosci.* 2008;28(30):7548–7554.
99. Nader K, Schafe G, LeDoux J. The labile nature of consolidation theory. *Nat Rev Neurosci.* 2000;1(3):216–219.
100. Woloszyn L, Sheinberg DL. Effects of long-term visual experience on responses of distinct classes of single units in inferior temporal cortex. *Neuron.* 2012;74(1):193–205.

CHAPTER EIGHT

Molecular Influences on Working Memory Circuits in Dorsolateral Prefrontal Cortex

Amy F.T. Arnsten, Lu E. Jin
Department of Neurobiology, Yale University School of Medicine, New Haven, Connecticut, USA

Contents

Abstract

The working memory circuits of the primate dorsolateral prefrontal cortex (dlPFC) are modulated in a unique manner, often opposite to the molecular mechanisms needed for long-term memory consolidation. Working memory, our "mental sketch pad" is an ephemeral process, whereby transient, mental representations form the foundation for abstract thought. The microcircuits that generate mental representations are found in deep layer III of the dlPFC, where pyramidal cells excite each other to keep information "in mind" through NMDA receptor synapses on spines. The catecholaminergic and cholinergic arousal systems have rapid and flexible influences on the strength of these connections, thus allowing coordination between arousal and cognitive states. These modulators can rapidly weaken connectivity, for example, as occurs during uncontrollable stress, via feedforward calcium–cAMP signaling opening potassium (K^+) channels near synapses on spines. Lower levels of calcium–cAMP–K^+ channel signaling provide negative feedback within recurrent excitatory circuits, and help to gate inputs to shape the contents of working memory. There are also explicit mechanisms to inhibit calcium–cAMP signaling and strengthen connectivity, for example, postsynaptic

Progress in Molecular Biology and Translational Science, Volume 122
ISSN 1877-1173
http://dx.doi.org/10.1016/B978-0-12-420170-5.00008-8

α2A-adrenoceptors on spines. This work has led to the development of the α2A agonist, guanfacine, for the treatment of a variety of dlPFC disorders. In mental illness, there are a variety of genetic insults to the molecules that normally serve to inhibit calcium–cAMP signaling in spines, thus explaining why so many genetic insults can lead to the same phenotype of impaired dlPFC cognitive function. Thus, the molecular mechanisms that provide mental flexibility may also confer vulnerability when dysregulated in cognitive disorders.

Understanding molecular influences on higher cognitive function is essential for illuminating how genetic insults lead to cognitive dysfunction in mental illness, and how environmental insults lead to age-related cognitive dysfunction, including vulnerability to neurodegeneration. Higher cognitive disorders such as schizophrenia and Alzheimer's disease preferentially afflict the highly evolved association cortices, including the dorsolateral prefrontal cortex (dlPFC).[1–3] The dlPFC subserves mental representations, the foundation of abstract thought, and recent research has begun to discover molecular influences on dlPFC function. This work has shown that the dlPFC is modulated in a fundamentally different manner than primary sensory cortex, hippocampus, or subcortical structures, and that small changes in molecular signaling can produce large changes in cognitive function.[4]

1. WORKING MEMORY, OUR "MENTAL SKETCH PAD"

Mental representations are the foundation of abstract thought: the ability to keep information in mind without sensory stimulation.[5,6] This cognitive operation is referred to as working memory or cognitive control when the representation is used to guide thought, action, and emotion.[5,7,8] This process can utilize the sustained representation of an event that has just occurred, or the reactivation of a memory from long-term stores,[9,10] and is particularly important for the inhibition of inappropriate responses.[11] The PFC is topographically organized whereby the dlPFC performs these functions for thought and action,[12] while the ventromedial PFC is involved with the representation and regulation of emotion.[13] In humans, inhibitory functions are particularly enriched in the right hemisphere.[11] Working memory abilities evolve greatly from rodent to monkey to humans, in keeping with the vast evolutionary expansion of the PFC and its connections.[14] The great expansion in the numbers of neurons, numbers of connections and numbers of dendritic spines in primate dlPFC allows dedicated circuits for the representation of distinct modalities,[15,16] for example, a precise location in visual space,[17,18] while it is likely that animals with fewer neurons in PFC, such as rodents, must rely on more generalized strategies.[19] Importantly, working memory is an ephemeral process, opposite to long-term memory consolidation, and thus it should not be surprising that the circuitry and modulation of working memory should be so different from that of long-term memory consolidation.

The physiology of spatial working memory has been studied extensively in monkeys, using tasks such as the one shown in Fig. 8.1A, in which the subject briefly views a spatial cue, and must remember the position over a delay period of several seconds. At the end of the delay period, the monkey can make a hand or eye movement (i.e., a saccade) to the remembered location, and if correct, receive a reward. The position of the cue randomly changes on every trial, creating extensive proactive interference. The dlPFC region needed to perform this task is area 46 surrounding the principal sulcus (Fig. 8.1B), where even small lesions produce permanent deficits.[21,22] This region receives highly processed visuospatial information from area 7 of the parietal association cortex (Fig. 8.1B), thus forming a circuit for visuospatial cognition.[12] Neuronal recordings from area 46 of dlPFC in monkeys performing a spatial working memory task have found neurons with spatially tuned, persistent firing across the delay period when the spatial position must be held in working memory (Ref. 23; Fig. 8.1C). Thus, these are the neurons that can represent visual space in the absence of sensory stimulation, the foundation of abstract thought.[23] The dlPFC also contains neurons with firing patterns specifically related to the Cue, Response, and Fixation, but the modulation of these neurons is less well understood and thus will not be the focus of the current review.

2. MICROCIRCUITS FOR WORKING MEMORY

Goldman-Rakic and colleagues revealed the dlPFC microcircuitry that underlies working memory-related neuronal firing,[20] summarized in Fig. 8.1D. The persistent firing of Delay cells appears to be generated by the recurrent excitation of glutamatergic pyramidal cell microcircuits in deep layer III (and possibly layer V as well[24]). Neurons with similar spatial tuning excite each other via connections on spines to maintain firing without the need for bottom-up sensory stimulation.[20,25] This persistent firing relies on glutamate excitation of NMDA receptors (NMDARs) (see below). The spatial tuning of Delay cells is shaped through lateral inhibition from GABAergic, parvalbumin-containing interneurons.[20] GABAergic neurons

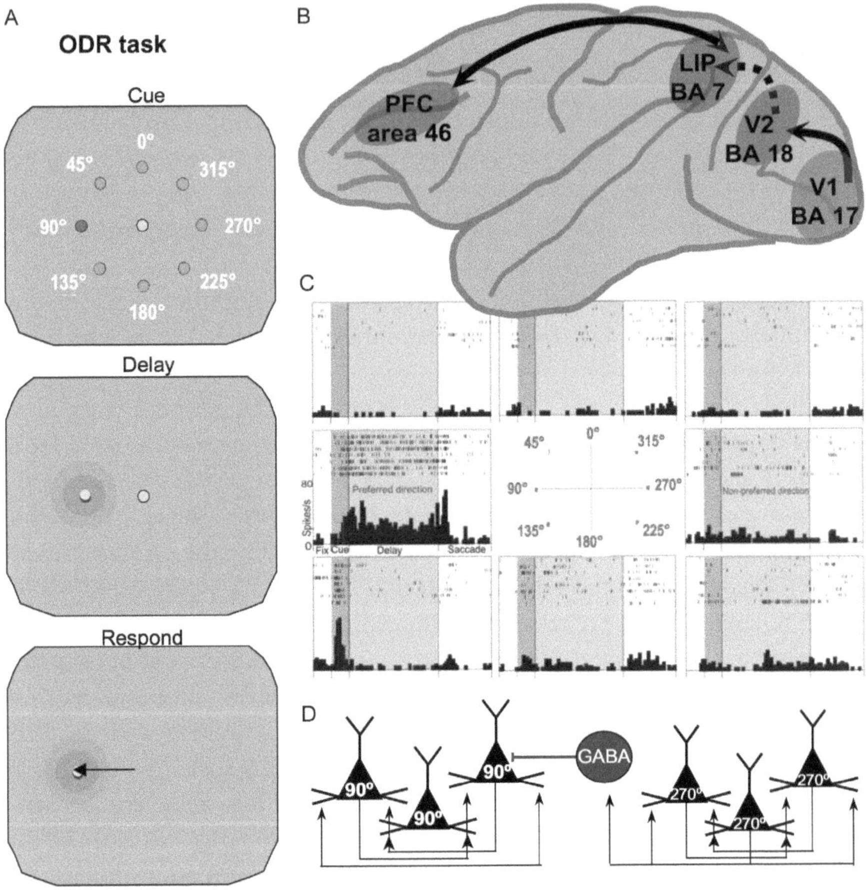

Figure 8.1 The neural circuitry underlying spatial working memory. (A) The oculomotor-delayed response (ODR) task, used to test visuospatial working memory in monkeys. The monkey fixates on the center spot, while a cue appears briefly in one of eight possible locations. The monkey needs to remember the cue location during the delay period (several seconds) using spatial working memory. At the end of the delay period, the monkey makes a quick eye movement to the remembered location to get a reward. The cue location changes from trial to trial, thus requiring constant updating of the contents of working memory. (B) Cortical areas involved in visuospatial working memory. The dlPFC (Walker's area 46) is essential for spatial working memory, and receives visuospatial inputs from the lateral intraparietal cortex (LIP) within Brodmann's Area 7. The parietal cortex receives visual information from earlier visual processing areas, originating from the primary visual cortex, area 17. Together, these areas form a circuit for visuospatial cognition. BA, Brodmann Area. (C) A representative neuron in dlPFC with spatially tuned, persistent activity during the delay period of the ODR task, which is able to represent spatial information in the absence of external cue. The neuron fires across the delay period on trials where the cue appeared at 90° (preferred direction), but not on trials where the cue appeared at other spatial locations. (D) Microcircuits for working memory in dlPFC as revealed by Goldman-Rakic.[20] Persistent firing is generated by recurrent excitation among pyramidal neurons with similar memory fields. The spatial tuning is created via lateral inhibition from GABAergic interneurons such as basket cells and chandelier cells. (See the color plate.)

are excited by pyramidal cell microcircuits with dissimilar tuning; these connections appear to rely on AMPA receptors (AMPARs) in the adult.[26] These layer III microcircuits subserve mental representation, but are also the focus of pathology in mental disease.

Layer III dlPFC pyramidal cells fill with neurofibrillary tangles and die early in Alzheimer's disease,[1] and these layer III dlPFC microcircuits are especially afflicted in schizophrenia, where there is loss of neuropil,[3,27] loss of dendrites and spines,[2,28] and weakening of GABAergic inhibition from parvalbumin-containing interneurons.[29] Taken together, these data suggest that patients with schizophrenia would have impaired ability to maintain information in mind, and eroded definition of information held in working memory stores. This speculation has been confirmed by functional imaging studies showing weaker dlPFC activation during working memory,[30–33] which correlates with symptoms of thought disorder.[33]

3. THE ROLE OF NMDARs

We have recently shown that the persistent firing of dlPFC neurons in monkeys is highly dependent on NMDARs, including those with NR2B subunits.[34] NMDAR–NR2B are localized exclusively within the postsynaptic density, and are not extrasynaptic as they are in adult sensory cortex or hippocampus.[34] Delay cell persistent firing was abolished by local NR2B NMDAR blockade, or by systemic ketamine administration. AMPARs had more subtle effects, contributing a small degree of background depolarization to sustain network firing. In contrast, many Response cells—which likely predominate in rodent PFC—were sensitive to AMPAR blockade and increased firing following systemic ketamine.[34] The physiological data showing loss of persistent, Delay cell firing following NMDAR–NR2B blockade are consistent with computational models predicting that persistent neuronal firing requires the slower kinetics of the NR2B receptor.[35] The reliance of Delay cells on NMDAR may explain why insults to NMDAR in schizophrenia or Alzheimer's disease profoundly impair cognition.

4. THE AROUSAL SYSTEMS PROJECT TO dlPFC

The arousal systems include the catecholamines, acetylcholine, serotonin, orexins, and histamine, all of which project to the PFC, for example, Refs. 36–39. The arousal systems also interact, for example, orexins activate noradrenergic neurons in the locus coeruleus (LC) that project to PFC.[40,41]

In general, these modulators are released during alert waking and not during deep sleep.[42,43] However, there are striking individual differences, for example, noradrenergic neurons are silent during rapid eye movement sleep while cholinergic neurons are active.[36,43] Recordings from monoamine and cholinergic neurons in monkeys indicate that these neurons often fire in response to salient events in the environment.[44–48] However, under conditions of danger or uncontrollable stress, there are very high levels of catecholamines released in the PFC which may involve increased tonic firing, for example, as seen in the LC.[49,50] As norepinephrine (NE) has high affinity for $\alpha 2$ adrenergic receptors and lower affinity for $\alpha 1$ and β adrenergic receptors, varying levels of NE release engage different types of receptors and thus can have qualitatively different actions.[51] As described below, these varying actions can coordinate arousal state with cognitive state. In general, the catecholamines[52] and acetylcholine[53] are particularly important for dlPFC function in primates, while serotonin is especially important for the flexible mediation of emotional behavior by the orbital PFC.[54]

5. DYNAMIC NETWORK CONNECTIVITY: RAPID CHANGES IN NETWORK STRENGTH IN dlPFC

Most studies of molecular modulation of synaptic events have focused on hippocampus, where synaptic inputs onto mushroom-shaped spines can enlarge and stabilize a connection via increases in Ca^{2+}–cAMP signaling (Fig. 8.2A). In contrast, most synaptic connections of pyramidal cells in deep layer III of dlPFC reside on long, thin dendritic spines,[55] an architecture appropriate for dynamic gating of network inputs.[4,56] As summarized in Fig. 8.2B, connections can be rapidly and reversibly weaken by Ca^{2+}–cAMP opening potassium (K^+) channels on spines, and strengthened by closing these channels and/or opening channels that depolarize the spine, for example, nicotinic $\alpha 7$ receptors (nic-α7Rs). We have called this process Dynamic Network Connectivity (DNC).[56] This dynamic gating (on a timescale of seconds) can weaken or strengthen network inputs and neuronal firing without architectural changes, and thus contrasts with classic neuroplasticity, for example, long-term potentiation (LTP) where the goal is enduring changes in circuit strength. DNC provides a mechanism for coordinating cognitive state with arousal state, and for sculpting the subset of network inputs that shape mental activity.[4,56] It is logical that the highly evolved dlPFC would be modulated in a manner different from sensory cortex and subcortical structures, whereby Ca^{2+}–cAMP signaling weakens

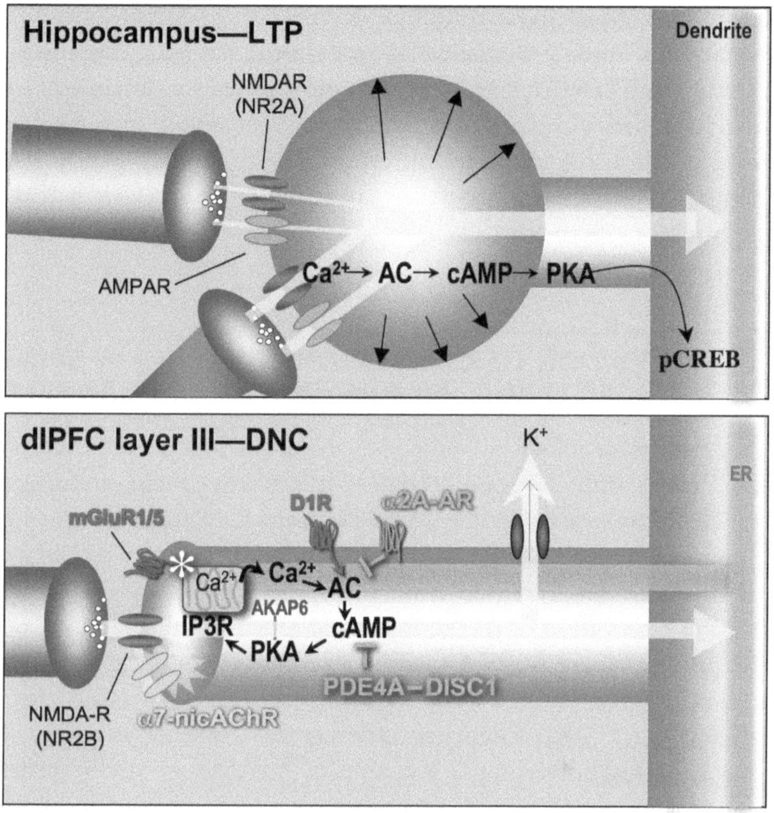

Figure 8.2 Comparison of long-term potentiation (LTP) in hippocampus and dynamic network connectivity in dlPFC. (A) Synaptic inputs onto mushroom-shaped spines in hippocampus can potentiate connections by increasing Ca^{2+}–cAMP phosphorylation of CREB (cAMP response element-binding protein), generating transcription events. The change in circuit strength is enduring, due to architectural changes such as enlarging of the spine head. Hippocampal glutamate synapses in the adult are mediated primarily by AMPAR excitation of NMDA receptors containing NR2A subunits; NR2B-containing NMDA receptors can be seen in extrasynaptic locations (not shown). (B) In contrast to the hippocampus, excitatory synaptic inputs onto long, thin spines in dlPFC layer III utilize NMDA receptors with NR2B subunits which are found exclusively in the postsynaptic density and are not extrasynaptic. Cholinergic α7-nicotinic receptors are also found in the postsynaptic density and are permissive for NMDA receptor actions. These synaptic connections can be rapidly and reversibly weakened through potassium channel opening by increased Ca^{2+}–cAMP signaling. The circuit strength can be weakened or strengthened via different molecular pathways on a timescale of seconds without long-term architectural changes, a process termed dynamic network connectivity (DNC). See text for more detailed description of these mechanisms. (See the color plate.)

dlPFC and strengthens more primitive circuits, as this allows a switch from slow, thoughtful dlPFC regulation to more reflexive brain circuits during danger. Although DNC normally acts on short timescale, sustained changes in arousal state, for example, chronic stress, can lead to architectural changes, for example, spine loss with chronic stress. The following section reviews some of the molecular pathways that we have discovered to modulate dlPFC working memory circuits.

6. ACTIVATION OF CA^{2+}–cAMP SIGNALING IN SPINES REDUCES FIRING VIA OPENING OF K^{+} CHANNELS

Synergistic, feedforward processes are needed to rapidly take dlPFC "off-line" and switch control of behavior to primitive brain circuits during danger. We have seen evidence of feedforward Ca^{2+}–cAMP signaling in dlPFC spines, whereby Ca^{2+} increases cAMP generation, and cAMP facilitates intracellular Ca^{2+} release from the spine apparatus (white asterisk), which is the elaboration of the endoplasmic reticulum in the spine. cAMP activates protein kinase A (PKA), which is tethered near the spine apparatus by an anchoring protein, AKAP6 (Ref. 57 and C. Paspalas and A. Arnsten, unpublished data). PKA facilitates internal Ca^{2+} release from the spine apparatus by phosphorylating IP3 receptors,[58] which are concentrated on the spine apparata in dlPFC.[59] Calcium in turn potentiates adenylyl cyclase (AC) generation of cAMP.[60] Calcium may additionally potentiate this process by activating protein kinase C (PKC), which can uncouple α2A-adrenergic receptors (α2A-ARs) which normally serve to inhibit cAMP signaling (Ref. 61; not shown in Fig. 8.2B). Calcium can build up in spines through a number of mechanisms, for example, through NMDARs (especially those with NR2B subunits)[62] and through IP3-mediated internal Ca^{2+} release. Internal Ca^{2+} release is stimulated by NE α1-AR, and by Gq-coupled metabotropic glutamate receptors (mGluR1α and/or mGluR5), which have been localized near the synapse in primate dlPFC spines.[63] mGluR generation of IP3 increases internal Ca^{2+} release from the spine apparatus where it can engage in a variety of actions, including reducing neuronal excitability by opening SK channels. SK channels provide negative feedback for NMDAR excitation,[64] reduce PFC cell firing,[65] and impair working memory in rats.[66] SK channels have not yet been mapped in primate dlPFC, but are likely to reside near the spine apparatus as well as on dendrites.

These mechanisms have clinical relevance to stress disorders. Studies in animals have shown that high levels of NE α1-AR stimulation contribute to PFC cognitive dysfunction during stress exposure.[67,68] NE α1-AR stimulation reduces dlPFC neuronal firing through a Ca^{2+}-PKC signaling mechanism.[69] This mechanism contributes to dendritic spine loss during chronic stress exposure,[70] perhaps via phosphorylation of MARCKS and collapse of the actin cytoskeleton in the spine.[71] This mechanism may contribute to PFC gray matter loss in humans exposed to repeated adversity.[72] As PKC signaling is overactive in bipolar disorder,[73] and in lead poisoning (which potently mimics Ca^2), this mechanism may also contribute to PFC gray matter loss in these disorders as well as posttraumatic stress disorder (PTSD).[74,75] Based, in part, on our work in animals, the NE α1-AR antagonist, prazosin, is now in use for the treatment of PTSD in veterans, soldiers, and civilians.[76,77]

cAMP signaling also influences network connectivity and neuronal excitability by altering the open state of a variety of ion channels: it directly increases the open probability of HCN (hyperpolarization-activated cyclic nucleotide-gated) channels,[78] while cAMP–PKA signaling increases the open probability of KCNQ channels.[79] KCNQ channels are also known as "M" channels, as they are closed by muscarinic receptor stimulation. In hippocampus, HCN channels are concentrated on distal apical pyramidal dendrites, where they gate distal inputs, for example, Ref. 80, and modulate excitability and plasticity, for example, Ref. 81. HCN channels are also on the distal apical dendrites of layer V dlPFC pyramidal cells.[82,83] However, in deep layer III of dlPFC, HCN channels are enriched in long, thin spines,[82] both in the spine neck and next to the synapse. These are likely HCN1–HCN2 heteromers, which rapidly respond to cAMP.[78,84] A variety of cAMP-related signaling proteins can be observed in deep layer III long, thin spines near the HCN channels. The phosphodiesterase PDE4A is commonly found in the spine neck and in the spine head near HCN channels, positioned to regulate the amount of cAMP and thus the degree of HCN channel opening. Inhibiting PDE4 regulation of cAMP by iontophoresis of etazolate onto dlPFC neurons induces a rapid collapse in dlPFC.[83] Delay cell firing is also reduced by agents that mimic cAMP, stimulate the generation of cAMP, or block the inhibition of cAMP generation, indicating the power of this mechanism.[83,85,86] Under conditions of high cAMP signaling, Delay cell firing can be restored by simultaneously blocking HCN channels with co-iontophoresis of ZD7288, demonstrating physiological interactions.[83] Similar effects have been observed at the behavioral level, where increased cAMP signaling in PFC impairs working memory,[87] and low dose blockade of HCN channels in rat PFC can restore working memory performance.[83] In some neurons, PKA phosphorylation of HCN channels can lead to sustained increases in channel opening[88]; if this occurs in dlPFC, it could contribute to prolonged cognitive impairment, for example, as occurs with fatigue and/or stress. We have also documented KCNQ channels on layer III dlPFC spines. KCNQ channels are present in many other cellular compartments as well, where they influence neuronal excitability and action potential generation, for example, Ref. 89, but may have a gating function in spines in dlPFC. KCNQ channels are of special interest to neuromodulation, as their open state is regulated by a variety of modulatory systems, including cAMP–PKA, muscarinic, and endocannabanoid/arachidonic acid signaling, as well as pH.[79] There are likely additional ionic mechanisms that contribute to rapid weakening of synaptic efficacy, but existing data already demonstrate powerful ionic mechanisms that can rapidly disconnect dlPFC neuronal networks and reduce neuronal firing.

7. INHIBITION OF CA^{2+}–cAMP–K^{+} SIGNALING IN SPINES ENHANCES TASK-RELATED FIRING

Research in nonhuman primates has identified mechanisms that can enhance working memory-related neuronal firing in dlPFC either by inhibiting Ca^{2+}–cAMP signaling in spines or by directly depolarizing the spine compartment (Fig. 8.2B). A variety of mechanisms regulate cAMP signaling in layer III dlPFC spines. As mentioned above, the phosphodiesterase PDE4A, which catabolizes cAMP, is often localized next to HCN channels in spines, and near the spine apparatus to regulate cAMP effects on internal Ca^{2+} release.[82] PDE4A is anchored to the correct location in the spine by the scaffolding protein, DISC1 (disrupted in schizophrenia), as DISC1 tethers a variety of PDE4s,[90] and colocalizes with PDE4A in layer III spines in monkey dlPFC.[82] DISC1 is also found next to HCN channels[82] and near the spine apparatus in monkey dlPFC layer III spines,[4] suggesting that a DISC1–PDE4A interactome is positioned to regulate both network gating by ion channels as well as cAMP–PKA facilitation of internal Ca^{2+} release. Both DISC1 and PDE4A are genetically altered in families with mental illness,[91–93] which may lead to dysregulated cAMP signaling and impaired dlPFC function. Indeed, knockdown of DISC1 lowers the threshold for stress-induced PFC dysfunction in rats.[94]

Another key regulator of cAMP signaling in spines is the α2A-AR, which inhibit cAMP generation. Although the presynaptic role of these receptors is often emphasized, the majority of these receptors in brain are postsynaptic, and α2A-ARs are colocalized with HCN channels in layer III spines near the synapse and in the spine neck (Fig. 8.3A; Ref. 83). Stimulation of α2A-ARs, for example, with the α2A-AR agonist, guanfacine, specifically increases firing for the neuron's preferred direction, thus enhancing mental representation (Fig. 8.3B; Ref. 83). Conversely, blockade of α2A-ARs with yohimbine causes a complete collapse of dlPFC network firing[95] that can be restored by blocking HCN channels.[83] Parallel effects are seen on cognitive behavior, where infusion of guanfacine directly into dlPFC improves working memory,[96] and systemic administration of guanfacine improves a variety of PFC cognitive functions, including spatial working memory, behavioral inhibition, top-down regulation of attention, and rapid associative learning (reviewed in Ref. 97). A recent study has shown that guanfacine improves impulse control by inhibiting responses to an immediate, small reward in order to wait over a delay for a larger reward.[98] All of these tasks require behavior to be guided by mental representation. Conversely, the work of Bao-Ming Li has shown that infusion of the α2A-AR antagonist, yohimbine, into the dlPFC impairs working memory and impulse control, and induces locomotor hyperactivity in monkeys (reviewed in Ref. 97). Thus, α2A-AR stimulation strengthens the efficacy of dlPFC microcircuit connections, enhancing mental representation and top-down regulation of behavior.

Based on this research in animals, guanfacine is now being used to treat a variety of PFC disorders in human patients, including attention deficit hyperactivity disorder (extended release pediatric formulation Intuniv™)[99], Tourette's Syndrome[100], autism spectrum illness[101], substance abuse (S. McKee, R. Sinha, and A. Arnsten, in preparation), and traumatic brain injury that involves the frontal lobe.[102] Guanfacine is also being used to help children who have been traumatized, for example, from experiencing a tragedy or from physical abuse.[103] This is one of the few examples where a treatment for cognitive disorders in humans has arisen from uncovering neural mechanisms in animals, a translational success and validation of this important research strategy.

There are also mechanisms that likely regulate internal Ca^{2+} release from the spine apparatus. For example, RGS4 (regulator of G protein signaling 4) flanks the synapse where it is positioned to inhibit perisynaptic mGluR1α/5-Gq signaling and reduce internal Ca^{2+} release.[104] RGS4 is of particular interest, as it is greatly reduced in the dlPFC of patients with schizophrenia,[105,106] and can be the target of genetic insults in mental illness.[107–110]

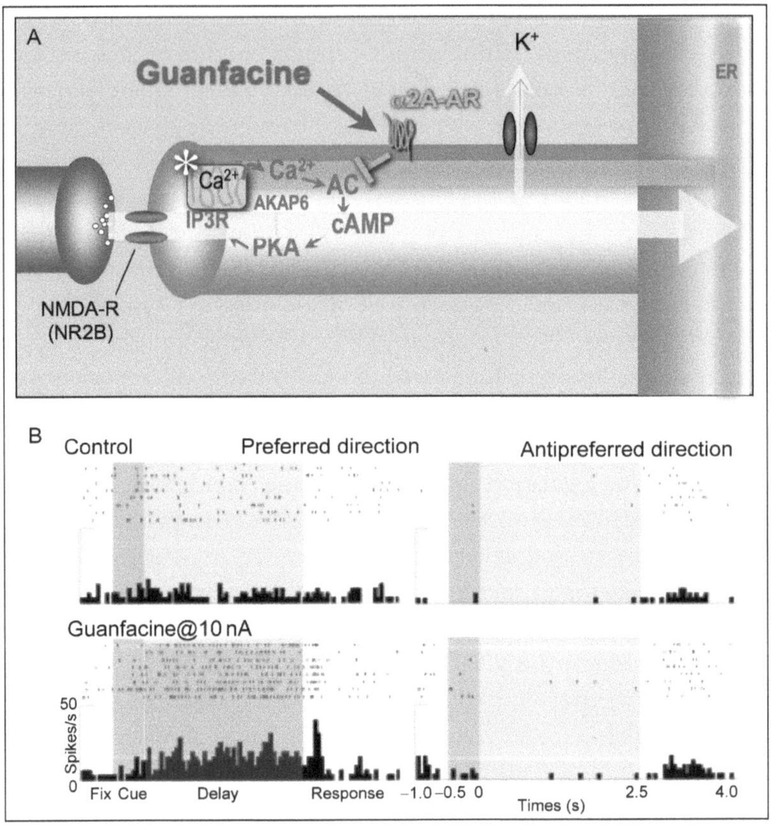

Figure 8.3 Guanfacine inhibits Ca^{2+}–cAMP–K^{+} signaling in spines and enhances task-related firing. (A) Schematic representation of α2A-AR therapeutic actions in primate PFC. α2A-AR are localized on dendritic spines near potassium channels in the spine neck and near the synapse (e.g., near HCN channels). Stimulation of these receptors by the α2A-AR agonist, guanfacine, inhibits cAMP signaling, thus closing potassium channels and strengthening network connectivity. (B) Stimulation of α2A-AR by guanfacine specifically increases persistent firing for the neuron's preferred direction, thus enhancing working memory representation. These effects can be reversed by activation of cAMP signaling, or mimicked by inhibition of cAMP signaling or by potassium channel blockade. Systemic administration of guanfacine in monkeys improves working memory performance, strengthens impulse control, and enhances the regulation of emotional behavior. Based on these animal data, guanfacine is now in use to treat PFC disorders in humans. See text for details. (See the color plate.)

Synaptic efficacy also may be strengthened by ion channels that directly depolarize the spine, for example, nic-α7Rs[111] and TRPC channels.[65] TRPC channels have not been studied in the primate dlPFC, but data from rodent PFC suggest these mechanisms also play an important role. Recent data indicate that nic-α7R may have a particularly powerful role in dlPFC. Nic-α7Rs have been localized in the postsynaptic density in the rodent medial PFC[111] and monkey dlPFC.[112] Systemic administration of nic-α 7R agonists improves working memory in monkeys, including those impaired by NMDAR antagonists.[113] Similar effects have been seen at the physiological level, where iontophoresis of nic-α7R agonists onto dlPFC neurons increases Delay cell task-related firing and reverses firing deficiencies from NMDAR blockade.[112] The physiological data are consistent with nic-α7R playing a permissive role in NMDAR actions,[112] similar to AMPAR permissive effects in sensory and hippocampal circuits. These effects may be particularly important for coordinating dlPFC network firing with sleep-waking states. Cholinergic stimulation also can strengthen connectivity and increase firing via stimulation of muscarinic M1 receptors, which close KCNQ channels.[114] Thus, cholinergic stimulation may strengthen network firing through both muscarinic and nicotinic mechanisms. These physiological data are consistent with recent behavioral studies in monkeys showing a profound impairment in working memory abilities following cholinergic depletion from the dlPFC.[53]

8. DOPAMINE D1R STIMULATION REGULATES THE BREADTH OF NETWORK INPUTS

The pioneering discovery of Brozoski *et al.* in 1979 revealed that dopamine (DA) is essential to dlPFC working memory function, whereby depletion of catecholamines from the dlPFC was as devastating as removal of the cortex itself.[52] Although the original work emphasized the beneficial effects of DA, we now know that DA stimulation of D1 receptors (D1Rs) has an inverted U dose–response influence on dlPFC neuronal firing[85] and on working memory performance,[115] with high doses decreasing firing and impairing working memory. *In vitro* recordings from PFC slices have been ideal preparations for examining the excitatory effects of very low dose D1R stimulation, as there is no endogenous DA in the slice. PFC neurons are also hyperpolarized in the slice, without the constant excitation from neighbors that occurs *in vivo*. It should be noted that most of these studies are done on layer V pyramidal cells; however, as some of layer V neurons may "migrate" into layer III in the more differentiated primate dlPFC,[16] these data may also be relevant to the recurrent layer III neurons. The *in vitro* studies have revealed excitatory effects of D1R stimulation in both rat medial PFC[116] and monkey dlPFC,[117] for example, by enhancing persistent sodium currents[118] and NMDAR actions, for example, Ref. 116. These data are echoed *in vivo*, where high doses of D1R antagonist lead to loss of dlPFC Delay cell firing[119] and to working memory impairment.[120] In contrast, moderate levels of D1R stimulation have sculpting actions on the pattern of task-related neuronal firing.[85] Iontophoresis of low doses of D1R agonists onto noisy dlPFC Delay cells can selectively decrease neuronal firing for the neurons' nonpreferred directions while leaving firing for the neurons' preferred direction intact.[85] These sculpting effects likely involve cAMP–K^+ channel gating actions, but may also involve facilitation of lateral inhibition from GABAergic interneurons[121,122] and presynaptic inhibition of glutamate release.[123] Finally, very high doses of DA D1R stimulation, as occurs during uncontrollable stressors, reduce all neuronal firing and impair working memory.[85] The deleterious effects of D1R agonists on neuronal firing and working memory performance are prevented by cAMP inhibition,[85] or by HCN or KCNQ channel blockade (N. Gamo, Y. Yang, M. Wang, L. Jin, and A. Arnsten, unpublished data) but are often not reversed once the D1R agonist has taken effect. These irreversible actions may involve cAMP–PKA phosphorylation of HCN channels maintaining channels in open state.[88]

Thus, in summary, optimal levels of DA D1R stimulation may sculpt network inputs to define the contents of working memory, while at higher levels, they take dlPFC "off-line" to switch control of behavior to more primitive circuits.[4] The finding of increased D1R expression in the dlPFC of unmedicated patients with schizophrenia[124] suggests that these suppressive actions may contribute to cognitive deficits in this devastating disorder. This may be especially problematic in adolescence when there is additional DA innervation of layer III in the primate dlPFC,[125,126] and particularly under stressful conditions such as going away to college or the military which often precipitate psychotic episodes.[127] Thus, understanding the mechanisms by which stress weakens dlPFC may help to illuminate the etiology of schizophrenia.

9. CONCLUSION

The dlPFC is modulated in a manner that is very different from classic neuroplasticity in sensory cortex and in subcortical circuits. Thus, while increases in calcium and cAMP signaling enhance long-term consolidation, they erode working memory in dlPFC by opening potassium channels near synapses on spines. These molecular differences may have evolved along with the PFC to allow the flexible connectivity needed for fluid intelligence. Understanding these mechanisms may help us to understand why so many genetic insults can lead to the same phenotype of impaired dlPFC cognitive function. Thus, the very mechanisms that have evolved to provide mental flexibility may also confer vulnerability to cognitive disorders.

ACKNOWLEDGMENTS

This research was supported by grants PO1 AG030004 and 1RL1AA017536 to A. F. T. A. and by a donation to L. E. J. in honor of Percy Sanguinetti Arnsten, whose optimism and courage continue in this work.

REFERENCES

1. Bussière T, Giannakopoulos P, Bouras C, Perl DP, Morrison JH, Hof PR. Progressive degeneration of nonphosphorylated neurofilament protein-enriched pyramidal neurons predicts cognitive impairment in Alzheimer's disease: stereologic analysis of prefrontal cortex area 9. *J Comp Neurol*. 2003;463:281–302.
2. Glantz LA, Lewis DA. Decreased dendritic spine density on prefrontal cortical pyramidal neurons in schizophrenia. *Arch Gen Psychiatry*. 2000;57:65–73.
3. Selemon LD, Rajkowska G, Goldman-Rakic PS. Abnormally high neuronal density in the schizophrenic cortex: a morphometric analysis of prefrontal area 9 and occipital area 17. *Arch Gen Psychiatry*. 1995;52:805–818.
4. Arnsten AFT, Wang MJ, Paspalas CD. Neuromodulation of thought: flexibilities and vulnerabilities in prefrontal cortical network synapses. *Neuron*. 2012;76:223–239.
5. Goldman-Rakic PS. The prefrontal landscape: implications of functional architecture for understanding human mentation and the central executive. *Phil Trans R Soc London*. 1996;351:1445–1453.
6. Jacobsen CF. Studies of cerebral function in primates. *Comp Psychol Monogr*. 1936;13:1–68.
7. Miller EK. The prefrontal cortex and cognitive control. *Nat Rev Neurosci*. 2000;1:59–65.
8. Robbins TW. Dissociating executive functions of the prefrontal cortex. *Phil Trans R Soc London*. 1996;351:1463–1471.
9. Fuster JM. The prefrontal cortex, mediator of cross-temporal contingencies. *Hum Neurobiol*. 1985;4:169–179.
10. Fuster JM. Network memory. *Trends Neurosci*. 1997;20:451–459.
11. Aron AR. From reactive to proactive and selective control: developing a richer model for stopping inappropriate responses. *Biol Psychiatry*. 2011;69:e55–e68.
12. Goldman-Rakic PS. Circuitry of the primate prefrontal cortex and the regulation of behavior by representational memory. In: Plum F, ed. *Handbook of Physiology, The Nervous System, Higher Functions of the Brain*. Bethesda: American Physiological Society; 1987:373–417.
13. Price JL, Carmichael ST, Drevets WC. Networks related to the orbital and medial prefrontal cortex; a substrate for emotional behavior? *Prog Brain Res*. 1996;107:523–536.
14. Elston GN, Benavides-Piccione R, Elston A, et al. Specializations of the granular prefrontal cortex of primates: implications for cognitive processing. *Anat Rec A: Discov Mol Cell Evol Biol*. 2006;288:26–35.
15. Elston GN. Pyramidal cells of the frontal lobe: all the more spinous to think with. *J Neurosci*. 2000;20:RC95.

16. Elston GN. Cortex, cognition and the cell: new insights into the pyramidal neuron and prefrontal function. *Cereb Cortex*. 2003;13:1124–1138.
17. O Scalaidhe SP, Wilson FA, Goldman-Rakic PS. Areal segregation of face-processing neurons in prefrontal cortex. *Science*. 1997;278:1135–1138.
18. Wilson FA, Scalaidhe SP, Goldman-Rakic PS. Dissociation of object and spatial processing domains in primate prefrontal cortex. *Science*. 1993;260:1955–1958.
19. Preuss T. Do rats have prefrontal cortex? The Rose-Woolsey-Akert program reconsidered. *J Cognit Neurosci*. 1995;7:1–26.
20. Goldman-Rakic PS. Cellular basis of working memory. *Neuron*. 1995;14:477–485.
21. Funahashi S, Bruce CJ, Goldman-Rakic PS. Dorsolateral prefrontal lesions and oculomotor delayed-response performance: evidence for mnemonic "scotomas" *J Neurosci*. 1993;13:1479–1497.
22. Goldman PS, Rosvold HE. Localization of function within the dorsolateral prefrontal cortex of the rhesus monkey. *Exp Neurol*. 1970;27:291–304.
23. Funahashi S, Bruce CJ, Goldman-Rakic PS. Mnemonic coding of visual space in the monkey's dorsolateral prefrontal cortex. *J Neurophysiol*. 1989;61:331–349.
24. Kritzer MF, Goldman-Rakic PS. Intrinsic circuit organization of the major layers and sublayers of the dorsolateral prefrontal cortex in the rhesus monkey. *J Comp Neurol*. 1995;359:131–143.
25. González-Burgos G, Barrionuevo G, Lewis DA. Horizontal synaptic connections in monkey prefrontal cortex: an in vitro electrophysiological study. *Cereb Cortex*. 2000;10:82–92.
26. Rotaru DC, Yoshino H, Lewis DA, Ermentrout GB, Gonzalez-Burgos G. Glutamate receptor subtypes mediating synaptic activation of prefrontal cortex neurons: relevance for schizophrenia. *J Neurosci*. 2011;31:142–156.
27. Selemon LD, Rajkowska G, Goldman-Rakic PS. Elevated neuronal density in prefrontal area 46 in brains from schizophrenic patients: application of a three-dimensional, stereologic counting method. *J Comp Neurol*. 1998;392:402–412.
28. Glantz LA, Lewis DA. Reduction of synaptophysin immunoreactivity in the prefrontal cortex of subjects with schizophrenia. Regional and diagnostic specificity. *Arch Gen Psychiatry*. 1997;54:943–952.
29. Lewis DA, Hashimoto T, Volk DW. Cortical inhibitory neurons and schizophrenia. *Nat Rev Neurosci*. 2005;6:312–324.
30. Barch DM. The cognitive neuroscience of schizophrenia. *Annu Rev Clin Psychol*. 2005;1:321–353.
31. Driesen NR, Leung HC, Calhoun VD, et al. Impairment of working memory maintenance and response in schizophrenia: functional magnetic resonance imaging evidence. *Biol Psychiatry*. 2008;64:1026–1034, PMCID: PMC2650279.
32. Goldman-Rakic PS. Prefrontal cortical dysfunction in schizophrenia: the relevance of working memory. In: Carroll BJ, Barrett JE, eds. *Psychopathology and the Brain*. New York, NY: Raven Press; 1991:1–23.
33. Perlstein WM, Carter CS, Noll DC, Cohen JD. Relation of prefrontal cortex dysfunction to working memory and symptoms in schizophrenia. *Am J Psychiatry*. 2001;158:1105–1113.
34. Wang MJ, Yang Y, Wang CJ, et al. NMDA receptors subserve working memory persistent neuronal firing In dorsolateral prefrontal cortex. *Neuron*. 2013;77:736–749.
35. Wang XJ. Synaptic basis of cortical persistent activity: the importance of NMDA receptors to working memory. *J Neurosci*. 1999;19:9587–9603.
36. Hobson JA. Sleep and dreaming: induction and mediation of REM sleep by cholinergic mechanisms. *Curr Opin Neurobiol*. 1992;2:759–763.
37. Levitt P, Rakic P, Goldman-Rakic P. Region-specific distribution of catecholamine afferents in primate cerebral cortex: a fluorescence histochemical analysis. *J Comp Neurol* 1984;227:23–36.

38. Lewis DA, Morrison JH. Noradrenergic innervation of monkey prefrontal cortex: a dopamine-beta-hydroxylase immunohistochemical study. *J Comp Neurol.* 1989;282:317–330.
39. Lewis DA, Cambell MJ, Foote SL, Goldstein M, Morrison JH. The distribution of tyrosine hydroxylase-immunoreactive fibers in primate neocortex is widespread but regionally specific. *J Neurosci.* 1987;282:317–330.
40. Del Cid-Pellitero E, Garzón M. Hypocretin1/OrexinA-containing axons innervate locus coeruleus neurons that project to the Rat medial prefrontal cortex. Implication in the sleep-wakefulness cycle and cortical activation. *Synapse.* 2011;65:843–857.
41. Horvath TL, Peyron C, Diano S, et al. Hypocretin (orexin) activation and synaptic innervation of the locus coeruleus noradrenergic system. *J Comp Neurol.* 1999;415:145–159.
42. Foote SL, Bloom FE, Aston-Jones G. Nucleus locus coeruleus: new evidence of anatomical and physiological specificity. *Physiol Rev.* 1983;63:844–914.
43. Jones BE. The role of noradrenergic locus coeruleus neurons and neighboring cholinergic neurons of the pontomesencephalic tegmentum in sleep-wake states. *Prog Brain Res.* 1991;88:533–543.
44. Aston-Jones G, Rajkowski J, Cohen J. Role of locus coeruleus in attention and behavioral flexibility. *Biol Psychiatry.* 1999;46:1309–1320.
45. Bromberg-Martin ES, Matsumoto M, Hikosaka O. Dopamine in motivational control: rewarding, aversive, and alerting. *Neuron.* 2010;68:815–834.
46. Foote SL, Aston-Jones G, Bloom FE. Impulse activity of locus coeruleus neurons in awake rats and monkeys is a function of sensory stimulation and arousal. *Proc Natl Acad Sci.* 1980;77:3033–3037.
47. Richardson RT, DeLong MR. Nucleus basalis of Meynert neuronal activity during a delayed response task in a monkey. *Brain Res.* 1986;399:364–368.
48. Schultz W. The phasic reward signal of primate dopamine neurons. *Adv Pharmacol.* 1998;42:686–690.
49. Rajkowski J, Kubiak P, Inanova S, Aston-Jones G. State-related activity, reactivity of locus coeruleus neurons in behaving monkeys. *Adv Pharmacol.* 1998;42:740–744.
50. Rajkowski J, Majczynski H, Clayton E, Aston-Jones G. Activation of monkey locus coeruleus neurons varies with difficulty and performance in a target detection task. *J Neurophysiol.* 2004;92:361–371.
51. Arnsten AFT. Through the looking glass: differential noradrenergic modulation of prefrontal cortical function. *Neural Plast.* 2000;7:133–146.
52. Brozoski T, Brown RM, Rosvold HE, Goldman PS. Cognitive deficit caused by regional depletion of dopamine in prefrontal cortex of rhesus monkey. *Science.* 1979;205:929–931.
53. Croxson PL, Kyriazis DA, Baxter MG. Cholinergic modulation of a specific memory function of prefrontal cortex. *Nat Neurosci.* 2011;14:1510–1512.
54. Clarke HF, Walker SC, Dalley JW, Robbins TW, Roberts AC. Cognitive inflexibility after prefrontal serotonin depletion is behaviorally and neurochemically specific. *Cereb Cortex.* 2007;17:18–27.
55. Dumitriu D, Hao J, Hara Y, et al. Selective changes in thin spine density and morphology in monkey prefrontal cortex correlate with aging-related cognitive impairment. *J Neurosci.* 2010;30:7507–7515.
56. Arnsten AFT, Paspalas CD, Gamo NJ, Yang Y, Wang M. Dynamic network connectivity: a new form of neuroplasticity. *Trends Cog Sci.* 2010;14:365–375, PMCID: PMC2914830.
57. McCartney S, Little BM, Langeberg LK, Scott JD. Cloning and characterization of A-kinase anchor protein 100 (AKAP100). A protein that targets A-kinase to the sarcoplasmic reticulum. *J Biol Chem.* 1995;270:9327–9333.

58. Soulsby MD, Wojcikiewicz RJ. The type III inositol 1,4,5-trisphosphate receptor is phosphorylated by cAMP-dependent protein kinase at three sites. *Biochem J.* 2005;392:493–497.
59. Paspalas CD, Goldman-Rakic PS. Microdomains for dopamine volume neurotransmission in primate prefrontal cortex. *J Neurosci.* 2004;24:5292–5300.
60. Ferguson GD, Storm DR. Why calcium-stimulated adenylyl cyclases? *Physiology.* 2004;19:271–276.
61. Wang Q, Limbird LL. Regulation of α2AR trafficking and signaling by interacting proteins. *Biochem Pharmacol.* 2007;73:1135–1145.
62. Liu Y, Wong TP, Aarts M, et al. NMDA receptor subunits have differential roles in mediating excitotoxic neuronal death both in vitro and in vivo. *J Neurosci.* 2007;27:2846–2857.
63. Muly EC, Maddox M, Smith Y. Distribution of mGluR1alpha and mGluR5 immunolabeling in primate prefrontal cortex. *J Comp Neurol.* 2003;467:521–535.
64. Faber ES. Functional interplay between NMDA receptors, SK channels and voltage-gated Ca2+ channels regulates synaptic excitability in the medial prefrontal cortex. *J Physiol.* 2010;588:1281–1292.
65. Hagenston AM, Fitzpatrick JS, Yeckel MF. mGluR-mediated calcium waves that invade the soma regulate firing in layer V medial prefrontal cortical pyramidal neurons. *Cereb Cortex.* 2008;18:407–423.
66. Brennan AR, Dolinsky B, Vu MA, Stanley M, Yeckel MF, Arnsten AF. Blockade of IP3-mediated SK channel signaling in the rat medial prefrontal cortex improves spatial working memory. *Learn Mem.* 2008;15:93–96.
67. Arnsten AFT, Jentsch JD. The alpha-1 adrenergic agonist, cirazoline, impairs spatial working memory performance in aged monkeys. *Pharmacol Biochem Behav.* 1997;58:55–59.
68. Birnbaum SG, Gobeske KT, Auerbach J, Taylor JR, Arnsten AFT. A role for norepinephrine in stress-induced cognitive deficits: alpha-1-adrenoceptor mediation in prefrontal cortex. *Biol Psychiatry.* 1999;46:1266–1274.
69. Birnbaum SB, Yuan P, Wang M, et al. Protein kinase C overactivity impairs prefrontal cortical regulation of working memory. *Science.* 2004;306:882–884.
70. Hains AB, Vu MA, Maciejewski PK, van Dyck CH, Gottron M, Arnsten AF. Inhibition of protein kinase C signaling protects prefrontal cortex dendritic spines and cognition from the effects of chronic stress. *Proc Natl Acad Sci U S A.* 2009;106:17957–17962.
71. Calabrese B, Halpain S. Essential role for the PKC target MARCKS in maintaining dendritic spine morphology. *Neuron.* 2005;48:77–90.
72. Ansell EB, Rando K, Tuit K, Guarnaccia J, Sinha R. Cumulative adversity and smaller gray matter volume in medial prefrontal, anterior cingulate, and insula regions. *Biol Psychiatry.* 2012;72:57–64.
73. Arnsten AFT, Manji HK. Mania: a rational neurobiology. *Future Neurol.* 2008;3:125–131.
74. Blumberg HP, Krystal JH, Bansal R, et al. Age, rapid-cycling, and pharmacotherapy effects on ventral prefrontal cortex in bipolar disorder: a cross-sectional study. *Biol Psychiatry.* 2006;59:611–618.
75. Cecil KM, Brubaker CJ, Adler CM, et al. Decreased brain volume in adults with childhood lead exposure. *PLoS Med.* 2008;5:e112.
76. Raskind MA, Peskind ER, Kanter ED, et al. Prazosin reduces nightmares and other PTSD symptoms in combat veterans: a placebo-controlled study. *Am J Psychiatry.* 2003;160:371–373.
77. Taylor FB, Lowe K, Thompson C, et al. Daytime prazosin reduces psychological distress to trauma specific cues in civilian trauma posttraumatic stress disorder. *Biol Psychiatry.* 2006;59:577–581.

78. Ulens C, Tytgat J. Functional heteromerization of HCN1 and HCN2 pacemaker channels. *J Biol Chem*. 2001;276:6069–6072.
79. Delmas P, Brown DA. Pathways modulating neural KCNQ/M (Kv7) potassium channels. *Nat Rev Neurosci*. 2005;6:850–862.
80. Nolan MF, Malleret G, Dudman JT, et al. A behavioral role for dendritic integration: HCN1 channels constrain spatial memory and plasticity at inputs to distal dendrites of CA1 pyramidal neurons. *Cell*. 2004;119:719–732.
81. Fan Y, Fricker D, Brager DH, et al. Activity-dependent decrease of excitability in rat hippocampal neurons through increases in I(h). *Nat Neurosci*. 2005;8:1542–1551.
82. Paspalas CD, Min Wang M, Arnsten AFT. Constellation of HCN Channels and cAMP regulating proteins in dendritic spines of the primate prefrontal cortex—potential substrate for working memory deficits in schizophrenia. *Cereb Cortex*. 2013;23:1643–1654.
83. Wang M, Ramos B, Paspalas C, et al. Alpha2A-adrenoceptor stimulation strengthens working memory networks by inhibiting cAMP-HCN channel signaling in prefrontal cortex. *Cell*. 2007;129:397–410.
84. Chen S, Wang J, Siegelbaum SA. Properties of hyperpolarization-activated pacemaker current defined by coassembly of HCN1 and HCN2 subunits and basal modulation by cyclic nucleotide. *J Gen Physiol*. 2001;117:491–504.
85. Vijayraghavan S, Wang M, Birnbaum SG, Bruce CJ, Williams GV, Arnsten AFT. Inverted-U dopamine D1 receptor actions on prefrontal neurons engaged in working memory. *Nat Neurosci*. 2007;10:376–384.
86. Wang M, Gamo NJ, Yang Y, et al. Neuronal basis of age-related working memory decline. *Nature*. 2011;476:210–213, PMCID: PMC3193794.
87. Taylor JR, Birnbaum SG, Ubriani R, Arnsten AFT. Activation of cAMP-dependent protein kinase A in prefrontal cortex impairs working memory performance. *J Neurosc (Online)*. 1999;19:RC23.
88. Vargas G, Lucero MT. Modulation by PKA of the hyperpolarization-activated current (Ih) in cultured rat olfactory receptor neurons. *J Membr Biol*. 2002;188:115–125.
89. Devaux JJ, Kleopa KA, Cooper EC, Scherer SS. KCNQ2 is a nodal K+ channel. *J Neurosci*. 2004;24:1236–1244.
90. Murdoch H, Mackie S, Collins DM, et al. Isoform-selective susceptibility of DISC1/phosphodiesterase-4 complexes to dissociation by elevated intracellular cAMP levels. *J Neurosci*. 2007;27:9513–9524.
91. Chubb JE, Bradshaw NJ, Soares DC, Porteous DJ, Millar JK. The DISC locus in psychiatric illness. *Mol Psychiatry*. 2008;13:36–64.
92. Deng X, Takaki H, Wang L, et al. Positive association of phencyclidine-responsive genes, PDE4A and PLAT, with schizophrenia. *Am J Med Genet B Neuropsychiatr Genet* 2011;156B:850–858.
93. Millar JK, Wilson-Annan JC, Anderson SL, et al. Disruption of two novel genes by a translocation co-segregating with schizophrenia. *Hum Mol Genet*. 2000;9:1415–1423.
94. Gamo NJ, Duque A, Paspalas CD, et al. Role of disrupted in schizophrenia 1 (DISC1) in stress-induced prefrontal cognitive dysfunction. *Transl Psychiatry*. 2013;3:e328.
95. Li B-M, Mao Z-M, Wang M, Mei Z-T. Alpha-2 adrenergic modulation of prefrontal cortical neuronal activity related to spatial working memory in monkeys. *Neuropsychopharmacol*. 1999;21:601–610.
96. Mao Z-M, Arnsten AFT, Li B-M. Local infusion of alpha-1 adrenergic agonist into the prefrontal cortex impairs spatial working memory performance in monkeys. *Biol Psychiatry*. 1999;46:1259–1265.
97. Arnsten AF. The use of alpha2A adrenergic agonists for the treatment of attention-deficit/hyperactivity disorder. *Expert Rev Neurother*. 2010;10:1595–1605.
98. Kim S, Bobeica I, Gamo NJ, Arnsten AF, Lee D. Effects of α-2A adrenergic receptor agonist on time and risk preference in primates. *Psychopharmacology*. 2012;219:363–375.

99. Biederman J, Melmed RD, Patel A, et al. A randomized, double-blind, placebo-controlled study of guanfacine extended release in children and adolescents with attention-deficit/hyperactivity disorder. *Pediatrics*. 2008;121:e73–e84.
100. Scahill L, Chappell PB, Kim YS, et al. A placebo-controlled study of guanfacine in the treatment of children with tic disorders and attention deficit hyperactivity disorder. *Am J Psychiatry*. 2001;158:1067–1074.
101. McCracken JT, Aman MG, McDougle CJ, et al. Possible influence of variant of the P-glycoprotein gene (MDR1/ABCB1) on clinical response to guanfacine in children with pervasive developmental disorders and hyperactivity. *J Child Adolesc Psychopharmacol*. 2010;20:1–5, PMCID: PMC2835389.
102. McAllister TW, McDonald BC, Flashman LA, et al. Alpha-2 adrenergic challenge with guanfacine one month after mild traumatic brain injury: altered working memory and BOLD response. *Int J Psychophysiol*. 2011;82:107–114, PMCID: PMC3210922.
103. Connor DF, Grasso DJ, Slivinsky MD, Pearson GS, Banga A. An open-label study of guanfacine extended release for traumatic stress related symptoms in children and adolescents. *J Child Adolesc Psychopharmacol*. 2013;23:244–251.
104. Paspalas CD, Selemon LD, Arnsten AF. Mapping the Regulator of G Protein Signaling 4 (RGS4): presynaptic and postsynaptic substrates for neuroregulation in prefrontal cortex. *Cereb Cortex*. 2009;19:2145–2155, PMCID: PMC2722426.
105. Mirnics K, Middleton FA, Stanwood GD, Lewis DA, Levitt P. Disease-specific changes in regulator of G-protein signaling 4 (RGS4) expression in schizophrenia. *Mol Psychiatry*. 2001;6:293–301.
106. Volk DW, Eggan SM, Lewis DA. Alterations in metabotropic glutamate receptor 1α and Regulator of G Protein Signaling 4 in the prefrontal cortex in schizophrenia. *Am J Psychiatry*. 2010;167:1489–1498, PMCID: PMC2997877.
107. Chowdari KV, Mirnics K, Semwal P, et al. Association and linkage analyses of RGS4 polymorphisms in schizophrenia. *Hum Mol Genet*. 2002;11:1373–1380.
108. Levitt P, Ebert PJ, Mirnics K, Nimgaonkar VL, Lewis DA. Making the case for a candidate vulnerability gene in schizophrenia: convergent evidence for regulator of G-protein signaling 4 (RGS4). *Biol Psychiatry*. 2006;60:534–537.
109. Prasad KM, Chowdari KV, Nimgaonkar VL, Talkowski ME, Lewis DA, Keshavan MS. Genetic polymorphisms of the RGS4 and dorsolateral prefrontal cortex morphometry among first episode schizophrenia patients. *Mol Psychiatry*. 2005;10:213–219.
110. Talkowski ME, Seltman H, Bassett AS, et al. Evaluation of a susceptibility gene for schizophrenia: genotype based meta-analysis of RGS4 polymorphisms from thirteen independent samples. *Biol Psychiatry*. 2006;60:152–162.
111. Duffy AM, Zhou P, Milner TA, Pickel VM. Spatial and intracellular relationships between the alpha7 nicotinic acetylcholine receptor and the vesicular acetylcholine transporter in the prefrontal cortex of rat and mouse. *Neuroscience*. 2009;161:1091–1103.
112. Yang Y, Paspalas CD, Jin LE, Picciotto MR, Arnsten AFT, Wang M. Nicotinic α7 receptors enhance NMDA cognitive circuits in dorsolateral prefrontal cortex. *Proc Natl Acad Sci U S A*. 2013;110:12078–12083.
113. Buccafusco JJ, Terry AVJ. A reversible model of the cognitive impairment associated with schizophrenia in monkeys: potential therapeutic effects of two nicotinic acetylcholine receptor agonists. *Biochem Pharmacol*. 2009;78:852–862.
114. Oldfield S, Hancock J, Mason A, et al. Receptor-mediated suppression of potassium currents requires colocalization within lipid rafts. *Mol Pharmacol*. 2009;76:1279–1289.
115. Arnsten AFT, Cai JX, Murphy BL, Goldman-Rakic PS. Dopamine D1 receptor mechanisms in the cognitive performance of young adult and aged monkeys. *Psychopharmacology*. 1994;116:143–151.

116. Seamans JK, Durstewitz D, Christie BR, Stevens CF, Sejnowski TJ. Dopamine D1/D5 receptor modulation of excitatory synaptic inputs to layer V prefrontal cortex neurons. *Proc Natl Acad Sci U S A*. 2001;98:301–306.
117. Henze DA, Gonzalez-Burgos GR, Urban NN, Lewis DA, Barrionuevo G. Dopamine increases excitability of pyramidal neurons in primate prefrontal cortex. *J Neurophysiol*. 2000;84:2799–2809.
118. Gorelova NA, Yang CR. Dopamine D1/D5 receptor activation modulates a persistent sodium current in rat prefrontal cortical neurons in vitro. *J Neurophysiol*. 2000;84:75–87.
119. Williams GV, Goldman-Rakic PS. Blockade of dopamine D1 receptors enhances memory fields of prefrontal neurons in primate cerebral cortex. *Nature*. 1995;376:572–575.
120. Sawaguchi T, Goldman-Rakic PS. D1 dopamine receptors in prefrontal cortex: involvement in working memory. *Science*. 1991;251:947–950.
121. Kroner S, Krimer LS, Lewis DA, Barrionuevo G. Dopamine increases inhibition in the monkey dorsolateral prefrontal cortex through cell type-specific modulation of interneurons. *Cereb Cortex*. 2007;17:1020–1032.
122. Seamans JK, Gorelova N, Daniel D, Yang CR. Bidirectional dopamine modulation of GABAergic inhibition in prefrontal cortical pyramidal neurons. *J Neurosci*. 2001;21:3628–3638.
123. Gao WJ, Krimer LS, Goldman-Rakic PS. Presynaptic regulation of recurrent excitation by D1 receptors in prefrontal circuits. *Proc Natl Acad Sci U S A*. 2001;98:295–300.
124. Abi-Dargham A, Xu X, Thompson JL, et al. Increased prefrontal cortical D1 receptors in drug naive patients with schizophrenia: a PET study with [11C]NNC112. *J Psychopharmacol*. 2012;26:794–805.
125. Rosenberg DR, Lewis DA. Changes in the dopaminergic innervation of monkey prefrontal cortex during late postnatal development: a tyosine hydroxylase immunohistochemical study. *Biol Psychiatry*. 1994;36:272–277.
126. Rosenberg DR, Lewis DA. Postnatal maturation of the dopaminergic innervation of monkey prefrontal cortices: a tyrosine hydroxylase immunohistochemical analysis. *J Comp Neurol*. 1995;358:383–400.
127. Breier A, Wolkowitz O, Pickar D. Stress and schizophrenia: advances in neuropsychiatry and psychopharmacology. In: Tamminga C, Schult S, eds. *Schizophrenia Research*. New York, NY: Raven Press, Ltd.; 1991.

CHAPTER NINE

Cost–Benefit Decision Circuitry: Proposed Modulatory Role for Acetylcholine

Wambura C. Fobbs, Sheri J.Y. Mizumori
Laboratory of Neural Systems, Decision Science, Learning and Memory, Neurobiology and Behavior Program, Psychology Department, University of Washington, Seattle, Washington, USA

Contents

Progress in Molecular Biology and Translational Science, Volume 122
ISSN 1877-1173
http://dx.doi.org/10.1016/B978-0-12-420170-5.00009-X

Abstract

In order to select which action should be taken, an animal must weigh the costs and benefits of possible outcomes associate with each action. Such decisions, called cost–benefit decisions, likely involve several cognitive processes (including memory) and a vast neural circuitry. Rodent models have allowed research to begin to probe the neural basis of three forms of cost–benefit decision making: effort-, delay-, and risk-based decision making. In this review, we detail the current understanding of the functional circuits that subserve each form of decision making. We highlight the extensive literature by detailing the ability of dopamine to influence decisions by modulating structures within these circuits. Since acetylcholine projects to all of the same important structures, we propose several ways in which the cholinergic system may play a local modulatory role that will allow it to shape these behaviors. A greater understanding of the contribution of the cholinergic system to cost–benefit decisions will permit us to better link the decision and memory processes, and this will help us to better understand and/or treat individuals with deficits in a number of higher cognitive functions including decision making, learning, memory, and language.

1. INTRODUCTION

Every day, animals are faced with numerous decisions, such as choosing what food to pursue or selecting an individual with whom to mate. Even the simplest decisions involve some form of cost–benefit analysis and engage a number of other high-level cognitive processes, including learning and memory, as well as motivational influences. For example, the decision to pick one food over another involves the recall of memories of past experiences with food targets, and whether the animal is hungry. Information about the outcome of a decision is used to update existing memories to ensure the continuation of adaptive choices in the future. Thus, an understanding of brain mechanisms of memory necessarily also includes an understanding of decision-making neural circuitry. While memory systems of the brain have been studied for decades (for excellent reviews, see Refs.[1–8]), only more recently have studies begun to identify functional neural circuits that subserve different forms of cost–benefit decision making. Those studies have primarily focused on interactions between prefrontal, amygdalar, and striatal systems, with considerable attention paid to the neuromodulatory role of the dopamine (DA) system. Yet, even though the cholinergic system is well positioned anatomically and functionally to directly influence the decision circuitry including the DA system and has been long studied for its role in learning and memory, few reports have considered its role in cost–benefit decision making. Therefore, in this review, we will present an updated view of the decision circuitry with a special emphasis on our proposal for a modulatory role of the cholinergic system in cost–benefit decisions.

2. GENERAL FEATURES OF COST–BENEFIT DECISION MAKING

Cost–benefit decision making, or value-based decision making, is a process by which an individual chooses between two or more options when each option represents an action linked to an outcome. The goal of cost–benefit decision making is to maximize rewarding outcomes by selecting actions that lead to the greatest subjective value. Subjective value (hereafter, simply referred to as value) is determined through cost–benefit analysis—an evaluation of the rewarding properties of an outcome in light of the costs that must be endured in order to obtain that outcome.

Here, we will focus on the three forms of cost–benefit decision making that have been studied most frequently in rodent models. They are effort-, delay-, and risk-based decision making, each named for the type of "cost" they involve: (1) a requirement of physical effort to pursue the reward, (2) a delay period that precedes the reward, or (3) reward delivery that is probabilistic (risky/uncertain). In behavioral tasks that model each form of decision making, animals are presented with choices between two options—a small food reward that is associated with little/no response cost or a large food reward that is linked to greater cost. The tasks have been performed in operant boxes and on T-mazes, with the actions taken to obtain the rewards differing by apparatus (Fig. 9.1A). In the box, animals press levers located on either the left or right side of a food cup, and depending on the lever selected, receive a small or large reward. Alternatively, animals run down the left or right arm in order to collect a small or large reward on the maze. While delays and risks are similarly implemented on both apparatuses, the type of effort required to obtain the large reward does differ between the two. In the box, animals must press the lever several times before the reward is delivered; while on the maze, they must climb over a barrier in order to access it.

Typically, within a single session, the cost to the large reward option is systematically increased. Such that when costs are initially held equal, animals reliably choose the large reward option over the small reward option, but, as the cost to the large reward increases, their preference for it decreases.

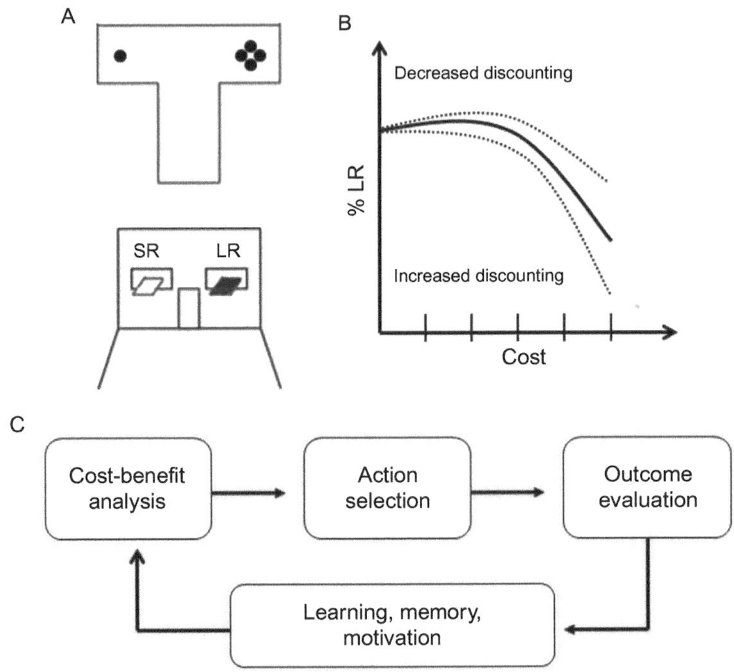

Figure 9.1 Features of cost–benefit decision making. (A) Diagram of a T-maze and an operant chamber typically used in rodent models of cost–benefit decision making. On the maze, animals are faced with choices between two arms. If they choose the left arm, they can retrieve one food pellet (small reward) without encountering a cost; but if they choose the right arm, they will face a cost (scalable barrier, delay, or uncertainty) before they can retrieve four food pellets (large reward). In the box version of the task, animals choose between pressing the left lever, which triggers the delivery of a small reward (SR), or pressing the right lever, which imposes a cost (multiple lever presses, delayed reward, uncertain reward) before the large reward (LR) is delivered. (B) Graphical representation of a cost–benefit decision choice profile. Specifically, the relationship between the amount of cost associated with the LR and the percentage of large, costly rewards (% LR) chosen is illustrated. The solid black line represents a typical discounting curve, while the dotted lines represent when the animal's curves are shifted up or down—increased or decreased cost discounting, respectively. (C) A schematic diagram of the general conceptual framework for cost–benefit decision making. When well-learned options are encountered, their value is determined by weighing their associated costs against their associated rewards (cost–benefit analysis). Next, the action linked to the greater value is selected, and the resultant outcome is compared to the expected outcome. Finally, learning, memory, and motivational systems help shape future behavior based on what, if anything, changed externally (actions, outcomes) or internally (hunger, thirst, cost sensitivity, reward sensitivity, etc.).

The decrement in preference reflects the fact that costs reduce or "discount" the value of an outcome, which explains why that pattern of choice is actually optimal (Fig. 9.1B). Animals are trained on these tasks until they display stable choice profiles across multiple days, and only then are neuroanatomical or neurochemical manipulations employed to test whether they can alter choice behavior by increasing or decreasing the overall preference for the large, costly reward. Greater preference for the large reward suggests reduced discounting, whereas diminished preference implies increased discounting.

Before considering the neural basis of each form of cost–benefit decision making, it is helpful to conceptualize the psychological or computational processes underlying them (Fig. 9.1C).[9–11] First, the animal must integrate cost and reward information to calculate the value of each option (cost–benefit analysis). Those values are then directly compared, and the animal selects the action with the greatest value. Once the outcome is obtained, it is evaluated to determine whether or not the experienced value matches the expected value, and that feedback is incorporated by learning, memory, and motivational processes to inform future decisions. Of note, though, is the fact that because animals are very well trained by the time they are tested, they likely already have mnemonic representations of the action–outcome associations for the options they will encounter in a given session.

Neuroscientists have only just begun to probe the neurobiological underpinnings of cost–benefit decision making. In Sections 3-6, we will highlight what is known about the role of the DA system, prefrontal cortex (PFC), nucleus accumbens (NAc), basolateral amygdala (BLA), and hippocampus (HPC) in effort-, delay-, and risk-based decision making. There is a breadth of decision-making data collected from many species using many different tasks, but we will only focus on the rodent literature that uses the tasks described in this section.

3. THE DA SYSTEM TRANSMITS REWARD AND COST INFORMATION THROUGHOUT THE BRAIN

The DA system is best known for its ability to broadcast reward information all over the brain, including to structures involved in cost–benefit analysis (Fig. 9.2). DA neurons fire in response to predictive cues and encode the magnitude of future rewards.[12] In fact, direct stimulation of DA neurons was shown to promote reward-seeking behavior,[13] just as inactivation of DA neurons impaired animals ability to respond to reward-predicting cues.[14]

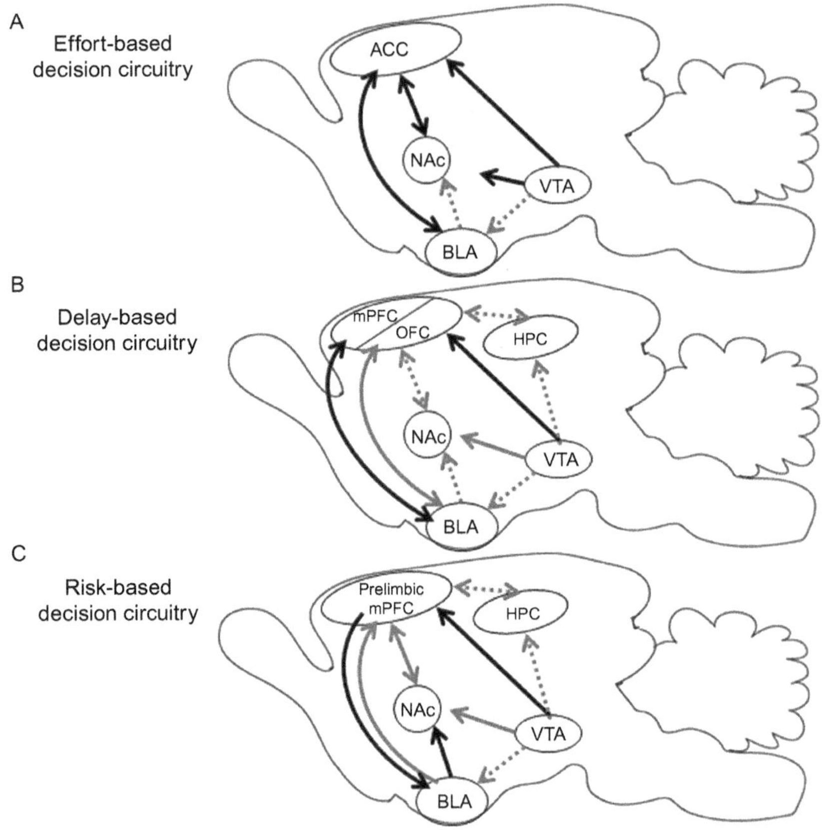

Figure 9.2 Anatomical circuit representations showing brain structures and pathways involved in cost–benefit decision making. (A) Effort-based decision circuit. (B) Delay-based decision circuit. (C) Risk-based decision circuit. *Solid black lines* represent pathways that support optimal performance; *solid gray lines* represent pathways that are not required for normal performance; and *dotted lines* indicate pathways that have yet to be tested. ACC, anterior cingulated cortex; NAc, nucleus accumbens; BLA, basolateral amygdala; VTA, ventral tegmental area; OFC, orbitofrontal cortex; HPC, hippocampus; mPFC, medial prefrontal cortex.

It is also accepted that even as DA neurons respond to cues, they continue to track reward information by firing at higher rates for greater than expected rewards and at lower rates for smaller than expected rewards, a phenomena called reward prediction error signaling.[12] However, it should be acknowledged that even though DA neurons signal reward information and promote reward seeking, DA is not required for animals to display subjective reward preference.[15]

Beyond signaling reward information, DA neurons have demonstrated the ability to encode value information as well. In several experiments, both cue- and reward-evoked DA activity recorded during Pavlovian and instrumental tasks exhibited sensitivity to costs associated with rewards.[16–19] When directly tested by different laboratories during all three cost–benefit decision making tasks, the importance of that sensitivity was revealed. In the moments prior to action selection when animals were presented with both options, the level of DA activity reflected the discounted value of the "better" option.[19–21] In other words, DA neurons were able to combine cost information with the reward information in a way that allowed them to signal relative value which may inform action selection. There is a caveat; however, a second report of DA activity during effort-based decision making found that the ability of DA to encode value depended on the specifics the trial, meaning DA might only signal effort-discounted values under certain conditions of choice.[22] Nevertheless, together the literature indicates that DA plays an important signaling role during cost–benefit decision making.

In fact, systemic pharmacological manipulations have confirmed that DA transmission shapes performance on cost–benefit decision making tasks (Table 9.1).[23–28] Blocking DA receptors with the antagonist flupenthixol caused animals to reduce their preference for the large, costly reward in

Table 9.1 Effects of systemic pharmacological manipulations of the DA and ACh systems on choice of the large, costly reward during each form of cost–benefit decision making

	Receptor blockade			**Receptor stimulation**		
Decision type	**DA receptor antagonist**	**mAChR antagonist**	**nAChR antagonist**	**DA receptor agonist**	**mAChR agonist**	**nAChR agonist**
Effort	⇩	?	?	Mixed	?	?
Delay	⇩	⇩	NE	⇧	NE	Mixed
Risk*	⇩	⇩	NE	⇩	NE	⇩
Risk**	⇩	?	?	⇧	?	⇧

⇧ increased choice of the large, costly reward; ⇩ decreased choice of the large, costly reward; NE—no significant effect; ?—unknown; mixed-findings are mixed; *probability of the large reward ascended; **probability of the large reward descended.

all three forms of cost–benefit decision making.[23,28] However, stimulating DA release altered each type of decision differently. In effort-based decision making, a low dose of amphetamine increased the choice of the large, high-effort option, whereas a high dose of amphetamine biased choice toward the small, low-effort option.[23] The effect on delay-based decision making did not depend on dose, as both high and low doses of amphetamine were shown to increase selection of the large, delayed reward.[24,25] Likewise, the impact of amphetamine on risk-based decision making did not vary by dose, but it did vary by session design. Amphetamine elevated the preference for the large, risky option when the probability of the large reward descended during the session, and reduced the preference for the large, risky option when the large reward probability ascended.[28,29] Primarily, these findings suggest that DA activity is necessary for optimal choice during cost–benefit decision making, but they also highlight the potential for DA transmission to sculpt choice in subtle ways depending on the level of DA activity and order of choice presentation. Further, they raise two questions about whether the nature of the modulation depends on (1) activation of specific receptors and/or (2) where in the brain the DA transmission occurs.

Answers to the first question can be gained through the use of systemic pharmacological agents that target specific DA receptors. Even though five different DA receptors have been indentified, drugs typically target either those called D_1 receptors (D_1 and D_5) or D_2 receptors (D_2, D_3, and D_4).[30] D_2 blockade mimicked the effect of flupenthixol when the large reward was associated with effort or risk but not delay.[25,29,120] D_1 blockade also reproduced the impairment but for a different pair of costs—delay and risk (its impact on effort has yet to be tested).[25,29] Indeed, those observations do suggest that DA's role in different forms of cost–benefit decision making may be mediated by different receptors, with D_1 activation promoting choice of delayed and risky options and D_2 activation encouraging effortful and risky choices. But, such a conclusion does not explain the mechanism by which DA influences these types of decisions. It could be the case that it does so globally through its influence on entire networks or it could contribute important information to computations that occur locally within single structures.

In the following three sections, we will present a case for the later. As we discuss each structure that is implicated in a given form of cost–benefit analysis, we will highlight the evidence concerning DA involvement in that structure.

4. EFFORT-BASED DECISION CIRCUITRY

4.1. Anterior cingulate cortex

The anterior cingulate cortex (ACC) is the only region of the PFC known to play a role in effort-based decision making.[32] In multiple studies, animals with ACC lesions exhibited suboptimal choice behavior. Instead of pursuing the high return, high-effort option, ACC-lesioned animals repeatedly selected the small reward that was linked to little effort. Interestingly, these effects depended on the magnitude of the difference between reward alternatives. They were observed when the discrepancy was small (four vs. two pellets), but not when it was bigger (five vs. one pellets).[32–34] Given that the effects cannot be attributed to deficits in spatial or magnitude discrimination or motor abilities, the lesion studies suggest that the ACC integrates effort and reward information and notifies the animal when an investment of effort will maximize value gains.[33] In fact, a subsequent single unit study directly demonstrated that the ACC encodes effort-discounted value. In that study, ACC neurons exhibited higher firing rates throughout the high-effort, high-reward trajectories than during the low-effort, low-reward trajectories; and further analysis revealed that the activity was not driven by the size of the reward or effort alone but by the combination of the two.[35] Taken together, the lesion and recording data suggest that ACC neurons are able to signal which action is "better" because it will lead to larger value.

Reports are mixed about whether or not DA transmission in the ACC plays a role in effort-based decision making. One group observed that selective DA lesion of the ACC failed to impact choice in one experiment,[31] while another reported that both DA lesion of the ACC and D_1, but not D_2, receptor antagonism in the ACC reduced preference for the large reward.[31,36,37] The discrepant results can probably be attributed to the differences in the amount of neurotoxin used, as the effect was seen following the larger dose. Therefore, a parsimonious interpretation is that DA signaling through D_1 receptors in the ACC is important for effort-related cost–benefit analysis.

4.2. Nucleus accumbens

Like the ACC, the NAc is important for biasing choice toward the large, effortful reward during effort-based decision making. When the NAc core was lesioned or reversibly inactivated, animals shifted their preference to the smaller, low-effort reward.[38–40] The effect was specific to the NAc core, as NAc shell manipulations were unable to induce similar shifts.[39] Additionally, two control experiments ruled out the possibility that the impairments were driven by deficits in reward discrimination or delay-discounting. The latter was demonstrated by presenting the NAc-core-lesioned animals with a modified version of the task in which the delays to reward delivery for the large and small reward were made equivalent instead of being determined by the amount of time it took the animal to fulfill the effort requirement. Since animals continued to bias their choice toward the small, low-effort option, it strengthened the conclusion that the NAc core is able to specifically process the impact of effort-related cost separate from delay (see Section 5.3).[39]

DA transmission in the NAc seems to play a vital role in selection during effort-based decision making. Multiple laboratoratories have reported that following DA depletion in the NAc, animals become less likely to choose the high-effort alternative.[40,41] When considered in light of the established body of literature that shows DA in the NAc is also essential for maintaining high levels of operant responding for reinforcement, it seems reasonable to conclude that the choice profile changed because DA was unable to encourage effort expenditure for the more valuable reward.[42–44] As a result, one would expect that DA release would signal effort information in the NAc. Surprisingly, that is not the case. Instead, electrochemically recorded DA release in the NAc reflected reward size and not the amount of effort associated with the options. However, the signal was able to encode the relative effort-based value of the best option under certain conditions.[20,22]

4.3. Basolateral amygdala

The BLA is another structure that has been implicated in effort-based decision making. Just like the other two, the BLA is necessary for integrating effort and reward information to support optimal performance. A study found that reversibly inactivating the BLA causes animals to exhibit the same impaired choice profile—diminished preference for the high return, high-effort option—without affecting their general motor abilities, their spatial discrimination, or their sensitivity to reward.[45] That finding does not help pinpoint the specific contribution of the structure to the analysis of effort and reward that underlie effort-based decision making, so we turn to other literature to speculate about its contributions. Previous work has demonstrated a propensity of the BLA to signal reward magnitude and calculate the potential value of actions/outcomes, so the BLA may be important for emphasizing the benefits of the reward, rather than accounting for effort specifically.[46–49]

4.4. Neural systems view of effort-based decision circuitry

It is clear from the earlier discussion that the ACC, NAc core, and BLA are all involved in effort-based decision making, with DA exerting its critical modulatory influence in the ACC and NAc (Fig. 9.2A). It is also known from lesion studies that two other regions of the PFC, the orbitofrontal cortex (OFC) and prelimbic medial PFC (mPFC), are not necessary for these decisions.[32,34] Further, two disconnection studies have revealed that the structures do not act in isolation because serial transfer of information between the ACC and both subcortical structures is also essential. One group demonstrated that unilateral inactivation of the ACC with contralateral inactivation of the BLA reduces preference for the large, high-effort option,[45]and a second group observed that asymmetric lesions of the ACC and NAc caused the same impairments.[38] Together, the data suggest that cortico-limbic and cortico-striatal communication are both necessary for animals to exert more effort to obtain a more profitable reward. In the future, it will be important to clarify the direction of information transfer between the ACC and BLA and the ACC and NAc as well as to determine whether signal transfer from the BLA to the NAc is required. Moreover, given the ventral pallidum has been implicated in effort-related processes, its role in effort discounting should be investigated using lesion.[50] Finally, additional work should be directed at pinpointing the exact contribution of each structure. To this end, single-unit recordings of neurons in the NAc core and BLA should be conducted during effort-based decision making.

5. DELAY-BASED DECISION CIRCUITRY

5.1. Orbitofrontal cortex

Within the PFC, the OFC is the region that has been most extensively studied for its role in delay-based decision making.[32] However, because there are discrepancies between studies, there remains some confusion about the precise function of the OFC. Following OFC lesion or inactivation, several laboratories reported decreased preference for the large, delayed reward (or more impulsive choice),[32,51,52] while another laboratory consistently observed the opposite effect,[53,54] and yet others found no effect.[55,56] It is possible that the discrepancies are due to differences in task design.[46,57] Nevertheless, the data collectively indicate that the OFC is important for combining delay and reward information for action selection during delay-based decision making. Single-unit data have since corroborated that interpretation. When OFC neurons were recorded during a delay-based decision-making task, reward delivery evoked firing that reflected the delay-discounted value.[58] Also, a small subset of the OFC recorded neurons exhibited firing patterns that ramped up throughout the delay period that preceded reward delivery.[58] Based on those observations, it seems likely that OFC neurons track delay information in order to accurately calculate the relative value of rewards and inform animals whether or not pursuing an option is a good choice.

The role of DA transmission in the OFC during delay-based decision making is currently ambiguous. Selectively lesioning the DAergic input to the OFC actually decreased impulsive choice,[59] even though pharmacologically blocking D_1 or D_2 receptors independently increased impulsive choice.[57] To further complicate the matter, an *in vivo* microdialysis study observed elevated levels of DOPAC, DA's main metabolite, while animals performed the task. Since the raised DOPAC levels could not be attributed to the physical actions performed, delays experienced, or rewards received, the observation indicated that tonic DA levels may be modulated by making delay-based decisions.[106] However, without more data, it is difficult to reconcile these findings and propose exactly how DA signaling in the OFC contributes to delay-based decision making.

5.2. Medial prefrontal cortex

The mPFC is another less-studied region of the PFC that may be important for delay-based decision making. When the mPFC was lesioned in one study, animals performed in a manner that was very suboptimal and could not be describes as delay-discounting: they reduced their choice of the large reward when it was preceded by a short delays and increased their choice of the large reward when it was preceded by longer delays.[60] The report did not follow up on the finding, so it is still unknown if the lesions were specific to choice and/or whether they were attributable to impairments in other aspects ofbehavior, such reward or delay sensitivity. However, since another recent report found that mPFC lesion decreased tolerance for delay during a reaction time task it is possible that the mPFC does play an important role in considering reward and delay information when selecting what and when to pursue rewards.[61] Much more work is needed to replicate the findings and clarify the interpretation.

5.3. Nucleus accumbens

Unlike the OFC, the importance of the NAc to delay-based decision making has not been called into question. In the only lesion study that has been published, lesioning the NAc core induced more impulsive choice—animals selected the large, delayed reward far less frequently than sham controls.[60] Since the ability to discriminate reward magnitudes remained intact, the choice profile indicates that NAc-core-lesioned animals discounted the value at each delay more than sham animals. In sum, the experiment suggests that the NAc core is needed to integrate delay and reward information and encourage animals to wait when it will maximize gains. Further, it turns out that DA is not critical to the processing of cost information within the NAc when the cost is delay instead of effort because eliminating DA projections to the NAc had no impact on the number of delayed rewards selected by animals.[26]

5.4. Basolateral amygdala

The BLA also likely plays a key role in integrating delay and reward information during delay-based decision making. Lesions of the BLA increase impulsive choice without affecting reward discrimination, which is identical to the effect of NAc lesions.[54] Additional evidence is needed to determine the exact contribution of the BLA to delay-based decision making, but it is possible that once again the structure's ability to encode reward information plays an integral role in directing behavior toward larger rewards, even when their delivery is delayed.

5.5. Hippocampus

The last structure currently implicated in delay-based decision making is one we have yet to discuss, the HPC. In three separate experiments, HPC lesions biased animals toward more impulsive choices.[55,56,62,63] The effect was not specific to specific portions of the structure, as lesions of both the dorsal and ventral HPC produced the same impairment.[62] Nor could the effect be interpreted as stemming from deficits in magnitude or spatial discrimination or, by extension, memory.[55] Instead, the results indicate that the HPC is essential for animals to properly integrate reward and delay information and to know when to tolerate delays in order to obtain a larger reward. One intriguing possibility that could explain the finding is that HPC lesions altered animal's perception of the delays. This is supported by previous work that identified a role for the HPC in temporal processing.[63–65] Among those studies, a simple choice task revealed that hippocampal lesions caused rats to switch their preference from a delayed certain option to an immediate uncertain alternative.[66] Of course, since the HPC may also be involved in processing uncertainty (see Section 6.5), the effect may have been driven by that aspect of the option rather than the delay associated with it.

5.6. Neural systems view of delay-based decision circuitry

Delay-based decision making also involves a cortico-limbic-striatal circuit, but the specific substructures within the circuit differ from effort-based decision making. Based on the lesion data, the OFC, NAc core, BLA, and HPC all seem to contribute to normal levels of delay-discounted choice while the function of the mPFC is not as clear (Fig. 9.2B). Additionally, while DA seems to exert influence in the OFC, its mechanism of involvement is not well understood. It may be the case that it plays a role within the BLA or HPC, but DA depletion of those areas during delay-based decision making has yet to be performed. To further complicate matters, in the only published disconnection study, it was demonstrated that serial transfer of information between the mPFC and BLA supports normal levels of choice while information transfer between the OFC and BLA is unnecessary.[67] In the future, it will be important to test all other projections between implicated structures as well as to carry out experiments which will better determine the contribution of each structure. Related to that last point, from now on, control experiments should test the possibility that the changes in choice profile are the result of changes in temporal sensitivity.

6. RISK-BASED DECISION CIRCUITRY

6.1. Prelimbic mPFC

The prelimbic mPFC is the region of PFC most implicated in risk-based decision making. Reversible inactivation of the prelimbic mPFC increased animal's preference for the larger, riskier option (risk seeking) when the probability of the large reward diminished across the session (i.e., 100%, 50%, 25%, or 12.5%), as is typically done in these experiments.[68] However, when the probability was increased across the session, the opposite effect was observed—prelimbic mPFC lesions decreased preference for the larger, riskier alternative (risk aversion).[68] The effects were not attributable to deficits in behavioral flexibility or an inability to make judgments of fixed probabilities, so it is likely that the prelimbic mPFC is involved in assessing any change in reward probabilities in order to update value representations and facilitate optimal decisions.

Interestingly, reports about DA in the mPFC during risk-based decision making seem to be ambiguous. Even though eliminating DA transmission through selective lesion in the mPFC did not impact risk-based decisions of one group of animals,[69] perturbations of the prelimbic mPFC DA system by D_1 or D_2 drugs altered choice profiles of another group of animals. Specifically, the results of pharmacologically manipulating signaling through D_1 and D_2 receptors were as follows: D_1 blockade caused risk aversion; D_1 stimulation had no effect; D_2 blockade increased the preference for the riskier option (risk seeking); and D_2 stimulation caused animals to perform extremely inefficiently, reducing choice of the large reward when it was less risky and increasing choice of the large reward when it was riskier.[70] Thus, DA transmission in the prelimbic mPFC is not required to accurately integrate reward information with the likelihood of receipt as long as the activity through D_1 and D_2 receptors is balanced.

6.2. Orbitofrontal cortex

The role of the OFC in risk-based discounting is also unclear. While one study observed risk-seeking behavior, or elevated choice of the large, risky option[56]; another reported no effect.[68] As was suggested for delay-based decision making, the discrepant results could be explained by differences in task design or training. Since the role of OFC is questionable for choices associated with delay and risk, it will be very important to further investigate the structure, especially through neural recordings.

6.3. Nucleus accumbens

In risk-based decision making, it is the NAc shell that plays a central role. An earlier experiment observed that lesioning the lateral NAc core and medial NAc shell caused animals to exhibit risk aversion, irrespective of whether the probability increased or decreased.[71] In a follow-up study, the data were clarified in two ways. First, it revealed that only the NAc shell was essential for normal choice performance. Since inactivating the NAc shell, not NAc core, induced risk aversion, it is likely that the NAc shell is necessary for the value of the risky reward to be properly discounted.[72] Second, through choice-by-choice analysis, the study was able to attribute the change in overall choice profile to reduced reward sensitivity when quantified as the tendency of the animals to pursue the large reward following reinforcement.[72] Combined, the experiment supports the interpretation that the NAc shell promotes tolerance to uncertainty associated with the larger reward-associated action during risk-based decision making.

The ability of the NAc to integrate magnitude and likelihood of rewards with different actions does not rely on intact DA transmission, even though it is altered by exogenous DAergic drugs.[69] The exact impact of the drug depended on the receptor it targeted: D_1 blockade resulted in risk aversion; D_1 stimulation made the selections more efficient by increasing choice of the large reward when it was less risky and decreasing choice of the large reward when it was riskier; D_2 drugs did not change behavior; and D_3 stimulation decreased risky choice.[73] Again choice-by-choice analysis helped explain these finding. In this case, the analysis revealed that the reduction of risky choices following D_1 blockade was likely due to elevated omission sensitivity, calculated as the propensity to bias choice toward the small, certain option after nonreinforced risky choices. Perhaps, the only explanation that can account for both the DA depletion and pharmacological data is the same one we have previously mentioned—DA transmission in the NAc may not contribute an essential signal to NAc processing during risk-based decision making, but it does support optimal choice through balanced activation of each DA receptor. However, there is another possibility, and that is that the DA activation of D_1 mediates the effect of reward omission to promote selection of larger reward options while D_3 may help discourage the selection of the risky reward when it need not be selected.

That is true despite the fact that DA signaling in the NAc is modulated by risk-based decision making. Microdialysis measurements revealed that tonic DA levels were adjusted by reward rates, uncertainty, and choice when animals performed risk-based judgments; and electrochemical recordings captured cue-evoked phasic DA signals that, prior to choice, reflected the value of the better option, irrespective of choice.[21,74] In other words, the DA signal in the NAc dynamically assesses task variables, even though it may or may not be important for sculpting optimal performance.

6.4. Basolateral amygdala

When the decreasing probability task was used, lesioning the BLA induced risk aversion, even though it did not affect magnitude discriminability.[75] Although these findings do little to assert a specific fundamental role for the BLA, they do confirm that the BLA is the only substructure known to be involved in facilitating an organism's ability to overcome all three costs (work, uncertainty, delays) to promote actions that may yield larger rewards.

6.5. Hippocampus

Our understanding of the role of HPC in risk-based decision making is still in its infancy. Recently, ventral HPC lesions were not shown to impair animals ability to discount reward value according to the expected probability of reward when tested in an enclosed chamber.[56] That was surprising in light of the fact that dorsal hippocampal place cells that were recorded from rats solving a maze task seemed to code reward value[76] and the probability of obtaining a large reward (uncertainty or risk[77]). In the latter experiment, place fields were differentially expressed depending on whether the expected probability of obtaining a large reward was 100%, 50%, 25%, or 12.5%. Further lesion and other place cell research show that the dorsal and ventral HPC likely play different roles in hippocampal-dependent memory.[78] The dorsal HPC is thought to preferentially process specific bits of information while the ventral HPC processes the emotional and motivational context of the remembered information. This sort of functional distinction is consistent with the seemingly discrepant finding of risk sensitive (dorsal) HPC neurons[77] and (ventral) HPC lesions that do not change risk-based decisions.[56] Therefore, future studies should compare dorsal versus ventral HPC contributions to risk-based decisions while rats are performing the same decision task.

6.6. Neural systems view of risk-based decision circuitry

Currently, it seems that the prelimbic mPFC, NAc shell, and BLA comprise the important circuitry that gives rise to optimal risk-based decision making, with the OFC and HPC possibly providing additional influences (Fig. 9.2C). Even though systemic blockade of DA by flupenthixol and D_1 and D_2 receptor antagonists changed risk-discounting, DA depletion of select structures (mPFC and NAc) has failed to replicate the findings. This may be because DA exerts its influence within structures not yet tested (BLA, HPC, OFC) or it might reveal that DA plays a role on a more global level. One laboratory conducted a series of disconnection studies, revealing which connections support normal choice. Specifically, they observed the following: disrupting communication between the prelimbic mPFC and the NAc shell had no effect; severing the connection between the BLA and NAc shell caused risk seeking; and disconnecting the mPFC and BLA increased risk seeking. The fact that the effect of PFC–BLA disconnection was identical to mPFC inactivation not BLA inactivation was striking and led the experimenters to wonder about the directionality of the signal. They found that by separately disrupting top-down communication (mPFC-to-BLA), they were able to induce the behavioral shift (risk seeking), while silencing bottom-up communication had no effect. Further, choice-by-choice analysis attributed the change in behavior to reduced omission sensitivity, which indicates that mPFC likely tracks and adjusts responding in the face of omissions, inhibiting responding relative to both downstream players, the BLA and NAc, which tend to drive response bias toward the larger reward.[79]

7. CHOLINERGIC MODULATION OF DECISION CIRCUITRY

It is clear from the literature that cholinergic neurotransmission is essential for certain behaviors. From selective cholinergic lesion studies, we know that while ACh is sometimes not required for performance on traditional learning and memory tasks[80–83] (exceptions[84–86]), it is necessary for performance on tasks that involve an attentional component.[71,74–81,112–119] It is also well established that ACh can influences arousal and information processing, including memory encoding, through direct modulation of target structures.[81,87] Therefore, it is highly likely that the cholinergic system plays a significant role in cost–benefit decisions because they are complex behaviors that rely on a number of psychological and computational properties.

In this section, we will highlight the little that is known about the cholinergic system in cost–benefit decision making and propose a few select roles for the ACh system to be investigated in future work.

7.1. Pharmacological evidence of cholinergic involvement in cost–benefit analysis

ACh signaling is ultimately mediated by activation of two different classes of receptors—nicotinic and muscarinic acetylcholine receptors (n/mAChRs). Often, the two receptor types are discussed separately because (1) nAChRs have an addictive exogenous ligand, nicotine and (2) there is evidence to suggest that the two receptors regulate different functions. Here, we will consider both receptors.

Recently, a comprehensive pharmacological study suggested that endogenous activation of mAChRs not nAChRs is necessary for optimal choice during delay- and risk-based decision making, but activation of nAChRs has the potential to shape decisions as well (Table 9.1). In order to draw those conclusions, the paper directly tested the effects of acute, systemic injections of muscarinic and nicotinic agonists and antagonists on delay- and risk-based decision making. For both the delay and risk task, blocking mAChRs with antagonists scopolamine and atropine induced suboptimal choice, while stimulating mAChRs with oxotremorine and blocking nAChRs with mecamylamine had no effect on behavior.[88] Yet, when nAChRs were stimulated with nicotine, different effects were seen for delay- and risk-based decision making. Nicotine was not shown to alter delay discounting but was reported to induce a perseverative choice profile in the risk task. The lack of nicotine effects on delay-discounting is contrary to other reports that observed either increased discounting[89,90] or decreased discounting[91] following both acute and chronic administration of nicotine. However, nicotine elevated the preference for the large, risky option when the probability of the large reward descended during the session and reduced the preference for the large, risky alternative when the large reward probability ascended, a pattern which reflects a sustained bias toward whichever option is initially more valuable and an inability to asses changing probabilities. Since, the experiment did not investigate effort-based decision making, it is still important to determine if the effects generalize to effort-discounting.

Overall, those pharmacological observations were the first to suggest that endogenous ACh plays a role in different forms of cost–benefit decision making and that its effects are mediated by different receptors. But, they do not explain the mechanism by which ACh exerts such influence on these decisions. ACh probably influences these behaviors via local activity within important structures, but since there is no direct evidence concerning its potential role at that level, we can only speculate.

7.2. Central cholinergic circuitry

The central cholinergic system, like the DA system, projects widely throughout the brain. Relevant to our discussion, three sources of cholinergic output directly innervate structures in the cost–benefit decision circuitries outlined in this review (Fig. 9.3).[92,93] Cholinergic cells of the midbrain laterodorsal tegmental nucleus (LDTg) and pedunculopontine tegmental nucleus (PPTg) project to DA neurons in the ventral tegmental area (VTA) and substantia nigra (SNc). In the striatum, cholinergic interneurons are found in both the core and shell subdivisions of the NAc. Lastly, in the basal forebrain, neurons of the medial septum (ms) project to the HPC, while the nucleus basalis magnocellularis (nbm) provides the major cholinergic input to the PFC. Thus, ACh is well positioned to influence cost–benefit decisions through its activity at any of the currently implicated structures.

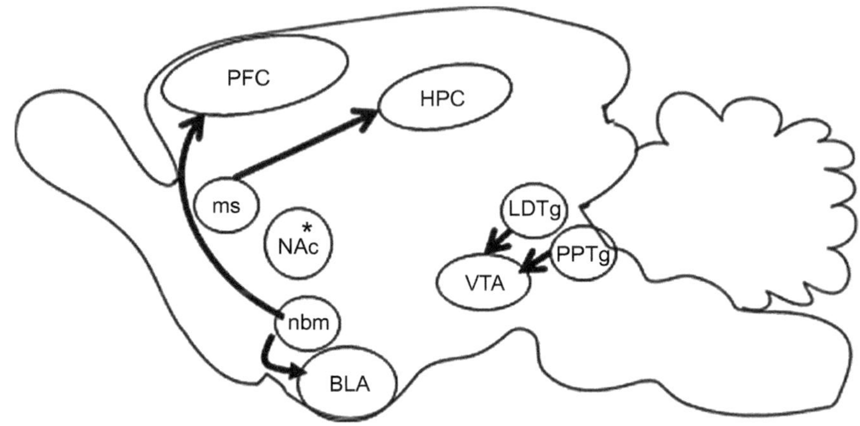

Figure 9.3 Schematic diagram showing cholinergic projections to structures implicated in cost–benefit decision making. Solid black lines represent cholinergic projections and the asterisk represents cholinergic interneurons. PFC, prefrontal cortex; NAc, nucleus accumbens; BLA, basolateral amygdala; VTA, ventral tegmental area; HPC, hippocampus; ms, medial septum; nbm, nucleus basalis magnocellularis; LDTg, laterodorsal tegmentum; PPTg, pedunculopontine tegmentum.

We will not detail all of the studies that have demonstrated the ability of the nAChRs and mAChRs to modulate those structures. Instead, we direct the readers to other reviews which highlight the exact distribution and known function of each distinct m/nAChR subtype[87,94,95] within the target structures.[81,87,94–100] From those reviews, it is clear that ACh can either directly modulate the activity of a target structure or mediate the release of neuromodulators within a structure. Only through selective cholinergic lesions of each structures and/or administration of n/mAChR antagonists within each structure, will the site of ACh action that contributes to cost–benefit decision making be determined. Until that and other work is done to clarify the role of ACh neurotransmission in cost–benefit decision making, we can only speculate about its role. Below, we present three testable hypotheses.

1. *Pharmacological effects of cholinergic drugs may be attributable to ACh regulation of midbrain DA neurons*: There is a breadth of evidence suggesting that ACh regulates midbrain DA activity and function.[99,100] mAChR agonists caused DA neurons in the VTA and SNc to increase their firing rates and DA release, and identical results were observed when endogenous ACh input was stimulated in the LDTg.[101,102] Similarly, nAChR activation also increased excitability and DA release of midbrain DA neurons.[103] Indeed, knockouts that lacked an mAChR (M_5) or an nAChR ($\beta 2$) throughout the

brain exhibited decreased dopaminergic function.[100,104] Of course, the latter effect could be driven by changes to non-DA systems.[100,104] Still, together the evidence suggests that the ACh activation of both receptors stimulates DA firing and release while blockade of both cholinergic receptors reduces DA activity.

Therefore, our first hypothesis is that the effects of cholinergic drugs on delay- and risk-based decision making can be explained in terms of how they alter the activity of the DA system (Table 9.1). When mAChRs, were blocked by antagonists, choice was affected in the same manner as if DA signaling was blocked directly. So, it is possible that the change in preference is driven by a reduction of DA signaling that is caused by the absence of mAChR activation. In order to test this hypothesis, future work should determine (1) if the same effects are observed following intra-VTA and intra-SNC injections of mAChR antagonists and (2) whether they can be counteracted by DA agonists.

This hypothesis could also explain the effects of nAChR stimulation on delay- and risk-based decision making. Even though the pharmacological evidence suggested that signaling through nAChRs was not required for cost–benefit analysis (perhaps because nAChR blockade does not dampen DA activity profoundly enough to impact choice) the same evidence also revealed that excess stimulation can change behavior in a manner that parallels the effect of elevated DA release. In fact, remarkably, nAChR stimulation caused the exact same perseverative choice as DA stimulation that depended on how the large reward probability changed during the session.

2. *Cholinergic modulation in the NAc may be important for effort-based decision making*: A recent paper suggests that ACh transmission in the NAc may shape effortful responding in a manner opposite to DA. In that paper, an operant task was used in which animals chose between a highly palatable food option that required five lever presses or a less palatable food (standard lab chow) that required no effort to obtain because it was freely available. When the muscarinic agonist pilocarpine was injected into the NAc, not the dorsal striatum, it caused animals to reduce their exertion of effort (lever pressing) and increase consumption of the standard chow.[50] Further, they demonstrated that the effect could be eliminated by coadministration of the muscarinic antagonist scopolamine. However, the significance of the results remains unclear. The authors suggest that the effect was not simply due to the animals' becoming sated, a state known to be accompanied by elevate endogenous ACh tone in the NAc,[99] because the animals did not simply stop eating all together. Rather they shifted their preference to the less palatable lab chow.

Therefore, we hypothesize that ACh activity in the NAc may be able to shape behavior by biasing choices away from the large effortful reward during cost–benefit decision making. These effects of mAChR stimulation are striking because they mirror those induced by DA antagonism during effort-based discounting. Before one can definitively interpret these effects, pilocarpine and scopolamine should be tested in effort-based decision tasks in which the effort is systematically changed. Such an experiment will better allow us to determine if ACh is involved in integrating reward and cost information as well as discouraging effortful actions.

3. *Cholinergic innervations of the PFC and HPC may be essential for delay-based decision making*: There was one lesion study that sought to determine the long-term effects of prolonged cholinergic hypofunction on impulsivity. In that experiment, ACh neurons of the basal forebrain were selectively lesioned in neonates prior to participation in a delay-based decision task.[105] When tested as adults, ACh lesioned animals exhibited greater impulsive choice than control animals, suggesting that early cholinergic hypofunction increased delay-discounting and caused lesioned animals to be less tolerant of delays. However, there are a couple aspects of the experiment which make it difficult to compare to other lesion studies we have covered in previous sections. First, since the lesions were made when the animals were neonates, the effect may be attributable to altered neuronal development. Second, the lesions were made prior to task exposure, meaning that the effects could be driven by impaired action–outcome learning rather than cost–benefit analysis. Thus, in order to directly test whether eliminating ACh neurotransmission impacts delay-discounting (and effort and risk discounting), cholinergic lesions should be performed in adult animals that have learned the tasks.

However, based on the fact that ACh is known to be involved in temporal processing, we suggest that basal forebrain depletions could alter delay discounting by impairing delay sensitivity. With respect to temporal processing, lesioning the nbm and ms (which project to the PFC and HPC, respectively) alters the performance on both temporal memory and temporal perception tasks[107–110]; while ms lesions have also been shown to shift animals preferences for delayed rewards—ms-lesioned animal began preferring an immediate uncertain reward over a delayed certain alternative.[66] Further, computational models demonstrate how altered estimations of time could easily impair cost–benefit computations by increasing or decreasing value estimations, even if reward magnitude sensitivity is spared.[111] In order to test this hypothesis, the effects of cholinergic depletions and pharmacological manipulations in the PFC and HPC on delay-based decision making should be determined, with attention paid to controls that can asses temporal processing.

8. CONCLUSION

A current and popular view of memory systems of the brain is that it is intimately tied to neural circuitry that mediates decisions and adaptive responding. The literature described in this review reveals that different neural systems mediate different types of cost–benefit decision making (i.e., those based on assessments of effort, delay, or risk). Yet common to these decision systems is the involvement of DA, which in turn is regulated in part by significant cholinergic afferent input. Interestingly, acetylcholine has long been considered central for normal memories. A challenge, then, for future neurobiological research on memory is to better understand its link to decision processing. Conversely, a challenge for the decision making field is to better understand its link to studies of memory systems of the brain. Here, we suggest that investigating the cholinergic involvement in both memory and decision should prove an effective strategy to approach both challenges.

ACKNOWLEDGMENTS

Supported by NIGMS grant GM007108 (W. C. F.) and NIMH grant MH58755 (S. J. Y. M.).

REFERENCES

1. Boyden ES, Katoh A, Raymond JL. Cerebellum-dependent learning: the role of multiple plasticity mechanisms. *Annu Rev Neurosci.* 2004;27:581–609.
2. Eichenbaum H, Cohen NJ. *From Conditioning to Conscious Recollection: Memory Systems of the Brain.* New York, NY: Oxford University Press; 2001.
3. Kesner RP. Tapestry of memory. *Behav Neurosci.* 2009;123(1):1–13.
4. Mizumori SJ, Jo YS. Homeostatic regulation of memory systems and adaptive decisions. *Hippocampus.* 2013;23(11):1103–1124.
5. Nadel L, Hupbach A, Gomez R, Newman-Smith K. Memory formation, consolidation and transformation. *Neurosci Biobehav Rev.* 2012;36(7):1640–1645.
6. Tulving E. Episodic memory: from mind to brain. *Annu Rev Psychol.* 2002;53:1–25.
7. Yin HH, Knowlton BJ. The role of the basal ganglia in habit formation. *Nat Rev Neurosci.* 2006;7(6):464–476.
8. Fuster JM. *The Prefrontal Cortex.* 4th ed. Boston, MA: Academic Press; 2008.
9. Rangel A, Camerer C, Montague PR. A framework for studying the neurobiology of value-based decision making. *Nat Rev Neurosci.* 2008;9(7):545–556.
10. Assadi SM, Yucel M, Pantelis C. Dopamine modulates neural networks involved in effort-based decision-making. *Neurosci Biobehav Rev.* 2009;33(3):383–393.
11. Penner MR, Mizumori SJ. Neural systems analysis of decision making during goal-directed navigation. *Prog Neurobiol.* 2012;96(1):96–135.

12. Schultz W. Behavioral theories and the neurophysiology of reward. *Annu Rev Psychol.* 2006;57:87–115.
13. Phillips PE, Stuber GD, Heien ML, Wightman RM, Carelli RM. Subsecond dopamine release promotes cocaine seeking. *Nature.* 2003;422(6932):614–618.
14. Yun IA, Wakabayashi KT, Fields HL, Nicola SM. The ventral tegmental area is required for the behavioral and nucleus accumbens neuronal firing responses to incentive cues. *J Neurosci.* 2004;24(12):2923–2933.
15. Phillips PE, Walton ME, Jhou TC. Calculating utility: preclinical evidence for cost-benefit analysis by mesolimbic dopamine. *Psychopharmacology (Berl).* 2007;191(3):483–495.
16. Kobayashi S, Schultz W. Influence of reward delays on responses of dopamine neurons. *J Neurosci.* 2008;28(31):7837–7846.
17. Fiorillo CD, Tobler PN, Schultz W. Discrete coding of reward probability and uncertainty by dopamine neurons. *Science.* 2003;299(5614):1898–1902.
18. Tobler PN, Fiorillo CD, Schultz W. Adaptive coding of reward value by dopamine neurons. *Science.* 2005;307(5715):1642–1645.
19. Roesch MR, Calu DJ, Schoenbaum G. Dopamine neurons encode the better option in rats deciding between differently delayed or sized rewards. *Nat Neurosci.* 2007;10(12):1615–1624.
20. Day JJ, Jones JL, Wightman RM, Carelli RM. Phasic nucleus accumbens dopamine release encodes effort- and delay-related costs. *Biol Psychiatry.* 2010;68(3):306–309.
21. Sugam JA, Day JJ, Wightman RM, Carelli RM. Phasic nucleus accumbens dopamine encodes risk-based decision-making behavior. *Biol Psychiatry.* 2012;71(3):199–205.
22. Gan JO, Walton ME, Phillips PE. Dissociable cost and benefit encoding of future rewards by mesolimbic dopamine. *Nat Neurosci.* 2010;13(1):25–27.
23. Floresco SB, Tse MT, Ghods-Sharifi S. Dopaminergic and glutamatergic regulation of effort- and delay-based decision making. *Neuropsychopharmacology.* 2008;33(8):1966–1979.
24. Winstanley CA, Dalley JW, Theobald DE, Robbins TW. Global 5-HT depletion attenuates the ability of amphetamine to decrease impulsive choice on a delay-discounting task in rats. *Psychopharmacology (Berl).* 2003;170(3):320–331.
25. van Gaalen MM, van Koten R, Schoffelmeer AN, Vanderschuren LJ. Critical involvement of dopaminergic neurotransmission in impulsive decision making. *Biol Psychiatry.* 2006;60(1):66–73.
26. Winstanley CA, Theobald DE, Dalley JW, Robbins TW. Interactions between serotonin and dopamine in the control of impulsive choice in rats: therapeutic implications for impulse control disorders. *Neuropsychopharmacology.* 2005;30(4):669–682.
27. Ghods-Sharifi S, St Onge JR, Floresco SB. Dopaminergic modulation of effort and risk-based decision making. *Neuropsychopharmacology.* 2006;31:S207.
28. St Onge JR, Chiu YC, Floresco SB. Differential effects of dopaminergic manipulations on risky choice. *Psychopharmacology (Berl).* 2010;211(2):209–221.
29. St Onge JR, Floresco SB. Dopaminergic modulation of risk-based decision making. *Neuropsychopharmacology.* 2009;34(3):681–697.
30. Civelli O. Molecular biology of dopamine receptor subtype. In: Bloom FE, Kupfer DJ, eds. *Psychopharmacology: The Fourth Generation of Progress.* Philadelphia, PA: Lippincott Williams & Wilkins; 1995:155–161.
31. Walton ME, Croxson PL, Rushworth MF, Bannerman DM. The mesocortical dopamine projection to anterior cingulate cortex plays no role in guiding effort-related decisions. *Behav Neurosci.* 2005;119(1):323–328.
32. Rudebeck PH, Walton ME, Smyth AN, Bannerman DM, Rushworth MF. Separate neural pathways process different decision costs. *Nat Neurosci.* 2006;9(9):1161–1168.

33. Walton ME, Bannerman DM, Rushworth MF. The role of rat medial frontal cortex in effort-based decision making. *J Neurosci*. 2002;22(24):10996–11003.
34. Walton ME, Bannerman DM, Alterescu K, Rushworth MF. Functional specialization within medial frontal cortex of the anterior cingulate for evaluating effort-related decisions. *J Neurosci*. 2003;23(16):6475–6479.
35. Hillman KL, Bilkey DK. Neurons in the rat anterior cingulate cortex dynamically encode cost-benefit in a spatial decision-making task. *J Neurosci*. 2010;30(22):7705–7713.
36. Schweimer J, Hauber W. Dopamine D1 receptors in the anterior cingulate cortex regulate effort-based decision making. *Learn Mem*. 2006;13(6):777–782.
37. Schweimer J, Saft S, Hauber W. Involvement of catecholamine neurotransmission in the rat anterior cingulate in effort-related decision making. *Behav Neurosci*. 2005;119(6):1687–1692.
38. Hauber W, Sommer S. Prefrontostriatal circuitry regulates effort-related decision making. *Cereb Cortex*. 2009;19(10):2240–2247.
39. Ghods-Sharifi S, Floresco SB. Differential effects on effort discounting induced by inactivations of the nucleus accumbens core or shell. *Behav Neurosci*. 2010;124(2):179–191.
40. Salamone JD, Cousins MS, Bucher S. Anhedonia or anergia? Effects of haloperidol and nucleus accumbens dopamine depletion on instrumental response selection in a T-maze cost/benefit procedure. *Behav Brain Res*. 1994;65(2):221–229.
41. Cousins MS, Wei W, Salamone JD. Pharmacological characterization of performance on a concurrent lever pressing/feeding choice procedure: effects of dopamine antagonist, cholinomimetic, sedative and stimulant drugs. *Psychopharmacology (Berl)*. 1994;116(4):529–537.
42. Salamone JD, Correa M, Mingote S, Weber SM. Nucleus accumbens dopamine and the regulation of effort in food-seeking behavior: implications for studies of natural motivation, psychiatry, and drug abuse. *J Pharmacol Exp Ther*. 2003;305(1):1–8.
43. Caine SB, Koob GF. Effects of mesolimbic dopamine depletion on responding maintained by cocaine and food. *J Exp Anal Behav*. 1994;61(2):213–221.
44. Ikemoto S, Panksepp J. The role of the nucleus accumbens dopamine in motivated behavior: a unifying interpretation with special reference to reward-seeking. *Brain Res Rev*. 1996;31:6–41.
45. Floresco SB, Ghods-Sharifi S. Amygdala-prefrontal cortical circuitry regulates effort-based decision making. *Cereb Cortex*. 2007;17(2):251–260.
46. Balleine BW, Killcross S. Parallel incentive processing: an integrated view of amygdala function. *Trends Neurosci*. 2006;29(5):272–279.
47. Baxter MG, Murray EA. The amygdala and reward. *Nat Rev Neurosci*. 2002;3(7):563–573.
48. Pratt WE, Mizumori SJ. Characteristics of basolateral amygdala neuronal firing on a spatial memory task involving differential reward. *Behav Neurosci*. 1998;112(3):554–570.
49. Schoenbaum G, Chiba AA, Gallagher M. Orbitofrontal cortex and basolateral amygdala encode expected outcomes during learning. *Nat Neurosci*. 1998;1(2):155–159.
50. Nunes EJ, Randall PA, Podurgiel S, Correa M, Salamone JD. Nucleus accumbens neurotransmission and effort-related choice behavior in food motivation: effects of drugs acting on dopamine, adenosine, and muscarinic acetylcholine receptors. *Neurosci Biobehav Rev*. 2013;37(9):2015–2025.
51. Mobini S, Body S, Ho MY, et al. Effects of lesions of the orbitofrontal cortex on sensitivity to delayed and probabilistic reinforcement. *Psychopharmacology (Berl)*. 2002;160(3):290–298.
52. Jo YS, Fobbs W, Mizumori SJ. *Effects of Orbitofrontal Inactivation on Dopamine Cell Activity During a Delay Based Decision Task*. San Diego, CA: Society for Neuroscience; 2013.

53. Floresco SB, St Onge JR, Ghods-Sharifi S, Winstanley CA. Cortico-limbic-striatal circuits subserving different forms of cost-benefit decision making. *Cogn Affect Behav Neurosci*. 2008;8(4):375–389.
54. Winstanley CA, Theobald DE, Cardinal RN, Robbins TW. Contrasting roles of basolateral amygdala and orbitofrontal cortex in impulsive choice. *J Neurosci*. 2004;24(20):4718–4722.
55. Mariano TY, Bannerman DM, McHugh SB, et al. Impulsive choice in hippocampal but not orbitofrontal cortex-lesioned rats on a nonspatial decision-making maze task. *Eur J Neurosci*. 2009;30(3):472–484.
56. Abela AR, Chudasama Y. Dissociable contributions of the ventral hippocampus and orbitofrontal cortex to decision-making with a delayed or uncertain outcome. *Eur J Neurosci*. 2013;37(4):640–647.
57. Zeeb FD, Floresco SB, Winstanley CA. Contributions of the orbitofrontal cortex to impulsive choice: interactions with basal levels of impulsivity, dopamine signalling, and reward-related cues. *Psychopharmacology (Berl)*. 2010;211(1):87–98.
58. Roesch MR, Taylor AR, Schoenbaum G. Encoding of time-discounted rewards in orbitofrontal cortex is independent of value representation. *Neuron*. 2006;51(4):509–520.
59. Kheramin S, Body S, Ho MY, et al. Effects of orbital prefrontal cortex dopamine depletion on inter-temporal choice: a quantitative analysis. *Psychopharmacology (Berl)*. 2004;175(2):206–214.
60. Cardinal RN, Pennicott DR, Sugathapala CL, Robbins TW, Everitt BJ. Impulsive choice induced in rats by lesions of the nucleus accumbens core. *Science*. 2001;292(5526):2499–2501.
61. Narayanan NS, Horst NK, Laubach M. Reversible inactivations of rat medial prefrontal cortex impair the ability to wait for a stimulus. *Neuroscience*. 2006;139(3):865–876.
62. McHugh SB, Campbell TG, Taylor AM, Rawlins JN, Bannerman DM. A role for dorsal and ventral hippocampus in inter-temporal choice cost-benefit decision making. *Behav Neurosci*. 2008;122(1):1–8.
63. Cheung TH, Cardinal RN. Hippocampal lesions facilitate instrumental learning with delayed reinforcement but induce impulsive choice in rats. *BMC Neurosci*. 2005;6:36.
64. Meck WH, Church RM, Olton DS. Hippocampus, time, and memory. *Behav Neurosci*. 1984;98(1):3–22.
65. Yin B, Troger AB. Exploring the 4th dimension: hippocampus, time, and memory revisited. *Front Integr Neurosci*. 2011;5:36.
66. Rawlins JN, Feldon J, Butt S. The effects of delaying reward on choice preference in rats with hippocampal or selective septal lesions. *Behav Brain Res*. 1985;15(3):191–203.
67. Churchwell JC, Morris AM, Heurtelou NM, Kesner RP. Interactions between the prefrontal cortex and amygdala during delay discounting and reversal. *Behav Neurosci* 2009;123(6):1185–1196.
68. St Onge JR, Floresco SB. Prefrontal cortical contribution to risk-based decision making. *Cereb Cortex*. 2010;20(8):1816–1828.
69. Mai B, Hauber W. Intact risk-based decision making in rats with prefrontal or accumbens dopamine depletion. *Cogn Affect Behav Neurosci*. 2012;12(4):719–729.
70. St Onge JR, Abhari H, Floresco SB. Dissociable contributions by prefrontal D1 and D2 receptors to risk-based decision making. *J Neurosci*. 2011;31(23):8625–8633.
71. Cardinal RN, Howes NJ. Effects of lesions of the nucleus accumbens core on choice between small certain rewards and large uncertain rewards in rats. *BMC Neurosci*. 2005;6:37.

72. Stopper CM, Floresco SB. Contributions of the nucleus accumbens and its subregions to different aspects of risk-based decision making. *Cogn Affect Behav Neurosci.* 2011;11(1):97–112.
73. Stopper CM, Khayambashi S, Floresco SB. Receptor-specific modulation of risk-based decision making by nucleus accumbens dopamine. *Neuropsychopharmacology.* 2013;38(5):715–728.
74. St Onge JR, Ahn S, Phillips AG, Floresco SB. Dynamic fluctuations in dopamine efflux in the prefrontal cortex and nucleus accumbens during risk-based decision making. *J Neurosci.* 2012;32(47):16880–16891.
75. Ghods-Sharifi S, St Onge JR, Floresco SB. Fundamental contribution by the basolateral amygdala to different forms of decision making. *J Neurosci.* 2009;29(16):5251–5259.
76. Lee H, Ghim JW, Kim H, Lee D, Jung M. Hippocampal neural correlates for values of experienced events. *J Neurosci.* 2012;32(43):15053–15065.
77. Penner MR, Larkin J, Tryon V, Mizumori SJ. *Hippocampal pyramidal cell activity is modulated by changes in reward context on a decision-making maze.* New Orleans, LA: Society for Neuroscience; 2012.
78. Fanselow MS, Dong HW. Are the dorsal and ventral hippocampus functionally distinct structures? *Neuron.* 2010;65(1):7–19.
79. St Onge JR, Stopper CM, Zahm DS, Floresco SB. Separate prefrontal-subcortical circuits mediate different components of risk-based decision making. *J Neurosci.* 2012;32(8):2886–2899.
80. Everitt BJ, Robbins TW. Central cholinergic systems and cognition. *Annu Rev Psychol.* 1997;48:6490–6684.
81. Hasselmo ME, Sarter M. Modes and models of forebrain cholinergic neuromodulation of cognition. *Neuropsychopharmacology.* 2011;36(1):52–73.
82. Deiana S, Platt B, Riedel G. The cholinergic system and spatial learning. *Behav Brain Res.* 2011;221(2):389–411.
83. Mesulam M. The cholinergic lesion of Alzheimer's disease: pivotal factor or side show? *Learn Mem.* 2004;11(1):43–49.
84. Croxson PL, Kyriazis DA, Baxter MG. Cholinergic modulation of a specific memory function of prefrontal cortex. *Nat Neurosci.* 2011;14(12):1510–1512.
85. Butt AE, Bowman TD. Transverse patterning reveals a dissociation of simple and configural association learning abilities in rats with 192 IgG-saporin lesions of the nucleus basalis magnocellularis. *Neurobiol Learn Mem.* 2002;77(2):211–233.
86. Steckler T, Keith AB, Wiley RG, Sahgal A. Cholinergic lesions by 192 IgG-saporin and short-term recognition memory: role of the septohippocampal projection. *Neuroscience.* 1995;66(1):101–114.
87. dos Santos Coura R, Granon S. Prefrontal neuromodulation by nicotinic receptors for cognitive processes. *Psychopharmacology (Berl).* 2012;221(1):1–18.
88. Mendez IA, Gilbert RJ, Bizon JL, Setlow B. Effects of acute administration of nicotinic and muscarinic cholinergic agonists and antagonists on performance in different cost-benefit decision making tasks in rats. *Psychopharmacology (Berl).* 2012;224(4): 489–499.
89. Locey ML, Dallery J. Nicotine and the behavioral mechanisms of intertemporal choice. *Behav Processes.* 2011;87(1):18–24.
90. Kolokotroni KZ, Rodgers RJ, Harrison AA. Acute nicotine increases both impulsive choice and behavioural disinhibition in rats. *Psychopharmacology (Berl).* 2011;217(4):455–473.
91. Anderson KG, Diller JW. Effects of acute and repeated nicotine administration on delay discounting in Lewis and Fischer 344 rats. *Behav Pharmacol.* 2010;21:754–764.

92. Mesulam M, Mufson EJ, Wainer BH, Levey AI. Central cholinergic pathways in the rate: an overview based on alternative nomenclature (Ch1-Ch6). *Neuroscience*. 1983;10(4):1185–1201.
93. Woolf NJ, Butcher LL. Cholinergic system in the rat brain: III. Projections from the pontomesencephalic tegmentum to the thalmus, tectum, basal ganglia, and basal forebrain. *Brain Res Bull*. 1986;16:603–637.
94. Changeux JP. Allosteric receptors: from electric organ to cognition. *Annu Rev Pharmacol Toxicol*. 2010;50:1–38.
95. Gotti C, Zoli M, Clementi F. Brain nicotinic acetylcholine receptors: native subtypes and their relevance. *Trends Pharmacol Sci*. 2006;27(9):482–491.
96. Wallace TL, Bertrand D. Importance of the nicotinic acetylcholine receptor system in the prefrontal cortex. *Biochem Pharmacol*. 2013;85(12):1713–1720.
97. Levin ED. Nicotinic receptor subtypes and cognitive functions. *J Neurobiol*. 2002;53(4):633–640.
98. Barak S, Weiner I. Differential role of muscarinic transmission within the entorhinal cortex and basolateral amygdala in the processing of irrelevant stimuli. *Neuropsychopharmacology*. 2010;35(5):1073–1082.
99. Mark GP, Shabani S, Dobbs LK, Hansen ST. Cholinergic modulation of mesolimbic dopamine function and reward. *Physiol Behav*. 2011;104(1):76–81.
100. Zhou FM, Wilson C, Dani JA. Muscarinic and nicotinic cholinergic mechanisms in the mesostriatal dopamine systems. *Neuroscientist*. 2003;9(1):23–36.
101. Gronier B, Rasmussen K. Activation of midbrain presumed dopaminergic neurones by muscarinic cholinergic receptors: an in vivo electrophysiological study in the rat. *Br J Pharmacol*. 1998;124(3):455–464.
102. Forster GL, Blaha CD. Laterodorsal tegmental stimulation elicits dopamine efflux in the rat nucleus accumbens by activation of acetylcholine and glutamate receptors in the ventral tegmental area. *Eur J Neurosci*. 2000;12(10):3596–3604.
103. Mansvelder HD, McGehee DS. Cellular and synaptic mechanisms of nicotine addiction. *J Neurobiol*. 2002;53(4):606–617.
104. Forster GL, Yeomans JS, Takeuchi J, Blaha CD. M5 muscarinic receptors are required for prolonged accumbal dopamine release after electrical stimulation of the pons in mice. *J Neurosci*. 2002;22(1):RC190.
105. Scattoni ML, Adriani W, Calamandrei G, Laviola G, Ricceri L. Long-term effects of neonatal basal forebrain cholinergic lesions on radial maze learning and impulsivity in rats. *Behav Pharmacol*. 2006;17(5–6):517–524.
106. Winstanley CA, Theobald DE, Dalley JW, Cardinal RN, Robbins TW. Double dissociation between serotonergic and dopaminergic modulation of medial prefrontal and orbitofrontal cortex during a test of impulsive choice. *Cereb Cortex*. 2006;16(1):106–114.
107. Meck WH, Church RM, Wenk GL, Olton DS. Nucleus basalis magnocellularis and medial septal area lesions differentially impair temporal memory. *J Neurosci*. 1987;7(11):3505–3511.
108. Meck WH, Church RM. Nutrients that modify the speed of internal clock and memory storage processes. *Behav Neurosci*. 1987;101(4):465–475.
109. Meck WH, Church RM. Cholinergic modulation of the content of temporal memory. *Behav Neurosci*. 1987;101(4):457–464.
110. Olton DS, Meck WH, Church RM. Separation of hippocampal and amygdaloid involvement in temporal memory dysfunctions. *Brain Res*. 1987;404(1–2):180–188.
111. Ho MY, Mobini S, Chiang TJ, Bradshaw CM, Szabadi E. Theory and method in the quantitative analysis of "impulsive choice" behaviour: implications for psychopharmacology. *Psychopharmacology (Berl)*. 1999;146(4):362–372.

112. Sarter M, Hasselmo ME, Bruno JP, Givens B. Unraveling the attentional functions of cortical cholinergic inputs: interactions between signal-driven and cognitive modulation of signal detection. *Brain Res Brain Res Rev*. 2005;48(1):98–111.
113. Baxter MG, Bucci DJ, Holland PC, Gallagher M. Impairments in conditioned stimulus processing and conditioned responding after combined selective removal of hippocampal and neocortical cholinergic input. *Behav Neurosci*. 1999;113(3):486–495.
114. Chiba AA, Bucci DJ, Holland PC, Gallagher M. Basal forebrain cholinergic lesions disrupt increments but not decrements in conditioned stimulus processing. *J Neurosci*. 1995;15(11):7315–7322.
115. Dalley JW, Theobald DE, Bouger P, Chudasama Y, Cardinal RN, Robbins TW. Cortical cholinergic function and deficits in visual attentional performance in rats following 192 IgG-saporin-induced lesions of the medial prefrontal cortex. *Cereb Cortex*. 2004;14(8):922–932.
116. McGaughy J, Everitt BJ, Robbins TW, Sarter M. The role of cortical cholinergic afferent projections in cognition: impact of new selective immunotoxins. *Behav Brain Res*. 2000;115(2):251–263.
117. McGaughy J, Sarter M. Sustained attention performance in rats with intracortical infusions of 192 IgG-saporin-induced cortical cholinergic deafferentation: effects of physostigmine and FG 7142. *Behav Neurosci*. 1998;112(6):1519–1525.
118. Newman LA, McGaughy J. Cholinergic deafferentation of prefrontal cortex increases sensitivity to cross-modal distractors during a sustained attention task. *J Neurosci*. 2008;28(10):2642–2650.
119. Turchi J, Sarter M. Cortical acetylcholine and processing capacity: effects of cortical cholinergic deafferentation on crossmodal divided attention in rats. *Brain Res Cogn Brain Res*. 1997;6(2):147–158.
120. Bardgett ME, Depenbrock M, Downs N, Points M, Green L. Dopamine modulates effort-based decision making in rats. *Behav Neurosci*. 2009;123(2):242–251.

CHAPTER TEN

Molecular Mechanisms of Threat Learning in the Lateral Nucleus of the Amygdala

Robert M. Sears[*], **Hillary C. Schiff**[*], **Joseph E. LeDoux**[*,†]
[*]Center for Neuroscience, New York University, New York, USA
[†]Emotional Brain Institute, Nathan Kline Institute for Psychiatric Research, Orangeburg, New York, USA

Contents

Abstract

Pavlovian threat conditioning is a behavioral paradigm that has been successfully utilized to define the mechanisms underlying threat (fear) memory formation. The amygdala is a temporal lobe structure required for the acquisition, consolidation, and expression of threat (fear) memories. In particular, the lateral nucleus of the amygdala (LA) is the major input structure of the amygdala and is required for all aspects of threat learning and memory. The LA expresses many neurotransmitter and neuromodulator receptors. This chapter covers the molecular mechanisms that occur downstream of these receptors and how they influence LA-dependent Pavlovian threat learning.

Progress in Molecular Biology and Translational Science, Volume 122
ISSN 1877-1173
http://dx.doi.org/10.1016/B978-0-12-420170-5.00010-6

1. INTRODUCTION

Pavlovian threat conditioning (fear conditioning) is a well-established behavioral paradigm to assess the formation, storage, and expression of threat memories (see Fig. 10.1; for review see Ref. 1). During training, animals learn to associate an aversive unconditioned stimulus (US), such as a footshock, with a neutral conditioned stimulus (CS), such as a tone, when both occur in close temporal proximity. After as little as one CS–US pairing, animals begin to show defensive reactions such as increased sympathetic activity (respiration and heart rate) and "freezing" behavior. Notably, the

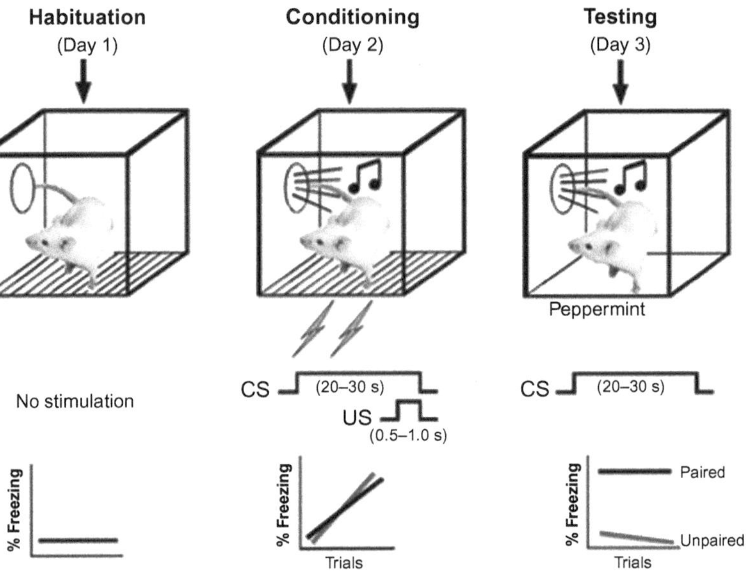

Figure 10.1 Depiction of a Pavlovian auditory threat conditioning paradigm and typical results obtained. Rats are first habituated to the conditioning chamber in the absence of stimuli, usually 24 h prior to conditioning. During the conditioning session, the unconditioned stimulus (US; a mild shock) is paired with the conditioned stimulus (CS; a tone), usually one to five times. After a maximum of 3 h (for STM) or at least 24 h (for LTM), animals are placed in a new context and exposed to multiple presentations of the CS alone. "Freezing" behavior, an innate defensive response elicited by the conditioned stimulus after conditioning, is then scored as total time spent in this state over the course of the CS presentation. To control for nonassociative effects of animal manipulations, for example, drug treatment, a control group in which the CS and US do not co-occur is often used. This unpaired control group shows very little CS-elicited freezing, indicating that associative learning processes are responsible for this defensive behavior. (See the color plate.)

memory can persist throughout the life of the animal.[2] The robustness of the paradigm has led to a detailed understanding of the neural circuitry of threat learning and the mechanisms of associative plasticity in general.[3]

Studies have designated the amygdala as a locus of threat memory formation, storage, and expression. Within the amygdala, the lateral nucleus (LA) has been described as the main input nucleus and a key site of associative plasticity. The plasticity mechanisms in LA are thought to involve a Hebbian process through which plasticity is induced at CS synapses by concurrent US input. Specifically, activation of LA neurons by CS inputs is enhanced by US-mediated depolarization and lead to strong cellular responses to future presentations of CS alone. The previously neutral CS therefore appears to take on emotional significance as assessed by the freezing response. Following LA-mediated learning, defensive reactions are believed to occur through CS activation of LA neurons which in turn project to the central nucleus of the amygdala via direct and indirect intra-amygdala pathways, and from there to target brain regions, to produce the integrated defense response.

Here, we focus on the molecular mechanisms in the amygdala important for threat conditioning. We will frame this in terms of "fast" synaptic transmission mediated by glutamate and gamma-aminobutyric acid (GABA) at ionotropic receptors and contrast these with "slow" synaptic transmission mediated by metabotropic receptor mechanisms. We will describe how these signaling mechanisms are involved in two stages of learning and memory: acquisition and consolidation, and, due to its essential involvement in these phases, we will focus on the LA. However, it is important to note that many studies covered here do not distinguish between the lateral and basal nuclei of the amygdala (BLA). Furthermore, we will occasionally refer to studies in other brain areas or cell systems to illustrate molecular events that may occur, but have not been explicitly studied in the amygdala. A review of the molecular mechanisms of reconsolidation and extinction is beyond the scope of this chapter and is covered in detail elsewhere.

2. HEBBIAN MECHANISMS IN LA: POSSIBLE CONTRIBUTIONS TO THREAT LEARNING

In the 1940s, the Canadian psychologist Donald Hebb proposed a theory to explain how neurons become intimately linked in learning, a phenomenon we now refer to as synaptic plasticity.[4–6] The phrase, "cells that fire together wire together" captures the essence of this idea: a neuron that is depolarized due to a given excitatory input will become more receptive to other co-occurring excitatory inputs. Consequently, this enhanced receptivity to a given input will be maintained long term. We now think of this as a cellular model for associative learning. A widely used and informative behavioral paradigm used to assess associative memory processes, auditory cued threat (fear) conditioning, has provided much support for Hebb's model (see Fig. 10.1; Refs. 1,3,7). The robust and replicable nature of this paradigm has allowed for mechanistic studies to understand the cellular and molecular processes underlying amygdala-dependent associative plasticity.[1,3,7,8]

Early cellular evidence in support of Hebb's model came from work in mammalian hippocampus slices[9,10] and in invertebrate preparations (see Refs. 11–13). These studies were the first to demonstrate associative plasticity in living tissue, and paved the way for the work described in this chapter. Here, we will focus solely on mammalian work, most of which was performed in rodents.

The induction of so-called long-term potentiation (LTP) of synaptic responses with high-frequency stimulation has also been demonstrated in many brain regions, including the amygdala,[14–19] and is currently the most popular cellular model for long-term memory (LTM) formation (Fig. 10.2). Detailed *in vitro* studies have shown enhanced plasticity of auditory thalamus to LA (thalamo-LA) and auditory cortex to LA (cortico-LA) following threat conditioning.[20–23] Moreover, LTP is occluded in LA slices of threat-conditioned animals.[23–25] *In vivo* extracellular recordings in the amygdala provide support for a causal link between LA plasticity and memory formation, showing enhancement of amygdala neuronal responses to the CS after pairing with US[26,27] reviews.[1,7,28] Moreover, when auditory thalamus (medial geniculate body) is artificially stimulated, LTP is induced at inputs to LA, and neural responses to auditory tones in the LA are enhanced following this stimulation.[16] Indeed, these artificially induced responses in LA mirror what is observed following auditory threat conditioning, which is strong evidence that LTP studied *ex vivo*, may represent learning and memory processes *in vivo*.[29] Advancing this idea, a recent study showed that replacing the foot shock stimulus with light activation of LA principal cells expressing excitatory opsins (channelrhodopsin-2) is sufficient for threat conditioning to an auditory tone.[30] Together, these data highlight the LA as a critical mediator of associative, Hebbian plasticity and memory formation and provided the foundation and impetus for detailed cellular and molecular studies discussed below.

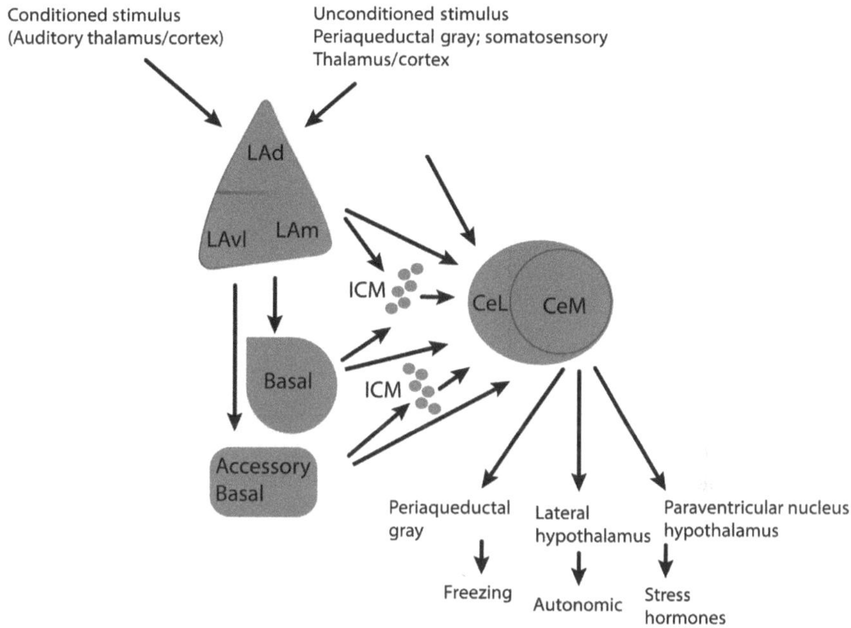

Figure 10.2 Amygdala circuitry involved in the acquisition, consolidation, and expression of threat memories. Convergence of the auditory CS (tone) and US (mild shock) in the lateral amygdala (LA) is required for synaptic plasticity and threat conditioning in LA. The LA communicates with Ce directly and indirectly by way of connections in the basal, accessory basal, and intercalated cell masses (ICM). The Ce, specifically the medial subnucleus (CeM), sends projections to hypothalamic and brainstem areas required for coordinated expression of the conditioned response, that is, the defensive response (freezing) as well as autonomic and hormonal responses. CeL, lateral nucleus of Ce. (See the color plate.)

3. MOLECULAR CORRELATES OF THREAT LEARNING

In cued threat conditioning, sensory stimuli in the form of an auditory tone and foot shock converge in the amygdala for threat memory formation.[31–33] This occurs rapidly, as defensive behaviors (i.e., freezing) are observed in response to the tone following the first tone-shock pairing and persist several hours following training, indicating that a form of learning has already occurred.[1,32] Such early learning may be referred to as short-term memory (STM), and is usually tested within 3 h following training.[34] Pretraining manipulations that affect STM, for example, drug microinfusions in the LA, generally also affect LTM, although the reverse is not always the case: pretraining manipulations may have little to no effect on STM, while significantly affecting LTM.[8,35] This suggests that distinct molecular processes are engaged during training (a) for the acquisition of a memory and (b) for the consolidation of the memory into long-term storage.[8] However, it is possible and even likely that some of these mechanisms share common molecular signaling cascades, with the distinction between early and late forms of memory reflected in early and late signaling events in the cell (see Fig. 10.3). Indeed, rapid events at the cell membrane may be responsible for short-term plasticity and STM, whereas the persistence of plasticity and associative memories likely requires synaptic and cellular changes that are triggered further downstream. Therefore, we present here the idea that early and late forms of plasticity and memory are defined by spatial and temporal factors, that is, signaling events cascading from the synapse to the nucleus.[36]

4. FAST VERSUS SLOW SYNAPTIC TRANSMISSION IN THE ACQUISITION OF THREAT MEMORIES

Events occurring at the synaptic membrane may be fast or slow. Presynaptic glutamate and GABA release contribute to the rapid excitatory and inhibitory (respectively) responses via the activation of postsynaptic ionotropic receptors, referred to here as "fast" synaptic transmission.[37] This rapid signaling leads to membrane depolarization or hyperpolarization in a matter of milliseconds, and directly influences downstream synaptic communication through action potential propagation. In contrast, slow synaptic transmission occurs on the order of seconds to minutes not only through the activity of biogenic amines and peptide signaling at metabotropic receptors but also through metabotropic glutamate and GABA receptors.[37] Many of the metabotropic receptors discussed in this chapter are G-protein-coupled receptors (GPCRs), 7-transmembrane receptors that upon ligand binding release guanosine nucleotide-binding proteins (G-proteins), triggering myriad short- and long-term cellular processes.[38] One important function of slow synaptic transmission in the brain is to modulate excitatory or inhibitory input either presynaptically (through affecting neurotransmitter release) or postsynaptically at the level of ionotropic receptors (through posttranslational modifications, trafficking, and localization), thus turning the gain up or down on short- and long-term plasticity mechanisms.[39]

Here, we discuss the mechanisms of fast synaptic transmission and contrast them with slow synaptic transmission as it relates to threat memory formation. Specifically, we present evidence to suggest that during acquisition of a threat memory, both fast and slow synaptic transmission mechanisms are

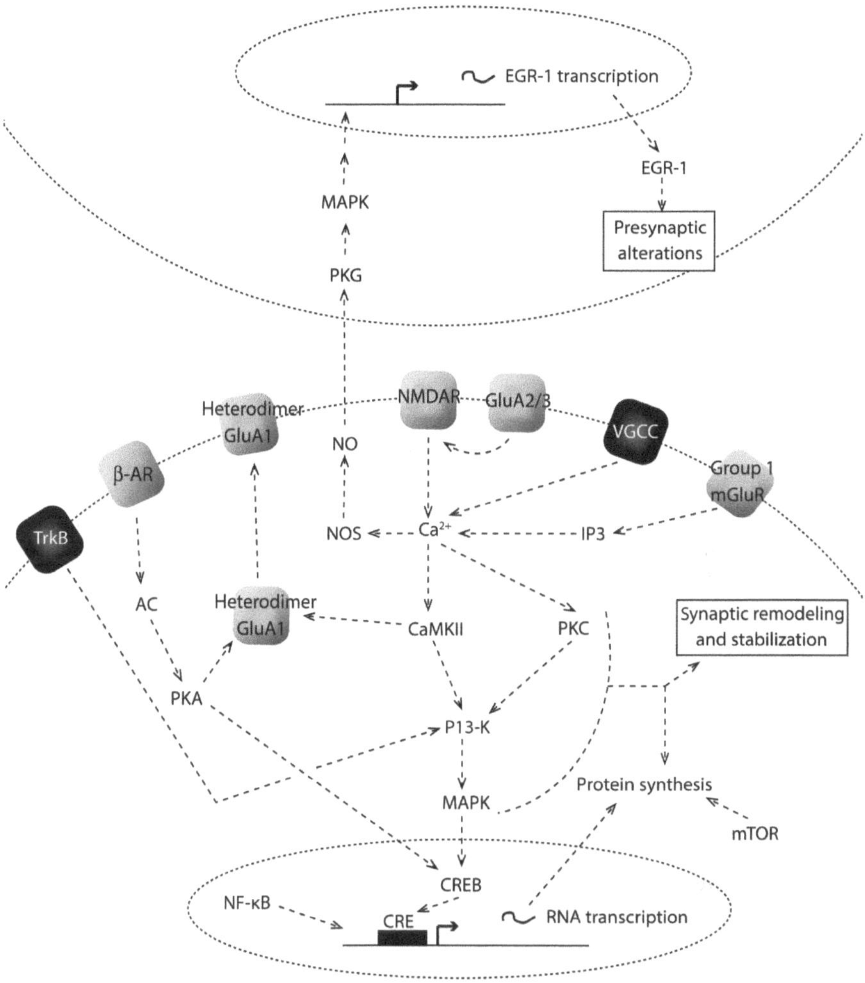

Figure 10.3 Molecular mechanisms of the acquisition and consolidation of threat memories in the B/LA. Dotted lines represent molecular pathways triggered during these two phases of learning. While all pathways are likely triggered during training, some molecular components may be brought online later during consolidation, and downstream events may be more important during the acquisition or consolidation phase. Molecular players known to be involved in the acquisition of threat conditioning are depicted as black, and those involved in the consolidation or maintenance of threat conditioning are gray. Abbreviations: AC, adenylyl cyclase; β-AR, β-adrenergic receptor; Ca^{2+}, calcium; CamKII, Ca^{2+}/calmodulin (Cam)-dependent protein kinase II; CREB, cAMP response element (CRE) binding protein; EGR-1, early growth response gene 1; GluA1, glutamate AMPA receptor subunit 1; GluA2/3, glutamate AMPA receptor subunit 2 and 3 heteromer; IP3, inositol 1,4,5-triphosphate; MAPK, mitogen-activated protein kinase; mGluR, metabotropic glutamate receptor; mTOR, mammalian target of rapamycin; NF-kB, nuclear factor kappa light-chain enhancer of activated B cells; NMDAR, *N*-methyl-D-aspartate glutamate receptor; NO, nitric oxide; NOS, nitric oxide synthase; PI3-K, phosphatidylinositol-3 kinase; PKA, protein kinase A; PKC, protein kinase C; PKG, cGMP-dependent protein kinase; RNA, ribonucleic acid; TrkB, tyrosine kinase B; VGCC, voltage-gated calcium channel.

engaged, leading to alterations in the activity of synaptic proteins, for example, ionotropic receptors that are already present at or near the synapse. Posttranslational modifications, trafficking, and localization of ionotropic receptors may underlie early forms of learning and memory as observed within trial (during training) and soon after training (STM). Further downstream of these relatively early signaling events, *de novo* protein synthesis is initiated and is responsible for the persistence of memory or LTM, which will be covered in the section on consolidation (Section 5) below.

4.1. Fast synaptic transmission through ionotropic receptors in LA

4.1.1 2-Amino-3-(3-hydroxy-5-methyl-isoxazol-4-yl)propanoic acid receptor

In vivo recordings demonstrate that LA neurons are largely inactive in the absence of external stimuli, and AMPA-type glutamate receptors (AMPARs) contribute to excitatory synaptic transmission in amygdala slices.[19,40] In fact, AMPARs mediate the majority of postsynaptic excitatory responses observed in the LA.[41] In the case of auditory threat conditioning, AMPARs are activated during glutamate release from terminals originating in the medial geniculate body of the thalamus (MGm/PIN) and the auditory cortex during auditory stimulus (CS) presentations.[42,43] These receptors are solely ligand gated and thus can rapidly pass current upon glutamate binding, contributing to excitatory synaptic transmission and membrane depolarization. Evidence suggests that AMPARs are necessary, but not sufficient for plasticity in amygdala slices,[19,40] and behavioral data show these receptors are required for learning in a fear potentiated startle task and cued threat conditioning.[44,45] Below, we will discuss how AMPAR activity and availability at the synapse are regulated downstream of slow synaptic transmission mechanisms, which are important for enhanced synaptic responses seen in early and late forms of memory.[8,35]

4.1.2 N-methyl-D-aspartate receptors

Unlike their AMPAR counterparts, *N*-methyl-D-aspartate receptors (NMDARs) or kainite receptors are the ideal candidates for mediating coincidence detection of pre- and postsynaptic events based on the mechanisms of their activation.[46] The activity of NMDARs requires a permissive step whereby postsynaptic membrane depolarization releases magnesium from the NMDAR pore region. As mentioned above, in auditory threat conditioning, it is believed that inputs MGm/PIN and auditory cortex, which are glutamatergic, depolarize the cell, likely through AMPAR activity and/or voltage-gated cation channel activity, and inputs carrying nociceptive (shock) information release glutamate to activate these voltage-primed NMDARs, allowing influx of cations including calcium.[35] Once activated, NMDARs contribute to normal synaptic transmission with a slowly decaying excitatory component.[47] Of relevance to their role in auditory threat conditioning, *in vivo* studies show that blockade of NMDARs in LA with the competitive NMDAR antagonist APV (or AP5) or the noncompetitive antagonist MK-801 reduces synaptic responses elicited from auditory thalamus stimulation.[42,48]

Behavioral studies show that NMDARs must be active in LA during the acquisition of threat memories.[49–54] APV impairs the acquisition of threat memories when infused in LA before training,[54] but does not affect memory consolidation when given posttraining,[51] suggesting that NMDAR activation occurs during acquisition and either does not persist or is not required, following training.

NMDAR contribution to fast synaptic transmission[47] may account for the observation that antagonism with general NMDAR antagonists has effects on both acquisition and expression.[35,51] However, blocking NMDARs in a subunit-specific manner has revealed a role beyond basic synaptic transmission. For example, LA infusion of the GluN2B-containing receptor blocker ifenprodil impairs the acquisition of threat memories while leaving normal synaptic transmission and expression of threat memories intact.[54–56] However, a recent study suggests that the involvement of GluN2B depends on the strength of conditioning (number of CS–US pairing trials), whereas the less well-studied GluN2A is necessary for all conditioning protocols tested.[57] The role of specific NMDAR subunits in threat learning and their regulation by second messenger signaling and protein–protein interactions will be discussed in detail below.

While most studies looking at NMDAR activity in threat conditioning are consistent with Hebb's hypothesis for associative plasticity mechanisms, a form of associative plasticity not requiring postsynaptic activity was reported in amygdala slices.[58] Plasticity was only observed at cortical inputs to the amygdala and required presynaptic NMDAR and calcium activity without a need for postsynaptic depolarization and calcium influx. Induction of this form of plasticity required simultaneous thalamic and cortical stimulation, leading the authors to speculate that synaptic release of glutamate from thalamic afferents activates presynaptic NMDARs at cortical inputs, thus inducing a non-Hebbian form of plasticity. This is perhaps consistent with *in vivo* studies showing that inputs from auditory thalamus reach the LA with shorter latency than those of auditory cortex (~15–20 ms before),[48,59] leading the authors to suggest that the slowly decaying responses elicited by rapid thalamus-glutamate release and NMDAR activation extends the window of coincidence detection at slower cortical auditory inputs. Under both scenarios, glutamate release from thalamic inputs activates both pre- and postsynaptic NMDARs, leading to pre- and postsynaptic plasticity at cortico-LA synapses. These data should be taken into consideration when interpreting behavioral results from pharmacological manipulations of NMDARs.[35]

4.1.3 GABA receptors

Spontaneous activity observed in LA principle cells is predominately inhibitory (inhibitory postsynaptic potentials, IPSPs), with a paucity of excitatory postsynaptic potentials (EPSPs),[60–63] suggesting these cells maintain tight control over plasticity and possibly memory formation and expression of defensive responses. Indeed, infusion of the GABA-A receptor antagonist bicuculline in LA revealed increases in evoked short-latency responses from MGm stimulation, supporting an important role in gating excitatory input.[64] Furthermore, *in vitro* studies show that local GABAergic interneurons negatively regulate LTP induction in the LA,[65–68] and behavioral pharmacology studies show that activation of ionotropic GABA-A receptors in LA disrupts the acquisition of threat learning.[69,70] Interestingly, GABA and GAD65 expression and LTP of inhibitory neurotransmission are transiently reduced following threat conditioning,[71–74] as are many GABA-A receptor subunits,[71] suggesting that temporary downregulation of GABA activity in LA may be permissive for memory formation.[75] Cellular responses to GABAergic signaling are highly dependent on GABA receptor subunit composition.

GABA-A receptors are composed of seven classes of subunits each with multiple isoforms differentially expressed throughout the amygdala, which can influence ligand affinity and subcellular localization.[75] Recent studies have attempted to understand the roles of different subunits in plasticity and memory formation in the amygdala. For example, genetic deletion of the $\alpha 1$ subunit of GABA-A receptors in LA/BLA enhances threat learning and LTP, not through direct effects in pyramidal cells, but possibly through dysregulation of GABAergic interneuron activity.[76] A recent study in slices showed that $\alpha 3$-containing GABA-A receptors play a role in tonic inhibition in the BLA, and although a role in threat memory was not explored, the authors suggest these receptors serve to prevent inappropriate defense states.[77] Another class of GABA-A-like receptors, GABA-A-ρ (formerly GABA(C)Rs), are composed entirely of ρ subunits, and receptor activation causes increases in evoked excitation in slices and enhanced threat memory formation when infused in LA, possibly through disinhibition of principle cells.[78]

Neuromodulators, in the form of biogenic amines and neuropeptides, play a large role in amygdala function due to effects on inhibitory drive, in some cases enhancing LTP and threat learning.[66,67,79,80] In other cases, neurotransmitters can increase GABAergic tone in the amygdala, ultimately inhibiting principle cell activity as is the case with serotonin.[81,82] These mechanisms will be discussed in greater detail below.

4.2. Slow synaptic transmission in LA

4.2.1 Biogenic amines

In addition to convergent CS and US signaling mediated through fast synaptic transmission mechanisms, neuromodulation through biogenic amines and peptide signaling has been shown to play an important role in a wide variety of species in memory formation for aversive paradigms and in synaptic plasticity in a plethora of brain systems including the amygdala (e.g., Ref. 83). In particular, norepinephrine (NE), serotonin, and dopamine (DA) act as neurotransmitters in the amygdala where they can induce either a form of non-Hebbian plasticity, for example, plasticity that is not dependent on postsynaptic plasticity, also known as heterosynaptic plasticity, or modulate of activity-dependent, Hebbian synaptic transmission, or homosynaptic plasticity.[83] Many neuromodulators are critical for learning and memory and contribute substantially to anxiety disorders in humans.[84,85]

4.2.1.1 Norepinephrine

NE release into the LA activates many subtypes of adrenergic receptors each of which couples to distinct G-protein signaling pathways. Specifically, the α1-ARs typically couple to Gq proteins, the α2-ARs to Gi, and the β-ARs to Gs.[86–92] Thus, due to differential coupling to adenylyl cyclase (AC) and cAMP or phospholipase C (PLC) signaling cascades, each receptors subtype plays a distinct role in modulating neural activity in response to NE release. NE in the LA likely originates in the locus coeruleus (LC).[93] LC neurons are activated in response to stressors including acute footshock, immobilization, and tail pinch,[94–96] leading to enhanced NE release in the amygdala and elsewhere in the brain.[94,97,98] Indeed, NE iontophoresis onto LA neurons causes excitation via β-ARs and a powerful inhibition via α1-ARs *in vivo*.[99] In threat conditioning, corresponding results have been found with β-ARs permitting acquisition[100] while α1-ARs inhibit acquisition.[101]

Furthermore, β-ARs have been shown to be permissive for thalamo-LA LTP induced by tetanus stimulation. In this induction protocol, β-ARs also appear to recruit PKA and extracellular-regulated kinase (ERK), two kinases required for LTP, suggesting that β-AR activation is responsible for recruiting the necessary kinases for the induction of plasticity.[102] PKA may also be involved in the ability of β-AR activity to phosphorylate ser845 GluA1[103] and in β-AR-mediated calcium-permeable AMPAR (CP-AMPAR) insertion in LA principal cells following threat conditioning.[104] Importantly, a mechanism has been described whereby β-ARs spatially restrict AMPAR trafficking to particular synapses through physical association in a structural complex.[105] β-AR activation has also been shown to enhance synaptic transmission at thalamo-LA synapses by reducing feed-forward inhibition.[67] β-AR-mediated phosphorylation or trafficking of GluA1, or its effect on feed-forward inhibition, may explain why β-AR blockade reduces threat memory acquisition.[100]

In addition, cellular excitability mediated by calcium-activated potassium channels (SK channels) is regulated by β-ARs. In this scenario, PKA, activated by β-ARs, phosphorylates SK2 channels, preventing their surface expression.[106,107] By reducing surface expression of SK channels, which normally shunt membrane excitability, possibly through opposing NMDAR-mediated potentials, β-ARs enhance synaptic excitability of LA pyramidal neurons.[108] Consistent with this finding, the β-AR agonist isoproterenol enhances the NMDAR-mediated component of the EPSP,[109] and this occurs through Gs-mediated activation of PKA,[110] although this was not shown to be through direct phosphorylation of NMDAR subunits.[111] More on the regulation of NMDARs is discussed below.

α1-AR activity appears to oppose β-AR activity in a number of ways. Activation of these receptors increases feed-forward inhibition,[112] which may serve to enhance signal-to-noise processing during CS–US pairing. In addition, blockade of α1-ARs enhances acquisition for threat conditioning and induction of LTP at thalamo-LA synapses through a reduction in feed-forward inhibition onto LA pyramidal cells.[101] α1-ARs have higher affinity for NE,[113] suggesting that they are tonically activated during periods of basal NE release such as during quiet wakefulness. This tonic activation of α1-ARs may prevent run-away plasticity from occurring. During periods of high, phasic NE release, β-ARs are activated, permitting plasticity in the system (e.g., Ref. 114).

4.2.1.2 Serotonin

A role for serotonin activity in threat memory formation has also been suggested. Indeed, serotonin is released in BLA during shock and conditioned threat stress,[115,116] and may therefore be critical for modulation of amygdala plasticity and learning during the acquisition period. In support of this, recent studies show that acute selective serotonin reuptake inhibitor (SSRI) treatment enhances threat memory formation, but chronic SSRI treatment reduces learning.[117,118] This may occur through Gs-coupled 5-HT2A receptors, which affect NMDAR activity and plasticity in BLA.[119] Chronic antidepressant treatment may reduce both 5-HT2 receptor activity and GluN2B in amygdala, which is a proposed mechanism for their antidepressant and/or anxiolytic efficacy.[118] Again, these data highlight the importance of GluN2B in threat memory formation.[54] Moreover, a recent study suggests that there may be an interaction with serotonin signaling and GluA1 levels in amygdala.[120]

4.2.1.3 Dopamine

DA is also an important neuromodulator released into the LA from projections originating in the ventral tegmental area. Like LC neurons, these neurons are activated by footshock,[121] and DA is released in amygdala during conditioned fear stress,[116] suggesting that DA and activation of its receptors in the LA may modulate Hebbian mechanisms recruited during CS–US pairings. Indeed, D1 (D1R) and D2 (D2R) receptors are expressed on LA pyramidal neurons.[122]

A couple of studies suggest a role of each receptor in acquisition of threat memories via action in LA. In one study, D1R inhibition and activation impaired and enhanced threat memory formation, respectively.[123] Another study showed that inhibition of D2 receptors impairs acquisition of threat conditioning.[124] The results should be interpreted with some reservation; however, given that consolidation effects were not explicitly tested, and in the Greba *et al.* study, cannula targeting was not always in LA (in fact, several cannulae appeared to be dorsal, in the caudate putamen).

D1Rs and D2Rs are GPCRs, coupling to Gs and Gi/o proteins, respectively. They are therefore thought to have opposing roles in cellular signaling.[125] However, in amygdala, D1 receptors can also couple to Gq, thus stimulating PLC and IP3 production.[126] Similar to the activity of β-ARs in LA, DA signaling through D1R reduces feed-forward inhibition,[65,66,127] suggesting that DA release into the amygdala may be permissive for plasticity.

4.2.2 Metabotropic glutamate receptors

Glutamate, in addition to initiating fast synaptic transmission mechanisms through ionotropic receptors also activates slow synaptic mechanisms mediated through metabotropic glutamate receptors (mGluRs). Evidence suggests that these receptors may be activated during training and mediate acquisition mechanisms. In brain slices, bath application of the mGluR5 antagonist 2-methyl-6-(phenylethynyl)-pyridine (MPEP) was shown to reduce plasticity in thalamic inputs to LA, and pretraining intra-LA infusion of MPEP reduced acquisition, but not consolidation or expression of threat memories.[128] Consistent with these findings, another study showed activation of group I mGluRs (mGluR1 and mGluR5) with (*R*,*S*)-3,5-dihydroxyphenylglycine (DHPG) before training enhanced threat memory formation in BLA, and this effect was blocked by coadministration of either the group I mGluR1 antagonist, (*R*,*S*)-1-aminoindan-1,5-dicarboxylic acid (AIDA) or MPEP, suggesting the observed effects were mediated by mGluR5.[129] Notably, however, unlike Rodrigues *et al.*,[128] this study did not look at posttraining infusions or expression. Two other studies provide evidence that group III mGluRs (mGluR7 and mGluR8) in LA may also be required for the acquisition of threat memories.[130,131] This may occur through enhancement of ionotropic receptor activity, particularly NMDAR activity, as is evidenced by their physical coupling and physiological data (see Ref. 128).

4.2.3 GABA-B receptors

Upon excitatory stimulation in BLA slices, an EPSP is followed by a fast GABA-A receptor-mediated IPSP, or by a combined fast IPSP and slow IPSP, the latter component likely mediated by $G_{i/o}$-coupled GABA-B receptors.[60,61] GABA-B receptors are metabotropic GPCRs that trigger membrane hyperpolarization, or shunt membrane excitation, through $G_{i/o}$-mediated activation of K^+ channels.[60,61] In the LA, these receptors act presynaptically to reduce both excitatory and inhibitory input to principle neurons in the BLA,[132] and another study shows that presynaptic GABA-B receptors reduce excitatory inputs and plasticity only on LA principal cells and not at inhibitory neurons.[133] Furthermore, a recent study uncovered a non-Hebbian and nonassociative form of NMDAR-independent LTP in GABA-B α1,2 KO mice, and behavioral data suggest that these receptors may serve to reduce generalization.[134]

4.2.4 Regulation of AMPARs by slow synaptic transmission mechanisms

Trafficking of AMPARs in and out of the synaptic membrane is required for memory formation for a variety of learning tasks including Pavlovian threat conditioning.[22,135,136] This process is regulated by neuromodulatory signaling through metabotropic receptors,[20,103] or calcium signaling from NMDAR activation (for reviews see Ref. 137) which will be discussed in greater detail in following sections. Early studies showed that synaptic activity can drive GluA1- and GluA4-containing AMPARs into the synaptic membrane, resulting in increased surface expression of AMPARs as well as increased conductance through these new channels.[22,136,138–145] Surface expression depends on phosphorylation at two serine residues on the long cytoplasmic carboxyl terminus of the GluA1 subunit, ser831 and ser845,[143,146,147] and phosphorylation of existing AMPARs at these serine residues can also influence conductance and open time allowing for stronger synaptic transmission.[148–150] These residues are phosphorylated by distinct kinase activity; specifically, ser831 is phosphorylated by calcium and calmodulin-dependent kinase II (CamKII)[143,151] or calcium-dependent protein kinase (PKC)[152,153] while ser845 is phosphorylated by PKA.[143,148]

Synaptic activity-induced AMPAR trafficking has mostly been shown to involve AMPARs lacking a GluA2 subunit, and the synaptic activity-induced changes tend to occur on the GluA1 subunit (i.e., phosphorylation). These are usually GluA1 homomeric receptors, but can occur in other combinations as well. The absence of the GluA2 subunit renders AMPARs permeable to calcium, and GluA2-lacking receptors are designated CP-AMPARs. In the absence of the GluA1 gene, LTP and long-term depression (LTD) are impaired or eliminated at thalamo-LA and cortico-LA synapses.[135] Furthermore, GluA1 knock out animals are impaired in acquisition of threat conditioning and conditioned freezing is abolished.[135] Conversely, in wild-type animals, threat conditioning induces synaptic insertion of CP-AMPARs, and this is required for memory formation.[20,22,154] Knockin animals with point mutations at the ser845 and ser831 sites of GluA1[155] have been used to show that insertion of CP-AMPARs in the LA following threat conditioning depends on phosphorylation at ser845 but not ser831.[20] However, plasticity and memory formation are normal in phospho-mutant animals. This suggests that phosphorylation at this site, though required for synaptic insertion of CP-AMPARs, may be dispensable for plasticity and memory formation.[20]

Phosphorylation of GluA1 subunits has also been implicated in a metaplasticity-like mechanism in the amygdala in which a subthreshold stimulus can be converted into a suprathreshold stimulus.[156] In this study, one pairing of a light CS with a mild footshock does not induce STM or LTM as indicated by a lack of fear potentiated startle response. However, when one pairing is followed by another pairing 45 min to 7 days apart, LTM for the CS is evident. This result suggests that the first pairing primes the system for learning to the second pairing provided it occurs within a specified time window (45 min to a week). Correspondingly, phosphorylation of ser845 GluA1 as well as ser133 cAMP-responsive element-binding protein (CREB) is further enhanced by the second pairing. Phosphorylation at these sites by the first trial may be important for learning during the second trial.

After learning, GluA2-containing AMPARs replace the newly inserted CP-AMPARs over the course of about a day.[154] Similarly, retrieval of a conditioned threat stimulus induces insertion of GluA1-containing AMPARs within 5 min, which are also replaced by GluA2-containing AMPARs over several hours. The exchange of GluA2-lacking CP-AMPARs to GluA2-containing calcium-impermeable AMPARs is thought to stabilize new plasticity into LTM because the GluA2 subunit appears to confer stability to synaptically expressed AMPARs.[136,138–142,144]

4.2.5 Regulation of NMDARs by slow synaptic transmission mechanisms

NMDARs are regulated by slow synaptic signaling mechanisms that influence channel function, coupling mechanisms, availability at the synapse, and consequently, plasticity and learning (for a review see Ref. 137). Regulation can occur through direct posttranslational modification of NMDAR subunits, for example, kinase-mediated phosphorylation, or indirectly through modification of targets that interact with and regulate NMDAR localization or function through protein–protein interactions.[137] The where, when, and how of these receptors depend in large part on their differential subunit composition.

NMDARs are heterotetrameric proteins composed of two GluN1 subunits, which are required for normal receptor function, and any combination of GluN2 (e.g., GluN2A or GluN2B) and/or GluN3 subunits, both of which impart different channel properties and localization signals.[56,157–159] GluN2B (NR2B)-containing receptors have been particularly well studied in amygdala. Rodrigues *et al.* showed that in contrast to broad-spectrum NMDAR antagonist, blockade of GluN2B-containing NMDARs in LA reduces acquisition but not expression of threat memories[54] and does not affect normal synaptic transmission,[54,55] whereas forebrain overexpression of GluN2B subunits enhances threat conditioning.[160] An earlier study in cell culture showed that GluN2B-containing receptors exhibit longer EPSPs than the GluN2A-containing receptors.[158] However, a recent study in BLA slices shows that RNA-interference (RNAi)-mediated knockdown of the GluN2B subunit has little effect on the kinetics of NMDAR activation and conductance but disrupts its association with signaling molecules that are essential for synaptic plasticity.[56] Notably, this is the first study to show that NMDARs exist as triheteromeric complexes of GluN1/GluN2A/GluN2B (as opposed to GluN1/GluN2A or GluN1/GluN2B diheteromers) in BLA. GluN2B subunits can also be regulated by posttranslational modification, that is, phosphorylation of C-terminal residues that can alter NMDAR localization and coupling to signaling molecules.

It is possible that Gs-coupled receptors, for example, β-ARs, 5-HT2A receptors, or D2Rs achieve their modulation of NMDARs through direct phosphorylation of GluN1 or GluN2 subunits. In fact, it is known that β-AR-mediated PKA activation can enhance NMDAR conductance to positively modulate LTP.[161,162] The subunit found in all NMDARs, NR1, is targeted by PKA (at ser897[163]), as is the newly described NR2C subunit (ser1244[164]), recently shown to be essential for fear acquisition and thalamo-amygdala LTP.[165]

In BLA, direct phosphorylation of GluN2B at a tyrosine residue (Tyr-1472) by Src family kinases (SFKs) is required for threat learning and plasticity, likely due to phosphorylation-dependent protein–protein interactions necessary for synaptic targeting.[166,167] One way this can occur is through activation of the receptor tyrosine kinase EphB, which leads to phosphorylation of this and two other tyrosine residues,[168,169] thus regulating NMDAR surface and synaptic localization.[170] Moreover, blockade of Src was shown to reduce surface expression of GluN2B-containing receptors and synaptic plasticity in amygdala neurons.[171] On the other hand, animals lacking Fyn, a SFK, show increased threat reactions,[172] and Fyn-overexpressing animals show the opposite behavior.[166]

Kinase activity initiated as a consequence of NMDAR activity can lead to phosphorylation of NMDAR residues that regulate its coupling to signaling complexes and the activity of such complexes. Calcium influx through NMDARs activates CamKII, which is essential for initiating molecular cascades that drive new protein synthesis and, it is assumed, consolidation processes (see Section 5 below). Protein–protein interactions between CamKII and NMDARs, particularly GluN2B (and to a much lesser extent GluN2A), are enhanced by calcium- and calmodulin-mediated autophosphorylation of CamKII.[173–175] This interaction restricts CamKII localization to synaptic sites and helps maintain CamKII activity, even after dissociation of calcium/calmodulin.[176] Eventually, phosphorylation of GluN2B ser1303 by CamKII leads to slow dissociation of the GluN2B–CamKII complex, thus terminating activity of the kinase.[175] Phosphorylation of NMDARs therefore serves to limit the time window of CamKII function.

As noted above, CamKII can phosphorylate the AMPAR subunit GluA1, which regulates receptor trafficking and function.[143,151] Enhanced AMPAR levels as a consequence of this signaling may be important during the acquisition phase, priming active synapses for further stimulation by subsequent CS–US pairings.

Calcium influx through NMDARs can also lead to the activation of other kinases, for example, through calcium-dependent ACs, possibly driving PKA-mediated enhancement in NMDAR function[177,178] although this has not been confirmed in amygdala processes. Again, this could be an alternative mechanism by which PKA-mediated phosphorylation of ionotropic glutamate receptors increases their availability and function at the synapse.[143,146,147]

An additional slow synaptic mechanism known to be important for regulating NMDAR activity involves mGluR signaling. mGluRs associate with NMDARs through protein–protein interactions and regulate NMDAR activity in LA (see Ref. 128). Although the exact mechanisms of regulation in LA are not known, it has been shown in hippocampus that Src and/or PKC may be involved.[179] Furthermore, a recent *in vitro* study shows that mGluRs may regulate expression of NMDAR trafficking and surface expression by way of SNARE proteins.[180] Indeed, group I and III mGluRs are required for the acquisition of threat memories,[128] and their interaction with NMDARs may account for this. The role of mGluRs in threat conditioning will be discussed in greater detail below.

NMDARs interact with membrane-bound molecules, many of which help localize NMDARs to the synapse.[137] In the amygdala, RNAi-mediated knockdown or knockout of the membrane-bound cell adhesion protein neuroligin-1 reduces NMDAR accumulation at the synapse and attenuates threat memory formation.[181–184] Phosphorylation of neuroligin-1 at a tyrosine residue (Y782) is likely required for the strengthening and/or stabilization of synapses during plasticity and possibly learning,[185] although this has not been shown in LA. Protein–protein interactions and phosphorylation are therefore essential for proper NMDAR trafficking, localization, and its association with different activity complexes.[137] In addition to being phosphorylated downstream of EphB activation, NMDARs also directly bind to EphB2, an interaction that is required for NMDAR trafficking and function.[186] This interaction is enhanced in the amygdala with stress-induced activation of the extracellular serine protease neuropsin, leading to behavioral expression of an anxiety-like phenotype[187] although a specific role in threat learning was not explored.

Taken together, these data highlight the importance of phosphorylation and protein–protein interactions in NMDAR regulation, which have implications for plasticity and threat memory formation in the amygdala. Upon NMDAR activation at the synapse, the second messenger calcium is admitted into the cell, thus regulating *de novo* protein synthesis and presumably memory consolidation. These mechanisms will be discussed in Section 5 below.

4.2.6 Neuropeptides

A heterologous population of GABAergic interneurons have been identified in LA and basal amygdala, each expressing unique combinations of calcium-binding proteins (parvalbumin and calbindin) or neuropeptides (neuropeptide Y (NPY), somatostatin (SOM), vasoactive intestinal peptide, or cholecystokinin (CCK).[188–201] Some neuropeptides expressed by these neurons regulate the excitability of LA principal cells through slow synaptic mechanisms and, consequently, influence amygdala plasticity and threat learning (see Ref. 202). For example, NPY and CCK activation have both been shown to hyperpolarize BLA principal neurons[201,203] and may therefore have an inhibitory effects on acquisition. While locally expressed peptides have shown to be involved in expression and extinction of threat memories,[204] little has been done to determine their potential role in acquisition. Therefore, further studies on the role of these peptides in the acquisition of threat memories are warranted.

4.3. Summary of acquisition

We have covered some of the molecular mechanisms that occur during the acquisition phase of Pavlovian threat conditioning. The first steps in this form of learning involve rapid activation of postsynaptic (and sometimes presynaptic) receptors by glutamate, triggering membrane depolarization, action potential generation, and the induction of Hebbian plasticity mechanisms thought to be required for memory stabilization. Activation of slow, metabotropic mechanisms also occurs during memory acquisition, influencing ionotropic receptor localization, coupling mechanisms, and function at the membrane. Such changes in ionotropic receptor activity possibly serve to "prime" relevant cells and synapses for subsequent CS–US trials, permitting calcium influx and other second messenger signaling to drive long-term changes required for memory stabilization. In addition to calcium-mediated mechanisms for memory stabilization, metabotropic signaling also triggers divergent pathways for the synthesis of new proteins for the same purpose. Such memory stabilization is referred to as consolidation and will be covered in Section 5 below.

5. MECHANISMS OF THREAT MEMORY CONSOLIDATION IN LA

Historically, acquisition processes have been shown to require cellular and biochemical processes that occur quickly and near the surface of the synaptic membrane through the fast and slow synaptic transmission mechanisms described above. Acquisition processes are sufficient to support only STM; however, they initiate signaling cascades that recruit consolidation processes. Consolidation processes, especially transcription and translation, are required for the persistence of memory.[205]

The consolidation phase begins once acquisition is complete and the learning task is over. In mammals, consolidation lasts for several hours during which time the memory is not stable, and memory formation can be interrupted by blocking the molecules described below. Thus, a molecule is determined to be involved in consolidation if manipulations to it (activating or inhibiting it) are effective when given after training. The manipulation should affect only LTM and leave STM intact because STMs do not depend on consolidation processes.[8]

We begin by examining the cell-surface expressed receptors including mGluRII and voltage-gated calcium channels (VGCCs) whose activity has been shown to be required for consolidation. We then review the intracellular molecules activated by the fast and slow synaptic transmission mechanisms reviewed above. Ultimately, these receptors and intracellular molecules modulate consolidation through their actions on translation and transcription. Together, these cellular processes stabilize STMs into LTMs

5.1. Surface expressed receptors involved in consolidation

5.1.1 Voltage-gated calcium channels

An additional mechanism permitting calcium influx and therefore calcium-mediated signaling in the cell occurs through VGCCs. As their name suggests, these channels are not ligand gated, but respond to membrane depolarization, thus allowing for local increases in intracellular calcium concentrations following activation of AMPARs. Indeed, these channels, L-type VGCCs (L-VGCCs) in particular, are required for certain forms of LTP induction at thalamo-LA or cortico-LA synapses[55,18] and LTM formation for threat conditioning.[55,206,207] Although certain forms of LTP require L-VGCCs, unlike GluN2B-containing NMDARs, L-VGCCs are not required for the acquisition of threat memories.[208,209] However, these channels are likely activated during training, and their passage of calcium can lead to molecular cascades that are important for consolidation processes, providing a link from activity at the synaptic membrane to changes in gene expression at the cell nucleus.

5.1.2 mGluRII

In contrast to group I and III mGluR function in amygdala, there is evidence that group II mGluRs (mGluR2 and mGluR3) regulate memory consolidation for threat-based memories using a fear potentiated startle task. Lin *et al.*[210] showed that activation of mGluRII with the agonist, DCG-IV, in the BLA attenuates the startle response, indicating a reduction in memory when the agonist is applied 1 h after training. This effect is completely reversed when the agonist is co-applied with an antagonist of mGluRII, LY341495. Using a slice preparation and examining plasticity at the cortico-LA pathway, this study delineated that the mechanism of action of mGluRII is an inhibition of presynaptic neurotransmitter release. This serves to reverse potentiation previously induced by tetanus stimulation. Indeed, in other studies at the thalamo-LA pathway, mGluRII activation has been shown to result in LTD via a similar mechanism of inhibiting presynaptic neurotransmitter release, though a postsynaptic mechanism of these receptors cannot be ruled out for this form of plasticity.[211,212] Together, these results suggest that normal threat memory formation occurs when glutamate reuptake from the synaptic cleft is maximal and prevents activation of presynaptic mGluRII. On the other hand, activation of presynaptic mGluRII by glutamate spillover may be a mechanism by which inappropriate plasticity is inhibited.

5.2. Kinase activity

Transcription and translation are achieved via kinase activation following synaptic activity.[213] These kinases, themselves, are often activated by increases in calcium concentrations following Hebbian synaptic activity, especially NMDAR activation, or by cAMP production following calcium or neuromodulatory signaling. Thus, acquisition processes, while insufficient to produce LTM, are responsible for recruiting the necessary processes for consolidation. A number of kinases have been implicated in the LA in underlying the consolidation of threat conditioning. Through their signaling pathways, each kinase plays a role in triggering translation, transcription, or both. Before reviewing the roles of translation and transcription in consolidation, we will review how the synaptic activity initiated during acquisition ultimately impinges on these cellular processes, focusing on diffusible intracellular kinases.

5.2.1 cAMP-dependent protein kinase (PKA)

PKA is activated by the production of cAMP via AC activity which can be stimulated by Gs signaling or calcium.[214] Both Gs signaling (via β-ARs) and calcium signaling (via Hebbian mechanisms) are known to occur following threat conditioning, suggesting the recruitment of PKA as well. Indeed, broad-spectrum inhibitors of PKA such as Rp-cAMPs impair consolidation of threat conditioning when administered into the LA before or after training.[215,216] In addition, a specific inhibitor of PKA binding to anchoring proteins near the synaptic membrane also impairs consolidation.[217] Activation of PKA results in nuclear signaling and PKA-dependent phosphorylation of a nuclear transcription factor, CREB,[218,219] suggesting that it mediates consolidation by changing gene expression.

Although PKA has mostly been implicated in consolidation, several of its targets are involved in acquisition, suggesting that PKA may regulate acquisition in subtle ways or that it is involved in a very rapid form of consolidation. These targets include ser845 GluA1,[148] NMDARs,[111] and potassium channels regulating cellular excitability, SK2 channels.[106] Detailed investigations will be required to parse out the role of these PKA targets in acquisition versus consolidation.

5.2.2 Extracellular-regulated kinases

ERK, also known as MAPK, is activated by MAP kinase kinase (MEK) which is regulated by PKA[220,221] and by increases in intracellular calcium concentrations.[222,223] Activation of ERK can occur via calcium signaling to CaMKs or G-protein signaling to PKA. Thus, ERK can be recruited in the absence of neuromodulation via calcium signaling or vice versa. However, optimal ERK activation may require signaling from both pathways, making ERK a point of convergence between Hebbian and neuromodulatory signaling. This convergent signaling may be required for ERK nuclear translocation and modulation of CREB and/or for ERK effects on protein synthesis. ERK is required for consolidation of threat conditioning and is activated 1 h following training.[224] ERK is also important for synaptic plasticity, specifically the induction of LTP at thalamic inputs to the LA.[102]

Many studies examining ERK's role in plasticity and learning and memory have shown that it is involved in recruiting cellular machinery involved in long-term stability of plasticity. Indeed, the interaction between ERK, protein synthesis, and gene transcription is well established.[225] Multiple studies link ERK signaling to CREB activity[226–228] via recruitment of the nuclear factors RSK and MSK.[225] This aspect of ERK function lends support to the notion that it is involved in LTM formation, but not required for STM.[224] Similarly, ERK activation following threat conditioning leads to the production of the immediate early gene Arc/Arg3.1,[229] further implicating ERK in the production of new gene products that are required for LTM formation.

5.2.3 Protein kinase C

PKC is typically activated by increases in intracellular calcium concentrations achieved following activation of NMDARs or VGCCs, or through Gq-protein-mediated activation of PLC and calcium release from intracellular stores. PKC, like other kinases, impinges on gene expression through modification of CREB. In this manner, it is required for LTM formation for threat memories.[216,230] Like PKA, however, it may also play a role in acquisition through its kinase activity on ser831 GluA1 as discussed above.

5.2.4 Phosphatidylinositol-3 kinase

Phosphatidylinositol-3 kinase (PI3K) is activated by increases in intracellular calcium concentrations and by calcium-induced stimulation of Ras,[231] known to occur during threat conditioning. PI3K is required for consolidation as well as in extinction of threat conditioning. Intra-LA blockade of PI3K with wortmannin impairs consolidation,[232,233] and this manipulation prevents activation of CREB. These results suggest that one of the main functions of PI3K may be to regulate transcription via CREB. Indeed, its main target, Akt, has also been shown to phosphorylate nuclear CREB in the LA.[232] PI3K may also affect consolidation by recruiting other nuclear factors such as nuclear factor kappa light-chain enhancer of activated B cells (NF-κB) and mammalian target of rapamycin (mTOR), which modulate gene expression and protein synthesis.

5.3. Neurotrophin signaling

Neurotrophin signaling has been implicated in LTM formation for threat-conditioned memories. Following training for threat conditioning, transcription of brain-derived neurotrophic factor (BDNF) mRNA occurs and BDNF protein levels increase. These increases are required for LTM formation. In addition, threat conditioning induces BDNF binding to its receptor, tyrosine kinase receptor B (TrkB), which results in its phosphorylation; blocking phosphorylation of TrkB in LA impairs consolidation.[234–237] TrkB phosphorylation may be critical for LTM formation because of its role in recruiting PI3-K and MAPK,[215,232] providing a link from BNDF activity to protein synthesis and gene expression.

5.4. Transcription and translation

Initial studies examined sensitization of the gill withdrawal reflex in *Aplysia* to show that long-term changes depend on transcription and translation.[238] In mammals, learning and memory for spatial, contextual, and cued associative memories were found to require new gene expression.[205,239] These basic findings have also been replicated in the amygdala for threat conditioning, showing that inhibition of protein synthesis with anisomycin restricted to the LA impairs LTM, but leaves STM intact.[215,240] Inhibitors of transcription such as actinomycin-D which prevent RNA synthesis have a similar pattern of effects on memory formation for threat conditioning.[240,241] These compounds are effective when administered before or after training, suggesting that translation and transcription are required only for the consolidation phase.

While transcription occurs in the cell nucleus, protein synthesis may occur either in the soma or in dendritic compartments, both of which appear to be necessary for LTM.[242–244] Indeed, recent studies show protein synthesis is upregulated in the LA[245] and that the number of polyribosomes, the site of protein synthesis, increases in the spines of LA pyramidal neurons following training for threat conditioning.[246] In addition, the protein, Arc/Arg3.1, which has been shown to be locally translated at synapses during learning,[247] is upregulated in the LA following fear learning and is also required for LTM formation.[229] Finally, the mTOR regulates protein synthesis and is also required for LTM formation for threat conditioning.[248] These studies show that both dendritic and somatic protein synthesis are required in the LA for LTM formation.

Protein degradation has also been shown to be important for normal memory formation,[249] and this process has recently been implicated in the amygdala during consolidation of threat conditioning.[250] This process occurs through the ubiquitin–proteasome system, which is upregulated 1 h following training for threat conditioning, as measured by an increase in polyubiquitinated proteins.[250] Interestingly, protein degradation was found to mirror protein synthesis in time course and in magnitude, suggesting that protein degradation and protein synthesis may go hand-in-hand.

5.4.1 Cyclic-AMP response element (CRE)-dependent gene transcription

Transcription of a subset of genes depends on the DNA sequence coded by the cAMP-responsive element (CRE). We are only beginning to understand which genes are transcribed in a CRE-dependent fashion, but studies show that two immediate early genes, *c-fos* and *zif268*, contain CRE sites.[251,252] Importantly, Tao *et al.*[253] show that CREB induces BDNF transcription, suggesting that it is transcribed in a CRE-dependent fashion. Thus, many of the gene products of CRE-dependent gene transcription appear to be involved in memory formation. CRE-dependent gene transcription is regulated by transcription factors and cofactors described below.

5.4.2 CRE-binding protein (CREB) and its cofactors

The transcription factor CREB regulates the transcription of a subset of activity-induced genes, which are transcribed in a CRE-dependent fashion. Activation of nuclear CREB is required for long-term synaptic plasticity and memory formation for many different paradigms.[239,254,255] Several upstream signaling pathways described above signal to CREB, placing it at an important point of convergent information processing following synaptic activity. Its activity is regulated by several posttranslational modifications, the most studied of which is phosphorylation at ser133. Phosphorylation of CREB (pCREB) at this site has been shown to be necessary but not sufficient for CRE-dependent gene transcription.[256–258]

Overexpression of CREB in LA pyramidal neurons facilitates memory formation for threat conditioning[259–262] and also determines which LA cells are incorporated into the memory trace,[260,263] possibly by enhancing cell excitability.[262] CREB phosphorylation by PKA at ser133 is also implicated in threat memory formation.[156,264] Basal levels of pCREB relative to total CREB may determine the strength of the memory formed and are thus involved in producing individual differences in the degree of memory formation.[264]

In addition to phosphorylation at ser133, CREB can be regulated by several cofactors including CREB-regulated transcriptional coactivator (CRTC1) and NF-κB. Upon synaptic activity, CRTC1 is dephosphorylated by calcineurin, leading to translocation from the synapse to the nucleus.[265,266] Once in the nucleus, CRTC1 binds CREB and drives transcription independent of CREB phosphorylation.[267] CRTC1 has been shown to be required for induction of synaptic plasticity[268] through its role in modulating CREB activity and thus transcription. Though these mechanisms have not been established in the LA, CRTC1 has recently been implicated in the LA in LTM formation for threat conditioning.[269] NF-κB is upregulated in the LA following threat conditioning, and this upregulation is required for LTM formation.[270]

5.5. Retrograde signaling by nitric oxide

In addition to molecular signaling events in the postsynaptic LA neurons, presynaptic changes in molecular composition have been shown to be important for memory consolidation in many systems, including LA-dependent threat conditioning.[25] One example is the soluble gas, nitric oxide (NO).[271–274] NO contributes to the consolidation of memory for threat conditioning as well as to LTP at thalamo-LA synapses.[275] It acts as a retrograde messenger and travels from the postsynaptic neuron, where it is produced, to the presynaptic neuron. Here, it acts on a signaling cascade involving guanylyl cyclase and cGMP-dependent protein kinase (PKG) to ultimately recruit ERK.[276,277] It appears that this signaling cascade is involved in changing gene expression in thalamic neurons projecting to the LA, probably due to ERK's known role in regulating translation and transcription.[225] Thus, plasticity underlying threat conditioning also depends on changes in gene expression within thalamic neurons, and this is at least partially achieved through NO signaling.

5.6. Summary of consolidation

Consolidation of threat memories ultimately depends on the recruitment of the basic cellular functions, transcription, and translation to produce new gene products and proteins involved in stabilizing early plasticity occurring during the acquisition phase. Coordinated activity of surface-expressed receptors (such as mGluRII) and intracellular molecules (including kinases and neurotrophic factors) mediates changes in translation, both locally and somatically, and transcription by impinging on transcription factors such as CREB. It is likely that convergence of many of these signaling pathways at the nucleus is important for memory formation. In this manner, slow and fast synaptic transmission induced during acquisition ultimately impinge on intracellular molecules and nuclear events which underlie consolidation.

6. OVERALL CONCLUSIONS

Detailed studies on the rodent amygdala identified the LA as a critical site for plasticity and learning during acquisition and consolidation of Pavlovian threat conditioning. The reliability and robustness of the paradigm have enabled detailed circuit studies in rodents to trace information flow to and from the LA.[1] Furthermore, cellular and molecular studies have uncovered several important molecular players that regulate information processing in the LA. Generally speaking, fast synaptic mechanisms are critical for all stages of memory formation and expression, and slow synaptic mechanisms modulate the information in a way such that only relevant or salient experiences are stored for later use.

So-called emotional memory formation is essential for survival in an uncertain environment, although in cases of pathological anxiety, such as PTSD, these mechanisms may be disrupted. It is therefore imperative to continue the study of molecular mechanisms of information processing in the amygdala so that we may better be able to treat or prevent these debilitating disorders in the future.

REFERENCES

1. LeDoux JE. Emotion circuits in the brain. *Ann Rev Neurosci*. 2000;23:155–184.
2. Gale GD, Anagnostaras SG, Godsil BP, et al. Role of the basolateral amygdala in the storage of fear memories across the adult lifetime of rats. *J Neurosci*. 2004;24(15):3810–3815.
3. Sah P, Westbrook RF, Luthi A. Fear conditioning and long-term potentiation in the amygdala: what really is the connection? *Ann NY Acad Sci*. 2008;1129:88–95.
4. Hebb DO. *The Organization of Behavior: A Neuropsychological Theory*. New York: Wiley; 1949.
5. Brown TH, Kairiss EW, Keenan CL. Hebbian synapses: biophysical mechanisms and algorithms. *Ann Rev Neurosci*. 1990;13:475–511.
6. Sejnowski TJ. The book of Hebb. *Neuron*. 1999;24(4):773–776.
7. Blair HT, Schafe GE, Bauer EP, Rodrigues SM, LeDoux JE. Synaptic plasticity in the lateral amygdala: a cellular hypothesis of fear conditioning. *Learn Mem*. 2001;8(5):229–242.
8. Rodrigues SM, Schafe GE, LeDoux JE. Molecular mechanisms underlying emotional learning and memory in the lateral amygdala. *Neuron*. 2004;44(1):75–91.
9. Bliss TV, Lomo T. Plasticity in a monosynaptic cortical pathway. *J Physiol*. 1970;207(2):61P.
10. Bliss TV, Collingridge GL. A synaptic model of memory: long-term potentiation in the hippocampus. *Nature*. 1993;361(6407):31–39.
11. Glanzman DL. Common mechanisms of synaptic plasticity in vertebrates and invertebrates. *Curr Biol*. 2010;20(1):R31–R36.
12. Walters ET, Carew TJ, Kandel ER. Associative learning in Aplysia: evidence for conditioned fear in an invertebrate. *Science*. 1981;211(4481):504–506.
13. Walters ET, Byrne JH. Associative conditioning of single sensory neurons suggests a cellular mechanism for learning. *Science*. 1983;219(4583):405–408.
14. Chapman PF, Kairiss EW, Keenan CL, Brown TH. Long-term synaptic potentiation in the amygdala. *Synapse*. 1990;6(3):271–278.
15. Clugnet MC, LeDoux JE. Synaptic plasticity in fear conditioning circuits: induction of LTP in the lateral nucleus of the amygdala by stimulation of the medial geniculate body. *J Neurosci*. 1990;10(8):2818–2824.
16. Rogan MT, LeDoux JE. LTP is accompanied by commensurate enhancement of auditory-evoked responses in a fear conditioning circuit. *Neuron*. 1995;15(1):127–136.
17. Huang YY, Kandel ER. Postsynaptic induction and PKA-dependent expression of LTP in the lateral amygdala. *Neuron*. 1998;21(1):169–178.
18. Weisskopf MG, Bauer EP, LeDoux JE. L-type voltage-gated calcium channels mediate NMDA-independent associative long-term potentiation at thalamic input synapses to the amygdala. *J Neurosci*. 1999;19(23):10512–10519.
19. Weisskopf MG, LeDoux JE. Distinct populations of NMDA receptors at subcortical and cortical inputs to principal cells of the lateral amygdala. *J Neurophysiol*. 1999;81(2):930–934.
20. Clem RL, Huganir RL. Calcium-permeable AMPA receptor dynamics mediate fear memory erasure. *Science*. 2010;330(6007):1108–1112.
21. McKernan MG, Shinnick-Gallagher P. Fear conditioning induces a lasting potentiation of synaptic currents in vitro. *Nature*. 1997;390(6660):607–611.
22. Rumpel S, LeDoux J, Zador A, Malinow R. Postsynaptic receptor trafficking underlying a form of associative learning. *Science*. 2005;308(5718):83–88.
23. Schroeder BW, Shinnick-Gallagher P. Fear learning induces persistent facilitation of amygdala synaptic transmission. *Eur J Neurosci*. 2005;22(7):1775–1783.

24. Schroeder BW, Shinnick-Gallagher P. Fear memories induce a switch in stimulus response and signaling mechanisms for long-term potentiation in the lateral amygdala. *Eur J Neurosci*. 2004;20(2):549–556.
25. Tsvetkov E, Carlezon WA, Benes FM, Kandel ER, Bolshakov VY. Fear conditioning occludes LTP-induced presynaptic enhancement of synaptic transmission in the cortical pathway to the lateral amygdala. *Neuron*. 2002;34(2):289–300.
26. Quirk GJ, Repa C, LeDoux JE. Fear conditioning enhances short-latency auditory responses of lateral amygdala neurons: parallel recordings in the freely behaving rat. *Neuron*. 1995;15(5):1029–1039.
27. Grace AA, Rosenkranz JA. Regulation of conditioned responses of basolateral amygdala neurons. *Physiol Behav*. 2002;77(4–5):489–493.
28. Maren S, Quirk GJ. Neuronal signalling of fear memory. *Nat Rev Neurosci*. 2004;5(11):844–852.
29. Rogan MT, Staubli UV, LeDoux JE. Fear conditioning induces associative long-term potentiation in the amygdala. *Nature*. 1997;390(6660):604–607.
30. Johansen JP, Hamanaka H, Monfils MH, et al. Optical activation of lateral amygdala pyramidal cells instructs associative fear learning. *Proc Natl Acad Sci USA*. 2010;107(28):12692–12697.
31. Pitkanen A, Savander V, LeDoux JE. Organization of intra-amygdaloid circuitries in the rat: an emerging framework for understanding functions of the amygdala. *Trends Neurosci*. 1997;20(11):517–523.
32. Fanselow MS, LeDoux JE. Why we think plasticity underlying Pavlovian fear conditioning occurs in the basolateral amygdala. *Neuron*. 1999;23(2):229–232.
33. Debiec J, Ledoux JE. Disruption of reconsolidation but not consolidation of auditory fear conditioning by noradrenergic blockade in the amygdala. *Neuroscience*. 2004;129(2):267–272.
34. Schafe GE, Nadel NV, Sullivan GM, Harris A, LeDoux JE. Memory consolidation for contextual and auditory fear conditioning is dependent on protein synthesis, PKA, and MAP kinase. *Learn Mem*. 1999;6(2):97–110.
35. Johansen JP, Cain CK, Ostroff LE, LeDoux JE. Molecular mechanisms of fear learning and memory. *Cell*. 2011;147(3):509–524.
36. Ch'ng TH, Martin KC. Synapse-to-nucleus signaling. *Curr Opin Neurobiol*. 2011;21(2):345–352.
37. Greengard P. Neuroscience—the neurobiology of slow synaptic transmission. *Science*. 2001;294(5544):1024–1030.
38. Lagerstrom MC, Schioth HB. Structural diversity of G protein-coupled receptors and significance for drug discovery. *Nat Rev Drug Discov*. 2008;7(4):339–357.
39. Li C, Dabrowska J, Hazra R, Rainnie DG. Synergistic activation of dopamine D1 and TrkB receptors mediate gain control of synaptic plasticity in the basolateral amygdala. *PLoS One*. 2011;6(10):e26065.
40. Tsvetkov E, Shin RM, Bolshakov VY. Glutamate uptake determines pathway specificity of long-term potentiation in the neural circuitry of fear conditioning. *Neuron*. 2004;41(1):139–151.
41. Gean PW, Chang FC. Pharmacological characterization of excitatory synaptic potentials in rat basolateral amygdaloid neurons. *Synapse*. 1992;11(1):1–9.
42. Li XF, Phillips R, LeDoux JE. NMDA and non-NMDA receptors contribute to synaptic transmission between the medial geniculate body and the lateral nucleus of the amygdala. *Exp Brain Res*. 1995;105(1):87–100.
43. Romanski LM, Clugnet MC, Bordi F, LeDoux JE. Somatosensory and auditory convergence in the lateral nucleus of the amygdala. *Behav Neurosci*. 1993;107(3):444–450.

44. Kim M, Campeau S, Falls WA, Davis M. Infusion of the non-NMDA receptor antagonist CNQX into the amygdala blocks the expression of fear-potentiated startle. *Behav Neural Biol.* 1993;59(1):5–8.
45. Rogan MT, Staubli UV, LeDoux JE. AMPA receptor facilitation accelerates fear learning without altering the level of conditioned fear acquired. *J Neurosci.* 1997;17(15):5928–5935.
46. Malenka RC, Nicoll RA. Long-term potentiation—a decade of progress? *Science.* 1999;285(5435):1870–1874.
47. Szinyei C, Stork O, Pape HC. Contribution of NR2B subunits to synaptic transmission in amygdaloid interneurons. *J Neurosci.* 2003;23(7):2549–2556.
48. Li XF, Stutzmann GE, LeDoux JE. Convergent but temporally separated inputs to lateral amygdala neurons from the auditory thalamus and auditory cortex use different postsynaptic receptors: in vivo intracellular and extracellular recordings in fear conditioning pathways. *Learn Mem.* 1996;3(2–3):229–242.
49. Gewirtz JC, Davis M. Second-order fear conditioning prevented by blocking NMDA receptors in amygdala. *Nature.* 1997;388(6641):471–474.
50. Lee H, Kim JJ. Amygdalar NMDA receptors are critical for new fear learning in previously fear-conditioned rats. *J Neurosci.* 1998;18(20):8444–8454.
51. Maren S, Aharonov G, Stote DL, Fanselow MS. *N*-methyl-D-aspartate receptors in the basolateral amygdala are required for both acquisition and expression of conditional fear in rats. *Behav Neurosci.* 1996;110(6):1365–1374.
52. Miserendino MJ, Sananes CB, Melia KR, Davis M. Blocking of acquisition but not expression of conditioned fear-potentiated startle by NMDA antagonists in the amygdala. *Nature.* 1990;345(6277):716–718.
53. Campeau S, Miserendino MJ, Davis M. Intra-amygdala infusion of the *N*-methyl-D-aspartate receptor antagonist AP5 blocks acquisition but not expression of fear-potentiated startle to an auditory conditioned stimulus. *Behav Neurosci.* 1992;106(3):569–574.
54. Rodrigues SM, Schafe GE, LeDoux JE. Intra-amygdala blockade of the NR2B subunit of the NMDA receptor disrupts the acquisition but not the expression of fear conditioning. *J Neurosci.* 2001;21(17):6889–6896.
55. Bauer EP, Schafe GE, LeDoux JE. NMDA receptors and L-type voltage-gated calcium channels contribute to long-term potentiation and different components of fear memory formation in the lateral amygdala. *J Neurosci.* 2002;22(12):5239–5249.
56. Delaney AJ, Sedlak PL, Autuori E, Power JM, Sah P. Synaptic NMDA receptors in basolateral amygdala principal neurons are triheteromeric proteins: physiological role of GluN2B subunits. *J Neurophysiol.* 2013;109(5):1391–1402.
57. Zhang XH, Liu F, Chen Q, et al. Conditioning-strength dependent involvement of NMDA NR2B subtype receptor in the basolateral nucleus of amygdala in acquisition of auditory fear memory. *Neuropharmacology.* 2008;55(2):238–246.
58. Humeau Y, Shaban H, Bissiere S, Luthi A. Presynaptic induction of heterosynaptic associative plasticity in the mammalian brain. *Nature.* 2003;426(6968):841–845.
59. Quirk GJ, Armony JL, LeDoux JE. Fear conditioning enhances different temporal components of tone-evoked spike trains in auditory cortex and lateral amygdala. *Neuron* 1997;19(3):613–624.
60. Rainnie DG, Asprodini EK, Shinnick-Gallagher P. Inhibitory transmission in the basolateral amygdala. *J Neurophysiol.* 1991;66(3):999–1009.
61. Washburn MS, Moises HC. Inhibitory responses of rat basolateral amygdaloid neurons recorded in vitro. *Neuroscience.* 1992;50(4):811–830.
62. Danober L, Pape HC. Mechanisms and functional significance of a slow inhibitory potential in neurons of the lateral amygdala. *Eur J Neurosci.* 1998;10(3):853–867.

63. Martina M, Royer S, Pare D. Cell-type-specific GABA responses and chloride homeostasis in the cortex and amygdala. *J Neurophysiol.* 2001;86(6):2887–2895.
64. Li XF, Armony JL, LeDoux JE. GABAA and GABAB receptors differentially regulate synaptic transmission in the auditory thalamo-amygdala pathway: an in vivo microiontophoretic study and a model. *Synapse.* 1996;24(2):115–124.
65. Marowsky A, Yanagawa Y, Obata K, Vogt KE. A specialized subclass of interneurons mediates dopaminergic facilitation of amygdala function. *Neuron.* 2005;48(6):1025–1037.
66. Bissiere S, Humeau Y, Luthi A. Dopamine gates LTP induction in lateral amygdala by suppressing feedforward inhibition. *Nat Neurosci.* 2003;6(6):587–592.
67. Tully K, Li Y, Tsvetkov E, Bolshakov VY. Norepinephrine enables the induction of associative long-term potentiation at thalamo-amygdala synapses. *Proc Natl Acad Sci USA.* 2007;104(35):14146–14150.
68. Morozov A, Sukato D, Ito W. Selective suppression of plasticity in amygdala inputs from temporal association cortex by the external capsule. *J Neurosci.* 2011;31(1):339–345.
69. Muller J, Corodimas KP, Fridel Z, LeDoux JE. Functional inactivation of the lateral and basal nuclei of the amygdala by muscimol infusion prevents fear conditioning to an explicit conditioned stimulus and to contextual stimuli. *Behav Neurosci.* 1997;111(4):683–691.
70. Wilensky AE, Schafe GE, LeDoux JE. Functional inactivation of the amygdala before but not after auditory fear conditioning prevents memory formation. *J Neurosci.* 1999;19(24):RC48.
71. Heldt SA, Ressler KJ. Training-induced changes in the expression of GABAA-associated genes in the amygdala after the acquisition and extinction of Pavlovian fear. *Eur J Neurosci.* 2007;26(12):3631–3644.
72. Stork O, Ji FY, Obata K. Reduction of extracellular GABA in the mouse amygdala during and following confrontation with a conditioned fear stimulus. *Neurosci Lett.* 2002;327(2):138–142.
73. Szinyei C, Narayanan RT, Pape HC. Plasticity of inhibitory synaptic network interactions in the lateral amygdala upon fear conditioning in mice. *Eur J Neurosci.* 2007;25(4):1205–1211.
74. Bergado-Acosta JR, Sangha S, Narayanan RT, Obata K, Pape HC, Stork O. Critical role of the 65-kDa isoform of glutamic acid decarboxylase in consolidation and generalization of Pavlovian fear memory. *Learn Mem.* 2008;15(3):163–171.
75. Makkar SR, Zhang SQ, Cranney J. Behavioral and neural analysis of GABA in the acquisition, consolidation, reconsolidation, and extinction of fear memory. *Neuropsychopharmacology.* 2010;35(8):1625–1652.
76. Wiltgen BJ, Godsil BP, Peng Z, et al. The alpha1 subunit of the GABA(A) receptor modulates fear learning and plasticity in the lateral amygdala. *Front Behav Neurosci.* 2009;3:37.
77. Marowsky A, Rudolph U, Fritschy JM, Arand M. Tonic inhibition in principal cells of the amygdala: a central role for alpha3 subunit-containing GABAA receptors. *J Neurosci.* 2012;32(25):8611–8619.
78. Cunha C, Monfils MH, Ledoux JE. GABA(C) receptors in the lateral amygdala: a possible novel target for the treatment of fear and anxiety disorders? *Front Behav Neurosci.* 2010;4:6.
79. Azad SC, Monory K, Marsicano G, et al. Circuitry for associative plasticity in the amygdala involves endocannabinoid signaling. *J Neurosci.* 2004;24(44):9953–9961.
80. Marsicano G, Wotjak CT, Azad SC, et al. The endogenous cannabinoid system controls extinction of aversive memories. *Nature.* 2002;418(6897):530–534.

81. Rainnie DG. Serotonergic modulation of neurotransmission in the rat basolateral amygdala. *J Neurophysiol.* 1999;82(1):69–85.
82. Stutzmann GE, LeDoux JE. GABAergic antagonists block the inhibitory effects of serotonin in the lateral amygdala: a mechanism for modulation of sensory inputs related to fear conditioning. *J Neurosci.* 1999;19(11):RC8.
83. Bailey CH, Giustetto M, Huang YY, Hawkins RD, Kandel ER. Is heterosynaptic modulation essential for stabilizing Hebbian plasticity and memory? *Nat Rev Neurosci.* 2000;1(1):11–20.
84. Krystal JH, Neumeister A. Noradrenergic and serotonergic mechanisms in the neurobiology of posttraumatic stress disorder and resilience. *Brain Res.* 2009;1293:13–23.
85. Geracioti Jr TD, Baker DG, Ekhator NN, et al. CSF norepinephrine concentrations in posttraumatic stress disorder. *Am J Psychiatry.* 2001;158(8):1227–1230.
86. Berridge CW, Waterhouse BD. The locus coeruleus-noradrenergic system: modulation of behavioral state and state-dependent cognitive processes. *Brain Res Brain Res Rev.* 2003;42(1):33–84.
87. Shreve PE, Toews ML, Bylund DB. Alpha 2A- and alpha 2C-adrenoceptor subtypes are differentially down-regulated by norepinephrine. *Eur J Pharmacol.* 1991;207(3):275–276.
88. Gu Q. Neuromodulatory transmitter systems in the cortex and their role in cortical plasticity. *Neuroscience.* 2002;111(4):815–835.
89. Philipp M, Hein L. Adrenergic receptor knockout mice: distinct functions of 9 receptor subtypes. *Pharmacol Ther.* 2004;101(1):65–74.
90. Ruffolo Jr RR, Nichols AJ, Stadel JM, Hieble JP. Structure and function of alpha-adrenoceptors. *Pharmacol Rev.* 1991;43(4):475–505.
91. Ruffolo Jr RR, Stadel JM, Hieble JP. Alpha-adrenoceptors: recent developments. *Med Res Rev.* 1994;14(2):229–270.
92. Minneman KP, Esbenshade TA. Alpha 1-adrenergic receptor subtypes. *Annu Rev Pharmacol Toxicol.* 1994;34:117–133.
93. Asan E. The catecholaminergic innervation of the rat amygdala. *Adv Anat Embryol Cell Biol.* 1998;142:1–118.
94. Galvez R, Mesches MH, McGaugh JL. Norepinephrine release in the amygdala in response to footshock stimulation. *Neurobiol Learn Mem.* 1996;66(3):253–257.
95. Tanaka T, Yokoo H, Mizoguchi K, Yoshida M, Tsuda A, Tanaka M. Noradrenaline release in the rat amygdala is increased by stress: studies with intracerebral microdialysis. *Brain Res.* 1991;544(1):174–176.
96. Hatfield T, Spanis C, McGaugh JL. Response of amygdalar norepinephrine to footshock and GABAergic drugs using in vivo microdialysis and HPLC. *Brain Res.* 1999;835(2):340–345.
97. Ennis M, Aston-Jones G. Activation of locus coeruleus from nucleus paragigantocellularis: a new excitatory amino acid pathway in brain. *J Neurosci.* 1988;8(10):3644–3657.
98. Ennis M, Aston-Jones G, Shiekhattar R. Activation of locus coeruleus neurons by nucleus paragigantocellularis or noxious sensory stimulation is mediated by intracoerulear excitatory amino acid neurotransmission. *Brain Res.* 1992;598(1–2): 185–195.
99. Buffalari DM, Grace AA. Noradrenergic modulation of basolateral amygdala neuronal activity: opposing influences of alpha-2 and beta receptor activation. *J Neurosci.* 2007;27(45):12358–12366.
100. Bush DE, Caparosa EM, Gekker A, Ledoux J. Beta-adrenergic receptors in the lateral nucleus of the amygdala contribute to the acquisition but not the consolidation of auditory fear conditioning. *Front Behav Neurosci.* 2010;4:154.

101. Lazzaro SC, Hou M, Cunha C, LeDoux JE, Cain CK. Antagonism of lateral amygdala alpha1-adrenergic receptors facilitates fear conditioning and long-term potentiation. *Learn Mem.* 2010;17(10):489–493.
102. Huang YY, Martin KC, Kandel ER. Both protein kinase A and mitogen-activated protein kinase are required in the amygdala for the macromolecular synthesis-dependent late phase of long-term potentiation. *J Neurosci.* 2000;20(17):6317–6325.
103. Hu H, Real E, Takamiya K, et al. Emotion enhances learning via norepinephrine regulation of AMPA-receptor trafficking. *Cell.* 2007;131(1):160–173.
104. Clem RL, Huganir RL. Norepinephrine enhances a discrete form of long-term depression during fear memory storage. *J Neurosci.* 2013;33(29):11825–11832.
105. Joiner ML, Lise MF, Yuen EY, et al. Assembly of a beta2-adrenergic receptor—GluR1 signalling complex for localized cAMP signalling. *EMBO J.* 2010;29(2):482–495.
106. Faber ES, Sah P. Functions of SK channels in central neurons. *Clin Exp Pharmacol Physiol.* 2007;34(10):1077–1083.
107. Faber ES, Delaney AJ, Power JM, Sedlak PL, Crane JW, Sah P. Modulation of SK channel trafficking by beta adrenoceptors enhances excitatory synaptic transmission and plasticity in the amygdala. *J Neurosci.* 2008;28(43):10803–10813.
108. Faber ES, Delaney AJ, Sah P. SK channels regulate excitatory synaptic transmission and plasticity in the lateral amygdala. *Nat Neurosci.* 2005;8(5):635–641.
109. Gean PW, Huang CC, Lin JH, Tsai JJ. Sustained enhancement of NMDA receptor-mediated synaptic potential by isoproterenol in rat amygdalar slices. *Brain Res.* 1992;594(2):331–334.
110. Huang CC, Tsai JJ, Gean PW. Enhancement of NMDA receptor-mediated synaptic potential by isoproterenol is blocked by Rp-adenosine 3′,5′-cyclic monophosphorothioate. *Neurosci Lett.* 1993;161(2):207–210.
111. Raman IM, Tong G, Jahr CE. Beta-adrenergic regulation of synaptic NMDA receptors by cAMP-dependent protein kinase. *Neuron.* 1996;16(2):415–421.
112. Braga MF, Aroniadou-Anderjaska V, Manion ST, Hough CJ, Li H. Stress impairs alpha(1A) adrenoceptor-mediated noradrenergic facilitation of GABAergic transmission in the basolateral amygdala. *Neuropsychopharmacology.* 2004;29(1):45–58.
113. Molinoff PB. Alpha- and beta-adrenergic receptor subtypes properties, distribution and regulation. *Drugs.* 1984;28(Suppl. 2):1–15.
114. Huang S, Trevino M, He K, et al. Pull–push neuromodulation of LTP and LTD enables bidirectional experience-induced synaptic scaling in visual cortex. *Neuron.* 2012;73(3):497–510.
115. Christianson JP, Ragole T, Amat J, et al. 5-Hydroxytryptamine 2C receptors in the basolateral amygdala are involved in the expression of anxiety after uncontrollable traumatic stress. *Biol Psychiatry.* 2010;67(4):339–345.
116. Yokoyama M, Suzuki E, Sato T, Maruta S, Watanabe S, Miyaoka H. Amygdalic levels of dopamine and serotonin rise upon exposure to conditioned fear stress without elevation of glutamate. *Neurosci Lett.* 2005;379(1):37–41.
117. Burghardt NS, Sullivan GM, McEwen BS, Gorman JM, LeDoux JE. The selective serotonin reuptake inhibitor citalopram increases fear after acute treatment but reduces fear with chronic treatment: a comparison with tianeptine. *Biol Psychiatry.* 2004;55(12):1171–1178.
118. Burghardt NS, Sigurdsson T, Gorman JM, McEwen BS, LeDoux JE. Chronic antidepressant treatment impairs the acquisition of fear extinction. *Biol Psychiatry.* 2013;73(11):1078–1086.
119. Chen A, Hough CJ, Li H. Serotonin type II receptor activation facilitates synaptic plasticity via *N*-methyl-D-aspartate-mediated mechanism in the rat basolateral amygdala. *Neuroscience.* 2003;119(1):53–63.

120. Tran L, Lasher BK, Young KA, Keele NB. Depletion of serotonin in the basolateral amygdala elevates glutamate receptors and facilitates fear-potentiated startle. *Transl Psychiatry*. 2013;3:e298.
121. Brischoux F, Chakraborty S, Brierley DI, Ungless MA. Phasic excitation of dopamine neurons in ventral VTA by noxious stimuli. *Proc Natl Acad Sci USA*. 2009;106(12):4894–4899.
122. Muly EC, Senyuz M, Khan ZU, Guo JD, Hazra R, Rainnie DG. Distribution of D1 and D5 dopamine receptors in the primate and rat basolateral amygdala. *Brain Struct Funct*. 2009;213(4–5):375–393.
123. Guarraci FA, Frohardt RJ, Kapp BS. Amygdaloid D1 dopamine receptor involvement in Pavlovian fear conditioning. *Brain Res*. 1999;827(1–2):28–40.
124. Greba Q, Gifkins A, Kokkinidis L. Inhibition of amygdaloid dopamine D2 receptors impairs emotional learning measured with fear-potentiated startle. *Brain Res*. 2001;899(1–2):218–226.
125. Greengard P. The neurobiology of dopamine signaling. *Biosci Rep*. 2001;21(3):247–269.
126. Leonard SK, Anderson CM, Lachowicz JE, Schulz DW, Kilts CD, Mailman RB. Amygdaloid D1 receptors are not linked to stimulation of adenylate cyclase. *Synapse*. 2003;50(4):320–333.
127. Rosenkranz JA, Grace AA. Dopamine-mediated modulation of odour-evoked amygdala potentials during Pavlovian conditioning. *Nature*. 2002;417(6886):282–287.
128. Rodrigues SM, Bauer EP, Farb CR, Schafe GE, LeDoux JE. The group I metabotropic glutamate receptor mGluR5 is required for fear memory formation and long-term potentiation in the lateral amygdala. *J Neurosci*. 2002;22(12):5219–5229.
129. Rudy JW, Matus-Amat P. DHPG activation of group 1 mGluRs in BLA enhances fear conditioning. *Learn Mem*. 2009;16(7):421–425.
130. Schmid S, Fendt M. Effects of the mGluR8 agonist (S)-3,4-DCPG in the lateral amygdala on acquisition/expression of fear-potentiated startle, synaptic transmission, and plasticity. *Neuropharmacology*. 2006;50(2):154–164.
131. Siegl S, Flor PJ, Fendt M. Amygdaloid metabotropic glutamate receptor subtype 7 is involved in the acquisition of conditioned fear. *Neuroreport*. 2008;19(11):1147–1150.
132. Yamada J, Saitow F, Satake S, Kiyohara T, Konishi S. GABA(B) receptor-mediated presynaptic inhibition of glutamatergic and GABAergic transmission in the basolateral amygdala. *Neuropharmacology*. 1999;38(11):1743–1753.
133. Pan BX, Dong Y, Ito W, Yanagawa Y, Shigemoto R, Morozov A. Selective gating of glutamatergic inputs to excitatory neurons of amygdala by presynaptic GABAb receptor. *Neuron*. 2009;61(6):917–929.
134. Shaban H, Humeau Y, Herry C, et al. Generalization of amygdala LTP and conditioned fear in the absence of presynaptic inhibition. *Nat Neurosci*. 2006;9(8):1028–1035.
135. Humeau Y, Reisel D, Johnson AW, et al. A pathway-specific function for different AMPA receptor subunits in amygdala long-term potentiation and fear conditioning. *J Neurosci*. 2007;27(41):10947–10956.
136. Malinow R, Malenka RC. AMPA receptor trafficking and synaptic plasticity. *Annu Rev Neurosci*. 2002;25:103–126.
137. Chen BS, Roche KW. Regulation of NMDA receptors by phosphorylation. *Neuropharmacology*. 2007;53(3):362–368.
138. Hayashi Y, Shi SH, Esteban JA, Piccini A, Poncer JC, Malinow R. Driving AMPA receptors into synapses by LTP and CaMKII: requirement for GluR1 and PDZ domain interaction. *Science*. 2000;287(5461):2262–2267.
139. Zhu JJ, Esteban JA, Hayashi Y, Malinow R. Postnatal synaptic potentiation: delivery of GluR4-containing AMPA receptors by spontaneous activity. *Nat Neurosci*. 2000;3(11):1098–1106.

140. Shi S, Hayashi Y, Esteban JA, Malinow R. Subunit-specific rules governing AMPA receptor trafficking to synapses in hippocampal pyramidal neurons. *Cell.* 2001;105(3):331–343.
141. Passafaro M, Piech V, Sheng M. Subunit-specific temporal and spatial patterns of AMPA receptor exocytosis in hippocampal neurons. *Nat Neurosci.* 2001;4(9):917–926.
142. Takahashi T, Svoboda K, Malinow R. Experience strengthening transmission by driving AMPA receptors into synapses. *Science.* 2003;299(5612):1585–1588.
143. Esteban JA, Shi SH, Wilson C, Nuriya M, Huganir RL, Malinow R. PKA phosphorylation of AMPA receptor subunits controls synaptic trafficking underlying plasticity. *Nat Neurosci.* 2003;6(2):136–143.
144. Song I, Huganir RL. Regulation of AMPA receptors during synaptic plasticity. *Trends Neurosci.* 2002;25(11):578–588.
145. Clem RL, Barth A. Pathway-specific trafficking of native AMPARs by in vivo experience. *Neuron.* 2006;49(5):663–670.
146. Ehlers MD. Reinsertion or degradation of AMPA receptors determined by activity-dependent endocytic sorting. *Neuron.* 2000;28(2):511–525.
147. Qin Y, Zhu Y, Baumgart JP, et al. State-dependent Ras signaling and AMPA receptor trafficking. *Genes Dev.* 2005;19(17):2000–2015.
148. Roche KW, O'Brien RJ, Mammen AL, Bernhardt J, Huganir RL. Characterization of multiple phosphorylation sites on the AMPA receptor GluR1 subunit. *Neuron.* 1996;16(6):1179–1188.
149. Derkach V, Barria A, Soderling TR. Ca2+/calmodulin-kinase II enhances channel conductance of alpha-amino-3-hydroxy-5-methyl-4-isoxazolepropionate type glutamate receptors. *Proc Natl Acad Sci USA.* 1999;96(6):3269–3274.
150. Banke TG, Bowie D, Lee H, Huganir RL, Schousboe A, Traynelis SF. Control of GluR1 AMPA receptor function by cAMP-dependent protein kinase. *J Neurosci.* 2000;20(1):89–102.
151. Yakel JL, Vissavajjhala P, Derkach VA, Brickey DA, Soderling TR. Identification of a Ca2+/calmodulin-dependent protein kinase II regulatory phosphorylation site in non-*N*-methyl-D-aspartate glutamate receptors. *Proc Natl Acad Sci USA.* 1995;92(5):1376–1380.
152. Mammen AL, Kameyama K, Roche KW, Huganir RL. Phosphorylation of the alpha-amino-3-hydroxy-5-methylisoxazole4-propionic acid receptor GluR1 subunit by calcium/calmodulin-dependent kinase II. *J Biol Chem.* 1997;272(51): 32528–32533.
153. Boehm J, Kang MG, Johnson RC, Esteban J, Huganir RL, Malinow R. Synaptic incorporation of AMPA receptors during LTP is controlled by a PKC phosphorylation site on GluR1. *Neuron.* 2006;51(2):213–225.
154. Hong I, Kim J, Lee S, et al. AMPA receptor exchange underlies transient memory destabilization on retrieval. *Proc Natl Acad Sci USA.* 2013;110(20):8218–8223.
155. Lee HK, Takamiya K, Han JS, et al. Phosphorylation of the AMPA receptor GluR1 subunit is required for synaptic plasticity and retention of spatial memory. *Cell.* 2003;112(5):631–643.
156. Parsons RG, Davis M. A metaplasticity-like mechanism supports the selection of fear memories: role of protein kinase a in the amygdala. *J Neurosci.* 2012;32(23):7843–7851.
157. Qiu S, Li XY, Zhuo M. Post-translational modification of NMDA receptor GluN2B subunit and its roles in chronic pain and memory. *Semin Cell Dev Biol.* 2011;22(5):521–529.
158. Monyer H, Sprengel R, Schoepfer R, et al. Heteromeric NMDA receptors: molecular and functional distinction of subtypes. *Science.* 1992;256(5060):1217–1221.
159. Nakanishi S. Molecular diversity of glutamate receptors and implications for brain function. *Science.* 1992;258(5082):597–603.

160. Tang YP, Shimizu E, Dube GR, et al. Genetic enhancement of learning and memory in mice. *Nature*. 1999;401(6748):63–69.
161. Stanton PK, Mody I, Heinemann U. A role for *N*-methyl-D-aspartate receptors in norepinephrine-induced long-lasting potentiation in the dentate gyrus. *Exp Brain Res*. 1989;77(3):517–530.
162. Westphal RS, Tavalin SJ, Lin JW, et al. Regulation of NMDA receptors by an associated phosphatase-kinase signaling complex. *Science*. 1999;285(5424):93–96.
163. Tingley WG, Ehlers MD, Kameyama K, et al. Characterization of protein kinase A and protein kinase C phosphorylation of the *N*-methyl-D-aspartate receptor NR1 subunit using phosphorylation site-specific antibodies. *J Biol Chem*. 1997;272(8):5157–5166.
164. Chen BS, Braud S, Badger 2nd JD, Isaac JT, Roche KW. Regulation of NR1/NR2C *N*-methyl-D-aspartate (NMDA) receptors by phosphorylation. *J Biol Chem*. 2006;281(24):16583–16590.
165. Hillman BG, Gupta SC, Stairs DJ, Buonanno A, Dravid SM. Behavioral analysis of NR2C knockout mouse reveals deficit in acquisition of conditioned fear and working memory. *Neurobiol Learn Mem*. 2011;95(4):404–414.
166. Kojima N, Sakamoto T, Endo S, Niki H. Impairment of conditioned freezing to tone, but not to context, in Fyn-transgenic mice: relationship to NMDA receptor subunit 2B function. *Eur J Neurosci*. 2005;21(5):1359–1369.
167. Nakazawa T, Komai S, Watabe AM, et al. NR2B tyrosine phosphorylation modulates fear learning as well as amygdaloid synaptic plasticity. *EMBO J*. 2006; 25(12):2867–2877.
168. Takasu MA, Dalva MB, Zigmond RE, Greenberg ME. Modulation of NMDA receptor-dependent calcium influx and gene expression through EphB receptors. *Science*. 2002;295(5554):491–495.
169. Antion MD, Christie LA, Bond AM, Dalva MB, Contractor A. Ephrin-B3 regulates glutamate receptor signaling at hippocampal synapses. *Mol Cell Neurosci*. 2010;45(4): 378–388.
170. Salter MW, Kalia LV. Src kinases: a hub for NMDA receptor regulation. *Nat Rev Neurosci*. 2004;5(4):317–328.
171. Sinai L, Duffy S, Roder JC. Src inhibition reduces NR2B surface expression and synaptic plasticity in the amygdala. *Learn Mem*. 2010;17(8):364–371.
172. Kubota O, Hattori K, Hashimoto K, et al. Auditory-conditioned-fear-dependent c-Fos expression is altered in the emotion-related brain structures of Fyn-deficient mice. *Brain Res Mol Brain Res*. 2004;130(1–2):149–160.
173. Leonard AS, Lim IA, Hemsworth DE, Horne MC, Hell JW. Calcium/calmodulin-dependent protein kinase II is associated with the *N*-methyl-D-aspartate receptor. *Proc Natl Acad Sci USA*. 1999;96(6):3239–3244.
174. Strack S, Colbran RJ. Autophosphorylation-dependent targeting of calcium/calmodulin-dependent protein kinase II by the NR2B subunit of the *N*-methyl-D-aspartate receptor. *J Biol Chem*. 1998;273(33):20689–20692.
175. Strack S, McNeill RB, Colbran RJ. Mechanism and regulation of calcium/calmodulin-dependent protein kinase II targeting to the NR2B subunit of the *N*-methyl-D-aspartate receptor. *J Biol Chem*. 2000;275(31):23798–23806.
176. Bayer KU, De Koninck P, Leonard AS, Hell JW, Schulman H. Interaction with the NMDA receptor locks CaMKII in an active conformation. *Nature*. 2001;411(6839):801–805.
177. Collingridge GL, Singer W. Excitatory amino acid receptors and synaptic plasticity. *Trends Pharmacol Sci*. 1990;11(7):290–296.
178. Wang H, Storm DR. Calmodulin-regulated adenylyl cyclases: cross-talk and plasticity in the central nervous system. *Mol Pharmacol*. 2003;63(3):463–468.

179. Benquet P, Gee CE, Gerber U. Two distinct signaling pathways upregulate NMDA receptor responses via two distinct metabotropic glutamate receptor subtypes. *J Neurosci*. 2002;22(22):9679–9686.
180. Cheng J, Liu W, Duffney LJ, Yan Z. SNARE proteins are essential in the potentiation of NMDA receptors by group II metabotropic glutamate receptors. *J Physiol*. 2013;591(Pt. 16):3935–3947.
181. Kim J, Jung SY, Lee YK, et al. Neuroligin-1 is required for normal expression of LTP and associative fear memory in the amygdala of adult animals. *Proc Natl Acad Sci USA*. 2008;105(26):9087–9092.
182. Jung SY, Kim J, Kwon OB, et al. Input-specific synaptic plasticity in the amygdala is regulated by neuroligin-1 via postsynaptic NMDA receptors. *Proc Natl Acad Sci USA*. 2010;107(10):4710–4715.
183. Jiang L, Emmetsberger J, Talmage DA, Role LW. Type III neuregulin 1 is required for multiple forms of excitatory synaptic plasticity of mouse cortico-amygdala circuits. *J Neurosci*. 2013;33(23):9655–9666.
184. Budreck EC, Kwon OB, Jung JH, et al. Neuroligin-1 controls synaptic abundance of NMDA-type glutamate receptors through extracellular coupling. *Proc Natl Acad Sci USA*. 2013;110(2):725–730.
185. Giannone G, Mondin M, Grillo-Bosch D, et al. Neurexin-1beta binding to neuroligin-1 triggers the preferential recruitment of PSD-95 versus gephyrin through tyrosine phosphorylation of neuroligin-1. *Cell Rep*. 2013;3(6):1996–2007.
186. Dalva MB, Takasu MA, Lin MZ, et al. EphB receptors interact with NMDA receptors and regulate excitatory synapse formation. *Cell*. 2000;103(6):945–956.
187. Attwood BK, Bourgognon JM, Patel S, et al. Neuropsin cleaves EphB2 in the amygdala to control anxiety. *Nature*. 2011;473(7347):372–375.
188. Jasnow AM, Banks MC, Owens EC, Huhman KL. Differential effects of two corticotropin-releasing factor antagonists on conditioned defeat in male Syrian hamsters (*Mesocricetus auratus*). *Brain Res*. 1999;846(1):122–128.
189. Kemppainen S, Pitkanen A. Distribution of parvalbumin, calretinin, and calbindin-D(28k) immunoreactivity in the rat amygdaloid complex and colocalization with gamma-aminobutyric acid. *J Comp Neurol*. 2000;426(3):441–467.
190. McDonald AJ, Betette RL. Parvalbumin-containing neurons in the rat basolateral amygdala: morphology and co-localization of Calbindin-D(28k). *Neuroscience*. 2001;102(2):413–425.
191. McDonald AJ, Mascagni F. Localization of the CB1 type cannabinoid receptor in the rat basolateral amygdala: high concentrations in a subpopulation of cholecystokinin-containing interneurons. *Neuroscience*. 2001;107(4):641–652.
192. McDonald AJ, Mascagni F. Colocalization of calcium-binding proteins and GABA in neurons of the rat basolateral amygdala. *Neuroscience*. 2001;105(3):681–693.
193. McDonald AJ, Mascagni F. Immunohistochemical characterization of somatostatin containing interneurons in the rat basolateral amygdala. *Brain Res*. 2002;943(2):237–244.
194. McDonald AJ, Mascagni F. Parvalbumin-containing interneurons in the basolateral amygdala express high levels of the alpha1 subunit of the GABAA receptor. *J Comp Neurol*. 2004;473(1):137–146.
195. Pitkanen A, Amaral DG. Distribution of calbindin-D28k immunoreactivity in the monkey temporal lobe: the amygdaloid complex. *J Comp Neurol*. 1993;331(2):199–224.
196. Pitkanen A, Amaral DG. Distribution of parvalbumin-immunoreactive cells and fibers in the monkey temporal lobe: the hippocampal formation. *J Comp Neurol*. 1993;331(1):37–74.

197. Pitkanen A, Amaral DG. Distribution of parvalbumin-immunoreactive cells and fibers in the monkey temporal lobe: the amygdaloid complex. *J Comp Neurol.* 1993; 331(1):14–36.
198. Pitkanen A, Amaral DG. The distribution of GABAergic cells, fibers, and terminals in the monkey amygdaloid complex: an immunohistochemical and in situ hybridization study. *J Neurosci.* 1994;14(4):2200–2224.
199. Pitkanen A, Amaral DG. Organization of the intrinsic connections of the monkey amygdaloid complex: projections originating in the lateral nucleus. *J Comp Neurol.* 1998;398(3):431–458.
200. Sorvari H, Soininen H, Paljarvi L, Karkola K, Pitkanen A. Distribution of parvalbumin-immunoreactive cells and fibers in the human amygdaloid complex. *J Comp Neurol.* 1995;360(2):185–212.
201. Sosulina L, Schwesig G, Seifert G, Pape HC. Neuropeptide Y activates a G-protein-coupled inwardly rectifying potassium current and dampens excitability in the lateral amygdala. *Mol Cell Neurosci.* 2008;39(3):491–498.
202. Pape HC, Pare D. Plastic synaptic networks of the amygdala for the acquisition, expression, and extinction of conditioned fear. *Physiol Rev.* 2010;90(2):419–463.
203. Meis S, Munsch T, Sosulina L, Pape HC. Postsynaptic mechanisms underlying responsiveness of amygdaloid neurons to cholecystokinin are mediated by a transient receptor potential-like current. *Mol Cell Neurosci.* 2007;35(2):356–367.
204. Gutman AR, Yang Y, Ressler KJ, Davis M. The role of neuropeptide Y in the expression and extinction of fear-potentiated startle. *J Neurosci.* 2008;28(48):12682–12690.
205. Alberini CM. Transcription factors in long-term memory and synaptic plasticity. *Physiol Rev.* 2009;89(1):121–145.
206. McKinney BC, Murphy GG. The L-type voltage-gated calcium channel Cav1.3 mediates consolidation, but not extinction, of contextually conditioned fear in mice. *Learn Mem.* 2006;13(5):584–589.
207. Shinnick-Gallagher P, McKernan MG, Xie J, Zinebi F. L-type voltage-gated calcium channels are involved in the in vivo and in vitro expression of fear conditioning. *Ann NY Acad Sci.* 2003;985:135–149.
208. Cain CK, Blouin AM, Barad M. L-type voltage-gated calcium channels are required for extinction, but not for acquisition or expression, of conditional fear in mice. *J Neurosci.* 2002;22(20):9113–9121.
209. Cain CK, Godsil BP, Jami S, Barad M. The L-type calcium channel blocker nifedipine impairs extinction, but not reduced contingency effects, in mice. *Learn Mem.* 2005;12(3):277–284.
210. Lin CH, Lee CC, Huang YC, Wang SJ, Gean PW. Activation of group II metabotropic glutamate receptors induces depotentiation in amygdala slices and reduces fear-potentiated startle in rats. *Learn Mem.* 2005;12(2):130–137.
211. Lin HC, Wang SJ, Luo MZ, Gean PW. Activation of group II metabotropic glutamate receptors induces long-term depression of synaptic transmission in the rat amygdala. *J Neurosci.* 2000;20(24):9017–9024.
212. Heinbockel T, Pape HC. Input-specific long-term depression in the lateral amygdala evoked by theta frequency stimulation. *J Neurosci.* 2000;20(7):RC68.
213. Schafe GE, Nader K, Blair HT, LeDoux JE. Memory consolidation of Pavlovian fear conditioning: a cellular and molecular perspective. *Trends Neurosci.* 2001;24(9): 540–546.
214. Chen J, Iyengar R. Interactions between the Gs/protein kinase A and the Ras/MAP-kinase signalling pathways. *Biochem Soc Trans.* 1995;23(1):129–133.
215. Schafe GE, LeDoux JE. Memory consolidation of auditory Pavlovian fear conditioning requires protein synthesis and protein kinase A in the amygdala. *J Neurosci.* 2000;20(18):RC96.

216. Goosens KA, Holt W, Maren S. A role for amygdaloid PKA and PKC in the acquisition of long-term conditional fear memories in rats. *Behav Brain Res*. 2000;114(1–2): 145–152.
217. Moita MA, Lamprecht R, Nader K, LeDoux JE. A-kinase anchoring proteins in amygdala are involved in auditory fear memory. *Nat Neurosci*. 2002;5(9):837–838.
218. Impey S, Wayman G, Wu Z, Storm DR. Type I adenylyl cyclase functions as a coincidence detector for control of cyclic AMP response element-mediated transcription: synergistic regulation of transcription by Ca2+ and isoproterenol. *Mol Cell Biol*. 1994;14(12):8272–8281.
219. Sassone-Corsi P. Coupling gene expression to cAMP signalling: role of CREB and CREM. *Int J Biochem Cell Biol*. 1998;30(1):27–38.
220. Roberson ED, English JD, Adams JP, Selcher JC, Kondratick C, Sweatt JD. The mitogen-activated protein kinase cascade couples PKA and PKC to cAMP response element binding protein phosphorylation in area CA1 of hippocampus. *J Neurosci*. 1999;19(11):4337–4348.
221. Sindreu CB, Scheiner ZS, Storm DR. Ca2+-stimulated adenylyl cyclases regulate ERK-dependent activation of MSK1 during fear conditioning. *Neuron*. 2007; 53(1):79–89.
222. Bading H, Greenberg ME. Stimulation of protein tyrosine phosphorylation by NMDA receptor activation. *Science*. 1991;253(5022):912–914.
223. Rosen LB, Ginty DD, Weber MJ, Greenberg ME. Membrane depolarization and calcium influx stimulate MEK and MAP kinase via activation of Ras. *Neuron*. 1994;12(6):1207–1221.
224. Schafe GE, Atkins CM, Swank MW, Bauer EP, Sweatt JD, LeDoux JE. Activation of ERK/MAP kinase in the amygdala is required for memory consolidation of Pavlovian fear conditioning. *J Neurosci*. 2000;20(21):8177–8187.
225. Thomas GM, Huganir RL. MAPK cascade signalling and synaptic plasticity. *Nat Rev Neurosci*. 2004;5(3):173–183.
226. Meitzen J, Luoma JI, Stern CM, Mermelstein PG. Beta1-adrenergic receptors activate two distinct signaling pathways in striatal neurons. *J Neurochem*. 2011; 116(6):984–995.
227. Dolmetsch RE, Pajvani U, Fife K, Spotts JM, Greenberg ME. Signaling to the nucleus by an L-type calcium channel-calmodulin complex through the MAP kinase pathway. *Science*. 2001;294(5541):333–339.
228. Wu GY, Deisseroth K, Tsien RW. Activity-dependent CREB phosphorylation: convergence of a fast, sensitive calmodulin kinase pathway and a slow, less sensitive mitogen-activated protein kinase pathway. *Proc Natl Acad Sci USA*. 2001;98(5):2808–2813.
229. Ploski JE, Pierre VJ, Smucny J, et al. The activity-regulated cytoskeletal-associated protein (Arc/Arg3.1) is required for memory consolidation of Pavlovian fear conditioning in the lateral amygdala. *J Neurosci*. 2008;28(47): 12383–12395.
230. Weeber EJ, Atkins CM, Selcher JC, et al. A role for the beta isoform of protein kinase C in fear conditioning. *J Neurosci*. 2000;20(16):5906–5914.
231. Kurig B, Shymanets A, Bohnacker T, et al. Ras is an indispensable coregulator of the class IB phosphoinositide 3-kinase p87/p110gamma. *Proc Natl Acad Sci USA*. 2009;106(48):20312–20317.
232. Lin CH, Yeh SH, Lu KT, Leu TH, Chang WC, Gean PW. A role for the PI-3 kinase signaling pathway in fear conditioning and synaptic plasticity in the amygdala. *Neuron*. 2001;31(5):841–851.
233. Lin CH, Lee CC, Gean PW. Involvement of a calcineurin cascade in amygdala depotentiation and quenching of fear memory. *Mol Pharmacol*. 2003;63(1):44–52.

234. Ou LC, Gean PW. Regulation of amygdala-dependent learning by brain-derived neurotrophic factor is mediated by extracellular signal-regulated kinase and phosphatidylinositol-3-kinase. *Neuropsychopharmacology*. 2006;31(2):287–296.
235. Ou LC, Yeh SH, Gean PW. Late expression of brain-derived neurotrophic factor in the amygdala is required for persistence of fear memory. *Neurobiol Learn Mem*. 2010;93(3):372–382.
236. Rattiner LM, Davis M, Ressler KJ. Differential regulation of brain-derived neurotrophic factor transcripts during the consolidation of fear learning. *Learn Mem*. 2004;11(6):727–731.
237. Rattiner LM, Davis M, French CT, Ressler KJ. Brain-derived neurotrophic factor and tyrosine kinase receptor B involvement in amygdala-dependent fear conditioning. *J Neurosci*. 2004;24(20):4796–4806.
238. Kandel ER. Genes, nerve cells, and the remembrance of things past. *J Neuropsychiatry Clin Neurosci*. 1989;1(2):103–125.
239. Bourtchuladze R, Frenguelli B, Blendy J, Cioffi D, Schutz G, Silva AJ. Deficient long-term memory in mice with a targeted mutation of the cAMP-responsive element-binding protein. *Cell*. 1994;79(1):59–68.
240. Duvarci S, Nader K, LeDoux JE. De novo mRNA synthesis is required for both consolidation and reconsolidation of fear memories in the amygdala. *Learn Mem*. 2008;15(10):747–755.
241. Bailey DJ, Kim JJ, Sun W, Thompson RF, Helmstetter FJ. Acquisition of fear conditioning in rats requires the synthesis of mRNA in the amygdala. *Behav Neurosci*. 1999;113(2):276–282.
242. Miller S, Yasuda M, Coats JK, Jones Y, Martone ME, Mayford M. Disruption of dendritic translation of CaMKIIalpha impairs stabilization of synaptic plasticity and memory consolidation. *Neuron*. 2002;36(3):507–519.
243. Sutton MA, Schuman EM. Dendritic protein synthesis, synaptic plasticity, and memory. *Cell*. 2006;127(1):49–58.
244. Helmstetter FJ, Parsons RG, Gafford GM. Macromolecular synthesis, distributed synaptic plasticity, and fear conditioning. *Neurobiol Learn Mem*. 2008;89(3):324–337.
245. Hoeffer CA, Cowansage KK, Arnold EC, et al. Inhibition of the interactions between eukaryotic initiation factors 4E and 4G impairs long-term associative memory consolidation but not reconsolidation. *Proc Natl Acad Sci USA*. 2011; 108(8):3383–3388.
246. Ostroff LE, Cain CK, Bedont J, Monfils MH, Ledoux JE. Fear and safety learning differentially affect synapse size and dendritic translation in the lateral amygdala. *Proc Natl Acad Sci USA*. 2010;107(20):9418–9423.
247. Guzowski JF, Lyford GL, Stevenson GD, et al. Inhibition of activity-dependent arc protein expression in the rat hippocampus impairs the maintenance of long-term potentiation and the consolidation of long-term memory. *J Neurosci*. 2000; 20(11):3993–4001.
248. Parsons RG, Gafford GM, Helmstetter FJ. Translational control via the mammalian target of rapamycin pathway is critical for the formation and stability of long-term fear memory in amygdala neurons. *J Neurosci*. 2006;26(50):12977–12983.
249. Lopez-Salon M, Alonso M, Vianna MR, et al. The ubiquitin-proteasome cascade is required for mammalian long-term memory formation. *Eur J Neurosci*. 2001; 14(11):1820–1826.
250. Jarome TJ, Werner CT, Kwapis JL, Helmstetter FJ. Activity dependent protein degradation is critical for the formation and stability of fear memory in the amygdala. *PLoS One*. 2011;6(9):e24349.
251. Hall J, Thomas KL, Everitt BJ. Fear memory retrieval induces CREB phosphorylation and Fos expression within the amygdala. *Eur J Neurosci*. 2001;13(7):1453–1458.

252. Hall J, Thomas KL, Everitt BJ. Cellular imaging of zif268 expression in the hippocampus and amygdala during contextual and cued fear memory retrieval: selective activation of hippocampal CA1 neurons during the recall of contextual memories. *J Neurosci.* 2001;21(6):2186–2193.
253. Tao X, Finkbeiner S, Arnold DB, Shaywitz AJ, Greenberg ME. Ca2+ influx regulates BDNF transcription by a CREB family transcription factor-dependent mechanism. *Neuron.* 1998;20(4):709–726.
254. Yin JC, Wallach JS, Del Vecchio M, et al. Induction of a dominant negative CREB transgene specifically blocks long-term memory in Drosophila. *Cell.* 1994;79(1):49–58.
255. Lonze BE, Ginty DD. Function and regulation of CREB family transcription factors in the nervous system. *Neuron.* 2002;35(4):605–623.
256. Ginty DD, Bonni A, Greenberg ME. Nerve growth factor activates a Ras-dependent protein kinase that stimulates c-fos transcription via phosphorylation of CREB. *Cell.* 1994;77(5):713–725.
257. Brindle P, Nakajima T, Montminy M. Multiple protein kinase A-regulated events are required for transcriptional induction by cAMP. *Proc Natl Acad Sci USA.* 1995;92(23):10521–10525.
258. Impey S, Mark M, Villacres EC, Poser S, Chavkin C, Storm DR. Induction of CRE-mediated gene expression by stimuli that generate long-lasting LTP in area CA1 of the hippocampus. *Neuron.* 1996;16(5):973–982.
259. Josselyn SA, Shi C, Carlezon Jr WA, Neve RL, Nestler EJ, Davis M. Long-term memory is facilitated by cAMP response element-binding protein overexpression in the amygdala. *J Neurosci.* 2001;21(7):2404–2412.
260. Han JH, Kushner SA, Yiu AP, et al. Neuronal competition and selection during memory formation. *Science.* 2007;316(5823):457–460.
261. Han JH, Yiu AP, Cole CJ, Hsiang HL, Neve RL, Josselyn SA. Increasing CREB in the auditory thalamus enhances memory and generalization of auditory conditioned fear. *Learn Mem.* 2008;15(6):443–453.
262. Zhou Y, Won J, Karlsson MG, et al. CREB regulates excitability and the allocation of memory to subsets of neurons in the amygdala. *Nat Neurosci.* 2009;12(11):1438–1443.
263. Han JH, Kushner SA, Yiu AP, et al. Selective erasure of a fear memory. *Science.* 2009;323(5920):1492–1496.
264. Cowansage KK, Bush DE, Josselyn SA, Klann E, Ledoux JE. Basal variability in CREB phosphorylation predicts trait-like differences in amygdala-dependent memory. *Proc Natl Acad Sci USA.* 2013;110(41):16645–16650.
265. Conkright MD, Canettieri G, Screaton R, et al. TORCs: transducers of regulated CREB activity. *Mol Cell.* 2003;12(2):413–423.
266. Ch'ng TH, Uzgil B, Lin P, Avliyakulov NK, O'Dell TJ, Martin KC. Activity-dependent transport of the transcriptional coactivator CRTC1 from synapse to nucleus. *Cell.* 2012;150(1):207–221.
267. Screaton RA, Conkright MD, Katoh Y, et al. The CREB coactivator TORC2 functions as a calcium- and cAMP-sensitive coincidence detector. *Cell.* 2004;119(1):61–74.
268. Kovacs KA, Steullet P, Steinmann M, et al. TORC1 is a calcium- and cAMP-sensitive coincidence detector involved in hippocampal long-term synaptic plasticity. *Proc Natl Acad Sci USA.* 2007;104(11):4700–4705.
269. Sekeres MJ, Mercaldo V, Richards B, et al. Increasing CRTC1 function in the dentate gyrus during memory formation or reactivation increases memory strength without compromising memory quality. *J Neurosci.* 2012;32(49):17857–17868.
270. Yeh SH, Lin CH, Lee CF, Gean PW. A requirement of nuclear factor-kappaB activation in fear-potentiated startle. *J Biol Chem.* 2002;277(48):46720–46729.
271. Schuman EM, Madison DV. A requirement for the intercellular messenger nitric oxide in long-term potentiation. *Science.* 1991;254(5037):1503–1506.

272. Arancio O, Kiebler M, Lee CJ, et al. Nitric oxide acts directly in the presynaptic neuron to produce long-term potentiation in cultured hippocampal neurons. *Cell.* 1996;87(6):1025–1035.
273. Arancio O, Lev-Ram V, Tsien RY, Kandel ER, Hawkins RD. Nitric oxide acts as a retrograde messenger during long-term potentiation in cultured hippocampal neurons. *J Physiol Paris.* 1996;90(5–6):321–322.
274. Chapman PF, Atkins CM, Allen MT, Haley JE, Steinmetz JE. Inhibition of nitric oxide synthesis impairs two different forms of learning. *Neuroreport.* 1992;3(7):567–570.
275. Schafe GE, Bauer EP, Rosis S, Farb CR, Rodrigues SM, LeDoux JE. Memory consolidation of Pavlovian fear conditioning requires nitric oxide signaling in the lateral amygdala. *Eur J Neurosci.* 2005;22(1):201–211.
276. Ota KT, Monsey MS, Wu MS, Schafe GE. Synaptic plasticity and NO-cGMP-PKG signaling regulate pre- and postsynaptic alterations at rat lateral amygdala synapses following fear conditioning. *PLoS One.* 2010;5(6):e11236.
277. Overeem KA, Ota KT, Monsey MS, Ploski JE, Schafe GE. A role for nitric oxide-driven retrograde signaling in the consolidation of a fear memory. *Front Behav Neurosci* 2010;4:2.

CHAPTER ELEVEN

Epigenetics of Memory and Plasticity

Bisrat T. Woldemichael, Johannes Bohacek, Katharina Gapp, Isabelle M. Mansuy

Brain Research Institute, Medical Faculty of the University of Zürich, and Department of Health Sciences and Technology, Swiss Federal Institute of Technology, Brain Research Institute Zürich, Switzerland

Contents

Abstract

Although all neurons carry the same genetic information, they vary considerably in morphology and functions and respond differently to environmental conditions. Such variability results mostly from differences in gene expression. Among the processes that regulate gene activity, epigenetic mechanisms play a key role and provide an additional layer of complexity to the genome. They allow the dynamic modulation of gene expression in a locus- and cell-specific manner. These mechanisms primarily involve DNA methylation, posttranslational modifications (PTMs) of histones and noncoding RNAs that together remodel chromatin and facilitate or suppress gene expression. Through these mechanisms, the brain gains high plasticity in response to experience and can

Progress in Molecular Biology and Translational Science, Volume 122
ISSN 1877-1173
http://dx.doi.org/10.1016/B978-0-12-420170-5.00011-8

integrate and store new information to shape future neuronal and behavioral responses. Dynamic epigenetic footprints underlying the plasticity of brain cells and circuits contribute to the persistent impact of life experiences on an individual's behavior and physiology ranging from the formation of long-term memory to the sequelae of traumatic events or of drug addiction. They also contribute to the way lifestyle, life events, or exposure to environmental toxins can predispose an individual to disease.

This chapter describes the most prominent examples of epigenetic marks associated with long-lasting changes in the brain induced by experience. It discusses the role of epigenetic processes in behavioral plasticity triggered by environmental experiences. A particular focus is placed on learning and memory where the importance of epigenetic modifications in brain circuits is best understood. The relevance of epigenetics in memory disorders such as dementia and Alzheimer's disease is also addressed, and promising perspectives for potential epigenetic drug treatment discussed.

1. OVERVIEW

We are currently in the midst of a revolution in genetics that is about to end a decade of "nature versus nurture" debate. This debate questioned the contribution of inherited genetic factors (nature) versus environmental influences (nurture) to individuals' development, features, personality, and disease susceptibility. The field of epigenetics has offered a novel and biologically relevant framework to explain how the genetic information contained in the DNA, which is static, can dynamically respond to environmental factors, and how stable changes in an organism can be induced with no change in the genetic code itself. Thus, epigenetics provides a molecular interface that allows integrate the interaction between genes and environment. This epigenetic revolution was brought about by progress in the understanding of chromatin, a dynamic and complex structure formed by DNA, histones, and nonhistone proteins in the cell nucleus. Chromatin can be modulated by multiple biochemical modifications triggered by environmental factors via complex intracellular signaling cascades. Identifying these modifications and understanding their mechanisms of regulation are essential steps to understand the interaction between an individual's genetic makeup and its environment. The genome is thus highly dependent on the environmental context in which it functions, and interacting genetic and environmental factors truly penetrate every aspect of life and every level of biology.

For many years, epigenetics was relegated to cellular identity during development and differentiation and was used to explain how the genome of a cell can be stably marked to gain and keep such identity. Recent advances have extended this concept and the additional layer of complexity and plasticity provided by epigenetic mechanisms also exists in other biological functions, including in the central nervous system. Because epigenetics offers a dynamic link between the genome and the environment, and adds plasticity to genetic programming, it is a particularly appealing concept in the realm of brain functions, where the ability to respond to environmental cues and demand is of utmost importance. A classic example is that of a child and a sleeping dog, where the emotional response of joy and fascination for the dog generated by the child's brain leads to the decision to extend a hand and sample the texture of the animal. If the dog, suddenly awakened, startles and bites the child, the child's brain will need to rapidly assess the danger of the situation, coordinate a flight response by activating the release of stress hormones that increases heart rate and blood pressure, and mobilize physical resources. At the same time, the brain also forms a long-lasting memory of the event that will commend the child to stay away from sleeping dogs or even perhaps all furry four-legged animals and will likely remain throughout the child's life. The instant formation of such long-lasting memory is remarkable when considering that molecular components in brain cells undergo constant turnover. Long-term memory traces are encoded by complex signal transduction cascades that involve gene transcription and translation.[1] Thus, their formation requires that these cascades be rapidly activated, which has been postulated to implicate epigenetic processes. The idea that "the epigenetic marking of the genome that confers cellular identity during early development is the ultimate example of long-term memory storage"[2] suggests that the same mechanisms have been coopted by the nervous system and its terminally differentiated, nonreplicating cells to achieve persistent long-term information storage. From the simple example of a child forming a life-long fear memory of dogs, it becomes evident how environmental factors in the form of daily experiences can permanently alter cellular processes. Such complex neuronal processes are usually taken for granted, but when they fail or malfunction, such as in devastating conditions like Alzheimer's disease (AD), their fundamental importance for basic cellular processes controlling our ability to learn and remember, perceive, interpret, and interact with our environment painfully appears to us. To place in perspective the recently appreciated importance of epigenetics in the functions of the nervous system, the current chapter first reviews general epigenetic processes and then describes the contribution of epigenetics to the integration of genetic and environmental information for brain functions in health and disease.

2. BACKGROUND

2.1. Definition of epigenetics

The term epigenetics was coined by Conrad Waddington in 1942 to address one of the fundamental problems of developmental biology: How can all cells in an organism carry the same genetic information, yet develop into different cells such as neurons, liver cells, or skin cells? Waddington conceptually defined epigenetics as "…the interactions of genes with their environment which brings the phenotype into being[3]" (Waddington, 1942). This captures two of the key features that will be discussed in this chapter: (1) the role of epigenetics as an interface between the genome and the environment and (2) the concept of long-lasting, stable, yet inducible (and reversible) changes within cells that determine cellular functions by altering the "interpretation" of the genetic information. Waddington's definition of epigenetics is the broadest, but more restricted definitions are used in the scientific literature. Another important definition characterizes epigenetics as all heritable changes in genome functions that occur without a change in DNA sequence.[4] This definition is more restricted because it places the focus not only on changes independent of the DNA sequence but includes the notion of heritability. When considering cells that no longer divide such as neurons, this definition excludes all epigenetic changes that accompany various neuronal functions. Thus, for this chapter, we will adopt a modified view of Waddington's original definition,[5] and define epigenetics "as the study of any potentially stable and, ideally, heritable change in gene expression or cellular phenotype that occurs without changes in DNA sequence."

2.2. Epigenetic mechanisms

2.2.1 DNA methylation (5mC)

Chromatin is a rigorously organized structure that can be locally modulated by epigenetic mechanisms that dynamically or stably alter the expression of the genes it carries. One of the best-known epigenetic mechanisms is DNA methylation. It stands out among epigenetic modifications because it modifies DNA directly and can be stable.[6] In mammals, DNA methylation consists in the transfer of a methyl group to the fifth position of the pyrimidine ring of cytosines (5mC), generally in dinucleotide CpG sequences. In mammalian genomes, about 1% of cytosines and 75% of CpG dinucleotides are methylated.[7–10] Because methylated cytosines are 10–50 times more likely to undergo mutations (by deamination of C to T), the representation of CpGs in the genome has decreased during evolution. However, clusters of CpG-dense regions called CpG islands are commonly found in the promoter region of genes, and promoter-associated CpG islands are typically minimally methylated. Increased 5mC generally results in transcriptional repression and is often associated with gene silencing[11,12] (see Fig. 11.1). 5mC is thought to interfere with gene expression via two general, not mutually exclusive mechanisms. CpG methylation can either directly block the DNA recognition sites of certain DNA-binding factors, or recruit binding proteins such as methyl-CpG-binding protein 2 (MeCP2) and transcription repressors to the CpG islands, which alters chromatin structure and makes it less accessible to the transcriptional machinery.[6,13]

The universal methyl donor *S*-adenosyl-L-methionine (SAM) necessary for DNA methylation is synthesized from methyl donors (methionine and choline) and cofactors (folic acid, vitamin B12, and pyridoxal phosphate). The methyl group from SAM is transferred to cytosine by DNA methyltransferases (DNMTs),[14] which exist in four major forms in mammals: DNMT1, 2, 3a and 3b.[6] DNMT1 is involved in the maintenance of

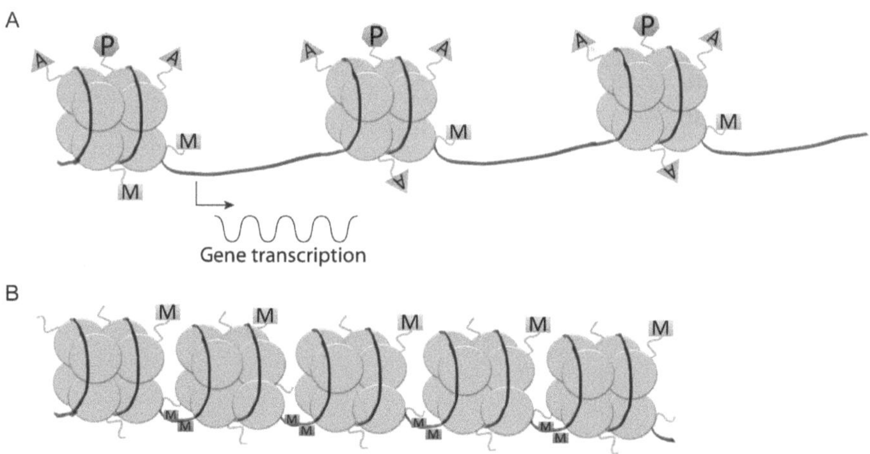

Figure 11.1 Epigenetic modifications and their influence on chromatin state. (A) Epigenetic modifications including acetylation, phosphorylation, and di/trimethylation of histone tail residues allow opening of the chromatin and recruitment of the transcriptional machinery. (B) Deacetylation, dephosphorylation, and demethylation of histone residues together with DNA methylation induce compaction of the chromatin and gene silencing. A, acetylation; P, phosphorylation; M, methylation. (For color version of this figure, the reader is referred to the online version of this chapter.)

DNA methylation, in particular across cell division. During replication, it methylates the newly synthesized DNA strand from the parent strand template. Thus although reversible, 5mC can stably mark DNA and be maintained across mitosis, as well as, in some cases across meiosis.[15,16] In contrast to maintenance DNMTs, DNMT3a and DNMT3b are *de novo* DNMTs. Their mechanisms of action are not fully understood, but they are known to involve domain-specific recognition sites, and be recruited to specific DNA sequences via protein–protein interaction and small RNAs.[6]

2.2.2 DNA hydroxymethylation (5hmC)

5mC was long thought to be an irreversible epigenetic mark, in part because no DNA demethylation mechanism or enzyme could be identified (for review, see Ref. 17). However, the fact that DNA in the zygote is first demethylated then remethylated during development always argued for the existence of an active demethylation process.[18] The recent discovery of an intermediate epigenetic mark between DNA methylation and demethylation in the form 5-hydroxymethylation (5hmC)[19] strongly suggests that DNA demethylation does occur in mammalian cells. 5hmC is generated by hydroxylation of 5mC by ten–eleven translocation (TET) proteins (TET1–3).[20] 5-Hydroxylation is thought to be the first step of a cascade of chemical reactions leading to the removal of 5mC (for review, see Ref. 21). Genome-wide analyses have revealed that 5mC and 5hmC are differently distributed on the genome. While 5mC occurs mostly in inter- and intragenic regions (CpG-islands surrounding promoter regions being largely unmethylated) and silences gene in most cases,[22,23] 5hmC is primarily confined to the 5′ end and correlates with gene transcription.[24,25] Interestingly, the level of 5hmc in the body is the highest in the brain, suggesting an important role for this modification in neural functions.[19,26] Although not much is known about the biological functions of 5hmC, the view that it is an epigenetic mark on its own that is associated with gene transcription is gaining momentum.

2.2.3 Histone posttranslational modifications

In addition to DNA methylation, histone posttranslational modifications (PTMs) play a critical role in chromatin remodeling. They form a histone code specific for each gene, and depending on their nature, they are associated with the activation or the repression of gene transcription. They are induced by a complex enzymatic machinery and occur on all histones (H2A, H2B, H3 and H4, and the H1 linker histone) in specific combinations.

Each histone can undergo a variety of PTMs, including acetylation, phosphorylation, methylation, ubiquitylation, sumoylation, ADP ribosylation, proline isomerization, and deamination, on the N-terminus tail protruding from the nucleosome, the C-terminus tail or the core. Some of these modifications are transient while others are stable and potentially heritable.

Together with DNA methylation, histone PTMs induce local and global structural changes in the chromatin. They can partition the genome into distinct domains of transcriptionally active chromatin (euchromatin) or transcriptionally inactive chromatin (heterochromatin).[27] They also alter the net electrical charge of nucleosomes and control the loosening or tightening of inter- and intra-nucleosomal DNA–histone interactions. For instance, acetylated histones (which are associated with transcriptionally active states) are more likely to be displaced from DNA, thus inducing a loosening of the chromatin.[28,29] Consistently, genome-wide studies have demonstrated that nucleosome density is typically lower at promoter regions that carry acetylated histones, than in the coding region.[30–32] Histone PTMs can also help recruit binding partners that can be positive or negative. Methylation of lysine 4 on histone 3 (H3K4me) can prevent the binding of histone deacetylases (HDACs), thus favor histone acetylation. In contrast, H3K18 acetylation can favor the binding of histone acetyl transferases (HATs) and transcription factors such as CREB-binding protein (CBP).[27,33,34] However, since PTMs are multiple and occur in combinations, it is difficult to define their individual impact (for a detailed review, see Ref. 35). Further, they have a different role depending on their location within a histone tail. For instance, the displacement of PTMs was shown to lead to the repression of usually transcribed genes.[34,36]

In addition to carrying complex combinations of PTMs, histones are also expressed as multiple sequence variants encoded by different genes that are associated with distinct transcriptional profiles. There is also increasing evidence that the composition of the nucleosome itself is plastic and carries important information about the transcriptional state of individual genes (for review, see Ref. 37).

2.2.4 Noncoding RNAs

Noncoding RNAs (ncRNAs), particularly small RNAs (sncRNAs), have recently emerged as key transcriptional and posttranscriptional regulators of gene expression that also contribute to non-genetic regulation. At least three classes of sncRNAs have so far been identified: microRNAs (miRNAs), small-interfering RNAs, and PIWI-interacting RNAs (piRNAs),[38,39] each with slightly different characteristics and modes of regulation and action. SncRNAs incorporate into an RNA-induced silencing complex and guide the silencing machinery by binding to specific sequences on target RNA(s).[38] Their role is two fold: they target chromatin-modifying enzymes, and recruit the silencing complex to specific genomic regions. Several miRNAs target chromatin-modifying enzymes. miR-1 and miR-140 regulate the level of HDAC4 during development,[40,41] while miR-290 targets transcriptional repressors of *de novo* DNMTs and maintains optimal level of DNA methylation in embryonic stem cells.[42] Further, the miR-29 family of miRNAs target DNMT3a, DNMT3b, and TET1-3, thus contributes to the balance between DNA methylation and hydroxymethylation.[43,44]

Components of the RNAi machinery have been shown to be essential for the formation of heterochromatin.[45] Studies in various model systems suggest that small RNAs can gain direct access to the chromatin and induce epigenetic silencing.[39] One mechanism involves incorporation of miRNAs to an RNA-induced transcriptional silencing complex and binding to an RNA transcript at the transcriptional machinery.[46–49] This mechanism is particularly exploited by piRNAs, which bind to several complementary regions in the genome and help to assemble the epigenetic machinery through their interaction with Piwi proteins. Indeed, lack of Piwi proteins causes dramatic changes to the epigenetic landscape and transcriptional states.[50,51]

3. BRAIN PLASTICITY THROUGH EPIGENETICS

One of the distinguishing features of the brain—in comparison to other organs—is its remarkable capacity to integrate information from the environment and adjust its activity accordingly, a property called plasticity. In the following section, we discuss two examples demonstrating how epigenetic processes contribute to brain plasticity in response to life experiences. The first is drug addiction, a condition for which a single exposure to an addictive substance can radically and permanently change the behavior of an individual. The second is the impact of early life stress on psychological development and health later in life. Addiction and early life stress are prime examples of the rapid and long-lasting changes that can be induced by environmental factors, and the complex epigenetic processes involved.

3.1. Drug addiction

Drug addiction is a chronic condition characterized by the compulsive seeking and usage of a substance even if it has injuring consequences.[52] It induces long-lasting changes in behavior in the form of craving and relapse. These changes are associated with structural alterations in reward circuits in the brain such as the ventral tegmental area, nucleus accumbens (NAc), amygdala, hippocampus, and prefrontal cortex (PFC) and damaged morphology and function of certain neurons in these regions.[53–55] Most drugs of abuse also activate major cellular signaling pathways such as those associated with the transcription factors deltaFosB and CREB.[56] They can affect hundreds of genes in different brain areas, with effects that persist long after cessation of treatment. Experimental evidence has provided insight into the epigenetic mechanisms that orchestrate these complex patterns of transcriptional regulation. From an epigenetic perspective, the lifelong vulnerability to relapse is particularly intriguing because it implies that drug exposure can activate molecular mechanisms that capture and maintain alterations in brain plasticity persistently.

Following chronic or acute cocaine administration, histone marks are globally changed in the NAc in adult rodents. Acetylated histone H4 and phosphoacetylated histone H3 are increased after a single cocaine injection,[57] while histone H3K9 dimethylation (H3K9me2), a repressive mark, is reduced after repeated cocaine injection.[58] Global level of histone H3 phosphorylation at serine 10 also increases in the striatum after acute cocaine treatment.[59] These histone modifications are associated with hundreds of different gene promoters and with differential expression of some genes.[60] But, the actual modes of gene regulation remain not fully understood. Many of the genes with altered promoter-associated histone marks do indeed not have any change in mRNA expression, and different histone modifications are altered on different genes and hardly overlap.[58] A particularly well-established molecular component linking epigenetic changes and drug addiction involves HDAC5 in the NAc.[61] Following chronic (but not acute) cocaine administration, HDAC5 gets phosphorylated by CaMKII, which triggers its nuclear export and results in a global increase in histone acetylation. Virus-mediated overexpression of HDAC5 in the NAc in adult mice attenuates the rewarding effects of cocaine, while a deficiency in HDAC5 leads to sensitization to cocaine reward following chronic administration.[61] Virus-mediated delivery of HDAC5 in the NAc of HDAC5-deficient animals normalizes the reward hypersensitivity in these mice, supporting a direct role of HDAC5. Histone methylation by cocaine is linked to the lysine methyltransferase G9a which is persistently downregulated in the NAc following drug administration.[62] Overexpression of G9a in mice reverses the global decrease in H3K9me2 induced by cocaine and reduces the animals' preference for the drug. The reduction in G9a and H3K9me2 depends on the upstream induction of deltaFosB, an immediate early gene necessary for the rewarding properties of the drug. When deltaFosB is turned off in the NAc, cocaine fails to reduce H3K9me2, while conditional deltaFosB overexpression reduces G9a and H3K9me2, and thereby mimics the effects of cocaine. A recent study further showed that alterations in histone marks also occur outside the brain. Histone acetylation is increased at the brain-derived neurotrophic factor (BDNF) gene in the sperm of cocaine-treated mice, and likewise in the brain of the offspring, raising the possibility that cocaine exerts transgenerational effects through epigenetic alterations in the germline.[63] Finally, DNA methylation is another epigenetic mark that is altered following cocaine exposure. The methylated DNA-binding protein MeCP2 has also been implicated and may involve small RNAs.[64,65] Future epigenomic analyses are expected to identify the ensemble of epigenetic changes induced by cocaine exposure. They are hoped to help design potential epigenetic drugs able to interfere with the long-lasting behavioral sequels of drug addiction and counteract drug craving and relapse.

3.2. Early life experiences

The exposure to traumatic and repeated stressful experiences has detrimental consequences on many physiological and psychological functions in humans, primates, and rodents.[66–68] Stressful events in early life in humans constitute a major risk factor for the development of emotional and cognitive disorders in adulthood, ranging from major depression to attention and anxiety disorders.[69,70] In rodents, early life stress has similarly dramatic and long-lasting effects on emotionality, depression-like behavior, and stress-responsiveness later in life.[70,71] The fact that early life experiences have persistent implications is widely accepted, yet the underlying mechanisms remain partially understood.[72] Rodent models have been instrumental to the study of these mechanisms and have revealed that epigenetic (re)programming in an important determinant of the response to early life experience.[72–75]

3.2.1 Maternal care

A naturally occurring form of early life stress is poor maternal care and neglect. Similar to human mothers, rodent dams show marked differences in the level and quality of maternal care they provide to their offspring.[76] Care is, however, highly consistent within an individual mother. Maternal care in rats and mice is characterized by the time mothers spend licking, grooming, and nursing their pups during the first week of life. Based on their maternal abilities, rat dams can be distinguished into high, mid, or low licking/grooming (LG) mothers. Such natural difference is associated with notable variability in stress responsiveness and emotionality in the offspring later in life. Rat pups raised by low-LG mothers have an increased responsiveness to stressful situations, associated with higher activity of the hypothalamic–pituitary–adrenal (HPA) stress axis compared to pups raised by high-LG dams. When adult, the neglected rats show prolonged ACTH and corticosterone elevation following restraint stress, reduced glucocorticoid receptor (GR) mRNA and protein in the hippocampus, and higher corticotrophin-releasing hormone mRNA in the hypothalamus.[77,78] They also have learning defects and increased anxiety in adulthood. Epigenetic mechanisms have been implicated in these long-term alterations, in particular in the compromised HPA axis. After the first week of life, pups of high-LG mothers have increased expression of the transcription factor NGFI-A and its binding to one of the GR promoters, leading to increased GR expression compared to low-LG offspring.[79] This increase is, however, transient and not observed in adult animals. A more persistent epigenetic alteration occurs through DNA methylation at a CpG site within the NGFI-A response element of the GR gene. Low-LG offspring have increased DNA methylation at this site starting 1 week after birth until adulthood.[80] These epigenetic changes have been postulated to result from the activation of a cascade of events involving the HAT CBP. In high-LG offspring, the change in NGFI-A expression during the first week of life increases NGFI-A binding to the GR promoter which recruits CBP. CBP enrichment at the GR promoter in turn increases the level of H3K9 acetylation, which activates GR expression and may also prevent DNA methylation in this region. In contrast, in low-LG offspring, the reduced NGFI-A binding to the GR promoter prevents GR expression and might favor the recruitment of the DNA methylation machinery and induce hypermethylation at this locus. Increased promoter methylation was indeed shown to prevent NGFI-A binding *in vitro* and may thus explain the reduced GR expression later in life.[79] Notably, increased GR promoter methylation can also be instated in the offspring by maternal behavior (high-LG).[80] However, when the offspring of low-LG mothers is cross-fostered to high-LG surrogate mothers within 12 h of birth, DNA methylation status at GR promoter is reversed and is similar to natural pups of high-LG dams. This indicates that the DNA methylation level is directly associated to the level of maternal care received by the pups, providing an example of transfer of DNA methylation profile through a behavioral mode of programming.

3.2.2 Early life stress

The fact that differences in maternal care can have a lasting impact throughout life suggests that more dramatic experiences might have even more dramatic and persistent consequences. Recent work has demonstrated that early life stress induced by repeated maternal separation during the first 2 weeks of life alters the stress response and involves epigenetic changes.[81] Early life stress induces hyperactivity of the HPA axis, corticosterone hypersecretion, and hyperresponsiveness to acute stressors later in life. These effects are linked to increased expression of the hormone arginine vasopressin (AVP) in a subnucleus of the hypothalamus, due to hypomethylation of a key *Avp* enhancer region at a high-affinity MeCP2 DNA-binding site. Since MeCP2 binding represses *Avp* expression when its site is methylated, lower DNA methylation releases this repression and leads to elevated *Avp*. These results overall provide strong evidence that early life stress can dynamically modify the epigenome persistently.

Compromised maternal care is another condition that induces long-lasting epigenetic changes. Rat pups raised by dams stressed for 30 min daily during the first postnatal week have significantly lower BDNF mRNA in PFC when adult.[82] This is associated with differential DNA methylation of an important regulatory region of the BDNF gene (exon IV). Methylation across 12 CpG sites in this region is higher in rats from stressed mothers, while there is no or only little DNA methylation in normally reared rats. Further, the offspring of maltreated rats have similarly increased DNA methylation at the BDNF promoter region, an effect that cannot be fully reversed by cross-fostering. Therefore, mechanisms seem to be in place that allow persistent changes in DNA methylation to be passed from one generation to the next, independent of postnatal experience of the affected individual.

When negative, early life experiences are particularly traumatic and can induce true transgenerational transmission of their effects. This means that exposure of one generation to stressful conditions can impact several following generations. In mice, chronic and unpredictable maternal separation combined with unpredictable maternal stress in early postnatal life is a severe condition that alters behavior across life. It induces depressive-like behaviors, social withdrawal, impaired cognition, and altered behavioral control in the animals when adult, but strikingly, it also severely affects the progeny across several generations.[83,74,84] Transmission occurs through both females and males, and is independent of maternal behaviors. It therefore involves the germline. Mechanistically, it in part implicates DNA methylation. Thus, methylation is altered at multiple genes in the brain of the stressed animals when adult, with some genes being hypermethylated on their promoter region, that is, MeCP2, and others, hypomethylated, that is, CRF receptor 2. Further to affecting the brain, methylation anomalies are also present in the germline of the stressed males, suggesting their potential implication in the inheritance of the traits induced by stress. This clear example of epigenetic inheritance, as well as other examples of transgenerational effects (i.e., diet and environmental toxicants), supports the idea that certain epigenetic marks are likely vectors of transgenerational transmission of the effects of environmental factors.[85–88] The impact of these marks at the chromatin is widespread and several genome- and epigenome-wide studies in rodents and humans have identified hundreds of genes affected in different brain regions.[89,90] This correlates with the complexity and multiplicity of the effects of, for instance, variations in maternal care or early life stress on behavior. Thus, the reductionist idea of associating single genes to complex behavioral phenotypes proves inadequate. More systematic epigenome-wide analyses will be essential in the future for a better understanding of the impact of adverse conditions early in life and the way they influence disease risk.

4. EPIGENETICS MECHANISMS OF LEARNING AND MEMORY FORMATION

Learning and memory are essential cognitive functions for mammals. Memory is a complex process that has several temporal phases, including short-, immediate-, and long-term, depending on the persistence of the stored information. Memory is also subdivided into explicit and implicit depending on the nature of the stored information. These phases and forms of memory implicate different regions and neural networks in the brain. However, they all depend on synaptic plasticity, a property of neuronal circuits to modulate their efficacy to transmit signals in an activity-dependent manner. Synaptic plasticity is a complex cellular process sustained by cascades of fine-tuned molecular events in individual neurons and synapses. It can be modeled experimentally *in vitro* or *in vivo* by electrophysiological means in different regions of the adult or developing brain. In the hippocampus, one of the major brain areas for memory formation, high-frequency stimulation of presynaptic neuronal fibers induces a sustained increase in the efficacy of synaptic transmission to postsynaptic neurons, a property known as long-term potentiation (LTP). In contrast, low-frequency stimulation of presynaptic fibers reduces the efficacy of synaptic transmission and induces long-term depression (LTD).[91,92]

While short-term memory is thought to recruit transient cellular and molecular changes such as covalent modifications of preexisting proteins, long-term memory requires long-lasting modifications and the synthesis of new proteins.[93–95] Many genes whose expression is upregulated by neuronal activity are essential for memory formation. They include immediate early genes such as c-Fos, structural proteins such as activity-regulated cytoskeleton-associated protein (Arc), transcription factors such as cyclic-AMP response element-binding protein (CREB), and other genes, that is, BDNF, major histocompatibility complex-1 and Homer 1.[93–95] These genes contribute to the cellular changes underlying synaptic plasticity and the acquisition and consolidation of memory traces like, for instance, the insertion of new AMPA receptors in postsynaptic membranes, the strengthening of synaptic contacts, and the modulation of dendritic spines.[96–98] Over the years, epigenetic mechanisms have emerged as key mechanisms of regulation of the molecular machinery necessary for learning and the formation and storage of memory.

4.1. DNA modifications in learning, memory, and synaptic plasticity

Learning and memory formation are accompanied by changes in the epigenetic landscape of the adult brain, in particular by DNA modifications and the associated machinery. Following contextual fear conditioning, a behavioral paradigm that induces the formation of a hippocampus-dependent associative memory between a neutral context and an aversive foot shock, DNMT3a and 3b, two enzymes necessary for *de novo* DNA methylation, increase in the hippocampus in rat.[99] This increase is paralleled by higher DNA methylation at some genes, but surprisingly, by hypomethylation at other genes. Thus, there is higher methylation and reduced expression of PP1γ, a memory suppressor, in the hippocampus but lower promoter methylation and increased expression of Reelin, a positive regulator of memory and synaptic plasticity.[100,101] Persistent hypomethylation of CpG sites at the BDNF promoter and increased BDNF expression has also been reported in the hippocampus following contextual fear conditioning.[102]

Further to the hippocampus, DNMT3a is also increased in the amygdala after cued fear conditioning (associative memory between a tone or light and an aversive foot shock that depends on the amygdala) in mice,[103] suggesting a global role for DNMTs in associative memory. This may be partly linked to the ability of some of these enzymes, such as DNMT3a2, to be rapidly upregulated by calcium-dependent neuronal activation.[104] Further insight on the importance of DNA methylation in memory processes was provided by genetic or pharmacologic manipulation of DNMTs in mice. Mice with a conditional deletion of DNMT1 and 3a in adult forebrain neurons have smaller hippocampus and impaired long-term spatial memory. Likewise, shRNA-based knockdown of DNMT3a2 in the mouse hippocampus induces long-term but not short-term, impairment in contextual fear memory and object location.[104] These effects can be largely reproduced by DNMT inhibitors. Infusion of 5-AZA in the lateral amygdala shortly after cued fear conditioning impairs long-term but not short-term memory, while infusion in medial PFC (mPFC) immediately after trace fear conditioning impairs long-term memory.[105]

DNMTs and DNA methylation are also modulated by synaptic plasticity, both *in vivo* and *in vitro*. *In vivo*, the induction of LTP in the rat mPFC increases the level of DNMTs,[105] while LTP in the hippocampus is blocked by DNMT inhibitors such as zebularin or 5-AZA.[101] Consistently, DNMT1 and 3a conditional deletion impairs LTP but enhances LTD in the hippocampus.[100] Likewise, the induction of synaptic plasticity by treatment with activators of PKC signaling increases DNMT3a in hippocampal slices *in vitro*. However, at the same time, a depolarizing stimulus can also reduce methylation at specific sites, for instance at some CpGs in one BDNF promoter, and induce BDNF expression in cultured hippocampal and cortical neurons.[106] This effect involves MeCP2, a methyl-DNA-binding protein that binds to the promoter when methylated, and its dissociation after promoter demethylation followed by CREB recruitment. This effect is increased by DNMTs inhibition[101] and oppositely, promoter activity after depolarization decreases when site-specific methylation at CREB sites in the BDNF exon IV promoter is induced, suggesting that activity-dependent change in DNA methylation is important for synaptic transmission.[101,106] Consistent with the requirement for activity, DNMT inhibitors produce an effect only when applied with behavioral training or synaptic activation. In the absence of training, zebularin or 5-AZA in the hippocampus *in vivo* does not affect the methylation of genes associated with learning, but it does following contextual fear conditioning. Likewise, zebularin or 5-AZA treatment of hippocampal slices impairs LTP but does not affect basal synaptic transmission.[101,107] Finally, epigenetic regulation linked to plasticity also occurs in the invertebrate Aplysia. In Aplysia neurons, stimulation of sensorimotor neurons by application of five pulses of serotonin enhances the response to subsequent stimuli, a form of long-lasting plasticity known as long-term facilitation (LTF) and is accompanied by increased methylation at several CpG sites in the promoter of CREB2, an important molecular suppressor of plasticity and memory formation. This epigenetic alteration is mediated by Piwi/piRNAs complexes that leads to a persistent down-regulation of CREB2 transcript.[108]

Although memory formation and synaptic plasticity are accompanied by a global increase in DNMTs, both DNA methylation and demethylation occur in a gene-specific manner. This in part is due to the role that DNMTs can play not only in active CpG methylation but also in demethylation of 5mCpGs through deamination.[109] But besides DNMTs, demethylating enzymes such as Gadd45, a member of a family of small (18 kDa) stress-inducible acidic nuclear proteins, may also be implicated. Thus, Gadd45 has been linked to active demethylation after learning and various forms of neuronal activation.[110,111] Indeed, the role of DNA hydroxymethylation (5-hydroxymethylation, 5-hmC), an epigenetic modification initially thought to be only a transition between 5-methylcytosine methylation and demethylation, is increasingly recognized as being important for brain functions. 5hmC is abundant in both the rodent and human brain and is particularly enriched at genes with synapse-related functions.[26] Viral-mediated overexpression of TET1, one of the enzymes that catalyze hydroxymethylation of cytosines, in the mouse hippocampus reduces CpG methylation at one of BDNF promoters (IX) and at a brain-specific promoter of Fgf1 and upregulates BDNF transcripts. In contrast, shRNA-mediated TET1 knockdown in the hippocampus increases CpG methylation.[112] TET1 knockout also alters short-term but not long-term spatial memory.[113] Overall, these studies suggest that a complex dynamics of DNA methylation/demethylation/hydroxymethylation operates during memory formation and synaptic plasticity.

4.2. Histone PTMs in learning and memory

Histone acetylation, phosphorylation, methylation, and poly-ADP ribosylation are PTMs that have been implicated in memory formation and synaptic plasticity. Acetylation of histone tails is one of the best-understood PTMs in the adult brain. This is in part because CBP, long known as an essential transcriptional regulator for synaptic plasticity and memory, also acts as a HAT that catalyzes the acetylation of histones and of transcription factors.[114] CBP is recruited by activity-dependent CaMKIV signaling and, together with CREB, regulates the transcription of many neuronal genes.[1,115,116] It operates as a scaffold within a transcription complex and favors the recruitment and modulation of different transcription factors to facilitate gene expression.[117] When the HAT activity of CBP is suppressed in the adult mouse brain, long-term memory for objects and space is impaired, and c-Fos expression is reduced. In hippocampal slices, it severely alters late phase of LTP, a phase associated with long-term memory, but spares basal synaptic transmission. Reversal of the HAT deficiency or treatment with the HDAC inhibitor TSA or SAHA rescues the memory and LTP deficits.[118,119] Further, activation of CBP and its homologue p300 by intraperitoneal injection of a small-molecule activator complex CSP-TTK21 facilitates the maturation and differentiation of adult neuronal progenitors in the dentate gyrus. This is accompanied by increased expression of genes such as BDNF, higher histone acetylation at the promoter of these genes, and prolonged spatial memory.[120] Likewise, mice conditionally expressing a truncated form of p300 lacking HAT activity have impaired long-term object and contextual fear memory, but normal spatial memory,[121] suggesting a role for acetylation in multiple forms of memory. Histone acetylation is indeed directly modulated by learning and memory formation in the adult brain. The level of H3K14 acetylation increases in different subregions of the hippocampus 1 h after associative learning in the adult rat.[122] Similarly, acetylation of H2B, H2AK9, and H4K12 increases at the promoter of activity-dependent genes such as cFos, Zif268, and BDNF exon-IV in the hippocampus after spatial learning on a water maze and is associated with upregulation of gene transcription. Consistently, the expression of several HATs including CBP, p300, and PCAF, and global HAT activity increases during the consolidation of spatial memory.[123] But such activation requires substantial training and does not occur when learning is weak. Thus, intense object recognition training that induces strong object memory enhances the global level of H3 acetylation but a weak training does not, again in line with the activity dependence of some epigenetic modifications in the brain.

Manipulation of histone acetylation by pharmacological drugs can modulate learning and memory performance. Intraperitoneal injection of HDAC inhibitors such as valproic acid, sodium butyrate, or TSA prior to training enhances long-term memory in mice.[124–126] However, HDACs inhibitors have different specificities and their effect on memory depends on which protein they target. For instance, HDAC2 but not HDAC1 impairs memory formation when overexpressed in neurons of adult mice, while it enhances memory when deficient.[127] HDAC2 knockout in adult forebrain neurons accelerates fear memory extinction after fear conditioning or conditioned taste aversion training, but HDAC1 knockout does not.[128] Incidentally, HDAC2 is enriched at the promoter of genes implicated in synaptic plasticity or that are regulated by neuronal activity such as BDNF, Egr1, CaMKIIα, and CREB1.[127] Besides HDAC2, deletion of HDAC3 in CA1 region of the dorsal hippocampus also improves long-term memory, specifically memory for object location for up to 7 days. This effect is in part mediated by increased expression of Nr4a2, a CREB-dependent gene implicated in long-term memory.[129] Overall, these findings are clinically relevant because they suggest the existence of specific epigenetic players that can be targeted pharmacologically and may limit the side effects typically associated with most drugs (for review, see Ref. 130).

In addition to acetylation, histone phosphorylation and methylation are also associated with learning and memory formation. H3S10 and H2K14 phosphorylation is increased in the hippocampus shortly after contextual fear conditioning, an effect that can be reproduced in hippocampal slices by activation of ERK, a protein kinase of signaling pathways downstream of the NMDAR.[131] Regulation of histone phosphorylation also depends on protein phosphatases. Protein phosphatase 1 (PP1), in particular, is a key phosphatase in the brain that controls the level of H3 phosphorylation on S10. When the nuclear pool of PP1 is selectively inhibited in excitatory forebrain neurons, H3S10 phosphorylation is significantly increased in the adult brain.[132] Further, since PP1 can associate with several components of the histone regulatory machinery including HDACs and histone demethylases at the chromatin, its inhibition also increases the acetylation of H2B, H3K14, and H4K5 and alters histone methylation on several specific residues. These combined PTMs are highly relevant for gene expression and affect CREB. They also enhance several forms of memory, and when present in the hippocampus, they improve spatial and object memory, while when present in the amygdala, they improve fear memory.[133–135] Further they contribute to different temporal phases of memory and are dynamically regulated in the hippocampus and cortex. While they first appear in the hippocampus and correlate with short- to long-term memory, they are later induced in the cortex and correlate with remote memory.[136] Such spatial and temporal regulation suggests that PP1 is a key regulator of the histone code in adult neurons in memory formation.[137]

H3K4 trimethylation and H3K9 dimethylation increase in the hippocampus 1 h after contextual fear conditioning in rat, and H3K4 trimethylation is present at the promoter of Zif-268 and BDNF genes.[138]

In the entorhinal cortex, a brain region where memory is processed for long-term storage, these PTMs are sequentially regulated. While both increase 1 h after training, H3K9 dimethylation is back to baseline after 24 h while H3K4 trimethylation is significantly decreased. Infusion of the G9a/GLP histone lysine dimethyltransferase complex inhibitor into the hippocampus 1 h before contextual fear conditioning impairs long-term memory while a similar treatment in entorhinal cortex enhances contextual fear memory for up to 7 days.[139] Consistent with enhanced methylation after behavioral training, conditional deletion of the histone methyltransferase MLL2 in adult excitatory forebrain neurons severely impairs long-term object, contextual fear, and spatial memory. MLL2 knockout is linked to downregulation of several genes involved in neuronal plasticity, specifically in the dentate gyrus.[140] These results demonstrate that epigenetic marks underlying learning and memory formation are dynamic and region specific.

Finally, poly(ADP)-ribosylation and poly(ADP)-ribose polymerase 1 (PARP-1), an enzyme that catalyzes this PTM, have been implicated in memory formation and in changes in synaptic plasticity underlying memory stabilization. Poly(ADP)-ribosylation increases on H1 in Aplysia neurons following LTF and in hippocampus and perirhinal cortex in mice trained on a novel object recognition task.[141] In mice, this is accompanied by a decrease in H1 expression after training. Consistently, intracerebroventricular injection of a PARP-1 inhibitor before training lowers H1 poly(ADP)-ribosylation and impairs long-term object memory and passive avoidance.[142,143] It also blocks LTP in the hippocampus. The decrease in H1 is linked to transcriptional activation and correlates with lower amount of H1 at the promoter of CREB target genes such as Egr-1, c-Jun, c-Fos, and i-Nos in the hippocampus. The resulting increase in the expression of these genes is consistent with the fact that H1 release from the chromatin is necessary for transcriptional activation and may be mediated by poly[ADP]-ribosylation.[144]

4.3. Epigenetic changes and the persistence and dynamics of memory

It was initially postulated that epigenetic marks in the nervous system, particularly DNA methylation, serve as stable molecular signatures of long-term memory.[145,146] However, experimental work has shown that most epigenetic marks induced by learning are not stable but are transiently regulated. Manipulating some of these marks during or after learning can alter the fate of memory traces. When mice are trained to recognize objects in just one session of 3 min, they do not form long-term memory for these objects. However, when this weak paradigm is combined with injection of an HDAC inhibitor immediately after training, long-term memory is formed and persists for several days.[129,147,148] Likewise, training on a socially transmitted food preference paradigm, which recruits the hippocampus and orbitofrontal cortex (OFC), combined with HDAC inhibitors in OFC immediately after learning, improves remote memory for up to 30 days. The treatment is, however, ineffective when applied later (i.e., 15 days after learning).[149] Further in fear extinction, while short reexposure (3 min) to a context after fear conditioning leads to poor extinction, combining it with systemic or intrahippocampal injection of HDAC inhibitors makes fear memory extinction as strong as with long reexposure (24 h) alone.[150–152] This is accompanied by H4 acetylation in PFC.[153] Paradoxically, however, infusion of p300/CBP inhibitors into infralimbic PFC shortly after fear extinction training also favors extinction but not if the inhibitors are administered 6 h after training or even during the initial step of conditioning. These results suggest that epigenetic marks involving acetylation operate during an early phase of memory consolidation and strongly influence memory persistence.[154]

The permanent storage of memory in the mammalian brain depends on the transfer of information from the hippocampus to the cortex. This involves memory consolidation, a process that allows memory traces to be strengthened and stored in the cortex in a way to become independent from the hippocampus.[155] Epigenetic changes play an important role in this process. During consolidation, DNA methylation appears on the promoter of the memory suppressor gene calcineurin in PFC, but only starting 1 day after training and not immediately like in the hippocampus. Consistently, inhibiting DNA methylation by DNMTs in PFC 30 days after learning impairs remote memory but has no effect if it occurs 1 day after training,[156] in line with the notion that memory consolidation is progressive. Likewise for histone PTMs, while rapidly activated after learning in the hippocampus, H3K4 phosphorylation, H3K14 acetylation, and H3K36 trimethylation increase only after 24 h in PFC. Further, they persist much longer than in the hippocampus and are still prominent 7 days after learning.[136] Such persistence correlates with strong memory at this time point, suggesting that the degree and possibly the extent of these PTMs may determine how well a memory trace is consolidated. When primed, by pharmacological (i.e., HDAC inhibitors), genetic (transgenic expression of a histone modifying enzyme or regulator) manipulation, epigenetic marks likely prompt preactivated gene expression programs, and act as an "epigenetic priming" signal.[157] Thus, only epigenetic activation within a certain time window can produce memory enhancement.[147,149,158,159]

Further to pharmacological and genetic manipulations, environmental conditions also influence brain plasticity and memory performance and implicate epigenetic mechanisms.[160,161] Similar to HDAC inhibition, voluntary exercise prior to learning favors long-lasting memory, even when training is minimal.[162,163] This implicates BDNF upregulation and epigenetic changes including H4K8 hyperacetylation at BDNF I and BDNF IV promoters, a global increase in H3 and H4 acetylation, and hypomethylation at BDNF IV promoter. These changes are accompanied by lower level of HDACs 5–8 and DNMTs 1, 3a, and 3b.[163–166] Likewise, environmental enrichment increases the level of BDNF and modulates H3K4, H3K9, and H3K27 trimethylation,[167] possibly through physical activity provided by enriched conditions.[162] Thus, large and dynamic epigenetic programs operate during learning and memory formation and determine the strength and persistence of memory traces. How individual modifications interact with each other and influence specific transcriptional programs for different aspect and phases of memory remain, however, to be elucidated.

Finally like in rodents, memory in insects like honeybees, also engages epigenetic mechanisms. In bees exposed to an appetitive Pavlovian olfactory discrimination task, where an odorant (neutral stimuli) is paired with a reward (sucrose), treatment with DNMT inhibitor alters discrimination when bees are reexposed to the conditioned odorant or a new odorant after training (discrimination task), but does not affect memory after initial conditioning. This impairment is observed 1 day after training, implying that DNA methylation mediates some aspects of long-term associative memory.[168,169]

5. EPIGENETICS AND COGNITIVE DYSFUNCTIONS

5.1. Age-associated cognitive decline

Cognitive decline is a normal aging process that affects a substantial portion of the aging population in human.[170–172] Likewise, cognitive alterations affect rodents during aging.[173–176] Cognitive aging is paralleled by substantial transcriptional reprogramming across the body in humans and rodents. In the human brain, transcriptional profiling across age reveals a decline in the expression of a set of genes in cerebral cortex, which starts after age 40. Most of these genes are important for synaptic plasticity, vesicular transport, and mitochondrial functions.[177–180] Likewise, in aged rodents, the expression of multiple genes essential for signaling, energy metabolism, and synaptic plasticity is altered. These age-related impairments partly derive from epigenetic dysregulation. In rodents, they have been associated with increased DNA damage at the promoter of the altered genes in the hippocampus.[175,178,181–183] This leads to the silencing of genes in the affected genomic region rather than apoptosis, possibly by the recruitment of epigenetic factors such as the HDAC SIRT1 at least in postmitotic neurons.[184,185]

Indeed, several epigenetic processes contribute to gene dysregulation in the aged brain. In the human and rat cerebral cortex during aging, methylation of several genes is increased, for instance at the promoter and intragenic regions of the immediate early gene Arc, or in Gabra5, Hspa5, and Syn1 genes.[164,183,186,47] This correlates with reduced gene expression and memory deficits. Likewise, histone PTMs are also altered in the aged brain. H4K12 acetylation at plasticity genes such as Prkca, Shank3, and Gsk3a is reduced in the aged brain compared to the young brain, and these genes are not differentially regulated following contextual fear conditioning unlike in young animals.[147,176] This dysregulation is reversed and H4K12 acetylation is restored by intra-hippocampal injection of SAHA before training, and they are associated with memory improvement. These findings indicate that age-related cognitive dysfunctions and epigenetic alterations are causally correlated, providing potential perspectives for the treatment of cognitive dysfunctions.

5.2. Epigenetics in the context of neurodegeneration-related cognitive decline

Neurodegenerative disorders such as AD, Parkinson´s disease (PD), and Huntington´s disease (HD) are characterized by progressive loss of neurons and to lead to cognitive decline. Genome-wide association studies examining the genetic basis of these disorders have not identified any specific marker, but have led to the recognition that several genes likely contribute, each a small part, and together with environmental factors, modulate the etiology of these diseases through complex interactions. The environment in early life, in particular, has a strong influence and has been proposed to mediate a latent early life-associated mode of regulation. Thus, environmental factors not only induce immediate but also delayed alterations in gene expression by modulating the epigenome. For delayed alterations, a secondary trigger following a delay (or latent period) is likely at play to perpetuate the effects of early stress later in life.[187] Epigenetic processes can provide such relay and explain the long-term effects of environmental factors on the genome that may ultimately contribute to the pathogenesis of neurodegenerative disorders in old age.[188]

5.2.1 Alzheimer's disease

AD is a common neurodegenerative disease and one the most common forms of dementia that affects more than 6% of people over 65.[189,190] The pathophysiology of AD is mainly characterized by a loss of neurons and synapses, and the deposition of neuritic plaques and neurofibrillary tangles in different brain regions.[50,190] Neurofibrillary tangles are composed of hyperphosphorylated tau protein,[191] and extracellular plaques contain amyloid-β fibrils. Plaques originate from the endoproteolysis of APP by β- and γ-secretases into different cleavage products including Aβ42. AD is associated with increased Aβ42 production and its accumulation and aggregation.[192] Aβ42 accumulation is caused by reduced amyloid degradation.[193] The biological functions of APP are not well understood but the protein is known to have a wide range of interaction partners that, like APP, can aggregate in plaques. Plaques and tangles have toxic effects on neurons and their synapses. They interfere with neuronal and synaptic functions and are thought to partly underlie the cognitive impairments associated with AD.

Gene variants have been described as predisposing factors of early forms of familial AD that comprise APP and presenilin (PS) 1 and 2, while variants of apolipoprotein (Apo) E4 are linked to late onset AD.[194] Besides these predisposing gene variants, environmental factors also play a significant role in the disease pathophysiology. This is supported by the high discordance rate of AD in monozygotic twins, and the fact that genetic risk factors diminish with age while environmental factors increase.[195–197] Environmental factors, such as exposure to metals, traumatic brain injury, and early life stress constitute a risk for AD and are associated with an ensemble of epigenetic alterations affecting DNA methylation, DNMTs expression, and histone PTMs.[14,187,198,199] A recent *postmortem* study in humans detected global DNA hypomethylation in the entorhinal cortex of AD patients when compared to age-matched controls.[200] However, studies on the methylation status of selected target genes, such as PS1, have reported hypermethylation in the dorsolateral PFC of some AD patients,[183,201–203] suggesting a region- and locus-specific DNA methylation pattern in AD. Altered DNMT1 expression or activity has been proposed as a possible mechanism for deregulated DNA methylation in AD. Consistently, DNMT1 activity is decreased and APP mRNA expression is increased in cortex in a primate model of AD.[204] However, APP itself might also be involved in the dysregulation of DNA methylation.[205] Effective clearance of Aβ requires crosslinking of the peptide by ApoE. ApoE4, an isoform of ApoE associated with AD, has reduced ability to crosslink Aβ.[206] This results in elevated Aβ concentrations leading to hypermethylation of the neprilysin gene that codes for the major Aβ-degrading enzyme in the brain.[207] Similarly, high concentration of Aβ applied to murine cerebral endothelial cells causes hypermethylation of the neprilysin gene. Further, neprilysin concentration is decreased in the hippocampus and midtemporal gyrus of AD patients.[208]

AD is also associated with an overall increase in histone acetylation. The mechanisms underlying such hyperacetylation are not known but may involve an AD-associated increase in APP C-terminal peptide (AICD), an APP cleavage product. AICD can interact with the HAT TIP60, directly or via a ligand, and lead to increased acetylation and transcriptional activation.[209] They may also involve decreased proteasome activity as AD-related mutations in PS1 inhibit proteasomal activity, leading to increased HAT CBP and CREB-mediated gene expression.[210] Lowering the level of acetylation by lentivirus-mediated overexpression of the HDAC SIRT1 provides neuroprotection in a mouse model of AD.[211] Likewise, in a mouse model of forebrain-specific neurodegeneration, increased SIRT1 activity resulting from caloric restriction diminishes acetylation, in particular at H3K56, in hippocampal CA1 and correlates with a correction of memory impairment in a cued fear conditioning task.[212]

However, histone acetylation has also been reported to be decreased in AD. In cultured cortical neurons, APP overexpression lowers H3 and H4 overall acetylation, and decreases CBP level.[213] Further in an APP/PS1 mouse model of AD, H4 acetylation is reduced in the hippocampus after fear conditioning, and the decrease is prevented by acute treatment with TSA.[214] Intracerebroventricular injection of sodium butyrate has been shown to reverse memory and plasticity deficits in a mouse model of AD.[174] Such reversal can also be obtained by environmental enrichment in old wild-type mice and is associated with increased H3K4 acetylation and methylation in hippocampus and cortex. Further, HDAC2 has been reported to be higher in the brain of AD patients and mouse models of AD and contributed to altered histone acetylation and expression of genes important for learning and memory associated with cognitive impairment. Inhibition of HDAC2 normalizes acetylation and can temporarily restore cognitive functions in mouse models.[215] Thus overall, bidirectional changes in histone acetylation in AD suggest a complex and gene-specific dysregulation of this epigenetic mark associated with the disease. But surprisingly, attempts targeting both global hyper- and hypoacetylation rescue cognitive impairments. Overall, the data available thus far support a primary involvement of DNA methylation and histone PTMs in AD pathophysiology. The precise role of these epigenetic modifications and their cross talk, however, still need to be better studied. Several pharmacological studies have revealed the promising potential of epigenetic therapies to reverse AD-associated cognitive decline. Cognitive decline in PD[216] and HD[217] may also benefit from epigenetic treatment as it has also been associated with altered DNA methylation, histone PTMs, and miRNAs,[218,219] but whether epigenetic dysregulation is a direct player in the cognitive pathology remains to be investigation.

6. CONCLUSIONS

Epigenetics is currently a subject of intense study in many disciplines including cancer research, immunology, and neuroscience. The underlying mechanisms are beginning to be clarified but a better understanding of how they exert control over the genome and how they are involved in health and disease still requires much research. In the brain, some epigenetic marks have been implicated in synaptic plasticity, and in complex brain functions such as learning and memory formation. These marks interact with each other to form a complex epigenetic code that bidirectionally affect gene expression, depending on the context and conditions of activation. A full understanding of how epigenetic mechanisms regulate plasticity and memory formation will require decoding the ensemble of epigenetic marks, and the language of their cross talk.[220] In this respect, high-throughput approaches will be instrumental to map global DNA modifications and histone PTMs, and computational modeling to determine the rules governing epigenetic cross talk.[221,222] Because the epigenome is very dynamic, it needs to be charted on multiple maps in different conditions to be identified in its entirety. And further to the DNA and histone code, the contribution of ncRNAs in epigenetic regulation is another important aspect that needs to be examined. How ncRNAs such as miRNAs affect the genome and its activity remain partly unresolved.[49,51,223] Despite these current limitations, drugs targeting epigenetic processes hold great promise in the clinic as potential cognitive enhancers. The beneficial effect of epigenetic drugs on cognition has been documented, but their use requires care due to nonspecific and secondary effects.[157,174] Improvements in the capacity to target and manipulate epigenetic marks in a selective manner are expected to accelerate their potential use in the clinics. As often in science, further technical advances will be required to provide answers to some of the currently most challenging questions in the field. Ultimately, a better understanding of epigenetic processes that govern health and disease during the life span and across generations will open many new novel perspectives for therapeutics.

ACKNOWLEDGMENTS

The lab of I. M. Mansuy is funded by the University Zürich, the Swiss Federal Institute of Technology, the Swiss National Science Foundation, and Roche.

REFERENCES

1. Kandel ER. The molecular biology of memory storage: a dialogue between genes and synapses. *Science*. 2001;294:1030–1038.
2. Levenson J, Sweatt J. Memory. *Cell Mol Life Sci*. 2006;63:1009–1016.
3. Liu Y. Like father like son. *EMBO Rep*. 2007;8:798–803.
4. Wolffe AP, Matzke MA. Epigenetics: regulation through repression. *Science*. 1999;286:481–486.
5. Goldberg AD, Allis CD, Bernstein E. Epigenetics: a landscape takes shape. *Cell*. 2007;128:635–638.
6. Klose RJ, Bird AP. Genomic DNA methylation: the mark and its mediators. *Trends Biochem Sci*. 2006;31:89–97.
7. Antequera F. Structure, function and evolution of CpG island promoters. *Cell Mol Life Sci*. 2003;60:1647–1658.
8. Ioshikhes IP, Zhang MQ. Large-scale human promoter mapping using CpG islands. *Nat Genet*. 2000;26:61–63.
9. Kim J, Samaranayake M, Pradhan S. Epigenetic mechanisms in mammals. *Cell Mol Life Sci*. 2009;66:596–612.
10. Tost J. DNA methylation: an introduction to the biology and the disease-associated changes of a promising biomarker. *Methods Mol Biol*. 2009;507:3–20.
11. Bird A. DNA methylation patterns and epigenetic memory. *Genes Dev*. 2002;16:6–21.
12. Esteller M. Cancer epigenomics: DNA methylomes and histone-modification maps. *Nat Rev Genet*. 2007;8:286–298.
13. Jones PA, Takai D. The role of DNA methylation in mammalian epigenetics. *Science*. 2001;293:1068–1070.
14. Ulrey CL, Liu L, Andrews LG, Tollefsbol TO. The impact of metabolism on DNA methylation. *Hum Mol Genet*. 2005;14:R139–R147.
15. Gräff J, Mansuy IM. Epigenetic codes in cognition and behaviour. *Behav Brain Res*. 2008;192:70–87.
16. Harper LV. Epigenetic inheritance and the intergenerational transfer of experience. *Psychol Bull*. 2005;131:340–360.
17. Ooi SKT, Bestor TH. The colorful history of active DNA demethylation. *Cell*. 2008;133:1145–1148.
18. Morgan HD, Ft Santos, Green K, Dean W, Reik W. Epigenetic reprogramming in mammals. *Hum Mol Genet*. 2005;14:R47–R58.
19. Kriaucionis S, Heintz N. The nuclear DNA base 5-hydroxymethylcytosine is present in Purkinje neurons and the brain. *Science*. 2009;324:929–930.

20. Tahiliani M, Koh KP, Shen Y, et al. Conversion of 5-methylcytosine to 5-hydroxymethylcytosine in mammalian DNA by MLL partner TET1. *Science.* 2009;324:930–935.
21. Branco MR, Ficz G, Reik W. Uncovering the role of 5-hydroxymethylcytosine in the epigenome. *Nat Rev Genet.* 2012;13:7–13.
22. Maunakea AK, Nagarajan RP, Bilenky M, et al. Conserved role of intragenic DNA methylation in regulating alternative promoters. *Nature.* 2010;466:253–257.
23. Weber M, Hellmann I, Stadler MB, et al. Distribution, silencing potential and evolutionary impact of promoter DNA methylation in the human genome. *Nat Genet.* 2007;39:457–466.
24. Mellen M, Ayata P, Dewell S, Kriaucionis S, Heintz N. MeCP2 binds to 5hmC enriched within active genes and accessible chromatin in the nervous system. *Cell.* 2012;151:1417–1430.
25. Song CX, Szulwach KE, Fu Y, et al. Selective chemical labeling reveals the genome-wide distribution of 5-hydroxymethylcytosine. *Nat Biotechnol.* 2011;29:68–72.
26. Khare T, Pai S, Koncevicius K, et al. 5-hmC in the brain is abundant in synaptic genes and shows differences at the exon-intron boundary. *Nat Struct Mol Biol.* 2012;19:1037–1043.
27. Kouzarides T. Chromatin modifications and their function. *Cell.* 2007;128:693–705.
28. Reinke H, Horz W. Histones are first hyperacetylated and then lose contact with the activated PHO5 promoter. *Mol Cell.* 2003;11:1599–1607.
29. Zhao J, Herrera-Diaz J, Gross DS. Domain-wide displacement of histones by activated heat shock factor occurs independently of Swi/Snf and is not correlated with RNA polymerase II density. *Mol Cell Biol.* 2005;25:8985–8999.
30. Bernstein BE, Liu CL, Humphrey EL, Perlstein EO, Schreiber SL. Global nucleosome occupancy in yeast. *Genome Biol.* 2004;5:R62.
31. Lee CK, Shibata Y, Rao B, Strahl BD, Lieb JD. Evidence for nucleosome depletion at active regulatory regions genome-wide. *Nat Genet.* 2004;36:900–905.
32. Sekinger EA, Moqtaderi Z, Struhl K. Intrinsic histone-DNA interactions and low nucleosome density are important for preferential accessibility of promoter regions in yeast. *Mol Cell.* 2005;18:735–748.
33. Jenuwein T, Allis CD. Translating the histone code. *Science.* 2001;293:1074–1080.
34. Strahl BD, Allis CD. The language of covalent histone modifications. *Nature.* 2000;403:41–45.
35. Young NL, Dimaggio PA, Garcia BA. The significance, development and progress of high-throughput combinatorial histone code analysis. *Cell Mol Life Sci.* 2010;67:3983–4000.
36. Landry J, Sutton A, Hesman T, et al. Set2-catalyzed methylation of histone H3 represses basal expression of GAL4 in Saccharomyces cerevisiae. *Mol Cell Biol.* 2003;23:5972–5978.
37. Maze I, Noh KM, Allis CD. Histone regulation in the CNS: basic principles of epigenetic plasticity. *Neuropsychopharmacology.* 2013;38:3–22.
38. Krol J, Loedige I, Filipowicz W. The widespread regulation of microRNA biogenesis, function and decay. *Nat Rev Genet.* 2010;11:597–610.
39. Moazed D. Small RNAs in transcriptional gene silencing and genome defence. *Nature.* 2009;457:413–420.
40. Chen JF, Mandel EM, Thomson JM, et al. The role of microRNA-1 and microRNA-133 in skeletal muscle proliferation and differentiation. *Nat Genet.* 2006;38:228–233.
41. Tuddenham L, Wheeler G, Ntounia-Fousara S, et al. The cartilage specific microRNA-140 targets histone deacetylase 4 in mouse cells. *FEBS Lett.* 2006;580:4214–4217.
42. Sinkkonen L, Hugenschmidt T, Berninger P, et al. MicroRNAs control de novo DNA methylation through regulation of transcriptional repressors in mouse embryonic stem cells. *Nat Struct Mol Biol.* 2008;15:259–267.

43. Fabbri M, Garzon R, Cimmino A, et al. MicroRNA-29 family reverts aberrant methylation in lung cancer by targeting DNA methyltransferases 3A and 3B. *Proc Natl Acad Sci USA*. 2007;104:15805–15810.
44. Zhang P, Huang B, Xu X, Sessa WC. Ten-eleven translocation (Tet) and thymine DNA glycosylase (TDG), components of the demethylation pathway, are direct targets of miRNA-29a. *Biochem Biophys Res Commun*. 2013;437:368–373.
45. Fukagawa T, Nogami M, Yoshikawa M, et al. Dicer is essential for formation of the heterochromatin structure in vertebrate cells. *Nat Cell Biol*. 2004;6:784–791.
46. Bao N, Lye KW, Barton MK. MicroRNA binding sites in Arabidopsis class III HD-ZIP mRNAs are required for methylation of the template chromosome. *Dev Cell*. 2004;7:653–662.
47. Khraiwesh B, Arif MA, Seumel GI, et al. Transcriptional control of gene expression by microRNAs. *Cell*. 2010;140:111–122.
48. Kim DH, Saetrom P, Snøve Jr O, Rossi JJ. MicroRNA-directed transcriptional gene silencing in mammalian cells. *Proc Natl Acad Sci USA*. 2008;105:16230–16235.
49. Volpe T, Martienssen RA. RNA interference and heterochromatin assembly. *Cold Spring Harb Perspect Biol*. 2011;3:a003731.
50. Huang HC, Jiang ZF. Accumulated amyloid-beta peptide and hyperphosphorylated tau protein: relationship and links in Alzheimer's disease. *J Alzheimers Dis*. 2009;16:15–27.
51. Peng JC, Lin H. Beyond transposons: the epigenetic and somatic functions of the Piwi-piRNA mechanism. *Curr Opin Cell Biol*. 2013;25:190–194.
52. Hyman SE, Malenka RC, Nestler EJ. Neural mechanisms of addiction: the role of reward-related learning and memory. *Annu Rev Neurosci*. 2006;29:565–598.
53. Kauer JA, Malenka RC. Synaptic plasticity and addiction. *Nat Rev Neurosci*. 2007;8:844–858.
54. Russo SJ, Dietz DM, Dumitriu D, Morrison JH, Malenka RC, Nestler EJ. The addicted synapse: mechanisms of synaptic and structural plasticity in nucleus accumbens. *Trends Neurosci*. 2010;33:267–276.
55. Volkow ND, Li TK. Drug addiction: the neurobiology of behaviour gone awry. *Nat Rev Neurosci*. 2004;5:963–970.
56. Nestler EJ. Molecular basis of long-term plasticity underlying addiction. *Nat Rev Neurosci*. 2001;2:119–128.
57. Kumar A, Choi KH, Renthal W, et al. Chromatin remodeling is a key mechanism underlying cocaine-induced plasticity in striatum. *Neuron*. 2005;48:303–314.
58. LaPlant Q, Vialou V, Covington 3rd HE, et al. Dnmt3a regulates emotional behavior and spine plasticity in the nucleus accumbens. *Nat Neurosci*. 2010;13:1137–1143.
59. Brami-Cherrier K, Valjent E, Herve D, et al. Parsing molecular and behavioral effects of cocaine in mitogen- and stress-activated protein kinase-1-deficient mice. *J Neurosci*. 2005;25:11444–11454.
60. Renthal W, Kumar A, Xiao G, et al. Genome-wide analysis of chromatin regulation by cocaine reveals a role for sirtuins. *Neuron*. 2009;62:335–348.
61. Renthal W, Maze I, Krishnan V, et al. Histone deacetylase 5 epigenetically controls behavioral adaptations to chronic emotional stimuli. *Neuron*. 2007;56:517–529.
62. Maze I, Covington 3rd HE, Dietz DM, et al. Essential role of the histone methyltransferase G9a in cocaine-induced plasticity. *Science*. 2010;327:213–216.
63. Vassoler FM, White SL, Schmidt HD, Sadri-Vakili G, Pierce RC. Epigenetic inheritance of a cocaine-resistance phenotype. *Nat Neurosci*. 2013;16:42–47.
64. Im HI, Hollander JA, Bali P, Kenny PJ. MeCP2 controls BDNF expression and cocaine intake through homeostatic interactions with microRNA-212. *Nat Neurosci*. 2010;13:1120–1127.

65. Pol Bodetto S, Carouge D, Fonteneau M, Dietrich JB, Zwiller J, Anglard P. Cocaine represses protein phosphatase-1Cbeta through DNA methylation and methyl-CpG binding protein-2 recruitment in adult rat brain. *Neuropharmacology*. 2013;73C:31–40.
66. de Kloet ER, Joels M, Holsboer F. Stress and the brain: from adaptation to disease. *Nat Rev Neurosci*. 2005;6:463–475.
67. McEwen BS. Physiology and neurobiology of stress and adaptation: central role of the brain. *Physiol Rev*. 2007;87:873–904.
68. Sapolsky RM, Romero LM, Munck AU. How do glucocorticoids influence stress responses? Integrating permissive, suppressive, stimulatory, and preparative actions. *Endocr Rev*. 2000;21:55–89.
69. Heim C, Nemeroff CB. The role of childhood trauma in the neurobiology of mood and anxiety disorders: preclinical and clinical studies. *Biol Psychiatry*. 2001;49:1023–1039.
70. McEwen BS. Understanding the potency of stressful early life experiences on brain and body function. *Metabolism*. 2008;57:S11–S15.
71. Holmes A, le Guisquet AM, Vogel E, Millstein RA, Leman S, Belzung C. Early life genetic, epigenetic and environmental factors shaping emotionality in rodents. *Neurosci Biobehav Rev*. 2005;29:1335–1346.
72. Weaver IC. Epigenetic programming by maternal behavior and pharmacological intervention. Nature versus nurture: let's call the whole thing off. *Epigenetics*. 2007;2:22–28.
73. Franklin TB, Mansuy IM. Epigenetic inheritance in mammals: evidence for the impact of adverse environmental effects. *Neurobiol Dis*. 2010;39:61–65.
74. Franklin TB, Russig H, Weiss IC, et al. Epigenetic transmission of the impact of early stress across generations. *Biol Psychiatry*. 2010;68:408–415.
75. Murgatroyd C, Wu Y, Bockmühl Y, Spengler D. Genes learn from stress: how infantile trauma programs us for depression. *Epigenetics*. 2010;5:194–199.
76. Champagne FA, Francis DD, Mar A, Meaney MJ. Variations in maternal care in the rat as a mediating influence for the effects of environment on development. *Physiol Behav*. 2003;79:359–371.
77. Champagne FA. Epigenetic mechanisms and the transgenerational effects of maternal care. *Front Neuroendocrinol*. 2008;29:386–397.
78. Meaney MJ. Maternal care, gene expression, and the transmission of individual differences in stress reactivity across generations. *Annu Rev Neurosci*. 2001;24:1161–1192.
79. Weaver IC, D'Alessio AC, Brown SE, et al. The transcription factor nerve growth factor-inducible protein a mediates epigenetic programming: altering epigenetic marks by immediate-early genes. *J Neurosci*. 2007;27:1756–1768.
80. Weaver IC, Cervoni N, Champagne FA, et al. Epigenetic programming by maternal behavior. *Nat Neurosci*. 2004;7:847–854.
81. Murgatroyd C, Patchev AV, Wu Y, et al. Dynamic DNA methylation programs persistent adverse effects of early-life stress. *Nat Neurosci*. 2009;12:1559–1566.
82. Roth TL, Lubin FD, Funk AJ, Sweatt JD. Lasting epigenetic influence of early-life adversity on the BDNF gene. *Biol Psychiatry*. 2009;65:760–769.
83. Franklin TB, Linder N, Russig H, Thöny B, Mansuy IM. Influence of early stress on social abilities and serotonergic functions across generations in mice. *PLoS One*. 2011;6:e21842.
84. Weiss IC, Franklin TB, Vizi S, Mansuy IM. Inheritable effect of unpredictable maternal separation on behavioral responses in mice. *Front Behav Neurosci*. 2011;5(3):1–11.
85. Bohacek J, Gapp K, Saab BJ, Mansuy IM. Transgenerational epigenetic effects on brain functions. *Biol Psychiatry*. 2013;73:313–320.
86. Bohacek J, Mansuy IM. Epigenetic inheritance of disease and disease risk. *Neuropsychopharmacology*. 2013;38:220–236.
87. Daxinger L, Whitelaw E. Understanding transgenerational epigenetic inheritance via the gametes in mammals. *Nat Rev Genet*. 2012;13:153–162.

88. Guerrero-Bosagna C, Skinner MK. Environmentally induced epigenetic transgenerational inheritance of phenotype and disease. *Mol Cell Endocrinol.* 2011;354(1–2):3–8.
89. Labonte B, Suderman M, Maussion G, et al. Genome-wide epigenetic regulation by early-life trauma. *Arch Gen Psychiatry.* 2012;69:722–731.
90. Suderman M, McGowan PO, Sasaki A, et al. Conserved epigenetic sensitivity to early life experience in the rat and human hippocampus. *Proc Natl Acad Sci USA.* 2012;109-(Suppl. 2):17266–17272.
91. Bliss TV, Lomo T. Long-lasting potentiation of synaptic transmission in the dentate area of the anaesthetized rabbit following stimulation of the perforant path. *J Physiol.* 1973;232:331–356.
92. Lynch MA. Long-term potentiation and memory. *Physiol Rev.* 2004;84:87–136.
93. Kandel ER. The molecular biology of memory storage: a dialogue between genes and synapses. *Science (New York, NY).* 2001;294:1030–1038.
94. Martin SJ, Grimwood PD, Morris RG. Synaptic plasticity and memory: an evaluation of the hypothesis. *Annu Rev Neurosci.* 2000;23:649–711.
95. Thompson RF. In search of memory traces. *Annu Rev Psychol.* 2005;56:1–23.
96. Caroni P, Donato F, Muller D. Structural plasticity upon learning: regulation and functions. *Nat Rev Neurosci.* 2012;13:478–490.
97. Dudai Y. The restless engram: consolidations never end. *Annu Rev Neurosci.* 2012;35:227–247.
98. Redondo RL, Morris RG. Making memories last: the synaptic tagging and capture hypothesis. *Nat Rev Neurosci.* 2011;12:17–30.
99. Miller CA, Sweatt JD. Covalent modification of DNA regulates memory formation. *Neuron.* 2007;53:857–869.
100. Feng J, Zhou Y, Campbell SL, et al. Dnmt1 and Dnmt3a maintain DNA methylation and regulate synaptic function in adult forebrain neurons. *Nat Neurosci.* 2010;13: 423–430.
101. Levenson JM, Roth TL, Lubin FD, et al. Evidence that DNA (cytosine-5) methyltransferase regulates synaptic plasticity in the hippocampus. *J Biol Chem.* 2006;281:15763–15773.
102. Mizuno K, Dempster E, Mill J, Giese KP. Long-lasting regulation of hippocampal Bdnf gene transcription after contextual fear conditioning. *Genes Brain Behav.* 2012;11:651–659.
103. Monsey MS, Ota KT, Akingbade IF, Hong ES, Schafe GE. Epigenetic alterations are critical for fear memory consolidation and synaptic plasticity in the lateral amygdala. *PLoS One.* 2011;6:e19958.
104. Oliveira AM, Hemstedt TJ, Bading H. Rescue of aging-associated decline in Dnmt3a2 expression restores cognitive abilities. *Nat Neurosci.* 2012;15:1111–1113.
105. Sui L, Wang Y, Ju L-H, Chen M. Epigenetic regulation of reelin and brain-derived neurotrophic factor genes in long-term potentiation in rat medial prefrontal cortex. *Neurobiol Learn Mem.* 2012;97:425–440.
106. Martinowich K, Hattori D, Wu H, et al. DNA methylation-related chromatin remodeling in activity-dependent BDNF gene regulation. *Science.* 2003;302:890–893.
107. Miller CA, Campbell SL, Sweatt JD. DNA methylation and histone acetylation work in concert to regulate memory formation and synaptic plasticity. *Neurobiol Learn Mem.* 2008;89:599–603.
108. Rajasethupathy P, Antonov I, Sheridan R, et al. A role for neuronal piRNAs in the epigenetic control of memory-related synaptic plasticity. *Cell.* 2012;149:693–707.
109. Metivier R, Gallais R, Tiffoche C, et al. Cyclical DNA methylation of a transcriptionally active promoter. *Nature.* 2008;452:45–50.
110. Ma DK, Jang MH, Guo JU, et al. Neuronal activity-induced Gadd45b promotes epigenetic DNA demethylation and adult neurogenesis. *Science.* 2009;323:1074–1077.

111. Sultan FA, Wang J, Tront J, Da Liebermann, Sweatt JD. Genetic deletion of Gadd45b, a regulator of active DNA demethylation, enhances long-term memory and synaptic plasticity. *J Neurosci.* 2012;32:17059–17066.
112. Guo JU, Su Y, Zhong C, G-l Ming, Song H. Hydroxylation of 5-methylcytosine by TET1 promotes active DNA demethylation in the adult brain. *Cell.* 2011;145: 423–434.
113. Zhang R-R, Cui Q-Y, Murai K, et al. Tet1 regulates adult Hippocampal Neurogenesis and Cognition. *Cell Stem Cell.* 2013;13:237–245.
114. Bannister AJ, Kouzarides T. The CBP co-activator is a histone acetyltransferase. *Nature.* 1996;384:641–643.
115. Impey S, Fong AL, Wang Y, et al. Phosphorylation of CBP mediates transcriptional activation by neural activity and CaM kinase IV. *Neuron.* 2002;34:235–244.
116. Wood MA, Attner MA, Oliveira AM, Brindle PK, Abel T. A transcription factor-binding domain of the coactivator CBP is essential for long-term memory and the expression of specific target genes. *Learn Mem.* 2006;13:609–617.
117. Janknecht R. The versatile functions of the transcriptional coactivators p300 and CBP and their roles in disease. *Histol Histopathol.* 2002;17:657–668.
118. Alarco JM, Touzani K, Vronskaya S, Ishii S, Kandel ER, Barco A. Chromatin acetylation, memory, and LTP are impaired in CBP / X mice: a model for the cognitive deficit in Rubinstein-Taybi syndrome and its amelioration. *Neuron.* 2004;42:947–959.
119. Korzus E, Rosenfeld MG, Mayford M. CBP histone acetyltransferase activity is a critical component of memory consolidation. *Neuron.* 2004;42:961–972.
120. Chatterjee S, Mizar P, Cassel R, et al. A novel activator of CBP/p300 acetyltransferases promotes neurogenesis and extends memory duration in adult mice. *J Neurosci.* 2013;33:10698–10712.
121. Oliveira AMM, Ma Wood, McDonough CB, Abel T. Transgenic mice expressing an inhibitory truncated form of p300 exhibit long-term memory deficits. *Learn Mem.* 2007;14:564–572.
122. Levenson JM, O'Riordan KJ, Brown KD, Ma Trinh, Molfese DL, Sweatt JD. Regulation of histone acetylation during memory formation in the hippocampus. *J Biol Chem* 2004;279:40545–40559.
123. Bousiges O, Vasconcelos APD, Neidl R, et al. Spatial memory consolidation is associated with induction of several lysine-acetyltransferase (histone acetyltransferase) expression levels and H2B/H4 acetylation-dependent transcriptional events in the rat hippocampus. *Neuropsychopharmacology.* 2010;35:2521–2537.
124. Bredy TW, Barad M. The histone deacetylase inhibitor valproic acid enhances acquisition, extinction, and reconsolidation of conditioned fear. *Learn Mem.* 2008;15:39–45.
125. Haettig J, Stefanko DP, Multani ML, Figueroa DX, McQuown SC, Ma Wood. HDAC inhibition modulates hippocampus-dependent long-term memory for object location in a CBP-dependent manner. *Learn Mem.* 2011;18:71–79.
126. Mahan AL, Mou L, Shah N, Hu J-H, Worley PF, Ressler KJ. Epigenetic modulation of Homer1a transcription regulation in amygdala and hippocampus with pavlovian fear conditioning. *J Neurosci.* 2012;32:4651–4659.
127. Guan J-S, Haggarty SJ, Giacometti E, et al. HDAC2 negatively regulates memory formation and synaptic plasticity. *Nature.* 2009;459:55–60.
128. Morris MJ, Mahgoub M, Na ES, Pranav H, Monteggia LM. Loss of histone deacetylase 2 improves working memory and accelerates extinction learning. *J Neurosci.* 2013;33:6401–6411.
129. McQuown SC, Barrett RM, Matheos DP, et al. HDAC3 is a critical negative regulator of long-term memory formation. *J Neurosci.* 2011;31:764–774.
130. Fischer AE, Sananbenesi F, Mungenast A, Tsai L-H. Targeting the correct HDAC(s) to treat cognitive disorders. *Trends Pharmacol Sci.* 2010;31:605–617.

131. Chwang WB, O'Riordan KJ, Levenson JM, Sweatt JD. ERK/MAPK regulates hippocampal histone phosphorylation following contextual fear conditioning. *Learn Mem.* 2006;13:322–328.
132. Koshibu K, Gra J, Beullens M, et al. Protein phosphatase 1 regulates the histone code for long-term memory. *October.* 2009;29:13079–13089.
133. Graff J, Koshibu K, Jouvenceau A, Dutar P, Mansuy IM. Protein phosphatase 1-dependent transcriptional programs for long-term memory and plasticity. *Learn Mem.* 2010;17:355–363.
134. Koshibu K, Graff J, Beullens M, et al. Protein phosphatase 1 regulates the histone code for long-term memory. *J Neurosci.* 2009;29:13079–13089.
135. Koshibu K, Graff J, Mansuy IM. Nuclear protein phosphatase-1: an epigenetic regulator of fear memory and amygdala long-term potentiation. *Neuroscience.* 2011;173:30–36.
136. Graff J, Woldemichael BT, Berchtold D, Dewarrat G, Mansuy IM. Dynamic histone marks in the hippocampus and cortex facilitate memory consolidation. *Nat Commun.* 2012;3:991.
137. Franklin TB, Mansuy IM. Kinases and phosphatases in the epigenetic regulation of cognitive functions. In: *Epigenomics: From Chromatin Biology to Therapeutics.* Cambridge, UK: Cambridge University Press; 2012, ISBN: 9781139533997.
138. Gupta S, Kim SY, Artis S, et al. Histone methylation regulates memory formation. *J Neurosci.* 2010;30:3589–3599.
139. Gupta-Agarwal S, Franklin AV, Deramus T, et al. G9a/GLP histone lysine dimethyltransferase complex activity in the hippocampus and the entorhinal cortex is required for gene activation and silencing during memory consolidation. *J Neurosci.* 2012;32:5440–5453.
140. Kerimoglu C, Agis-Balboa RC, Kranz A, et al. Histone-methyltransferase MLL2 (KMT2B) is required for memory formation in mice. *J Neurosci.* 2013;33:3452–3464.
141. Cohen-Armon M, Visochek L, Katzoff A, et al. Long-term memory requires polyADP-ribosylation. *Science.* 2004;304:1820–1822.
142. Fontan-Lozano A, Suarez-Pereira I, Horrillo A, del-Pozo-Martin Y, Hmadcha A, Carrion AM. Histone H1 poly[ADP]-ribosylation regulates the chromatin alterations required for learning consolidation. *J Neurosci.* 2010;30:13305–13313.
143. Goldberg S, Visochek L, Giladi E, Gozes I, Cohen-Armon M. PolyADP-ribosylation is required for long-term memory formation in mammals. *J Neurochem.* 2009;111:72–79.
144. Kraus WL, Lis JT. PARP goes transcription. *Cell.* 2003;113:677–683.
145. Crick F. Memory and molecular turnover. *Nature.* 1984;312:101.
146. Levenson JM, Sweatt JD. Epigenetic mechanisms in memory formation. *Nat Rev Neurosci.* 2005;6:108–118.
147. Federman N, de la Fuente V, Zalcman G, et al. Nuclear factor κB-dependent histone acetylation is specifically involved in persistent forms of memory. *J Neurosci.* 2013;33:7603–7614.
148. Stefanko DP, Barrett RM, Ly AR, Reolon GK, Ma Wood. Modulation of long-term memory for object recognition via HDAC inhibition. *Proc Natl Acad Sci USA.* 2009;106:9447–9452.
149. Lesburgueres E, Gobbo OL, Alaux-Cantin S, Hambucken A, Trifilieff P, Bontempi B. Early tagging of cortical networks is required for the formation of enduring associative memory. *Science.* 2011;331:924–928.
150. Lattal KM, Barrett RM, Ma Wood. Systemic or intrahippocampal delivery of histone deacetylase inhibitors facilitates fear extinction. *Behav Neurosci.* 2007;121:1125–1131.
151. Lattal KM, Wood MA. Epigenetics and persistent memory: implications for reconsolidation and silent extinction beyond the zero. *Nat Neurosci.* 2013;16:124–129.
152. Maren S. Seeking a spotless mind: extinction, deconsolidation, and erasure of fear memory. *Neuron.* 2011;70:830–845.

153. Bredy TW, Wu H, Crego C, Zellhoefer J, Sun YE, Barad M. Histone modifications around individual BDNF gene promoters in prefrontal cortex are associated with extinction of conditioned fear. *Learn Mem*. 2007;14:268–276.
154. Marek R, Coelho CM, Sullivan RKP, et al. Paradoxical enhancement of fear extinction memory and synaptic plasticity by inhibition of the histone acetyltransferase p300. *J Neurosci*. 2011;31:7486–7491.
155. Wang SH, Morris RG. Hippocampal-neocortical interactions in memory formation, consolidation, and reconsolidation. *Annu Rev Psychol*. 2010;61(49–79):C1–C4.
156. Miller CA, Gavin CF, White JA, et al. Cortical DNA methylation maintains remote memory. *Nat Neurosci*. 2010;13:664–666.
157. Graff J, Tsai LH. The potential of HDAC inhibitors as cognitive enhancers. *Annu Rev Pharmacol Toxicol*. 2013;53:311–330.
158. Dagnas M, Mons N. Region- and age-specific patterns of histone acetylation related to spatial and cued learning in the water maze. *Hippocampus*. 2013;23:581–591.
159. Lesburguères E, Gobbo OL, Se Alaux-Cantin, Hambucken A, Trifilieff P, Bontempi B. Early tagging of cortical networks is required for the formation of enduring associative memory. *Science*. 2011;331:924–928.
160. Hillman CH, Erickson KI, Kramer AF. Be smart, exercise your heart: exercise effects on brain and cognition. *Nat Rev Neurosci*. 2008;9:58–65.
161. Voss MW, Nagamatsu LS, Liu-Ambrose T, Kramer AF. Exercise, brain, and cognition across the life span. *J Appl Physiol (1985)*. 2011;111:1505–1513.
162. Bechara RG, Kelly AM. Exercise improves object recognition memory and induces BDNF expression and cell proliferation in cognitively enriched rats. *Behav Brain Res*. 2013;245:96–100.
163. Intlekofer KA, Berchtold NC, Malvaez M, et al. Exercise and sodium butyrate transform a subthreshold learning event into long-term memory via a brain-derived neurotrophic factor-dependent mechanism. *Neuropsychopharmacology*. 2013;38:2027–2034.
164. Abel JL, Rissman EF. Running-induced epigenetic and gene expression changes in the adolescent brain. *Int J Dev Neurosci*. 2013;31:382–390.
165. Gomez-Pinilla F, Zhuang Y, Feng J, Ying Z, Fan G. Exercise impacts brain-derived neurotrophic factor plasticity by engaging mechanisms of epigenetic regulation. *Eur J Neurosci*. 2011;33:383–390.
166. Lovatel GA, Elsner VR, Bertoldi K, et al. Treadmill exercise induces age-related changes in aversive memory, neuroinflammatory and epigenetic processes in the rat hippocampus. *Neurobiol Learn Mem*. 2013;101:94–102.
167. Kuzumaki N, Ikegami D, Tamura R, et al. Hippocampal epigenetic modification at the brain-derived neurotrophic factor gene induced by an enriched environment. *Hippocampus*. 2011;21:127–132.
168. Biergans SD, Jones JC, Treiber N, Galizia CG, Szyszka P. DNA methylation mediates the discriminatory power of associative long-term memory in honeybees. *PLoS One*. 2012;7:e39349.
169. Lockett GA, Helliwell P, Maleszka R. Involvement of DNA methylation in memory processing in the honey bee. *Neuroreport*. 2010;21:812–816.
170. Plassman BL, Langa KM, Fisher GG, et al. Prevalence of dementia in the United States: the aging, demographics, and memory study. *Neuroepidemiology*. 2007;29:125–132.
171. Driscoll I, Hamilton DA, Yeo RA, Brooks WM, Sutherland RJ. Virtual navigation in humans: the impact of age, sex, and hormones on place learning. *Horm Behav*. 2005;47:326–335.
172. Lanchini I, Lavarone A, Senese VP, Ruotolo F, Ruggiero G. Visuospatial memory in healthy elderly, AD and MCI: a review. *Curr Aging Sci*. 2009;2:43–59.
173. Bach ME, Barad M, Son H, et al. Age-related defects in spatial memory are correlated with defects in the late phase of hippocampal long-term potentiation in vitro and are

attenuated by drugs that enhance the cAMP signaling pathway. *Proc Natl Acad Sci USA*. 1999;96:5280–5285.
174. Fischer A, Sananbenesi F, Wang X, Dobbin M, Tsai LH. Recovery of learning and memory is associated with chromatin remodelling. *Nature*. 2007;447:178–182.
175. Rowe WB, Blalock EM, Chen KC, et al. Hippocampal expression analyses reveal selective association of immediate-early, neuroenergetic, and myelinogenic pathways with cognitive impairment in aged rats. *J Neurosci*. 2007;27:3098–3110.
176. Pelge S, Sananbenesi F, Zovoilis A, et al. Altered histone acetylation is associated with age-dependent memory impairment in mice. *Science*. 2010;328:753–756.
177. Kang HJ, Kawasawa YI, Cheng F, et al. Spatio-temporal transcriptome of the human brain. *Nature*. 2011;478:483–489.
178. Tao Lu, Pan Y, Kao SY, et al. Gene regulation and DNA damage in the ageing human brain. *Nature*. 2004;429:883–891.
179. Wood SH, Craig T, Li Y, Merry B, de Magalhaes JP. Whole transcriptome sequencing of the aging rat brain reveals dynamic RNA changes in the dark matter of the genome. *Age (Dordr)*. 2012;4:4.
180. Yankner BA, Lu T, Loerch P. The aging brain. *Annu Rev Pathol*. 2008;3:41–66.
181. Blalock Eric M, Chen KC, Sharrow Keith, et al. Gene microarrays in hippocampal aging: statistical profiling identifies novel processes correlated with cognitive impairment. *J Neurosci*. 2003;23:3807–3819.
182. Hernandez DG, Nalls MA, Gibbs JR, et al. Distinct DNA methylation changes highly correlated with chronological age in the human brain. *Hum Mol Genet*. 2011;20:1164–1172.
183. Siegmund KD, Connor CM, Campan M, et al. DNA methylation in the human cerebral cortex is dynamically regulated throughout the life span and involves differentiated neurons. *PLoS One*. 2007;2:e895.
184. Bishop NA, Lu T, Yankner BA. Neural mechanisms of aging and cognitive decline. *Nature*. 2010;464:529–535.
185. Oberdoerffer P, Michan S, McVay M, et al. SIRT1 redistribution on chromatin promotes genomic stability but alters gene expression during aging. *Cell*. 2008;135:907–918.
186. Haberman RP, Quigley CK, Gallagher M. Characterization of CpG island DNA methylation of impairment-related genes in a rat model of cognitive aging. *Epigenetics* 2012;7:1008–1019.
187. Lahiri DK, Maloney B, Zawia NH. The LEARn model: an epigenetic explanation for idiopathic neurobiological diseases. *Mol Psychiatry*. 2009;14:992–1003.
188. Rutten BP, Mill J. Epigenetic mediation of environmental influences in major psychotic disorders. *Schizophr Bull*. 2009;35:1045–1056.
189. Burns A, Iliffe S. Alzheimer's disease. *BMJ*. 2009;338:b158.
190. Dickson TC, Vickers JC. The morphological phenotype of beta-amyloid plaques and associated neuritic changes in Alzheimer's disease. *Neuroscience*. 2001;105:99–107.
191. Mudher A, Lovestone S. Alzheimer's disease-do tauists and baptists finally shake hands? *Trends Neurosci*. 2002;25:22–26.
192. Selkoe DJ. Alzheimer's disease results from the cerebral accumulation and cytotoxicity of amyloid beta-protein. *J Alzheimers Dis*. 2001;3:75–80.
193. Di Fede G, Catania M, Morbin M, et al. A recessive mutation in the APP gene with dominant-negative effect on amyloidogenesis. *Science*. 2009;323:1473–1477.
194. Bezprozvanny I, Mattson MP. Neuronal calcium mishandling and the pathogenesis of Alzheimer's disease. *Trends Neurosci*. 2008;31:454–463.
195. Brickell KL, Leverenz JB, Steinbart EJ, et al. Clinicopathological concordance and discordance in three monozygotic twin pairs with familial Alzheimer's disease. *J Neurol Neurosurg Psychiatry*. 2007;78:1050–1055.

196. Fraga MF, Ballestar E, Paz MF, et al. Epigenetic differences arise during the lifetime of monozygotic twins. *Proc Natl Acad Sci USA*. 2005;102:10604–10609.
197. Silverman JM, Ciresi G, Smith CJ, Marin DB, Schnaider-Beeri M. Variability of familial risk of Alzheimer disease across the late life span. *Arch Gen Psychiatry*. 2005;62:565–573.
198. Lundberg J, Karimi M, von Gertten C, Holmin S, Ekstrom TJ, Sandberg-Nordqvist-AC. Traumatic brain injury induces relocalization of DNA-methyltransferase 1. *Neurosci Lett*. 2009;457:8–11.
199. Sierksma AS, van den Hove DL, Steinbusch HW, Prickaerts J. Major depression, cognitive dysfunction and Alzheimer's disease: is there a link? *Eur J Pharmacol*. 2009;626:72–82.
200. Mastroeni D, Grover A, Delvaux E, Whiteside C, Coleman PD, Rogers J. Epigenetic changes in Alzheimer's disease: decrements in DNA methylation. *Neurobiol Aging*. 2010;31:2025–2037.
201. Bollati V, Galimberti D, Pergoli L, et al. DNA methylation in repetitive elements and Alzheimer disease. *Brain Behav Immun*. 2011;25:1078–1083.
202. Philip De Jager GS, Matthew Eaton, Lori Chibnik, Manolis Kellis and David Bennett. 2012. Genome-Wide Exploration of DNA Methylation in the Aging Brain and Its Relation to Alzheimer's Disease In *Neurology April 25, 2012; 78(Meeting Abstracts 1): P05.070*
203. Wang SC, Oelze B, Schumacher A. Age-specific epigenetic drift in late-onset Alzheimer's disease. *PLoS One*. 2008;3:e2698.
204. Wu J, Basha MR, Brock B, et al. Alzheimer's disease (AD)-like pathology in aged monkeys after infantile exposure to environmental metal lead (Pb): evidence for a developmental origin and environmental link for AD. *J Neurosci*. 2008;28:3–9.
205. Urdinguio RG, Sanchez-Mut JV, Esteller M. Epigenetic mechanisms in neurological diseases: genes, syndromes, and therapies. *Lancet Neurol*. 2009;8:1056–1072.
206. Gunzburg MJ, Perugini MA, Howlett GJ. Structural basis for the recognition and cross-linking of amyloid fibrils by human apolipoprotein E. *J Biol Chem*. 2007;282:35831–35841.
207. Iwata N, Tsubuki S, Takaki Y, et al. Identification of the major Abeta1-42-degrading catabolic pathway in brain parenchyma: suppression leads to biochemical and pathological deposition. *Nat Med*. 2000;6:143–150.
208. Yasojima K, Akiyama H, McGeer EG, McGeer PL. Reduced neprilysin in high plaque areas of Alzheimer brain: a possible relationship to deficient degradation of beta-amyloid peptide. *Neurosci Lett*. 2001;297:97–100.
209. Cao X, Sudhof TC. A transcriptionally [correction of transcriptively] active complex of APP with Fe65 and histone acetyltransferase Tip60. *Science*. 2001;293:115–120.
210. Marambaud P, Wen PH, Dutt A, et al. A CBP binding transcriptional repressor produced by the PS1/epsilon-cleavage of N-cadherin is inhibited by PS1 FAD mutations. *Cell*. 2003;114:635–645.
211. Kim D, Nguyen MD, Dobbin MM, et al. SIRT1 deacetylase protects against neurodegeneration in models for Alzheimer's disease and amyotrophic lateral sclerosis. *EMBO J*. 2007;26:3169–3179.
212. Graff J, Kahn M, Samiei A, et al. A dietary regimen of caloric restriction or pharmacological activation of SIRT1 to delay the onset of neurodegeneration. *J Neurosci*. 2013;33:8951–8960.
213. Rouaux C, Jokic N, Mbebi C, Boutillier S, Loeffler JP, Boutillier AL. Critical loss of CBP/p300 histone acetylase activity by caspase-6 during neurodegeneration. *EMBO J*. 2003;22:6537–6549.
214. Francis YI, Fa M, Ashraf H, et al. Dysregulation of histone acetylation in the APP/PS1 mouse model of Alzheimer's disease. *J Alzheimers Dis*. 2009;18:131–139.

215. Graff J, Rei D, Guan JS, et al. An epigenetic blockade of cognitive functions in the neurodegenerating brain. *Nature*. 2012;483:222–226.
216. Lang AE, Obeso JA. Challenges in Parkinson's disease: restoration of the nigrostriatal dopamine system is not enough. *Lancet Neurol*. 2004;3:309–316.
217. Martino D, Stamelou M, Bhatia KP. The differential diagnosis of Huntington's disease-like syndromes: 'red flags' for the clinician. *J Neurol Neurosurg Psychiatry*. 2013;84:650–656.
218. Gapp K, Woldemichael BT, Bohacek J, Mansuy IM. Epigenetic regulation in neurodevelopment and neurodegenerative diseases. *Neuroscience*. 2012;S0306-4522:1151–1157.
219. Ng CW, Yildirim F, Yap YS, et al. Extensive changes in DNA methylation are associated with expression of mutant huntington. *Proc Natl Acad Sci USA*. 2013;110:2354–2359.
220. Lee J-S, Smith E, Shilatifard A. The language of histone crosstalk. *Cell*. 2010;142:682–685.
221. Brunner AM, Tweedie-Cullen RY, Mansuy IM. Epigenetic modifications of the neuroproteome. *Proteomics*. 2012;12:2404–2420.
222. Tweedie-Cullen RY, Brunner AM, Grossmann J, et al. Identification of combinatorial patterns of post-translational modifications on individual histones in the mouse brain. *PLoS One*. 2012;7:e36980.
223. Wang X, Song X, Glass CK, Rosenfeld MG. The long arm of long noncoding RNAs: roles as sensors regulating gene transcriptional programs. *Cold Spring Harb Perspect Biol*. 2011;3:a003756.

CHAPTER TWELVE

Deciphering Memory Function with Optogenetics

Anna Beyeler, Christine A. Eckhardt, Kay M. Tye
Department of Brain and Cognitive Sciences, Picower Institute for Learning and Memory, Massachusetts Institute of Technology, Cambridge, Massachusetts, USA

Contents

Abstract

Optogenetics has accelerated the field of neuroscience by overcoming many of the spatial, genetic, and temporal limitations of previous techniques to control neural activity. The study of learning and memory has profoundly benefited from these tools mainly from their use in rodents. New insights have been made regarding the involvement of specific cell types or populations of synapses in the acquisition, consolidation, and retrieval of memories. The cellular specificity and temporal precision of optogenetic manipulations have also shown to be useful to study synaptic mechanisms supporting learning and memory including long-term synaptic plasticity. Recently, new light-sensitive molecules have been developed to control intracellular pathways or gene

Progress in Molecular Biology and Translational Science, Volume 122
ISSN 1877-1173
http://dx.doi.org/10.1016/B978-0-12-420170-5.00012-X

expression, which promise to enhance our understanding of the molecular mechanism of memory function.

Optogenetics is a branch of biotechnology that combines genetic engineering, optics, and light-sensitive molecules to achieve precise control of cell activity. In this method, illumination of natural or synthetic photoreceptor proteins changes their conformation, leading to activation or inhibition of the host cell. Although optogenetic tools have been applied to many cell types in the retina, heart, and peripheral and central nervous system, this chapter focuses on memory function in the mammalian brain and reviews the insights neuroscientists have achieved by utilizing optogenetic approaches. Learning and memory involve neural interactions occurring on the order of milliseconds at specific synapses. Before optogenetics, electrophysiology allowed recording of neural processing at a satisfying time resolution, but manipulation of inputs was spatially unspecific (electric stimulation) or extremely slow compared to neurotransmission (pharmacological activation or inhibition). Because optogenetics overcomes these two major limitations, this tool has revolutionized the study of learning and memory from the behavioral to molecular level.

In Section 1, we introduce the most common optogenetic tools used to probe the neural substrates of learning and memory. In Section 2, we review how optogenetics causally implicates brain structures and/or cell types in specific behaviors, followed by Section 3 presenting how optogenetics has helped advance our understanding of synaptic mechanisms involved in learning and memory. Finally, we briefly discuss how optogenetics creates unique opportunities for future investigations.

1. OPTOGENETICS

1.1. Light-sensitive molecules

Optical control of neuronal activity was initiated in 1971 when Richard Fork activated aplysia abdominal ganglion neurons with the light-sensitive protein bacteriorhodopsin (BR).[1] In 1983, Faber and Grinvald successfully initiated action potentials in neurons using a fluorescent dye and light.[2] Later, Schmucker *et al.* combined a chromophore and light amplification, via stimulated emission of radiation (laser), to inhibit cells in larval Drosophila.[3] In 2002, Zemelman and colleagues created an explosive combination termed "chARGe," which is composed of arrestin-2, rhodopsin, and the α-subunit of a heterotrimeric G protein, to influence neuronal activity with light.[4] These pioneering efforts accelerated the subsequent generations of optogenetic tools. Today, the major optogenetic tools used in the mammalian brain include light-activated nonspecific cation channels, chloride pumps, proton pumps, and membrane-bound G-protein-coupled receptors (GPCR, Fig. 12.1A). When expressed in neurons, photoactivation of these molecules can activate or inhibit neural activity and/or signaling. Here, we briefly survey the currently available light-sensitive molecules that have provided new insights into our understanding of learning and memory.

1.1.1 Modulation of neuron excitability

The light-sensitive cation channels, channelrhodopsins (ChRs), are a class of opsins used to depolarize neuronal membranes by allowing sodium and calcium cations to flow down their electrochemical gradient in response to light (Fig. 12.1A). A variety of ChR variants have been engineered to optimize different properties, such as the channel's photocurrent, temporal precision, peak wavelength activation, light sensitivity, and bistability.[6] Channelrhodopsin-2 (ChR2) was first isolated from the alga *Chlamydomonas reinhardtii*[7] and humanized to increase expression levels.[8–11] Subsequent mutations to the channel have produced higher photocurrents[12,13] and additional derivatives, including the ChETA family and ChIEF, provide high fidelity between optical stimulation and spiking activity at ultrafast firing frequencies.[14–16] While the majority of ChR variants respond to blue light (peak activation wavelength of ~470 nm), red-shifted opsins (C1V1 and VChR1) have also been created,[17,18] thereby permitting combinatorial perturbation of neural activity and improved integration of optogenetics with calcium imaging tools that overlap in spectra with ChR2, such as Fura-2 and the green fluorescent protein–Calmodulin-M13 fusion protein known as GCaMP. Another group of excitatory proteins, step-function opsins, have been developed to allow bistable, step-like control of membrane potential with varying deactivation time constants.[17,19,20] The variety of tools for fast excitation highlights the need to select the optimal channel depending on the goals of the experiment. Optical stimulation, similar to electrical and pharmacological stimulation, also does not exactly replicate physiological patterns of activity since all cells or synapses are artificially activated simultaneously. These channels also alter intracellular membranes,[21] which may affect studies of intracellular processes, common in research of plasticity and memory.

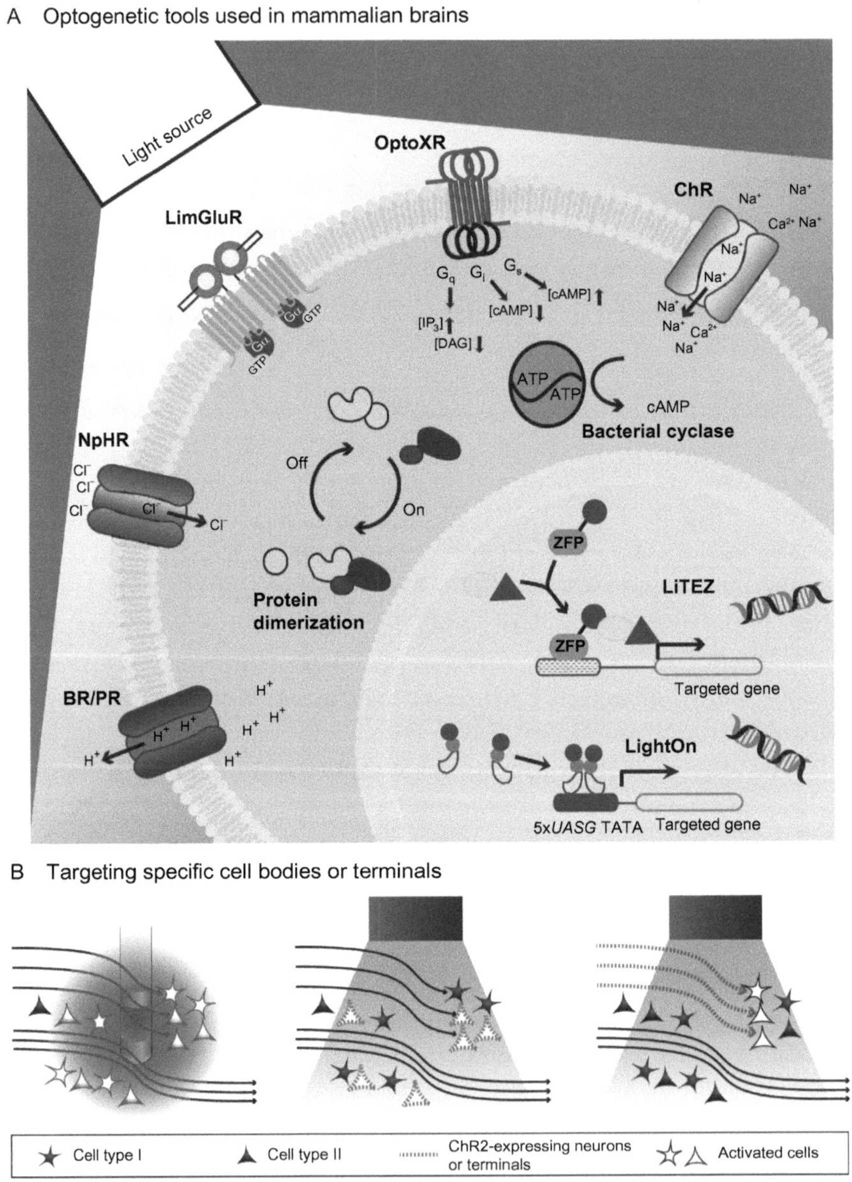

Figure 12.1 Optogenetic tools for modulation of cell activity in the mammalian brain. (A) Currently available light-activated molecules enable modulation of membrane potential, intracellular signaling, and genetic transcription. The cation-permeable channels, such as channelrhodopsin (ChR), produce fast membrane depolarization in response to light, while chloride pumps, such as halorhodopsin (NpHR), and proton pumps, such as bacteriorhodopsin (BR) or proteorhodopsin (PR), facilitate fast membrane hyperpolarization. Light-sensitive bacterial cyclases, protein dimerizers, and

Subsequent development of tools for fast inhibition has been an extremely important complement to excitatory channels, in testing for the necessity of specific cell types or projections in animal behavior. When expressed in a host cell, light-sensitive proton pumps, such as BR, proteorhodopsin, or archaerhodopsin (Arch),[22,23] and light-sensitive chloride pumps, such as halorhodopsin (NpHR), hyperpolarize the membrane when exposed to yellow light (~590 nm, Fig. 12.1A). The original NpHR has been enhanced to improve membrane localization and photocurrents, resulting in an enhanced third-generation variant, eNpHR3.0,[23] that has been proved to be effective in mammalian systems.[24,25] Unlike ChR, both proton and chloride light-sensitive pumps are generally subjected to constant light instead of pulses, in order to diminish rebound excitation or ionic or pH imbalances.[21]

1.1.2 Modulation of intracellular signaling

While fast excitation and inhibition provide direct control of neural electric activity, additional tools have been developed to perturb intracellular biochemical pathways, beginning with light-responsive GPCR. Researchers have developed chimeras between vertebrate rhodopsin and conventional ligand-gated GPCR, termed OptoXR.[26–28] Developed for adrenergic and serotoninergic pathways, these receptors initiate a G-protein-coupled signaling cascade in response to light. However, although these OptoXR can be targeted to specific cell types, they lack the signaling specificity of native GPCR, due to shortening of the complete native GPCR sequence in the creation of the chimera. OptoXR also do not fully recover following initial activation, thereby preventing sustained or reproducible perturbations.[29] Targeting synthetic photoswitchable tethered ligands to metabotropic glutamate receptors (mGluR)[29] avoided these issues and created

membrane-bound G-protein-coupled receptors, such as OptoXR and LimGluR, may be used to trigger or regulate intracellular signaling cascades. Light-switchable gene regulation systems, along with protein dimerizers, permit control of transcription of specific genes. ATP, adenosine triphosphate; cAMP, cyclic adenosine monophosphate; DAG, diacylglycerol; IP3, inositol trisphosphate; LiTEZ, light-inducible transcription using zinc finger proteins system, ZFP, zinc finger protein. Note: the light source is a schematic and is not drawn to scale. (B) Selectivity of optogenetic versus electric stimulation. Stimulation electrodes nonspecifically affect all nearby neurons and fibers (left), whereas blue light activation of cell bodies (center) or axon terminals (right) expressing ChR2, results in selective activation of a neural subpopulation defined by a promoter (center) or by its inputs (left). *Panel (B) is adapted from Ref. 5.* (See the color plate.)

light agonized and antagonized mGluR (LimGluR) for mGluR2, mGluR3, and mGluR6. Photostimulation turns LimGluR on and off in a rapid, reversible, reproducible manner. While ChR2 can certainly be used to activate GPCR, via depolarization of the presynaptic terminal, both LimGluR and OptoXR allow researchers to isolate the contribution of a specific ligand to a cell's response.

Extending the direct influence of light from the membrane to the cytoplasm and nucleus, optical methods also promise to enhance modulation of biochemical signaling and advance our knowledge of the molecular biology of memory. Many of these methods replicate existing strategies by replacing a chemical switch with light, thereby bestowing greater spatial resolution than previous chemical tools. Although not yet demonstrated in mammalian cells, researchers have used microbial adenylyl cyclases to optically regulate levels of the cyclic nucleotides cGMP and cAMP (Fig. 12.1A), which are important second messengers in numerous signaling pathways[30,31] involved in memory consolidation.[32]

Researchers have also created genetically encoded, light-activated dimerizers that enable control of protein–protein interactions in a reversible manner on a subsecond timescale (Fig. 12.1A). These dimerizers can be used to affect signaling cascades or genetic transcription. In one such system, researchers successfully induced modest protein translocation, transcription, and Cre-recombinase-mediated DNA recombination with light.[33] Other similar systems have been created but are essentially irreversible with a long time course of deactivation[34] or require addition of exogenous cofactors.[35–38]

Other notable light switchable gene regulation systems include the gene promoter system based on the optimized p65 activation domain (LightOn) and the light-inducible transcription using zinc finger proteins system (LITEZ).[39,40] In the LightOn system,[40] a transactivator binds a promoter in response to continuous light or pulses, thereby triggering transcription of targeted transgenes; cessation of light detaches the transactivator and halts transcription. In the LITEZ system,[39] light induces dimerization of two proteins that control synthetic zinc finger transcription factor activity in mammalian cells. Although not a single component system, the LITEZ system has, as of yet, faster kinetics of induction and silencing than any other reversible light-inducible gene regulation system. LITEZ also possesses customizable DNA-binding specificity, in which LITEZ can affect any promoter sequence by engineering an appropriate zinc finger protein, in contrast to LightOn and other systems,[33,34,37,40,41] which require that all targeted transgenes possess the same upstream binding sequence. However, LITEZ still shares drawbacks with other light-regulated gene expression systems, such as potential interference with endogenous signaling events and a need for further characterization in mammalian cells. While not reviewed here, additional optical tools for manipulation of genetic transcription and signaling cascades exist,[42,43] some of which may prove ideal, with further engineering, for mammalian experiments.

1.2. Opsin transduction of neurons with viral vectors

Optogenetic tools are continuously optimized to increase the specificity of experimental paradigms requiring *in vivo* targeting. In this section, we review viral delivery and targeting strategies, including lentiviral (LV), adeno-associated viral (AAV), and herpes simplex viral (HSV) vectors, as well as rabies virus vectors. Viral expression can be targeted with promoter-based approaches to distinguish the roles of specific neural types in the physiological or pathological functioning of the nervous system. Viral expression systems provide high gene copy number, high cellular transfection rate, and flexible genetic and anatomical specificity. Indeed, viral vectors currently represent the most popular means of delivering optogenetic tools to intact systems.

1.2.1 LV vectors

LV vectors[44] have been used to introduce opsins into the brains of mice,[8] rats,[9,45] and primates.[46] After transduction of neurons, LV vectors travel anterogradely to the axon terminals and achieve high expression levels over long periods of time. LV vectors possess the advantage of easy production with standard tissue culture techniques.[11,47] LV vectors have restricted diffusion *in vivo* and can be used to target subfields of a structure, such as the CA1 region of the mouse hippocampus. LV vectors can carry a genome of up to 9 kb[48] including the regulatory elements and opsin gene, a relatively large packaging capacity.

1.2.2 AAV vectors

AAV vectors enter the neuron soma and are generally transported anterogradely out to the axon terminals. As with LV, they have been used to express opsins in mouse,[49,50] rat,[51] and primate[46] neural tissue. AAV vectors are generally restricted to a genome size of 4.7 kb,[52] which limits the use of many large promoters. AAV vectors are challenging to produce within a standard laboratory but can be obtained through viral core facilities.

Interestingly, they display low immunogenicity and are safer than LV since currently available strains do not broadly integrate into the host genome. They also offer the advantage of larger transduced tissue volumes compared to LV. Many serotypes of AAV vectors exist, each having specific properties, such as volume of transduced tissue. For example, AAV5 injected into mouse hippocampus will diffuse and transduce neurons throughout most of the structure, whereas AAV2 can result in a relatively restricted expression pattern and thus may be used for experiments testing the role of specific local regions.[53] AAV2 is known to attach and infect neurons via proteoglycan molecules,[54] while AAV5 binds sialic acid and enters neurons through platelet-derived growth factor PDGF receptors.[55] There are more than 120 different AAV serotypes, which yet remain to be characterized.[56,57]

1.2.3 HSV vectors

Herpes simplex-based vectors transduce neurons and travel retrogradely from the axon terminals to the neuron cell body. HSV vectors can carry larger genetic material than LV and AAV vectors, allowing package of larger promoter fragments or transgenes (~150 kb). Although HSV presents consistency and toxicity concerns,[58] it provides strong and rapid (1–2 weeks) expression in all applied mammalian models.[59–62]

1.2.4 Rabies virus vectors

Rabies virus is particularly useful for the study of neuronal circuits because of its ability to spread transsynaptically in the retrograde direction.[63–65] Although rabies viruses are restricted to nonpromoter-based targeting because of their negative-strand RNA nature, infected cells remain viable for weeks[66] and the virus can amplify from even a single viral particle.[67] In 2007, Wickersham and his collaborators developed a rabies virus variant that spread only to the cells monosynaptically connected to the infected neurons, allowing tests of the functional role of neurons one synapse away from the initially infected cell.[66]

1.3. Transgenic lines

1.3.1 Transgenic lines natively expressing opsins

Although opsins can be targeted to specific regions and cell types via virus, the generation of transgenic mouse lines has yielded animals with reliable, homogenous opsin expression under specific promoters that do not require the use of viral transduction to achieve opsin expression. Moreover, promoter-based expression in transgenic line is not limited by the length of the promoter, compared to viral vector approaches. To control for gene expression, the opsin is combined or fused to a reporter fluorescent protein which is often an enhanced yellow fluorescent protein (eYFP). The development of mouse lines directly expressing opsins began with the Thy1-*ChR2-eYFP* line, which expresses ChR2 primarily in projection neurons.[68] Many other transgenic ChR2 mouse lines were subsequently developed, and some of them are commercially available, facilitating the technical requirements for optogenetic experiments. The following list, while not exhaustive, highlights the diversity of cell types that can be targeted *in vivo* through a simple optic fiber implant. Indeed, transgenic mouse lines have been developed to express ChR2 in olfactory sensory neurons (OMP-*ChR2-eYFP* line),[69] in glutamatergic cells of the spinal cord and hindbrain (Vglut-*ChR2-eYFP* line),[70] in striatal medium spiny neurons (MSN) (CaMKIIα-*tTA-BTR6* line),[71] in parvalbumin (PV) interneurons, GABAergic and glycinergic neurons, or serotonergic neurons (PV-*ChR2-eYFP*, Vgat-*ChR2-eYFP*, or TPH2-*ChR2-eYFP* mice).[72] Undoubtedly, the development of transgenic lines will continue, thereby enabling precise optical modulation of a variety of cell types and pathways possessing consistent expression of opsins.

1.3.2 Conditional expression systems

Although cell-specific promoters of transgenic lines are effective at restricting opsin gene expression to subsets of genetically defined neurons, some promoters have weak transcriptional activity. Indeed, many cell-specific promoters are unable to achieve the level of opsin gene expression necessary to mediate effective action potential firing or blockade.[11,47] To amplify the transcriptional activity in a cell-specific manner, conditional AAV vectors[73,74] have been developed to capitalize on the numerous cell-specific Cre-driver transgenic mouse lines available. In this strategy, AAV vectors carry the microbial opsins, driven by a strong promoter and followed by a floxed-stop sequence. The cell type of interest (expressing Cre) efficiently recombines the stop sequence out of the genome leading to opsin expression under the strong promoter.[8,9] Many studies have used conditional expression systems *in vivo* to express ChR2 in specific cell types in freely behaving mice.[61,75,76] Interestingly, some rat Cre-lines are also available, such as tyrosine hydroxylase (TH::Cre) and choline acetyl transferase (ChAT::Cre) lines,[25] allowing neuroscientists to take advantage of the substantial and flexible behavioral repertoire of rats.

1.4. Applying optogenetic tools to understand brain circuits

Microbial opsin gene products have been shown to traffic into neural dendrites and axons.[23,77,78] This property allows researchers to test the function of specific projections, by transducing the opsin into the cell bodies of the upstream region and illuminating the axonal projections in the downstream region.[12,24,45,79] Through this experimental design, researchers can test the role of specific projections *in vivo* by activating or inhibiting them during behavioral tasks involving learning or memory processing.[80,81] Compared to electrical stimulation, *in vivo* optogenetic stimulation offers two tremendous advantages: First, photostimulation does not activate passing fibers projecting to other parts of the brain (Fig. 12.1B). Second, photostimulation is selective for one cell type, depending on the promoter controlling ChR2 expression[5] (Fig. 12.1B). In *ex vivo* preparations, optogenetic stimulation offers the same benefits of cellular selectivity and, combined with pharmacology, enables exclusive recordings of monosynaptically evoked photocurrents.[50,82]

Optogenetics can also be used as a tool to control for the projection pattern of recorded cells, also known as "phototagging."[60] For example, in a recent study, Zhang *et al.* used this combined optogenetic-electrophysiological strategy to determine the functional identity of entorhinal neurons contacting place cells in the hippocampus.[81] A retrograde AAV vector carrying *ChR2* gene was injected into the hippocampus and an assembly of tetrodes and a laser-coupled optical fiber was implanted into the medial entorhinal cortex (MEC). Several weeks later, spike activity was recorded from MEC cells while the animals collected food morsels in a 1-m wide square enclosure. The MEC neurons projecting to hippocampal place cells were identified as the cells that fired at fixed minimal latencies in response to flashes of light at the recording site. Using this approach, they showed that hippocampal place cells receive inputs from a large number of grid cells, border cells, and head-direction cells, as well as cells from irregular or nonspatial firing correlates. These results suggest that place fields may be generated by convergence of signals from a broad spectrum of entorhinal functional cell types.

The reverse approach, consisting of "tagging" where the recorded cells are projecting, is also possible by injecting the viral construct expressing ChR2 in the recording site and photostimulating the terminals in a downstream region(s). High power stimulation of the terminals can trigger backpropagating action potentials that can be recorded in the upstream structure. To test for direct functional connection between two brain structures *in vivo*, electrophysiologists use collision tests where an electrically evoked orthodromic action potential blocks a simultaneously electrically evoked backpropagating (antidromic) action potential from a downstream structure. This strategy, broadly used in *in vivo* electrophysiology, was recently applied to optogenetic tools, adding the advantages of optogenetics including cellular specificity.[83]

2. OPTOGENETIC MANIPULATION OF MEMORY

Optogenetic strategies now allow neuroscientists to manipulate defined cell populations with high temporal precision during the different phases of memory: learning (acquisition), storage (consolidation), recall and extinction. In the acquisition phase of memory, an animal first learns to associate an unconditioned stimulus (US), an innately negative or positive stimulus, with a conditioned stimulus (CS) previously neutral. When a CS predicts a US, the animal adapts its behavior to optimize survival, most commonly by avoiding punishment and seeking reward, and the consolidation phase ensures long-term stabilization of the acquired association. When the CS no longer predicts the US, the extinction phase occurs, and the animal loses its previously acquired behavioral response to the CS.

2.1. Memory acquisition

2.1.1 Acquisition in fear conditioning

The study of aversive conditioning has provided significant insight into the dynamics of memory acquisition. In aversive conditioning, the animal first links the occurrence of a US, most often a footshock, to a CS, such as a cue or context. This association leads to behaviorally adaptive responses to the CS alone, such as freezing.[84,85] The lateral nucleus of the amygdala (LA) is considered a critical site for convergence of information about the CS and US in fear learning[86] although the association of US and CS most likely occurs in multiple regions.[87–89] The periaqueductal gray, as well as the thalamus and anterior cingulate cortex (ACC), are thought to relay information about the US to the LA,[85] while sensory information about the CS arrives from cortical areas and thalamic nuclei.[90] Research with electrophysiology, pharmacology, and brain lesions has elucidated numerous elements of the fear conditioning circuit; however, many aspects remain unknown.

2.1.1.1 Application of optogenetics in understanding cued aversive conditioning

Previous research has indicated that the association of US and CS likely occurs in the LA via synaptic plasticity in which the US strongly depolarizes LA pyramidal neurons and strengthens coincident inputs from the CS.[91–102] However, it was still puzzling whether LA neuron depolarization paired to CS presentations is sufficient to support learning. Johansen and colleagues employed optogenetics to directly test whether depolarization of LA pyramidal neurons guides fear conditioning.[103] The researchers targeted ChR2 to LA pyramidal neurons and conducted cued aversive conditioning, in which they replaced the aversive stimulus (US), normally a footshock, with laser stimulation of LA pyramidal neurons (laser-US). The researchers exposed the animals to repeated pairings of CS and laser-US, or presented the laser-US independently of the CS to control for laser stimulation aversive quality. And indeed, the animals exhibited small, but statistically significant freezing in response to the laser-US alone. Twenty four hours later, the animals exposed to pairings of laser-US and CS showed significantly more freezing in response to the CS alone, as compared to controls exposed to unpaired laser-US and CS. Even if the amount of freezing is relatively weak, and acquisition needed a large number of CS and laser-US pairings, this study shows for the first time that depolarization of pyramidal cells in the LA during a cue presentation is sufficient to generate an aversive memory. Moreover, a subsequent investigation by the same group showed more robust fear learning by combining the laser-US and CS association with pharmacological activation of noradrenergic β-receptors.[103] These complementary results indicate that both depolarization of projecting neurons and neuromodulatory processes are important for associative fear memory formation in the LA.

Downstream structure of the LA, the central nucleus of the amygdala (CeA), divided into medial and lateral parts (CeM and CeL), has been shown to affect acquisition of conditioned fear. Indeed, acute inactivation of CeA prevents fear acquisition,[104,105] and whole-cell recordings in slice have demonstrated activity-dependent plasticity in the CeM.[106] The CeM also receives significant inhibitory input from CeL[107–109] and projects to regions in the brainstem and hypothalamus.[110–112] However, the function of this intra-CeA inhibitory circuit in fear conditioning was previously unclear. Ciocchi and colleagues bilaterally induced ChR2 expression in CeM using a synapsin promotor in an AAV construct and showed that bilateral activation of the CeM triggered strong and reversible freezing.[49] Interestingly, pharmacological inhibition through injection of the $GABA_A$ receptor agonist muscimol into the CeL produced freezing, indicating that CeL tonically inhibits CeM and that activation of CeM output neurons induces a conditioned fear response. Altogether, these optogenetic and pharmacology experiments defined the functional architecture of CeA microcircuits and their role in the acquisition and regulation of conditioned fear behavior.

Disinhibitory circuits in regions besides the amygdala have been found to play a role in conditioned fear learning. Previous studies demonstrated plasticity in the auditory cortex in response to auditory fear conditioning,[113–115] which appears to depend upon pyramidal neuron disinhibition[116] and cholinergic input from the basal forebrain.[114,115,117] Nonetheless, a role for the auditory cortex in memory acquisition remained debatable, and the specific contribution of cortical interneurons unknown. Letzkus and colleagues first substantiated the involvement of auditory cortex through local injection of the $GABA_A$ receptor agonist muscimol during a differential fear conditioning paradigm with complex tones.[88] Testing 24 h later revealed significant memory deficits, as reflected by diminished freezing behavior. The US, a footshock, was found to depolarize the majority of interneurons in layer 1, via cholinergic afferents from the basal forebrain. L1 interneurons, in turn, were shown to inhibit fast-spiking PV interneurons in layers 2/3 and disinhibit L2/L3 pyramidal neurons during footshock. Under calcium imaging, L2/L3 neurons showed a significantly enhanced response to tone and shock together as compared to tone alone, whereas isolated shock evoked minimal activity. These data suggest that the auditory cortex, rather than simply relaying CS information, mediates the integration of CS and US information. To conclusively demonstrate the function of L1 disinhibition in fear learning, Letzkus and colleagues (2011) then implanted bilateral optic fibers in the auditory cortex of mice expressing ChR2 in PV interneurons. During the training phase of fear conditioning, the researchers optically stimulated PV interneurons when these neurons previously exhibited inhibition during and for 5 s after the footshock. Such activation significantly reduced fear responses in testing 24 h later, whereas reconditioning in the same animals, without laser stimulation, produced normal fear learning. These findings support the presence of a disinhibitory microcircuit in the auditory cortex, necessary for the acquisition of associative fear memories for complex auditory tones.

2.1.1.2 Utilizing optogenetics to understand contextual aversive conditioning

Acquisition of fear contextual memories depends upon the hippocampus,[118] a region essential for episodic memory formation[119] and affective processing.[120–124] Previous research suggested differential functions of the hippocampus along the dorsal–ventral axis, with the dorsal pole vital for spatial memory and the ventral pole for affect.[125–127] However, the extent of this functional organization within the subregions of the hippocampus (dentate gyrus (DG), CA3, CA1) was previously unknown. Kheirbek and colleagues (2013) investigated the effect of the dorsal and ventral DG on contextual fear learning.[118] The DG primarily consists of granule cells (GC), which are genetically defined by expression of proopiomelanocortin (POMC) and thus amenable to genetic targeting. By crossing a POMC Cre-recombinase mouse line[128] to either a conditional mouse line with NpHR[11] or ChR2,[129] the researchers created POMC-*NpHR* and POMC-*ChR2* mouse lines. In addition to cell-type specificity through genetics, implantation of an optic fiber in either the dorsal or the ventral DG imparted intrahippocampal specificity to their manipulations. Mice were then subjected to a contextual fear conditioning paradigm, where they were trained to associate a context (CS) with a footshock (US), followed by memory quantification 24 h later through measure of freezing in response to the context alone. Bilateral inhibition of dorsal GC during training substantially impaired acquisition of the contextual fear memory, as shown by reduced freezing in the testing phase. Importantly, this inhibition specifically altered acquisition of contextual fear memory since the same mice successfully learned to associate an auditory cue with the footshock. Surprisingly, bilateral activation of dorsal GC also impaired contextual fear memory encoding, most likely due to indiscriminate activation of GC, which are thought to be only sparsely activated by a specific context. In contrast, bilateral inhibition or activation of ventral GC had no effect on acquisition. Thus, the dorsal GC appears to be necessary for acquisition of contextual fear memories.

Fear memory generalization allows an animal to recognize resemblances between novel and previously encountered stimuli, in order to respond adaptively and avoid potential danger. However, overgeneralization can produce maladaptive responses to safe stimuli and excessive anxiety. Although memory generalization has primarily been discussed in connection to consolidation or retrieval,[130–132] the use of optogenetics has clarified the dynamics of generalization of contextual fear memories during acquisition.[133] Through genetic tracers and toxins, Xu and Südhof discovered a circuit, consisting of the medial prefrontal cortex (mPFC), nucleus reuniens

(NR), and hippocampus, that modulates fear memory generalization.[133] In order to directly observe the effect of the NR within this circuit, researchers perturbed NR activity in behaving mice using the ChR variant ChIEF.[16] Stimulation was delivered at 4 Hz either continuously or phasically through 15-pulse bursts at 30 Hz every 5 s, during the acquisition phase of a contextual fear conditioning paradigm. In this paradigm, mice were trained to associate a context and a tone with a footshock and were then tested for freezing in the same context, an altered context, and in response to the tone. Continuous 4 Hz stimulation of the NR during training produced a decrease in the freezing response to the altered context, indicative of reduced memory generalization. In contrast, phasic stimulation increased freezing to the altered context, indicative of memory overgeneralization. Importantly, both photostimulation parameters of the NR substantially affected memory generalization but did not alter contextual or cued conditioning. Given that NR neurons cause subthreshold depolarization in CA1 pyramidal neurons and firing of inhibitory interneurons,[134] the NR may regulate memory generalization by altering the excitability of hippocampal neurons. The researchers hypothesize that the contrasting effects of stimulation parameters may arise from varying excitation of these subclasses of neurons. Previous research also demonstrated a correlation between NR activity and hippocampal oscillations,[135] raising the possibility that NR stimulation may produce differential behavioral effects through modulation of these oscillations.[133]

During the past decade, optogenetics allowed researchers to test hypotheses through previously unfeasible experiments regarding aversive memory acquisition, as well as to dissect neural circuitry within and across regions. Focusing primarily on associative learning, these studies demonstrate new mechanisms of integration of aversive US and CS, isolating the contributions of specific cell types and brain structures.

2.1.2 Acquisition in appetitive conditioning

As with aversive conditioning, mammals can learn to associate a cue, context, or action with an appetitive stimulus. Such reward learning depends not only on the association of CS and US, but the recognition of the US as a reward and the motivation to seek such a reward. In the mammalian brain, the ventral tegmental area (VTA) located in the midbrain and the nucleus accumbens (NAc) located in the ventral striatum were shown to be extremely important for reward processing.

Thus far, the primary assays conducted with optogenetics to detect changes in reward-related behavior involve Pavlovian and instrumental conditioning. Conditioned place preference (CPP) paradigm is a variant of passive Pavlovian conditioning where the animal is placed in a multichamber apparatus and learns to associate each chamber with a form of stimulation, such as pharmacological administration, electrical stimulation, or optical manipulation. On the first day, before conditioning, the animal explores all chambers with no stimulation, to control for any preexisting biases toward a chamber. On the next day, the first conditioning session, the animal is confined to one chamber and receives the chamber's associated stimulation. The following day, the animal is confined to another chamber and receives an alternate treatment. On the final day, the animal can freely explore all chambers without stimulation, and conditioned preference (or avoidance, if the stimulations were aversive) for a chamber is quantified as the duration of time spent in each chamber, before and after conditioning. Because optogenetic activation or inhibition, and electric stimulation are reversible at the millisecond timescale, preference or avoidance measures can also be obtained through an abbreviated version of CPP, the real-time place preference or aversion (RTPP) paradigm. In RTPP experiments, the animal can freely explore a multichamber apparatus, and the entry into a chamber triggers the chamber's associated stimulation or optogenetic inhibition. If this manipulation affects the animal's behavioral state, the animal will learn an association between the chamber and the state. By the end of a single session, the animal spends most of its time in whichever chamber possesses the most appetitive (or least aversive) association. The main difference between these two place preference paradigms is that in CPP the different phases of memory formation are separated and testable, whereas in RTPP, the gradual transition from acquisition to recall cannot be quantified.

In contrast to Pavlovian conditioning tasks, in which an animal is passively presented with repeated pairings of CS and US, appetitive operant tasks require that the animal learns an association between an action, such as a nose poke or lever press, and a reward, such as sucrose, or, in the case of intracranial self-stimulation (ICSS), direct stimulation of a brain region. In optogenetic ICSS paradigm, an animal has access to one active lever or nosepoke, paired to optical stimulation, and one inactive lever or nosepoke, not paired to optical stimulation. If the targeted cell type or region plays a role in reward processes, the animal will learn to preferentially press the active lever or enter the active nosepoke. In this section, we discuss recent discoveries regarding reward learning, realized through the use of optogenetics during these behavioral assays.

2.1.2.1 Parsing the function of VTA in reward-related behavior

Via its projections throughout the forebrain, the VTA is thought to mediate reward-seeking behavior to guide learning.[136,137] A heterogeneous structure consisting of dopaminergic, GABAergic, and glutamatergic neurons,[138] the VTA connects to numerous structures, including the prefrontal cortex (PFC), the amygdala, and the NAc.[139] Early studies suggested that putative dopamine (DA) neurons of the VTA may encode reward prediction error, defined as the discrepancy between expectation and outcome.[140–142] Consistent with this hypothesis, DA neurons exhibit phasic increased firing in response to an unexpected reward, firing depression following omission of an expected reward, and minimal response when the outcome accords with expectation.[142] However, with such observational data, the causal role of this activity in reward-related behavior remained unknown, while lesion, electrical stimulation, and pharmacology did not permit both specific and temporally precise manipulation of DA neurons.

In 2009, Tsai and colleagues used ChR2 to activate VTA DA neurons and observe the perturbation's effect on behavior[143] (Fig. 12.2). Specific expression of ChR2 in DA neurons was achieved through viral infusion into the VTA of TH::Cre transgenic mice, and optic fiber implant above the VTA. Mice were then subjected to a series of CPP experiments to compare the effects of phasic and tonic DA neural activity. Phasic 50-Hz stimulation produced robust CPP, whereas 1-Hz tonic stimulation did not. *In vivo* fast-scan cyclic voltammetry revealed that 50-Hz phasic photostimulation of VTA DA neurons triggers large DA transient concentrations in the NAc, similar to those observed in response to natural reward and that 1 Hz tonic stimulation induced lower DA concentrations. These findings indicate that phasic, high-frequency firing of DA neurons in VTA is sufficient for passive behavioral conditioning.

Further research has demonstrated that VTA DA neurons support not only passive conditioning but also active, operant conditioning. Adamantidis and colleagues induced expression of ChR2 in DA neurons of the VTA and phasically activated DA neurons in freely behaving mice during different phases of a food-seeking operant task[144] (Fig. 12.2). In the acquisition phase of the task, optical stimulation enhanced positive reinforcing actions, producing significantly more lever presses for delivery of food and optical stimulation than food alone. After extinction of food-seeking

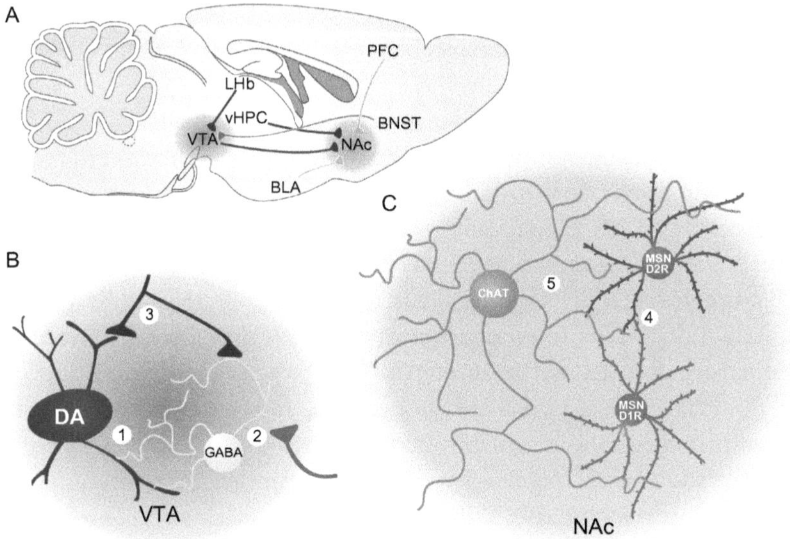

Figure 12.2 Limbic circuitry revealed by optogenetic study of reward-related behavior. (A) Ventral tegmental area (VTA) and nucleus accumbens (NAc) inputs involved in reward processing. Phasic activation of VTA dopaminergic (DA) neurons supports CPP[143] and operant responding[144] including intracranial self-stimulation (ICSS, Ref. 145,146). Animals actively nosepoke for activation of glutamatergic inputs from the ventral hippocampus (vHPC), basolateral amygdala (BLA), or PFC.[147,148] (B) GABAergic contributions to reward processing within the VTA. VTA GABAergic neurons, which provide inhibition to VTA DA neurons (1, Ref. 138), receive glutamatergic and GABAergic input from the vBNST (2) and are thought to contribute to the computation of reward prediction error.[149] Activation of this subpopulation produces conditioned place avoidance[150] and diminished reward seeking,[151] while inhibition results in opposite behavioral effects.[83] The lateral habenula (LHb) also projects to VTA GABAergic neurons (3, Ref. 152). (C) NAc microcircuitry dissected by optogenetics. Within the NAc, D1R-or D2R-expressing NAc neurons exhibit opposing effects on cocaine-CPP (CPP, 4, Ref. 61), while cholinergic interneurons expressing choline acetyltransferase (ChAT) affect the development of cocaine-CPP (5, Ref. 25). Stimulation of α_1 adrenergic signaling within NAc neurons also resulted in CPP.[26] *Adapted from Ref. 153.* (See the color plate.)

behavior, presentation of optical stimulation reactivated operant behavior, despite the absence of cues, stress, or primary reinforcers. Furthermore, in a single-session reversal paradigm, the majority of mice successfully switched their responses to press the appropriate lever for food and optical stimulation. These results demonstrate that contingent activation of dopaminergic neurons supports the development of positive reinforcement during reward learning.

Although Adamantidis and colleagues observed reinforcement learning only with the combination of food and optical stimulation,[144] subsequent optogenetic studies showed that activation of DA VTA neurons alone can support ICSS in rodents. Early studies of ICSS with electric stimulations or pharmacological infusions found that mice will learn to preferentially press a lever for stimulation in the medial forebrain bundle, among other regions, which helped fostered hypotheses regarding DA's potential role in reward signaling.[154–156] Optogenetic studies have improved upon these studies by specifically stimulating dopaminergic neurons. Rodents have been shown to nosepoke for a single 5-ms optical pulse delivered to DA VTA neurons, with response rates increasing with duration of stimulation, up to 1 s[145] (Fig. 12.2). Kim and colleagues also demonstrated ICSS with 200 ms activation of VTA DA neurons[146] (Fig. 12.2). With extinction, in which previously active nosepokes no longer deliver stimulation, the animal's response rate rapidly diminishes.[145,146] Thus, transient activation of dopaminergic neurons is sufficient to drive reinforcement learning and appears to provide a reward signal to other regions.

In addition to perturbing VTA activity to establish causal roles, optogenetics has also been used to uncover the cell type of VTA neurons involved in computing reward prediction error[149] (Fig. 12.2B). Although previous research had established that VTA GABAergic neurons inhibit VTA DA neurons,[138] their contribution to reward processing was unclear. Cohen and colleagues conducted extracellullar recordings in the VTA while mice performed a conditioning task, in which mice learned to associate different odor cues with appetitive and aversive outcomes.[149] Through the infusion of virus into two strains of transgenic mice, the researchers induced expression of ChR2 in the dopaminergic or GABAergic neurons of the VTA. ChR2 expression thus enabled optical tagging of dopaminergic and GABAergic neurons, such that researchers could identify a neuron's cell type during extracellular recordings by optically stimulating and observing a response. This technique improved upon previous methods to identify cell types via firing patterns, which have been called into question in respect to dopaminergic neurons.[157,158] With optical tags, researchers discovered activity in DA neurons consistent with reward prediction error coding. All identified GABAergic neurons exhibited persistent excitation following reward-predicting odors and no modulation in response to outcome. With aversive stimuli, the majority of dopaminergic neurons were inhibited, although some displayed excitation, and all GABAergic neurons were excited by the aversive stimuli. These data suggest that when reward is expected, VTA GABAergic neurons may provide inhibition to balance the excitation of primary reward. Following aversive stimuli, GABAergic neurons may also contribute to inhibition of dopaminergic neurons.

Subsequent optogenetic studies complemented these observations on the functional role of GABAergic neurons of the VTA. In a CPP paradigm, optical activation of VTA GABAergic neurons produced conditioned place aversion, as did the direct inhibition of VTA DA neurons[150] (Fig. 12.2B). In a cue-evoked reward-seeking paradigm, optical activation of GABAergic VTA neurons during reward delivery, but not cue presentation, diminished sucrose consummatory behavior, which accords with the reward prediction-error hypothesis[151] (Fig. 12.2B). Although stimulation had no effect on anticipatory behavior, this most likely occurred because of the task design, in which animals ultimately consumed the reward after cessation of light stimulation. Activation of VTA GABAergic inputs to NAc alone had no effect on behavior, suggesting that VTA GABAergic projections to other regions or other unknown alterations to DA signaling may mediate the cessation of consummatory behaviors instead. Given the conditioned place aversion and reduction in reward-seeking behavior, VTA GABAergic neurons may produce an acute anhedonia-like phenotype that impacts motivation and reward learning.[151] Although not a direct modulation of VTA GABAergic neurons, Stamatakis and Stuber found similar avoidance behavior with the activation of lateral habenula inputs to the ventral midbrain, which appeared to activate nondopaminergic neurons in the rostromedial tegmental nucleus, as well as VTA GABAergic neurons[152] (Fig. 12.2). This manipulation produced real-time and conditioned place avoidance, as well as active avoidance, in which the animal repeatedly nosepoked to terminate optical stimulation.

Complementing these results, a recent study showed that optical inhibition of VTA GABAergic neurons increases reward-related behaviors, and that the activity of these neurons is influenced by the ventral bed nucleus of the stria terminalis (vBNST, Fig. 12.2).[83] Previous studies demonstrated that subpopulations of vBNST neurons respond to aversive and appetitive stimuli[159–161] and send efferents to the VTA.[52,162–165] However, the constituent cell types of vBNST to VTA projections, as well as their effects, were unknown. Employing optogenetics, Jennings and colleagues discovered both glutamatergic and GABAergic neurons projecting from the vBNST to the VTA, primarily to nondopaminergic neurons, and targeted ChR2 to these neurons using the Vglut2::Cre and Vgat::Cre mouse lines, respectively.[83] Whereas all neurons displayed heterogeneous firing activity during aversive stimuli, Vglut2$^{vBNST-VTA}$-ChR2-expressing neurons primarily showed increased firing, in contrast to net inhibition among Vgat$^{vBNST-VTA}$-ChR2-expressing neurons. In an RTPP assay, optical stimulation of Vglut2$^{vBNST-VTA}$–ChR2-expressing neurons elicited place aversion, and additional assays revealed decreased reward-seeking and increased anxiety-like behavior with activation of this pathway. Conversely to the aversive effects produced by the activation of vBNST to VTA glutamatergic cells, activation of the GABAergic projections of this pathway evoked appetitive behaviors. In fact, optical activation of Vgat$^{vBNST-VTA}$ projections produced place preference in RTPP and supported reward-seeking in an ICSS assay. Given that GABAergic vBNST neurons synapse onto VTA GABAergic neurons, Jennings and colleagues hypothesized that these reward-related behaviors occurred through inhibition of VTA GABAergic neurons.[83] Indeed, optical inhibition of VTA GABAergic neurons via NpHR replicated these behavioral effects, with stimulation eliciting place preference, ICSS, and decreased anxiety-like behaviors. Together, these results indicate that distinct glutamatergic and GABAergic pathways from vBNST to VTA are involved in coordinating opposing motivational states. Considering the close relationship of motivation and learning, this circuitry would appear to modulate both appetitive and aversive learning. The contribution of the BNST to divergent behavioral states also emerged in a study by Kim and colleagues, in which researchers investigated the role of the oval nucleus of the BNST (ovBNST) and anterodorsal BNST (adBNST) in the regulation of anxiety levels.[166] Investigators targeted ChR2 to the adBNST and stimulated adBNST terminals in either the parabrachial nucleus (PB), lateral hypothalamus (LH), or VTA, all known targets of the adBNST.[167–171] In contrast to projections to PB or LH, activation of adBNST projections to VTA produced place preference in RTPP. The remaining pathways were found to modulate other aspects of anxiolysis, while the ovBNST contributed to anxiogenesis. By promoting reward-related behavior, the pathway from adBNST to VTA may influence reward learning.

2.1.2.2 Identifying the role of specific inputs to the NAc in reward-related behaviors

The NAc mediates goal-directed behaviors through integration of information from limbic and cortical structures and projections to basal ganglia nuclei. The ventral hippocampus (vHPC), basolateral amygdala (BLA), and mPFC send substantial glutamatergic afferents to the NAc.[172]

The VTA is also an important source of inputs to NAc, which are mainly dopaminergic (80%) with a subpopulation of these dopaminergic neurons coreleasing glutamate.[173] Less than 10% of the mesoaccumbal projections are from glutamatergic neurons only and 10% are GABAergic.[174,175] Through the combination of reinforcement signals and environmental information, the NAc is thought to regulate reward-related behavior.[176–180]

Previous studies indicated that the projection from BLA to NAc may regulate cue-evoked motivated behaviors,[181–185] but lack of specificity precluded definitive conclusions. Stuber and colleagues targeted this projection by expressing ChR2 in BLA glutamatergic neurons of mice and stimulating BLA fibers in the NAc[147] (Fig. 12.2). Photostimulation of these terminals evoked ICSS, which was blocked by a D1, but not D2, receptor antagonist, supporting that the reinforcing effect of BLA glutamatergic input to NAc depends upon postsynaptic modulation by D1 receptors. In contrast, activation of glutamatergic mPFC terminals in NAc did not elicit optical self-stimulation, although a subsequent study (discussed below) showed ICSS through photoactivation of mPFC to NAc glutamatergic inputs.[148] Finally, inhibition of the BLA to NAc projection with NpHR also decreased cue-triggered sucrose consumption. Together, these results demonstrate that the BLA to NAc glutamatergic pathway influences reward-related learning in a D1 receptor-dependent manner.

Another optogenetic study of glutamatergic inputs to NAc has indicated that the amount of glutamate released, rather than the specific pathway activated, may be most important in reinforcement learning (Fig. 12.2). Britt and colleagues used ChR2 to selectively activate and compare inputs from the vHPC, BLA, and PFC.[148] The strength and synaptic properties of the pathways differed, as well as the response of the pathways to cocaine use, which selectively strengthened vHPC synapses in the medial NAc shell. In a CPP paradigm, activation of each pathway induced a CPP for optical stimulation. Each group of mice also learned to optically self-stimulate, although progressively stronger, nonphysiological stimulation was required for BLA and PFC inputs. Consistent with the excitatory effect of glutamatergic input, mice with ChR2 expression in NAc projection neurons learned to optically self-stimulate. In contrast, an earlier study did not find CPP with ChR2 expression in NAc neurons,[26] possibly due to lower-frequency stimulation. Although previous research had suggested that inhibition of NAc neurons represents reward,[186–188] these data indicate that indiscriminate activation of

NAc neurons by any of its glutamatergic inputs can drive reinforcement learning.

Noradrenergic inputs to the NAc have also been involved in reinforcement learning. Using the OptoXR approach, Airan *et al.* showed that activation of opto-α_1 adrenergic receptors (AR) in the NAc induces place preference in freely behaving mice.[26] Interestingly, activation of opto-β_2 AR did not produce any preference.[26] Together with the demonstration that glutamatergic inputs to NAc are a key player in reinforcement learning, the results suggest that norepinephrine modulates the effect of glutamatergic inputs in NAc.

2.1.2.3 Determining the function of NAc subpopulations in reward-related behavior

In both the dorsal and ventral striatum, additional studies have demonstrated an influence of D1 receptors upon reward-seeking behavior.[61,76] Lobo and colleagues found that optical activation of MSN expressing D1 receptors enhances reward preference in a cocaine-CPP paradigm, while activation of MSN expressing D2 receptors suppresses the influence of cocaine reward[76] (Fig. 12.2C). This functional division of D1 and D2 receptors in ventral striatum resembles the organization of the dorsal striatum, in which the "direct," D1 receptor pathway enables body movements and the "indirect," D2 receptor pathway inhibits body movements.[189–192]

Despite its primary implication in motor function,[193,194] the dorsal striatum may also influence reward-related behavior in a similar manner to the NAc. With a Cre-dependent viral strategy, Kravitz and colleagues expressed ChR2 in the two populations of striatal neurons: D1-expressing direct pathway medium spiny neurons (dMSN) and D2-expressing indirect pathway medium spiny neurons (iMSN).[61] Activation of dMSN neurons produced persistent ICSS, while activation of iMSN evoked initial avoidance of the active trigger, followed by no preference for stimulation. Consistent with these findings, animals developed persistent place preference for stimulation of dMSN neurons and initial place avoidance for iMSN activation, which faded with continual stimulation. These data suggest that dMSN neurons support persistent appetitive reinforcement, in contrast to the transient punishment elicited by iMSN neurons. Both dorsal and ventral striatum appear to influence reinforcement learning.

Within the NAc, cholinergic interneurons comprise less than 1% of the NAc neural population,[195] yet they appear to influence reward-related learning (Fig. 12.2C). Prior to optogenetics, contradictory results had emerged from pharmacology and molecular ablation studies,[196–199] rendering the role of cholinergic interneurons in learning debatable. By expressing ChR2 or NpHR in cholinergic neurons in the NAc, Witten and colleagues investigated the effect of these neurons upon physiology and behavior.[25] Researchers found that cholinergic neurons inhibit MSN, primarily through nicotinic receptors, and that inhibition of cholinergic neurons increases MSN activity. During cocaine-CPP, in which cocaine administration is paired to a chamber, inhibition of cholinergic neurons through NpHR reduced cocaine-induced CPP, with no behavioral effect in the absence of cocaine. Inhibition did not influence other learning paradigms, such as contextual or cued fear conditioning. Activation of cholinergic neurons via ChR2 also did not affect cocaine place preference or optical place preference. These findings imply that appetitive stimuli, such as drugs, may inhibit MSN activity and, in turn, disinhibit target structures—a process that may have been blocked by inhibiting cholinergic interneurons. Thus, as with aversive conditioning, optogenetics has helped reveal the influence of local circuits upon behavior and the neural activity among regions.

While nonoptical methods had previously identified the essential role of NAc and VTA in reward processes, the studies featured here highlight the interactions of these regions with other areas of the brain, as well as the function of specific cell types within the NAc and VTA. By building upon knowledge obtained through electrophysiology, pharmacology, and lesions, optogenetic tools have helped resolve outstanding debates in the field and uncover new findings.

2.2. Memory consolidation

Following the acquisition of a memory, consolidation facilitates the long-lasting stabilization of a memory through changes on the synaptic and systems levels. Although numerous studies have suggested mechanisms for memory consolidation, the temporal and spatial precision of optogenetics has confirmed specific contributions to this process, such as the roles of the BLA and sleep continuity.

The BLA has been implicated in the consolidation of multiple forms of learning, either via posttraining electrical stimulation or pharmacological infusion to alter activity.[200–209] Previous studies also observed enhanced gamma frequency (35–45 Hz) coupling between BLA and other structures as learning progresses,[210,211] which raised the possibility that firing of BLA neurons in this range may contribute to consolidation. However, pharmacology lacks temporal precision to control firing frequency, while both pharmacological and electrical stimulation potentially modulate multiple cell types and fibers of passage. Thus, researchers used optogenetics to stimulate and inhibit the BLA following training in an inhibitory avoidance (IA) task and observe the manipulation's effect on consolidation.[212] ChETA or ArchT was expressed in the pyramidal neurons of the BLA via viral infusion, and the animals were subjected to IA training. Immediately after training, optical stimulation of rats expressing ChETA was applied at 20 or 40 Hz. Activation of BLA neurons enhanced retention in a frequency-dependent manner, with statistically significant retention observed with the 40 Hz gamma range stimulation, as compared to sham-controls. Conversely, inhibition of BLA pyramidal neurons for 15 min, but not 1 min, immediately after training impaired retention. Inhibition 3 h after training had no effect, indicating that optical inhibition did not decrease retention through a long-lasting impairment of BLA function. These results highlight the importance of the 15 min posttraining period in consolidation, potentially due to the influence of stress hormones or the noradrenergic system, which have both been shown to modulate memory consolidation. The findings also demonstrate a causal role for BLA gamma frequency firing in consolidation, where previously only correlations had been observed. Although the researchers did not investigate the effect of BLA manipulation upon downstream structures, future investigations could potentially use optogenetics to modulate axon terminals from BLA or those terminating in the BLA in the same task design.

Sleep continuity has been hypothesized to influence memory consolidation,[213–217] but attempts to study sleep, mostly through deprivation, have introduced confounds that prevent unambiguous conclusions. Most studies that alter sleep continuity inevitably affect other characteristics of sleep, such as duration, quality, and composition,[218–220] as well as a host of intracellular and synaptic processes.[219,221–224] However, optogenetic targeting of ChR2 to hypocretin/orexin neurons in the LH has been effectively used to isolate the effect of sleep continuity upon memory consolidation. Previous studies demonstrated that hypocretin/orexin neurons in mice influence sleep-to-wake transitions,[8,225] with seemingly minimal effects upon homeostatic sleep regulation.[225] This function rendered them an ideal candidate for fragmenting the sleep of mice. Thus, researchers stimulated hypocretin neurons in mice by expressing

ChR2 and observed the effect of light stimulation upon object recognition, a memory assay. Sixty-second interval stimulation during rest was found to disrupt nonrapid eye movement sleep without otherwise altering sleep architecture, total sleep time, or stress-related factors. However, this stimulation significantly impaired object recognition, with performance correlated to sleep continuity and not sleep duration. In contrast, stimulation every 120 s, which produces less sleep fragmentation, had no effect upon memory consolidation. As the authors propose, sleep fragmentation may interfere with consolidation by impairing replay phenomenon,[226,227] which may support memory consolidation via transfer of information from hippocampus to cortex.[228]

2.3. Probing mechanisms of memory recall

2.3.1 The memory engram

At the beginning of the twentieth century, Richard Semon introduced the word "engram" to describe a memory trace, the physical substrate of memory in the brain. Since then, modern memory research has attempted to locate this engram in the brain with recent aid from the use of optogenetics. In 2007, Reijmers *et al.* developed a transgenic mouse that enables the long-lasting genetic tagging of c-fos-active neurons.[229] The authors found neurons in the BLA that are activated during contextual fear conditioning and are reactivated during memory retrieval. Similarly, they found neurons in the lateral amygdala (LA) that were active during cued fear conditioning and reactivated during retrieval. The number of reactivated neurons correlated positively with the behavioral expression of the fear memory, indicating a stable neural correlate of associative memory. Two years later, a study showed that selective depletion of neurons of the LA, which were active during encoding of fear memory, is sufficient to erase this memory.[230] These data suggest that the memory trace is a neural ensemble active during learning. However, sufficiency of a sparse neural population to evoke memory-driven behavior was still puzzling. By creating and using new optogenetic tools, Liu *et al.* demonstrated, for the first time, that a sparse population of neurons is sufficient to support memory recall.[231] Their strategy combines a doxycycline (dox) system for inducible gene expression and viral expression of ChR2 in the dentage gyrus (DG) of dorsal hippocampus. Briefly, the author removed doxycycline (dox) from the diet of c-fos–tTA (tetracycline transactivator) transgenic mice for 3 days. During this time window, tTA was expressed specifically in the active cells (i.e., expressing c-fos) and bound to the tetracycline

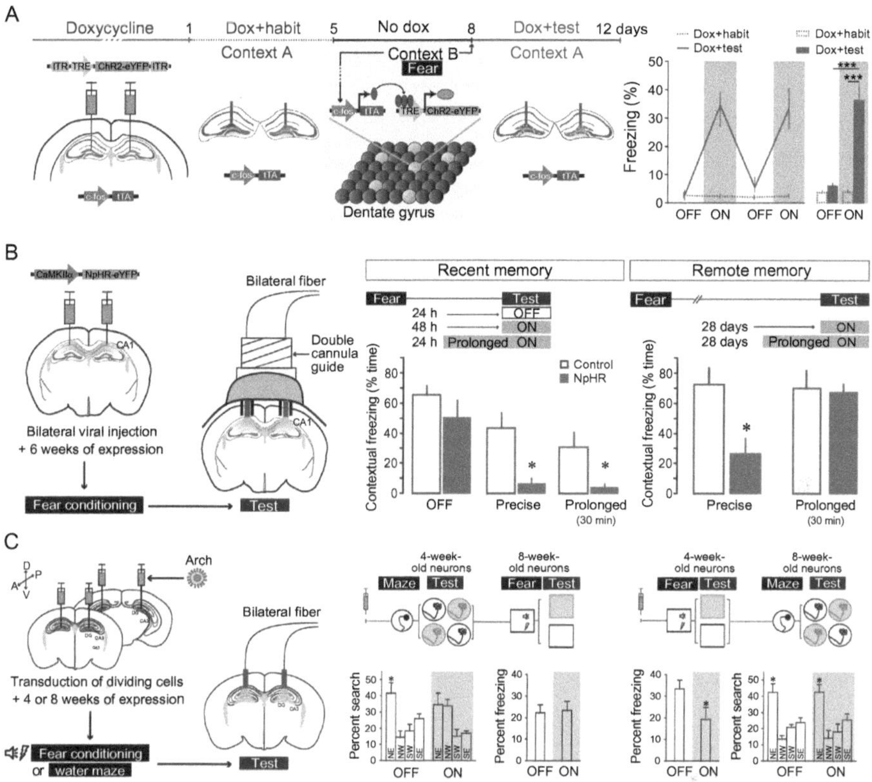

Figure 12.3 Parsing the role of hippocampal subregions in memory recall. (A) Targeting a memory engram. Selective labeling of dentate gyrus (DG) cells active during fear learning. The c-fos–tTA mice were bilaterally injected with AAV_9-TRE-ChR2-eYFP and implanted with an optical fiber over the DG. Mice were then habituated for 5 days in context A with light stimulation while on doxycycline (dox). After dox removal from the diet for 2 days (No dox), the mice were fear conditioned in context B. Without doxycycline, fear conditioning induces the expression of tTA, which binds to TRE and drives the expression of ChR2-eYFP, labeling a subpopulation of activated cells (lighter filling) in the DG. Mice were then put back on dox and tested for 5 days in context A with light stimulation. Bilateral light stimulation (ON, shaded rectangles) of the DG cells that were active during fear conditioning (dox + test, continuous lines) significantly increased freezing ~30% when compared to stimulation during the habituation period ($n=6$, $^{***}p<0.001$). Consistent with these results, the amount of freezing is significantly higher during the ON periods of light stimulation compared to the OFF periods ($n=6$, $^{***}p<0.001$). (B) Targeting recent and remote memory. After bilateral injections of lentiviral vector expressing NpHR under the control of CaMKIIα in CA1 neurons, optogenetic inhibition during the test phase of a fear conditioning protocol (Precise ON), prevented recall of recent fear memory ($n=4$ control, white bar and $n=5$ NpHR, filled bar, $p<0.05$), whereas without photoinhibition (OFF) the mice were still able to recall the memory as measured by freezing to the context ($n=4$ control, white bar

(Continued)

responsive element (TRE) of the AAV_9-TRE-*ChR2-eYFP* virus, previously injected into the DG of the mice. This cascade of activation led to ChR2 expression, specifically in the active cells of the DG (Fig. 12.3A). To create a memory engram, the authors fear conditioned the animals at the end of the 3-day window, before reintroducing dox in the mouse diet, in order to stop c-fos dependent-ChR2 expression. At that point, the animals were expressing ChR2 only in the DG cells that were active during the 3-day window and especially the cells activated during the fear conditioning. The reactivation of this specific population of neurons by shining light through an optic fiber implanted in the DG evoked freezing behavior, indicating recall of the fear memory. Along with providing the critical demonstration that an ensemble of neurons that were active during acquisition is sufficient to recall the encoded memory, this study brings a new strategy to identify and manipulate memory engrams in different paradigms and brain regions.

2.3.2 "Multiple trace theory" versus hippocampo-cortical transfer

Retrograde amnesia observed following hippocampal lesions in humans[234] and animals is temporally graded, with recent memory being impaired while remote memories remain intact, suggesting that the hippocampal formation

Figure 12.3—Cont'd and $n=5$ NpHR, filled bar; not significantly different $p>0.05$). Thirty minutes of prolonged optogenetic inhibition also prevented recall of recent fear memory ($n=7$ control, white bar; $n=3$ NpHR, filled bar; $^*p<0.05$). The same 30 min of prolonged photoinhibition did not alter remote fear memory recall (prolonged group, left: $n=3$ control, $n=4$ NpHR, $^*p>0.05$) even if precise inhibition blocked the recall of remote fear memory (precise group, right: $n=4$ control, $n=8$ NpHR, $^{**}p<0.01$). (C) Testing the role of adult born neurons. After bilateral injection into the DG of a retroviral vector that selectively transduces dividing neurons and expresses Arch, the mice were trained to associate a context to a foot shock or to find a platform in the northeast (NE) quarter of a circular watermaze. Inhibition of 4-week-old newborn neurons during the recall of these tasks significantly impaired the performance of the mice. In the first experimental group, photoinhibition of the 4-week neurons blocked the recall of the platform location, since the mice did not spent significantly more time in the platform quarter (first histogram $n=10$ mice). In the second experimental group, photoinhibition of 4-week-old neurons significantly decreased the amount of freezing (third histogram, $n=10$). Conversely, inhibition of 8-week-old newborn neurons did not have a significant effect on the recall of these two types of memory (second and fourth histograms $n=10$ mice). Four quarters of the water maze: north east (NE), north west (NW), south west (SW), south east (SE). Error bars show mean $\pm$ SEM for (A)–(C). $^*p<0.05$ *Panel (A) is adapted from Ref. 231, panel (B) is adapted from Ref. 232, and panel (C) is adapted from Ref. 233.* (See the color plate.)

has a time-limited role in memory storage. Pioneering studies on the circuitry of memory have shown that long-term spatial memory requires early involvement of the hippocampus, followed by a predominant role of the neocortex.[235,236] Specifically, pharmacological inactivation of the hippocampus impairs recent fear and spatial memory 1 day after training, but has no effect on remote memories several weeks after training.[236,237] This gradual decay in hippocampus dependency is proposed to play a role in clearing disused old memories once the memory has become dependent on cortical networks, in order to preserve the learning capacity of the hippocampus.[238] However, controversial results were obtained regarding contextual fear conditioning[239] leaving the hippocampo-cortical transfer theory elusive.

In 2011, Goshen *et al.* used optogenetics to bring new insights to this debate.[232] Capitalizing on the temporal precision of optogenetic inhibition, Goshen and colleagues showed that real-time inhibition of CA1 hippocampal excitatory neurons can reversibly abolish contextual fear memory recall, even weeks after training (Fig. 12.3B). Furthermore, when this inhibition was extended to match the typical time course of pharmacological inhibition, remote memory dependence on the hippocampus no longer existed, suggesting that long-term memory retrieval normally depends on the hippocampus but can adaptively shift to alternate structures (Fig. 12.3B). Consistently with previous work,[236,237] they also found that inhibition of neocortical areas also impairs remote memory recall, supporting the existence of multiple traces for remote memories. This study is an exemplar case showing how the temporal resolution of optogenetic tools enhances our knowledge regarding controversial points of memory physiology.

2.3.3 Hippocampal neurogenesis

The subgranular zone of the hippocampus gives rise to thousands of new neurons in the DG each day throughout the adulthood of mammals, including mice, rats, monkeys, and humans.[240–243] Evidence from studies using global perturbation or suppression of adult hippocampal neurogenesis has revealed deficits in some forms of hippocampal memory in rodents.[237,244] Although traditional manipulations of adult neurogenesis disrupt hippocampal memory function, we cannot rule out the possibility that the observed memory deficits are a result of a suppression of neurogenesis. Optogenetic manipulation of adult newborn cells allows researchers to test the involvement of neurons that developed at a specific time point for defined periods of time without killing the cells. By combining retroviral birth-dating and gene delivery with optogenetic stimulation, Gu *et al.* identified critical time points of adult-born neurons integration into the functional hippocampal trisynaptic circuit.[233] The authors used self-inactivating murine oncoretroviruses to deliver genes of interest (opsins) specifically to proliferating cells and their progeny.[245,246] When expressing ChR2, the authors characterized the integration time course of newborn neurons that establish functional synapses with CA3 pyramidal neurons as early as 2 weeks after birth and present stable synaptic responses by ~4 weeks of age. Interestingly, the newborn neurons also present a higher plasticity at this age (4 weeks) than at other stages. When expressing Arch,[22,23] bilateral silencing through yellow laser (589 nm) illumination of 4-week-old newborn neurons substantially decreased the retrieval of hippocampal-dependent spatial memory as well as contextual fear memory (Fig. 12.3C). Importantly, the same manipulation of 2- or 8-week-old neurons did not alter memory retrieval, suggesting that 4-week-old new neurons have a critical role in hippocampal memory recall. In contrast, inhibiting 4-week-old neurons did not alter acquisition of spatial memory nor contextual fear memory, showing that 4-week-old newborn neurons are selectively involved in the recall of memories. In this study, optogenetics allowed researchers to characterize the development of output circuit function of adult-born GC in the DG and revealed a precise time window in which newborn neurons exhibit enhanced plasticity, and are involved in the recall of hippocampal memories.

3. OPTOGENETIC-FACILITATED SYNAPTIC DISSECTION OF MEMORY FUNCTION

3.1. How optogenetics unraveled pathway-specific long-term synaptic plasticity

Donald Hebb (1949) speculated that psychological concepts could be represented by simultaneous activity of many nerve cells distributed throughout the brain, which he called a cell assembly. He postulated that cell assemblies are formed by an amplification process taking place in all synapses between active nerve cells during learning. This process of synaptic strengthening depends on coincident neural activity of pre- and postsynaptic neurons and was later called Hebbian learning. Decades after Hebb's postulate, the synaptic phenomena of long-term potentiation (LTP) and long-term depression (LTD) were discovered in rodents and together provided a possible mechanism for memory acquisition and storage in the nervous system.[247] These results were then replicated in many regions of the rodent brain where LTP and LTD can be induced experimentally (LTP[248,249] and LTD[250–252]). Moreover, learning has been shown to be correlated with synaptic plasticity.[97,253–257] The recent development of optogenetic tools has allowed us to probe the role of synaptic plasticity in learning and memory,[258] surpassing the limits of classical protocols of long-term plasticity induction and recording, in terms of projection and cell-type specificity as well as population spike-timing control.

In many areas of the brain, LTP is generally induced by electrically stimulating bundles of fibers (Schaffer collaterals in the hippocampus, internal capsule in the BLA) at high-frequency stimulation (~100 Hz), whereas LTD is induced by low-frequency stimulations (~1 Hz). The plasticity is then quantified from the postsynaptic cell response to a single electric stimulation, recorded either intracellularly through voltage-clamp experiments or from the extracellular space where "population spikes" can be recorded. After high- or low-frequency stimulation protocols, the postsynaptic response will present a stable increase or decrease in neural response amplitude, respectively. However, electrical stimulation presents a major limitation because it nonspecifically activates all the cell bodies and fibers traveling through the region surrounding the location of the stimulation electrode. One of the main advantages of optogenetic approaches is the ability to specifically activate specific circuit elements targeted either genetically or based on projection, leaving neighboring neurons and neural processes unaffected (Fig. 12.1B).

Pathway-specific synaptic plasticity[259–261] has been correlated with sensory manipulation or learning.[257,262] These studies have been limited to pathways where axons converge onto a structure via anatomically distinct tracts, such as the internal and the external capsule carrying thalamic and cortical afferents, respectively, to the LA. Optogenetic tools allow more precise pathway-specific manipulation of neural activity, even if axon fibers from different regions are intermingled (Fig. 12.1B). For example, during Pavlovian conditioning, if the CS and US are delivered through different sensory modalities (like an auditory cue and a gustatory outcome), they will recruit specific pathways arising from different thalamic nuclei and projecting to the LA.[263,264] Optogenetic tools offer the possibility to selectively manipulate CS and US pathways, to study the mechanisms underlying convergence of CS and US information that leads to synaptic plasticity and associative learning.

Because axonal fibers from different cortical areas are intermingled in the amygdala and cannot be activated selectively using conventional electrical stimulation, photostimulation of specific inputs can bring new insights regarding the connectivity and plasticity of these different inputs. For example, temporal association cortex (TAC) inputs to the LA transmits polymodal sensory information (McDonald, 1998), and plasticity in this TAC–LA path running through the external capsule is implicated in fear learning.[97,265] GABAergic neurons in the external capsule are known to provide feedforward inhibition in the LA,[266] but how they affect synaptic transmission and plasticity of inputs from specific cortical areas remained unknown. In 2011, Morozov *et al.* achieved selective activation of fibers from the TAC or the ACC through ChR2 photostimulation of either TAC or AAC terminals in the LA.[267] Activation of TAC fibers produced LTP in LA pyramidal cells recorded in whole-cell patch-clamp, but only after disconnecting the external capsule from the LA or applying a $GABA_A$ antagonist. This finding implies that paracapsular and/or local interneurons suppress plasticity of TAC to LA synapses. In contrast, optical activation of ACC to LA pathway, which bypasses the external capsule, produced LTP without addition of $GABA_A$ antagonist or external capsule–LA transection. Thus, GABAergic neurons appear to strongly gate sensory information conveyed by the external capsule. By combining optogenetic stimulation and fiber transections, the authors were able to show that the external capsule provides pathway-specific gating of amygdala plasticity.

One way to measure postsynaptic LTP or LTD in a single cell is to quantify the ratio between α-amino-3-hydroxy-5-methyl-4-isoxazolepropionate (AMPA) and *N*-methyl-D-aspartate (NMDA) glutamate receptors of this cell. Indeed after potentiation of a synapse, the AMPA/NMDA ratio has been shown to increase.[268] Interestingly, fear conditioning is known to induce long-term potentiation of thalamus to LA synapses through a stable increase in AMPA/NMDA ratio.[97,256] Similarly, reward-related learning also generates LTP in BLA neurons with a graded increase in synaptic potentiation in correlation with learning performance.[253–255,257] In previous studies, the potentiation was measured after extracellular electric stimulation of inputs to the recorded cell. By using optogenetics, Britt and colleagues examined the innervation patterns and synaptic properties of inputs from the vHPC, BLA, and PFC to the NAc.[148] After characterizing the three pathways, they assayed each of them for cocaine-induced synaptic plasticity, by quantifying AMPA/NMDA ratio in cells recorded *ex vivo* of cocaine-treated mice and drug-naïve control mice. They showed that vHPC inputs were uniquely localized and predominant in the medial shell of the NAc, and selectively potentiated after cocaine exposure.[148] In this study, optogenetics allowed researchers to show how learning induces synaptic plasticity in specific inputs in a structure of interest.

Another recent study exploited the power of optogenetic tools to study the mechanisms of drug-induced synaptic plasticity.[269] In this study, the authors used optogenetic tools to induce pathway-specific synaptic depression at the infralimbic cortex to NAc synapses. They expressed ChR2 in cell bodies of projection neurons coming from the infralimbic cortex and stimulated axon terminals of these neurons in the NAc shell to induce synaptic depression. The authors showed that this manipulation reduced cocaine-induced locomotor sensitization and showed that this is due to reversal of synaptic potentiation induced by cocaine.

3.2. Optogenetic study of monosynaptic and subcellular currents

Stimulating inputs can often trigger network activity that can lead to compound currents with monosynaptic as well as polysynaptic excitatory and/or inhibitory components. Using optogenetic approaches, two studies elegantly combined optogenetic activation of terminals with bath application of sodium and potassium channels blockers (1 μM tetrodotoxin, TTX, and 100 μM 4-aminopyridin, 4AP, respectively) to record exclusive monosynaptic currents evoked in postsynaptic cell recorded in whole-cell by photostimulation of presynaptic terminals expressing ChR2.[50,82] Because ChR2 is a nonspecific cation channel itself, it can permit the passage of both sodium and calcium ions leading to vesicle fusion even in the presence of TTX, which silences all but spontaneous vesicle release. However, ChR2 activation alone is not sufficient to trigger synaptic release in the presence of TTX, as measured by voltage clamp experiments.[50,82] After addition of 4AP, which blocks voltage-gated potassium channels,[270,271] stimulation of terminals expressing ChR2 then triggers monosynaptic currents. Whereas this approach has proved to be extremely useful to determine input circuit mechanism, it could also be useful for pathway-specific synaptic plasticity quantification.

LTP can be induced even at the level of single dendritic spines[272] using two photon glutamate uncaging[273–275] and is used to study molecular mechanisms underlying LTP.[276] Glutamate uncaging, however, limits study of LTP mechanisms to the postsynaptic site. Two-photon excitability of ChR2[277,278] allows manipulation of neural activity at an unprecedented spatial resolution and can possibly be used to study plasticity mechanisms at single presynaptic boutons, in addition to imaging single presynaptic boutons.[279,280]

The axon initial segment (AIS), where action potentials initiate, is a region rich in voltage-gated sodium channel (Nav).[281–285] Plasticity at the AIS, either through relocation[286] or change in size,[287] is a mechanism that regulates neural activity.[288] Optogenetic tools can be targeted to specific regions within the cell by adding a targeting signal to ChR2 construct, allowing us to selectively study the contribution of these regions to synaptic plasticity. The ability to target ChR2 selectively to the AIS[289] may help to elucidate mechanisms of homeostatic regulation of neural activity by the AIS.

3.3. Spike-timing-dependent plasticity

Spike-timing-dependent plasticity or STDP[290] involves a change in synaptic strength based on the difference in firing time of pre- and postsynaptic neurons. ChR2, in combination with the red-shifted opsin C1V1, allows remote induction of STDP by millisecond scale manipulation of neural activity of both the presynaptic and postsynaptic neurons. This can be achieved by selectively stimulating axon terminals of the presynaptic neuron expressing ChR2 with blue light and stimulating the postsynaptic neuron expressing C1V1 with yellow light.[17] Despite the advantages over previous methods, it is important to keep in mind that opsins activate or inhibit neurons through their permeability to ions, which can also regulate synaptic function. For example, ChR2 is a nonspecific cation channel and Ca^{2+} concentration changes influence many intracellular pathways affecting, for example, the probability of neurotransmitter release.[291] As well as for long-term plasticity, optogenetic tools also bring the possibility to study STDP in a pathway and cell-specific manner.

3.4. Homeostatic plasticity

Homeostatic processes are proposed to explain the discrepancy between synaptic plasticity and the stability of brain function.[292] Although network-wide excitation triggers compensatory homeostatic changes, it was unknown whether neurons initiate homeostatic synaptic changes in response to an increase in excitation.[293] In 2010, Goold and Nicoll used optogenetics to study homeostatic synaptic plasticity.[294] Using sparse transfection, the authors expressed ChR2 in single cells of organotypic hippocampal slice cultures from ~1-week-old rodents. Four to eight days after viral expression, the authors stimulated the transfected cell for 24 h (3 Hz, 50 ms pulses) and observed a postsynaptic depression of the synaptic inputs to the cell that require Ca^{2+} entry through L-type Ca^{2+} channels. The researchers also showed that spiking of the cell is irrelevant to the induction of this homeostatic plasticity. In this particular set of experiments, optogenetics allowed for chronic stimulation of a single-cell unraveling compensatory postsynaptic depression of glutamate receptor function.

Together, these recent studies of synaptic plasticity with optogenetics highlight how these tools can improve our understanding of the synaptic mechanisms underlying learning and memory through circuit dissection at millisecond time-scale with a cell-type-specific resolution.

4. CONCLUSIONS

Elucidation of the neural substrates underlying learning and memory is facilitated by tools that precisely control neural activity and compatible readout methods. Optogenetics can directly probe the causal role of circuit elements with millisecond timescale and cell-type-specific resolution and can be coupled with electrophysiology and freely moving behavior in mammals. This new way of probing the neural circuits underlying learning and memory has provided considerable insight into the function of specific neuron populations and synaptic inputs in many structures, such as the hippocampus, the amygdala, the NAc, and the VTA. Along with determining the function of specific cell ensembles in supporting fear or reward memories, optogenetic tools have been used to parse the involvement of neurons or populations of synapses in selective phases of memory. Optogenetic tools have also been extremely valuable in the study of synaptic plasticity during memory formation. However, understanding the molecular mechanisms of memory consolidation could also be enhanced through optogenetic control of intracellular signaling. Indeed, a complete panel of light-sensitive molecules interfering with biochemical pathways has been developed but still needs to be applied to mammalian neurons, in order to test the function of specific intracellular pathways or genes during memory formation.

ACKNOWLEDGMENTS

This work was supported by the JPB Foundation, Picower Institute Innovation Funds, The Whitehall Foundation, the Picower Institute and Department of Brain and Cognitive

Science startup funds (MIT), and the Klingenstein Foundation Award. Anna Beyeler is a postdoctoral fellow of the Swiss National Science Foundation (SNSF).

REFERENCES

1. Fork RL. Laser stimulation of nerve cells in Aplysia. *Science*. 1971;171(3974):907–908.
2. Farber IC, Grinvald A. Identification of presynaptic neurons by laser photostimulation. *Science*. 1983;222(4627):1025–1027.
3. Schmucker D, Su AL, Beermann A, Jäckle H, Jay DG. Chromophore-assisted laser inactivation of patched protein switches cell fate in the larval visual system of Drosophila. *Proc Natl Acad Sci USA*. 1994;91(7):2664–2668.
4. Zemelman BV, Lee GA, Ng M, Miesenböck G. Selective photostimulation of genetically chARGed neurons. *Neuron*. 2002;33(1):15–22.
5. Schneider MB, Gradinaru V, Zhang F, Deisseroth K. Controlling neuronal activity. *Am J Psychiatry*. 2008;165(5):562.
6. Mattis J, Tye KM, Ferenczi EA, et al. Principles for applying optogenetic tools derived from direct comparative analysis of microbial opsins. *Nat Methods*. 2011;9(2):159–172.
7. Nagel G, Szellas T, Huhn W, et al. Channelrhodopsin-2, a directly light-gated cation-selective membrane channel. *Proc Natl Acad Sci USA*. 2003;100(24):13940–13945.
8. Adamantidis AR, Zhang F, Aravanis AM, Deisseroth K, de Lecea L. Neural substrates of awakening probed with optogenetic control of hypocretin neurons. *Nature*. 2007;450(7168):420–424.
9. Aravanis AM, Wang L-P, Zhang F, et al. An optical neural interface: in vivo control of rodent motor cortex with integrated fiberoptic and optogenetic technology. *J Neural Eng*. 2007;4(3):S143–S156.
10. Boyden ES, Zhang F, Bamberg E, Nagel G, Deisseroth K. Millisecond-timescale, genetically targeted optical control of neural activity. *Nat Neurosci*. 2005; 8(9):1263–1268.
11. Zhang F, Wang L-P, Brauner M, et al. Multimodal fast optical interrogation of neural circuitry. *Nature*. 2007;446(7136):633–639.
12. Gradinaru V, Thompson KR, Zhang F, et al. Targeting and readout strategies for fast optical neural control in vitro and in vivo. *J Neurosci*. 2007;27(52):14231–14238.
13. Nagel G, Brauner M, Liewald JF, Adeishvili N, Bamberg E, Gottschalk A. Light activation of channelrhodopsin-2 in excitable cells of Caenorhabditis elegans triggers rapid behavioral responses. *Curr Biol*. 2005;15(24):2279–2284.
14. Berndt A, Schoenenberger P, Mattis J, et al. High-efficiency channelrhodopsins for fast neuronal stimulation at low light levels. *Proc Natl Acad Sci USA*. 2011; 108(18):7595–7600.
15. Gunaydin LA, Yizhar O, Berndt A, Sohal VS, Deisseroth K, Hegemann P. Ultrafast optogenetic control. *Nat Neurosci*. 2010;13(3):387–392.
16. Lin JY, Lin MZ, Steinbach P, Tsien RY. Characterization of engineered channelrhodopsin variants with improved properties and kinetics. *Biophys J*. 2009; 96(5):1803–1814.
17. Yizhar O, Fenno LE, Prigge M, et al. Neocortical excitation/inhibition balance in information processing and social dysfunction. *Nature*. 2011;477(7363):171–178.
18. Zhang F, Prigge M, Beyrière F, et al. Red-shifted optogenetic excitation: a tool for fast neural control derived from Volvox carteri. *Nat Neurosci*. 2008;11(6):631–633.
19. Bamann C, Gueta R, Kleinlogel S, Nagel G, Bamberg E. Structural guidance of the photocycle of channelrhodopsin-2 by an interhelical hydrogen bond. *Biochemistry*. 2010;49(2):267–278.
20. Berndt A, Yizhar O, Gunaydin LA, Hegemann P, Deisseroth K. Bi-stable neural state switches. *Nat Neurosci*. 2009;12(2):229–234.

21. Yizhar O, Fenno LE, Davidson TJ, Mogri M, Deisseroth K. Optogenetics in neural systems. *Neuron*. 2011;71(1):9–34.
22. Chow BY, Han X, Dobry AS, et al. High-performance genetically targetable optical neural silencing by light-driven proton pumps. *Nature*. 2010;463(7277):98–102.
23. Gradinaru V, Zhang F, Ramakrishnan C, et al. Molecular and cellular approaches for diversifying and extending optogenetics. *Cell*. 2010;141(1):154–165.
24. Tye KM, Prakash R, Kim S-Y, et al. Amygdala circuitry mediating reversible and bidirectional control of anxiety. *Nature*. 2011;471(7338):358–362.
25. Witten IB, Lin S-C, Brodsky M, et al. Cholinergic interneurons control local circuit activity and cocaine conditioning. *Science*. 2010;330(6011):1677–1681.
26. Airan RD, Thompson KR, Fenno LE, Bernstein H, Deisseroth K. Temporally precise in vivo control of intracellular signalling. *Nature*. 2009;458(7241):1025–1029.
27. Kim J-M, Hwa J, Garriga P, Reeves PJ, RajBhandary UL, Khorana HG. Light-driven activation of β2-adrenergic receptor signaling by a chimeric rhodopsin containing the β2-adrenergic receptor cytoplasmic loops†. *Biochemistry*. 2005;44(7):2284–2292.
28. Oh E, Maejima T, Liu C, Deneris E, Herlitze S. Substitution of 5-HT1A receptor signaling by a light-activated G protein-coupled receptor. *J Biol Chem*. 2010;285(40):30825–30836.
29. Levitz J, Pantoja C, Gaub B, et al. Optical control of metabotropic glutamate receptors. *Nat Neurosci*. 2013;16(4):507–516.
30. Ryu M-H, Moskvin OV, Siltberg-Liberles J, Gomelsky M. Natural and engineered photoactivated nucleotidyl cyclases for optogenetic applications. *J Biol Chem*. 2010;285(53):41501–41508.
31. Stierl M, Stumpf P, Udwari D, et al. Light modulation of cellular cAMP by a small bacterial photoactivated adenylyl cyclase, bPAC, of the soil bacterium Beggiatoa. *J Biol Chem*. 2011;286(2):1181–1188.
32. Kandel ER. The molecular biology of memory: cAMP, PKA, CRE, CREB-1, CREB-2, and CPEB. *Mol Brain*. 2012;5:14.
33. Kennedy MJ, Hughes RM, Peteya LA, Schwartz JW, Ehlers MD, Tucker CL. Rapid blue-light-mediated induction of protein interactions in living cells. *Nat Methods*. 2010;7(12):973–975.
34. Yazawa M, Sadaghiani AM, Hsueh B, Dolmetsch RE. Induction of protein-protein interactions in live cells using light. *Nat Biotechnol*. 2009;27(10):941–945.
35. Leung DW, Otomo C, Chory J, Rosen MK. Genetically encoded photoswitching of actin assembly through the Cdc42-WASP-Arp2/3 complex pathway. *Proc Natl Acad Sci USA*. 2008;105(35):12797–12802.
36. Levskaya A, Weiner OD, Lim WA, Voigt CA. Spatiotemporal control of cell signalling using a light-switchable protein interaction. *Nature*. 2009;461(7266):997–1001.
37. Shimizu-Sato S, Huq E, Tepperman JM, Quail PH. A light-switchable gene promoter system. *Nat Biotechnol*. 2002;20(10):1041–1044.
38. Tyszkiewicz AB, Muir TW. Activation of protein splicing with light in yeast. *Nat Methods*. 2008;5(4):303–305.
39. Polstein LR, Gersbach CA. Light-inducible spatiotemporal control of gene activation by customizable zinc finger transcription factors. *J Am Chem Soc*. 2012;134(40):16480–16483.
40. Wang X, Chen X, Yang Y. Spatiotemporal control of gene expression by a light-switchable transgene system. *Nat Methods*. 2012;9(3):266–269.
41. Ye H, Daoud-El Baba M, Peng R-W, Fussenegger M. A synthetic optogenetic transcription device enhances blood-glucose homeostasis in mice. *Science*. 2011;332(6037):1565–1568.
42. Cong L, Ran FA, Cox D, et al. Multiplex genome engineering using CRISPR/Cas systems. *Science*. 2013;339(6121):819–823.

43. Tucker CL. Manipulating cellular processes using optical control of protein–protein interactions. [chapter 6]. In: Knöpfel Thomas, Boyden Edward S, eds. Elsevier, 2012:95–117. *Progress in Brain Research*. Vol. 196. Available at: http://www.sciencedirect.com/science/article/pii/B9780444594266000069. Accessed April 25, 2013.
44. Dittgen T, Nimmerjahn A, Komai S, et al. Lentivirus-based genetic manipulations of cortical neurons and their optical and electrophysiological monitoring in vivo. *Proc Natl Acad Sci USA*. 2004;101(52):18206–18211.
45. Gradinaru V, Mogri M, Thompson KR, Henderson JM, Deisseroth K. Optical deconstruction of parkinsonian neural circuitry. *Science*. 2009;324(5925):354–359.
46. Diester I, Kaufman MT, Mogri M, et al. An optogenetic toolbox designed for primates. *Nat Neurosci*. 2011;14(3):387–397.
47. Zhang F, Gradinaru V, Adamantidis AR, et al. Optogenetic interrogation of neural circuits: technology for probing mammalian brain structures. *Nat Protoc*. 2010;5(3):439–456.
48. Kumar M, Keller B, Makalou N, Sutton RE. Systematic determination of the packaging limit of lentiviral vectors. *Hum Gene Ther*. 2001;12(15):1893–1905.
49. Ciocchi S, Herry C, Grenier F, et al. Encoding of conditioned fear in central amygdala inhibitory circuits. *Nature*. 2010;468(7321):277–282.
50. Petreanu L, Mao T, Sternson SM, Svoboda K. The subcellular organization of neocortical excitatory connections. *Nature*. 2009;457(7233):1142–1145.
51. Lee JH, Durand R, Gradinaru V, et al. Global and local fMRI signals driven by neurons defined optogenetically by type and wiring. *Nature*. 2010;465(7299):788–792.
52. Dong B, Nakai H, Xiao W. Characterization of genome integrity for oversized recombinant AAV vector. *Mol Ther*. 2010;18(1):87–92.
53. Burger C, Gorbatyuk OS, Velardo MJ, et al. Recombinant AAV viral vectors pseudotyped with viral capsids from serotypes 1, 2, and 5 display differential efficiency and cell tropism after delivery to different regions of the central nervous system. *Mol Ther*. 2004;10(2):302–317.
54. Summerford C, Samulski RJ. Membrane-associated heparan sulfate proteoglycan is a receptor for adeno-associated virus type 2 virions. *J Virol*. 1998;72(2):1438–1445.
55. Di Pasquale G, Davidson BL, Stein CS, et al. Identification of PDGFR as a receptor for AAV-5 transduction. *Nat Med*. 2003;9(10):1306–1312.
56. Broekman MLD, Comer LA, Hyman BT, Sena-Esteves M. Adeno-associated virus vectors serotyped with AAV8 capsid are more efficient than AAV-1 or -2 serotypes for widespread gene delivery to the neonatal mouse brain. *Neuroscience*. 2006;138(2):501–510.
57. Lawlor PA, Bland RJ, Mouravlev A, Young D, During MJ. Efficient gene delivery and selective transduction of glial cells in the mammalian brain by AAV serotypes isolated from nonhuman primates. *Mol Ther*. 2009;17(10):1692–1702.
58. Fink DJ, DeLuca NA, Goins WF, Glorioso JC. Gene transfer to neurons using herpes simplex virus-based vectors. *Annu Rev Neurosci*. 1996;19:265–287.
59. Lilley CE, Branston RH, Coffin RS. Herpes simplex virus vectors for the nervous system. *Curr Gene Ther*. 2001;1(4):339–358.
60. Lima SQ, Hromádka T, Znamenskiy P, Zador AM. PINP: a new method of tagging neuronal populations for identification during in vivo electrophysiological recording. *PLoS One*. 2009;4(7):e6099.
61. Lobo MK, Covington III HE, Chaudhury D, et al. Cell type-specific loss of BDNF signaling mimics optogenetic control of cocaine reward. *Science*. 2010;330(6002):385–390.
62. Neve RL. Overview of gene delivery into cells using HSV-1-based vectors. *Curr Protoc Neurosci*. 2012;61:4.12.1–4.12.7.

63. Callaway EM. Transneuronal circuit tracing with neurotropic viruses. *Curr Opin Neurobiol.* 2008;18(6):617–623.
64. Ugolini G. Specificity of rabies virus as a transneuronal tracer of motor networks: transfer from hypoglossal motoneurons to connected second-order and higher order central nervous system cell groups. *J Comp Neurol.* 1995;356(3):457–480.
65. Ugolini G. Advances in viral transneuronal tracing. *J Neurosci Methods.* 2010;194(1):2–20.
66. Wickersham IR, Lyon DC, Barnard RJO, et al. Monosynaptic restriction of transsynaptic tracing from single, genetically targeted neurons. *Neuron.* 2007;53(5): 639–647.
67. Coulon P, Rollin P, Blancou J, Flamand A. Avirulent mutants of the CVS strain of rabies virus. *Comp Immunol Microbiol Infect Dis.* 1982;5(1–3):117–122.
68. Arenkiel BR, Peca J, Davison IG, et al. In vivo light-induced activation of neural circuitry in transgenic mice expressing channelrhodopsin-2. *Neuron.* 2007;54(2): 205–218.
69. Dhawale AK, Hagiwara A, Bhalla US, Murthy VN, Albeanu DF. Non-redundant odor coding by sister mitral cells revealed by light addressable glomeruli in the mouse. *Nat Neurosci.* 2010;13(11):1404–1412.
70. Hägglund M, Borgius L, Dougherty KJ, Kiehn O. Activation of groups of excitatory neurons in the mammalian spinal cord or hindbrain evokes locomotion. *Nat Neurosci.* 2010;13(2):246–252.
71. Chuhma N, Tanaka KF, Hen R, Rayport S. Functional connectome of the striatal medium spiny neuron. *J Neurosci.* 2011;31(4):1183–1192.
72. Zhao S, Ting JT, Atallah HE, et al. Cell-type specific optogenetic mice for dissecting neural circuitry function. *Nat Methods.* 2011;8(9):745.
73. Atasoy D, Aponte Y, Su HH, Sternson SM. A FLEX switch targets Channelrhodopsin-2 to multiple cell types for imaging and long-range circuit mapping. *J Neurosci.* 2008;28(28):7025–7030.
74. Kuhlman SJ, Huang ZJ. High-resolution labeling and functional manipulation of specific neuron types in mouse brain by Cre-activated viral gene expression. *PLoS One.* 2008;3(4):e2005.
75. Haubensak W, Kunwar PS, Cai H, et al. Genetic dissection of an amygdala microcircuit that gates conditioned fear. *Nature.* 2010;468(7321):270–276.
76. Kravitz AV, Tye LD, Kreitzer AC. Distinct roles for direct and indirect pathway striatal neurons in reinforcement. *Nat Neurosci.* 2012;15(6):816–818.
77. Lewis Jr TL, Mao T, Svoboda K, Arnold DB. Myosin-dependent targeting of transmembrane proteins to neuronal dendrites. *Nat Neurosci.* 2009;12(5):568–576.
78. Lewis Jr TL, Mao T, Arnold DB. A role for myosin VI in the localization of axonal proteins. *PLoS Biol.* 2011;9(3):e1001021.
79. Petreanu L, Huber D, Sobczyk A, Svoboda K. Channelrhodopsin-2-assisted circuit mapping of long-range callosal projections. *Nat Neurosci.* 2007;10(5):663–668.
80. Huber D, Gutnisky DA, Peron S, et al. Multiple dynamic representations in the motor cortex during sensorimotor learning. *Nature.* 2012;484(7395):473–478.
81. Zhang S-J, Ye J, Miao C, et al. Optogenetic dissection of entorhinal-hippocampal functional connectivity. *Science.* 2013;340(6128):1232627.
82. Hull C, Adesnik H, Scanziani M. Neocortical disynaptic inhibition requires somatodendritic integration in interneurons. *J Neurosci.* 2009;29(28):8991–8995.
83. Jennings JH, Sparta DR, Stamatakis AM, et al. Distinct extended amygdala circuits for divergent motivational states. *Nature.* 2013;496(7444):224–228.
84. Blair HT, Sotres-Bayon F, Moita MAP, Ledoux JE. The lateral amygdala processes the value of conditioned and unconditioned aversive stimuli. *Neuroscience.* 2005;133(2):561–569.

85. Johansen JP, Tarpley JW, LeDoux JE, Blair HT. Neural substrates for expectation-modulated fear learning in the amygdala and periaqueductal gray. *Nat Neurosci.* 2010;13(8):979–986.
86. Johansen JP, Wolff SBE, Lüthi A, LeDoux JE. Controlling the elements: an optogenetic approach to understanding the neural circuits of fear. *Biol Psychiatry.* 2012;71(12):1053–1060.
87. Helmstetter FJ, Parsons RG, Gafford GM. Macromolecular synthesis, distributed synaptic plasticity, and fear conditioning. *Neurobiol Learn Mem.* 2008;89(3):324–337.
88. Letzkus JJ, Wolff SBE, Meyer EMM, et al. A disinhibitory microcircuit for associative fear learning in the auditory cortex. *Nature.* 2011;480(7377):331–335.
89. Weinberger NM. The medial geniculate, not the amygdala, as the root of auditory fear conditioning. *Hear Res.* 2011;274(1–2):61–74.
90. Li XF, Stutzmann GE, LeDoux JE. Convergent but temporally separated inputs to lateral amygdala neurons from the auditory thalamus and auditory cortex use different postsynaptic receptors: in vivo intracellular and extracellular recordings in fear conditioning pathways. *Learn Mem.* 1996;3(2–3):229–242.
91. Barot SK, Chung A, Kim JJ, Bernstein IL. Functional imaging of stimulus convergence in amygdalar neurons during Pavlovian fear conditioning. *PLoS One.* 2009;4(7): e6156.
92. Bauer EP, Schafe GE, LeDoux JE. NMDA receptors and L-type voltage-gated calcium channels contribute to long-term potentiation and different components of fear memory formation in the lateral amygdala. *J Neurosci.* 2002;22(12):5239–5249.
93. Collins DR, Paré D. Differential fear conditioning induces reciprocal changes in the sensory responses of lateral amygdala neurons to the CS(+) and CS(-). *Learn Mem.* 2000;7(2):97–103.
94. Goosens KA, Hobin JA, Maren S. Auditory-evoked spike firing in the lateral amygdala and Pavlovian fear conditioning: mnemonic code or fear bias? *Neuron.* 2003;40(5):1013–1022.
95. Huang YY, Kandel ER. Postsynaptic induction and PKA-dependent expression of LTP in the lateral amygdala. *Neuron.* 1998;21(1):169–178.
96. Humeau Y, Herry C, Kemp N, et al. Dendritic spine heterogeneity determines afferent-specific Hebbian plasticity in the amygdala. *Neuron.* 2005;45(1):119–131.
97. McKernan MG, Shinnick-Gallagher P. Fear conditioning induces a lasting potentiation of synaptic currents in vitro. *Nature.* 1997;390(6660):607–611.
98. Muller J, Corodimas KP, Fridel Z, LeDoux JE. Functional inactivation of the lateral and basal nuclei of the amygdala by muscimol infusion prevents fear conditioning to an explicit conditioned stimulus and to contextual stimuli. *Behav Neurosci.* 1997;111(4):683–691.
99. Quirk GJ, Repa C, LeDoux JE. Fear conditioning enhances short-latency auditory responses of lateral amygdala neurons: parallel recordings in the freely behaving rat. *Neuron.* 1995;15(5):1029–1039.
100. Rodrigues SM, Schafe GE, LeDoux JE. Intra-amygdala blockade of the NR2B subunit of the NMDA receptor disrupts the acquisition but not the expression of fear conditioning. *J Neurosci.* 2001;21(17):6889–6896.
101. Romanski LM, Clugnet MC, Bordi F, LeDoux JE. Somatosensory and auditory convergence in the lateral nucleus of the amygdala. *Behav Neurosci.* 1993;107(3): 444–450.
102. Rosenkranz JA, Grace AA. Dopamine-mediated modulation of odour-evoked amygdala potentials during pavlovian conditioning. *Nature.* 2002;417(6886):282–287.
103. Johansen J, Hamanka H, Diaz-Mataix L, LeDoux J. Hebbian and neuromodulatory mechanisms act synergistically to instruct associative memory formation. *Soc Neurosci Abst.* 2010;914:15.

104. Goosens KA, Maren S. Pretraining NMDA receptor blockade in the basolateral complex, but not the central nucleus, of the amygdala prevents savings of conditional fear. *Behav Neurosci.* 2003;117(4):738–750.
105. Wilensky AE, Schafe GE, Kristensen MP, LeDoux JE. Rethinking the fear circuit: the central nucleus of the amygdala is required for the acquisition, consolidation, and expression of Pavlovian fear conditioning. *J Neurosci.* 2006;26(48):12387–12396.
106. Samson RD, Paré D. Activity-dependent synaptic plasticity in the central nucleus of the amygdala. *J Neurosci.* 2005;25(7):1847–1855.
107. Cassell MD, Freedman LJ, Shi C. The intrinsic organization of the central extended amygdala. *Ann N Y Acad Sci.* 1999;877(1):217–241.
108. Huber D, Veinante P, Stoop R. Vasopressin and oxytocin excite distinct neuronal populations in the central amygdala. *Science.* 2005;308(5719):245–248.
109. Sun N, Yi H, Cassell MD. Evidence for a GABAergic interface between cortical afferents and brainstem projection neurons in the rat central extended amygdala. *J Comp Neurol.* 1994;340(1):43–64.
110. Krettek JE, Price JL. A description of the amygdaloid complex in the rat and cat with observations on intra-amygdaloid axonal connections. *J Comp Neurol.* 1978;178(2):255–280.
111. LeDoux JE, Iwata J, Cicchetti P, Reis DJ. Different projections of the central amygdaloid nucleus mediate autonomic and behavioral correlates of conditioned fear. *J Neurosci.* 1988;8(7):2517–2529.
112. Veening JG, Swanson LW, Sawchenko PE. The organization of projections from the central nucleus of the amygdala to brainstem sites involved in central autonomic regulation: a combined retrograde transport-immunohistochemical study. *Brain Res.* 1984;303(2):337–357.
113. Quirk GJ, Armony JL, LeDoux JE. Fear conditioning enhances different temporal components of tone-evoked spike trains in auditory cortex and lateral amygdala. *Neuron* 1997;19(3):613–624.
114. Suga N, Ma X. Multiparametric corticofugal modulation and plasticity in the auditory system. *Nat Rev Neurosci.* 2003;4(10):783–794.
115. Weinberger NM. Auditory associative memory and representational plasticity in the primary auditory cortex. *Hear Res.* 2007;229(1–2):54–68.
116. Froemke RC, Merzenich MM, Schreiner CE. A synaptic memory trace for cortical receptive field plasticity. *Nature.* 2007;450(7168):425–429.
117. Ji W, Suga N, Gao E. Effects of agonists and antagonists of NMDA and ACh receptors on plasticity of bat auditory system elicited by fear conditioning. *J Neurophysiol.* 2005;94(2):1199–1211.
118. Kheirbek MA, Drew LJ, Burghardt NS, et al. Differential control of learning and anxiety along the dorsoventral axis of the dentate gyrus. *Neuron.* 2013;77(5):955–968.
119. Burgess N, Maguire EA, O'Keefe J. The human hippocampus and spatial and episodic memory. *Neuron.* 2002;35(4):625–641.
120. Campbell S, Marriott M, Nahmias C, MacQueen GM. Lower hippocampal volume in patients suffering from depression: a meta-analysis. *Am J Psychiatry.* 2004;161(4):598–607.
121. Dannlowski U, Stuhrmann A, Beutelmann V, et al. Limbic scars: long-term consequences of childhood maltreatment revealed by functional and structural magnetic resonance imaging. *Biol Psychiatry.* 2012;71(4):286–293.
122. Gilbertson MW, Shenton ME, Ciszewski A, et al. Smaller hippocampal volume predicts pathologic vulnerability to psychological trauma. *Nat Neurosci.* 2002;5(11):1242–1247.
123. Irle E, Ruhleder M, Lange C, et al. Reduced amygdalar and hippocampal size in adults with generalized social phobia. *J Psychiatry Neurosci.* 2010;35(2):126–131.

124. Kitayama N, Vaccarino V, Kutner M, Weiss P, Bremner JD. Magnetic resonance imaging (MRI) measurement of hippocampal volume in posttraumatic stress disorder: a meta-analysis. *J Affect Disord.* 2005;88(1):79–86.
125. Bannerman DM, Yee BK, Good MA, Heupel MJ, Iversen SD, Rawlins JN. Double dissociation of function within the hippocampus: a comparison of dorsal, ventral, and complete hippocampal cytotoxic lesions. *Behav Neurosci.* 1999;113(6):1170–1188.
126. Kjelstrup KG, Tuvnes FA, Steffenach H-A, Murison R, Moser EI, Moser M-B. Reduced fear expression after lesions of the ventral hippocampus. *Proc Natl Acad Sci USA.* 2002;99(16):10825–10830.
127. Richmond MA, Yee BK, Pouzet B, et al. Dissociating context and space within the hippocampus: effects of complete, dorsal, and ventral excitotoxic hippocampal lesions on conditioned freezing and spatial learning. *Behav Neurosci.* 1999;113(6):1189–1203.
128. McHugh TJ, Jones MW, Quinn JJ, et al. Dentate gyrus NMDA receptors mediate rapid pattern separation in the hippocampal network. *Science.* 2007;317(5834): 94–99.
129. Madisen L, Mao T, Koch H, et al. A toolbox of Cre-dependent optogenetic transgenic mice for light-induced activation and silencing. *Nat Neurosci.* 2012;15(5):793–802.
130. Perkins Jr CC, Weyant RG. The interval between training and test trials as a determiner of the slope of generalization gradients. *J Comp Physiol Psychol.* 1958;51(5):596–600.
131. Wiltgen BJ, Silva AJ. Memory for context becomes less specific with time. *Learn Mem.* 2007;14(4):313–317.
132. Zhou Riccio. Manipulation of components of context: the context shift effect and forgetting of stimulus attributes. *Learn Motiv.* 1996;27(4):400–407.
133. Xu W, Südhof TC. A neural circuit for memory specificity and generalization. *Science.* 2013;339(6125):1290–1295.
134. Dolleman-Van der Weel MJ, Lopes da Silva FH, Witter MP. Nucleus reuniens thalami modulates activity in hippocampal field CA1 through excitatory and inhibitory mechanisms. *J Neurosci.* 1997;17(14):5640–5650.
135. Zhang Y, Yoshida T, Katz DB, Lisman JE. NMDAR antagonist action in thalamus imposes δ oscillations on the hippocampus. *J Neurophysiol.* 2012;107(11):3181–3189.
136. Fields HL, Hjelmstad GO, Margolis EB, Nicola SM. Ventral tegmental area neurons in learned appetitive behavior and positive reinforcement. *Annu Rev Neurosci.* 2007;30:289–316.
137. Wise RA. Dopamine, learning and motivation. *Nat Rev Neurosci.* 2004;5(6):483–494.
138. Dobi A, Margolis EB, Wang H-L, Harvey BK, Morales M. Glutamatergic and non-glutamatergic neurons of the ventral tegmental area establish local synaptic contacts with dopaminergic and nondopaminergic neurons. *J Neurosci.* 2010;30(1):218–229.
139. Swanson LW. The projections of the ventral tegmental area and adjacent regions: a combined fluorescent retrograde tracer and immunofluorescence study in the rat. *Brain Res Bull.* 1982;9(1–6):321–353.
140. Bayer HM, Glimcher PW. Midbrain dopamine neurons encode a quantitative reward prediction error signal. *Neuron.* 2005;47(1):129–141.
141. Schultz W. Behavioral theories and the neurophysiology of reward. *Annu Rev Psychol.* 2006;57:87–115.
142. Schultz W, Dayan P, Montague PR. A neural substrate of prediction and reward. *Science.* 1997;275(5306):1593–1599.
143. Tsai H-C, Zhang F, Adamantidis A, et al. Phasic firing in dopaminergic neurons is sufficient for behavioral conditioning. *Science.* 2009;324(5930):1080–1084.
144. Adamantidis AR, Tsai H-C, Boutrel B, et al. Optogenetic interrogation of dopaminergic modulation of the multiple phases of reward-seeking behavior. *J Neurosci.* 2011;31(30):10829–10835.

145. Witten IB, Steinberg EE, Lee SY, et al. Recombinase-driver rat lines: tools, techniques, and optogenetic application to dopamine-mediated reinforcement. *Neuron.* 2011;72(5):721–733.
146. Kim KM, Baratta MV, Yang A, Lee D, Boyden ES, Fiorillo CD. Optogenetic mimicry of the transient activation of dopamine neurons by natural reward is sufficient for operant reinforcement. *PLoS One.* 2012;7(4):e33612.
147. Stuber GD, Sparta DR, Stamatakis AM, et al. Excitatory transmission from the amygdala to nucleus accumbens facilitates reward seeking. *Nature.* 2011;475(7356):377–380.
148. Britt JP, Benaliouad F, McDevitt RA, Stuber GD, Wise RA, Bonci A. Synaptic and behavioral profile of multiple glutamatergic inputs to the nucleus accumbens. *Neuron.* 2012;76(4):790–803.
149. Cohen JY, Haesler S, Vong L, Lowell BB, Uchida N. Neuron-type-specific signals for reward and punishment in the ventral tegmental area. *Nature.* 2012;482(7383):85–88.
150. Tan KR, Yvon C, Turiault M, et al. GABA neurons of the VTA drive conditioned place aversion. *Neuron.* 2012;73(6):1173–1183.
151. Van Zessen R, Phillips JL, Budygin EA, Stuber GD. Activation of VTA GABA neurons disrupts reward consumption. *Neuron.* 2012;73(6):1184–1194.
152. Stamatakis AM, Stuber GD. Activation of lateral habenula inputs to the ventral midbrain promotes behavioral avoidance. *Nat Neurosci.* 2012;15(8):1105–1107.
153. Tye KM, Deisseroth K. Optogenetic investigation of neural circuits underlying brain disease in animal models. *Nat Rev Neurosci.* 2012;13(4):251–266.
154. Gallistel CR, Shizgal P, Yeomans JS. A portrait of the substrate for self-stimulation. *Psychol Rev.* 1981;88(3):228–273.
155. Redgrave P. Modulation of intracranial self-stimulation behaviour by local perfusions of dopamine, noradrenaline and serotonin within the caudate nucleus and nucleus accumbens. *Brain Res.* 1978;155(2):277–295.
156. Wise RA, Bozarth MA. Brain reward circuitry: four circuit elements "wired" in apparent series. *Brain Res Bull.* 1984;12(2):203–208.
157. Lammel S, Hetzel A, Häckel O, Jones I, Liss B, Roeper J. Unique properties of mesoprefrontal neurons within a dual mesocorticolimbic dopamine system. *Neuron.* 2008;57(5):760–773.
158. Margolis FL, Grillo M, Brown CE, Williams TH, Pitcher RG, Elgar GJ. Enzymatic and immunological evidence for two forms of carnosinase in the mouse. *Biochim Biophys Acta.* 1979;570(2):311–323.
159. Briand LA, Vassoler FM, Pierce RC, Valentino RJ, Blendy JA. Ventral tegmental afferents in stress-induced reinstatement: the role of cAMP response element-binding protein. *J Neurosci.* 2010;30(48):16149–16159.
160. Christianson JP, Jennings JH, Ragole T, et al. Safety signals mitigate the consequences of uncontrollable stress via a circuit involving the sensory insular cortex and bed nucleus of the stria terminalis. *Biol Psychiatry.* 2011;70(5):458–464.
161. Mahler SV, Aston-Jones GS. Fos activation of selective afferents to ventral tegmental area during cue-induced reinstatement of cocaine seeking in rats. *J Neurosci.* 2012;32(38):13309–13326.
162. Dumont EC, Williams JT. Noradrenaline triggers GABAA inhibition of bed nucleus of the stria terminalis neurons projecting to the ventral tegmental area. *J Neurosci.* 2004;24(38):8198–8204.
163. Geisler S, Zahm DS. Afferents of the ventral tegmental area in the rat-anatomical substratum for integrative functions. *J Comp Neurol.* 2005;490(3):270–294.
164. Jalabert M, Aston-Jones G, Herzog E, Manzoni O, Georges F. Role of the bed nucleus of the stria terminalis in the control of ventral tegmental area dopamine neurons. *Prog Neuropsychopharmacol Biol Psychiatry.* 2009;33(8):1336–1346.

165. Kudo T, Uchigashima M, Miyazaki T, et al. Three types of neurochemical projection from the bed nucleus of the stria terminalis to the ventral tegmental area in adult mice. *J Neurosci*. 2012;32(50):18035–18046.
166. Kim S-Y, Adhikari A, Lee SY, et al. Diverging neural pathways assemble a behavioural state from separable features in anxiety. *Nature*. 2013;496(7444):219–223.
167. Dabrowska J, Hazra R, Ahern TH, et al. Neuroanatomical evidence for reciprocal regulation of the corticotrophin-releasing factor and oxytocin systems in the hypothalamus and the bed nucleus of the stria terminalis of the rat: implications for balancing stress and affect. *Psychoneuroendocrinology*. 2011;36(9):1312–1326.
168. Dong H-W, Swanson LW. Organization of axonal projections from the anterolateral area of the bed nuclei of the stria terminalis. *J Comp Neurol*. 2004;468(2): 277–298.
169. Dong H-W, Swanson LW. Projections from bed nuclei of the stria terminalis, anteromedial area: cerebral hemisphere integration of neuroendocrine, autonomic, and behavioral aspects of energy balance. *J Comp Neurol*. 2006;494(1):142–178.
170. Georges F, Aston-Jones G. Activation of ventral tegmental area cells by the bed nucleus of the stria terminalis: a novel excitatory amino acid input to midbrain dopamine neurons. *J Neurosci*. 2002;22(12):5173–5187.
171. Sartor GC, Aston-Jones G. Regulation of the ventral tegmental area by the bed nucleus of the stria terminalis is required for expression of cocaine preference. *Eur J Neurosci*. 2012;36(11):3549–3558.
172. Groenewegen HJ, Wright CI, Beijer AV, Voorn P. Convergence and segregation of ventral striatal inputs and outputs. *Ann N Y Acad Sci*. 1999;877:49–63.
173. Stuber GD, Hnasko TS, Britt JP, Edwards RH, Bonci A. Dopaminergic terminals in the nucleus accumbens but not the dorsal striatum corelease glutamate. *J Neurosci*. 2010;30(24):8229–8233.
174. Morales M, Pickel VM. Insights to drug addiction derived from ultrastructural views of the mesocorticolimbic system. *Ann N Y Acad Sci*. 2012;1248:71–88.
175. Yamaguchi T, Wang H-L, Li X, Ng TH, Morales M. Mesocorticolimbic glutamatergic pathway. *J Neurosci*. 2011;31(23):8476–8490.
176. Brown HD, McCutcheon JE, Cone JJ, Ragozzino ME, Roitman MF. Primary food reward and reward-predictive stimuli evoke different patterns of phasic dopamine signaling throughout the striatum. *Eur J Neurosci*. 2011;34(12):1997–2006.
177. Day JJ, Roitman MF, Wightman RM, Carelli RM. Associative learning mediates dynamic shifts in dopamine signaling in the nucleus accumbens. *Nat Neurosci*. 2007;10(8):1020–1028.
178. Flagel SB, Clark JJ, Robinson TE, et al. A selective role for dopamine in stimulus-reward learning. *Nature*. 2011;469(7328):53–57.
179. Phillips PEM, Stuber GD, Heien MLAV, Wightman RM, Carelli RM. Subsecond dopamine release promotes cocaine seeking. *Nature*. 2003;422(6932):614–618.
180. Stuber GD, Klanker M, de Ridder B, et al. Reward-predictive cues enhance excitatory synaptic strength onto midbrain dopamine neurons. *Science*. 2008;321(5896): 1690–1692.
181. Ambroggi F, Ishikawa A, Fields HL, Nicola SM. Basolateral amygdala neurons facilitate reward-seeking behavior by exciting nucleus accumbens neurons. *Neuron*. 2008;59(4):648–661.
182. Cador M, Robbins TW, Everitt BJ. Involvement of the amygdala in stimulus-reward associations: interaction with the ventral striatum. *Neuroscience*. 1989;30(1):77–86.
183. Di Ciano P, Everitt BJ. Direct interactions between the basolateral amygdala and nucleus accumbens core underlie cocaine-seeking behavior by rats. *J Neurosci*. 2004;24(32):7167–7173.

184. Setlow B, Holland PC, Gallagher M. Disconnection of the basolateral amygdala complex and nucleus accumbens impairs appetitive pavlovian second-order conditioned responses. *Behav Neurosci.* 2002;116(2):267–275.
185. Shiflett MW, Balleine BW. At the limbic-motor interface: disconnection of basolateral amygdala from nucleus accumbens core and shell reveals dissociable components of incentive motivation. *Eur J Neurosci.* 2010;32(10):1735–1743.
186. Carlezon Jr WA, Thomas MJ. Biological substrates of reward and aversion: a nucleus accumbens activity hypothesis. *Neuropharmacology.* 2009;56(Suppl. 1):122–132.
187. Carlezon Jr WA, Wise RA. Microinjections of phencyclidine (PCP) and related drugs into nucleus accumbens shell potentiate medial forebrain bundle brain stimulation reward. *Psychopharmacology (Berl).* 1996;128(4):413–420.
188. Taha SA, Fields HL. Inhibitions of nucleus accumbens neurons encode a gating signal for reward-directed behavior. *J Neurosci.* 2006;26(1):217–222.
189. Albin RL, Young AB, Penney JB. The functional anatomy of basal ganglia disorders. *Trends Neurosci.* 1989;12(10):366–375.
190. Gerfen CR, Engber TM, Mahan LC, et al. D1 and D2 dopamine receptor-regulated gene expression of striatonigral and striatopallidal neurons. *Science.* 1990;250(4986):1429–1432.
191. Hikida T, Kimura K, Wada N, Funabiki K, Nakanishi S. Distinct roles of synaptic transmission in direct and indirect striatal pathways to reward and aversive behavior. *Neuron.* 2010;66(6):896–907.
192. Kravitz AV, Freeze BS, Parker PRL, et al. Regulation of parkinsonian motor behaviours by optogenetic control of basal ganglia circuitry. *Nature.* 2010;466(7306):622–626.
193. DeLong MR, Wichmann T. Circuits and circuit disorders of the basal ganglia. *Arch Neurol.* 2007;64(1):20–24.
194. Graybiel AM. The basal ganglia. *Curr Biol.* 2000;10(14):R509–R511.
195. Rymar VV, Sasseville R, Luk KC, Sadikot AF. Neurogenesis and stereological morphometry of calretinin-immunoreactive GABAergic interneurons of the neostriatum. *J Comp Neurol.* 2004;469(3):325–339.
196. Crespo JA, Sturm K, Saria A, Zernig G. Activation of muscarinic and nicotinic acetylcholine receptors in the nucleus accumbens core is necessary for the acquisition of drug reinforcement. *J Neurosci.* 2006;26(22):6004–6010.
197. Hikida T, Kaneko S, Isobe T, et al. Increased sensitivity to cocaine by cholinergic cell ablation in nucleus accumbens. *Proc Natl Acad Sci USA.* 2001;98(23):13351–13354.
198. Pratt WE, Kelley AE. Nucleus accumbens acetylcholine regulates appetitive learning and motivation for food via activation of muscarinic receptors. *Behav Neurosci.* 2004;118(4):730–739.
199. Pratt WE, Spencer RC, Kelley AE. Muscarinic receptor antagonism of the nucleus accumbens core causes avoidance to flavor and spatial cues. *Behav Neurosci.* 2007;121(6):1215–1223.
200. Campolongo P, Roozendaal B, Trezza V, et al. Endocannabinoids in the rat basolateral amygdala enhance memory consolidation and enable glucocorticoid modulation of memory. *Proc Natl Acad Sci USA.* 2009;106(12):4888–4893.
201. Hatfield T, McGaugh JL. Norepinephrine infused into the basolateral amygdala post-training enhances retention in a spatial water maze task. *Neurobiol Learn Mem.* 1999;71(2):232–239.
202. Huff NC, Rudy JW. The amygdala modulates hippocampus-dependent context memory formation and stores cue-shock associations. *Behav Neurosci.* 2004;118(1):53–62.
203. Lalumiere RT, Nguyen LT, McGaugh JL. Post-training intrabasolateral amygdala infusions of dopamine modulate consolidation of inhibitory avoidance memory: involvement of noradrenergic and cholinergic systems. *Eur J Neurosci.* 2004;20(10):2804–2810.

204. LaLumiere RT, Buen T-V, McGaugh JL. Post-training intra-basolateral amygdala infusions of norepinephrine enhance consolidation of memory for contextual fear conditioning. *J Neurosci.* 2003;23(17):6754–6758.
205. LaLumiere RT, Nawar EM, McGaugh JL. Modulation of memory consolidation by the basolateral amygdala or nucleus accumbens shell requires concurrent dopamine receptor activation in both brain regions. *Learn Mem.* 2005;12(3):296–301.
206. Miranda MI, LaLumiere RT, Buen TV, Bermudez-Rattoni F, McGaugh JL. Blockade of noradrenergic receptors in the basolateral amygdala impairs taste memory. *Eur J Neurosci.* 2003;18(9):2605–2610.
207. Packard MG, Cahill L, McGaugh JL. Amygdala modulation of hippocampal-dependent and caudate nucleus-dependent memory processes. *Proc Natl Acad Sci USA.* 1994;91(18):8477–8481.
208. Roozendaal B, Castello NA, Vedana G, Barsegyan A, McGaugh JL. Noradrenergic activation of the basolateral amygdala modulates consolidation of object recognition memory. *Neurobiol Learn Mem.* 2008;90(3):576–579.
209. Roozendaal B, Schelling G, McGaugh JL. Corticotropin-releasing factor in the basolateral amygdala enhances memory consolidation via an interaction with the beta-adrenoceptor-cAMP pathway: dependence on glucocorticoid receptor activation. *J Neurosci.* 2008;28(26):6642–6651.
210. Bauer EP, Paz R, Paré D. Gamma oscillations coordinate amygdalo-rhinal interactions during learning. *J Neurosci.* 2007;27(35):9369–9379.
211. Popescu AT, Popa D, Paré D. Coherent gamma oscillations couple the amygdala and striatum during learning. *Nat Neurosci.* 2009;12(6):801–807.
212. Huff ML, Miller RL, Deisseroth K, Moorman DE, Lalumiere RT. Posttraining optogenetic manipulations of basolateral amygdala activity modulate consolidation of inhibitory avoidance memory in rats. *Proc Natl Acad Sci USA.* 2013;110(9):3597–3602.
213. Frank MG, Benington JH. The role of sleep in memory consolidation and brain plasticity: dream or reality? *Neuroscientist.* 2006;12(6):477–488.
214. Graves L, Pack A, Abel T. Sleep and memory: a molecular perspective. *Trends Neurosci.* 2001;24(4):237–243.
215. Stickgold R, Hobson JA, Fosse R, Fosse M. Sleep, learning, and dreams: off-line memory reprocessing. *Science.* 2001;294(5544):1052–1057.
216. Wagner U, Born J. Memory consolidation during sleep: interactive effects of sleep stages and HPA regulation. *Stress.* 2008;11(1):28–41.
217. Walker MP, Stickgold R. Sleep-dependent learning and memory consolidation. *Neuron.* 2004;44(1):121–133.
218. Van Der Werf YD, Altena E, Schoonheim MM, et al. Sleep benefits subsequent hippocampal functioning. *Nat Neurosci.* 2009;12(2):122–123.
219. Mackiewicz M, Naidoo N, Zimmerman JE, Pack AI. Molecular mechanisms of sleep and wakefulness. *Ann N Y Acad Sci.* 2008;1129:335–349.
220. Tartar JL, Ward CP, McKenna JT, et al. Hippocampal synaptic plasticity and spatial learning are impaired in a rat model of sleep fragmentation. *Eur J Neurosci.* 2006;23(10):2739–2748.
221. Franken P, Gip P, Hagiwara G, Ruby NF, Heller HC. Glycogen content in the cerebral cortex increases with sleep loss in C57BL/6J mice. *Neurosci Lett.* 2006;402(1–2):176–179.
222. Knutson KL, Spiegel K, Penev P, Van Cauter E. The metabolic consequences of sleep deprivation. *Sleep Med Rev.* 2007;11(3):163–178.
223. Ramanathan L, Gulyani S, Nienhuis R, Siegel JM. Sleep deprivation decreases superoxide dismutase activity in rat hippocampus and brainstem. *Neuroreport.* 2002;13(11):1387–1390.

224. Tononi G, Cirelli C. Sleep and synaptic homeostasis: a hypothesis. *Brain Res Bull.* 2003;62(2):143–150.
225. Carter ME, Adamantidis A, Ohtsu H, Deisseroth K, de Lecea L. Sleep homeostasis modulates hypocretin-mediated sleep-to-wake transitions. *J Neurosci.* 2009;29(35):10939–10949.
226. Ego-Stengel V, Wilson MA. Disruption of ripple-associated hippocampal activity during rest impairs spatial learning in the rat. *Hippocampus.* 2010;20(1):1–10.
227. Girardeau G, Benchenane K, Wiener SI, Buzsáki G, Zugaro MB. Selective suppression of hippocampal ripples impairs spatial memory. *Nat Neurosci.* 2009;12(10):1222–1223.
228. Diekelmann S, Born J. The memory function of sleep. *Nat Rev Neurosci.* 2010;11(2):114–126.
229. Reijmers LG, Perkins BL, Matsuo N, Mayford M. Localization of a stable neural correlate of associative memory. *Science.* 2007;317(5842):1230–1233.
230. Han J-H, Kushner SA, Yiu AP, et al. Selective erasure of a fear memory. *Science.* 2009;323(5920):1492–1496.
231. Liu X, Ramirez S, Pang PT, et al. Optogenetic stimulation of a hippocampal engram activates fear memory recall. *Nature.* 2012;484(7394):381–385.
232. Goshen I, Brodsky M, Prakash R, et al. Dynamics of retrieval strategies for remote memories. *Cell.* 2011;147(3):678–689.
233. Gu Y, Arruda-Carvalho M, Wang J, et al. Optical controlling reveals time-dependent roles for adult-born dentate granule cells. *Nat Neurosci.* 2012;15(12):1700–1706.
234. Scoville WB, Milner B. Loss of recent memory after bilateral hippocampal lesions. *J Neurol Neurosurg Psychiatry.* 1957;20(1):11–21.
235. Bontempi B, Laurent-Demir C, Destrade C, Jaffard R. Time-dependent reorganization of brain circuitry underlying long-term memory storage. *Nature.* 1999; 400(6745):671–675.
236. Maviel T, Durkin TP, Menzaghi F, Bontempi B. Sites of neocortical reorganization critical for remote spatial memory. *Science.* 2004;305(5680):96–99.
237. Kitamura T, Saitoh Y, Takashima N, et al. Adult neurogenesis modulates the hippocampus-dependent period of associative fear memory. *Cell.* 2009;139(4): 814–827.
238. Willshaw DJ, Buckingham JT. An assessment of Marr's theory of the hippocampus as a temporary memory store. *Philos Trans R Soc Lond B Biol Sci.* 1990;329(1253):205–215.
239. Wiltgen BJ, Zhou M, Cai Y, et al. The hippocampus plays a selective role in the retrieval of detailed contextual memories. *Curr Biol.* 2010;20(15):1336–1344.
240. Altman J, Das GD. Post-natal origin of microneurones in the rat brain. *Nature.* 1965;207(5000):953–956.
241. Cameron HA, McKay RD. Adult neurogenesis produces a large pool of new granule cells in the dentate gyrus. *J Comp Neurol.* 2001;435(4):406–417.
242. Eriksson PS, Perfilieva E, Björk-Eriksson T, et al. Neurogenesis in the adult human hippocampus. *Nat Med.* 1998;4(11):1313–1317.
243. Gould E, Reeves AJ, Graziano MS, Gross CG. Neurogenesis in the neocortex of adult primates. *Science.* 1999;286(5439):548–552.
244. Arruda-Carvalho M, Sakaguchi M, Akers KG, Josselyn SA, Frankland PW. Post-training ablation of adult-generated neurons degrades previously acquired memories. *J Neurosci.* 2011;31(42):15113–15127.
245. Ge S, Goh ELK, Sailor KA, Kitabatake Y, Ming G, Song H. GABA regulates synaptic integration of newly generated neurons in the adult brain. *Nature.* 2006;439(7076):589–593.
246. Van Praag H, Schinder AF, Christie BR, Toni N, Palmer TD, Gage FH. Functional neurogenesis in the adult hippocampus. *Nature.* 2002;415(6875):1030–1034.

247. Bliss TV, Lomo T. Long-lasting potentiation of synaptic transmission in the dentate area of the anaesthetized rabbit following stimulation of the perforant path. *J Physiol.* 1973;232(2):331–356.
248. Teyler TJ, DiScenna P. Long-term potentiation. *Annu Rev Neurosci.* 1987;10:131–161.
249. Malenka RC, Bear MF. LTP and LTD: an embarrassment of riches. *Neuron.* 2004;44(1):5–21.
250. Ito M. Long-term depression. *Annu Rev Neurosci.* 1989;12:85–102.
251. Dudek SM, Bear MF. Homosynaptic long-term depression in area CA1 of hippocampus and effects of N-methyl-D-aspartate receptor blockade. *Proc Natl Acad Sci USA.* 1992;89(10):4363–4367.
252. Artola A, Singer W. Long-term depression of excitatory synaptic transmission and its relationship to long-term potentiation. *Trends Neurosci.* 1993;16(11):480–487.
253. Grant SG, O'Dell TJ, Karl KA, Stein PL, Soriano P, Kandel ER. Impaired long-term potentiation, spatial learning, and hippocampal eevelopment in fyn mutant mice. *Science* 1992;258(5090):1903–1910.
254. Ichise T, Kano M, Hashimoto K, et al. mGluR1 in cerebellar purkinje cells essential for long-term depression, synapse elimination, and motor coordination. *Science.* 2000;288(5472):1832–1835.
255. Morris RG, Anderson E, Lynch GS, Baudry M. Selective impairment of learning and blockade of long-term potentiation by an N-methyl-D-aspartate receptor antagonist, AP5. *Nature.* 1986;319(6056):774–776.
256. Rumpel S, LeDoux J, Zador A, Malinow R. Postsynaptic receptor trafficking underlying a form of associative learning. *Science.* 2005;308(5718):83–88.
257. Tye KM, Stuber GD, de Ridder B, Bonci A, Janak PH. Rapid strengthening of thalamo-amygdala synapses mediates cue-reward learning. *Nature.* 2008;453(7199): 1253–1257.
258. Liewald JF, Brauner M, Stephens GJ, et al. Optogenetic analysis of synaptic function. *Nat Methods.* 2008;5(10):895–902.
259. Lynch GS, Dunwiddie T, Gribkoff V. Heterosynaptic depression: a postsynaptic correlate of long-term potentiation. *Nature.* 1977;266(5604):737–739.
260. Zalutsky RA, Nicoll RA. Comparison of two forms of long-term potentiation in single hippocampal neurons. *Science.* 1990;248(4963):1619–1624.
261. Schacher S, Wu F, Sun Z-Y. Pathway-specific synaptic plasticity: activity-dependent enhancement and suppression of long-term heterosynaptic facilitation at converging inputs on a single target. *J Neurosci.* 1997;17(2):597–606.
262. Deeg KE, Aizenman CD. Sensory modality-specific homeostatic plasticity in the developing optic tectum. *Nat Neurosci.* 2011;14(5):548–550.
263. LeDoux JE. Emotion circuits in the brain. *Annu Rev Neurosci.* 2000;23:155–184.
264. LeDoux JE, Farb C, Ruggiero DA. Topographic organization of neurons in the acoustic thalamus that project to the amygdala. *J Neurosci.* 1990;10(4):1043–1054.
265. Tsvetkov E, Carlezon Jr WA, Benes FM, Kandel ER, Bolshakov VY. Fear conditioning occludes LTP-induced presynaptic enhancement of synaptic transmission in the cortical pathway to the lateral amygdala. *Neuron.* 2002;34(2):289–300.
266. Marowsky A, Yanagawa Y, Obata K, Vogt KE. A specialized subclass of interneurons mediates dopaminergic facilitation of amygdala function. *Neuron.* 2005;48(6): 1025–1037.
267. Morozov A, Sukato D, Ito W. Selective suppression of plasticity in amygdala inputs from temporal association cortex by the external capsule. *J Neurosci.* 2011;31(1): 339–345.
268. Hayashi Y, Shi SH, Esteban JA, Piccini A, Poncer JC, Malinow R. Driving AMPA receptors into synapses by LTP and CaMKII: requirement for GluR1 and PDZ domain interaction. *Science.* 2000;287(5461):2262–2267.

269. Pascoli V, Turiault M, Lüscher C. Reversal of cocaine-evoked synaptic potentiation resets drug-induced adaptive behaviour. *Nature*. 2012;481(7379):71–75.
270. Schechter LE. The potassium channel blockers 4-aminopyridine and tetra-ethylammonium increase the spontaneous basal release of [3H]5-hydroxytryptamine in rat hippocampal slices. *J Pharmacol Exp Ther*. 1997;282(1):262–270.
271. Shu Y, Yu Y, Yang J, McCormick DA. Selective control of cortical axonal spikes by a slowly inactivating K^+ current. *Proc Natl Acad Sci USA*. 2007;104(27):11453–11458.
272. Harvey CD, Svoboda K. Locally dynamic synaptic learning rules in pyramidal neuron dendrites. *Nature*. 2007;450(7173):1195–1200.
273. Adams SR, Tsien RY. Controlling cell chemistry with caged compounds. *Annu Rev Physiol*. 1993;55(1):755–784.
274. Pettit DL, Wang SS, Gee KR, Augustine GJ. Chemical two-photon uncaging: a novel approach to mapping glutamate receptors. *Neuron*. 1997;19(3):465–471.
275. Nikolenko V, Poskanzer KE, Yuste R. Two-photon photostimulation and imaging of neural circuits. *Nat Methods*. 2007;4(11):943–950.
276. Harvey CD, Yasuda R, Zhong H, Svoboda K. The spread of ras activity triggered by activation of a single dendritic spine. *Science*. 2008;321(5885):136–140.
277. Rickgauer JP, Tank DW. Two-photon excitation of channelrhodopsin-2 at saturation. *Proc Natl Acad Sci USA*. 2009;106(35):15025–15030.
278. Papagiakoumou E, Anselmi F, Bègue A, et al. Scanless two-photon excitation of channelrhodopsin-2. *Nat Methods*. 2010;7(10):848–854.
279. Ryan TA, Reuter H, Wendland B, Schweizer FE, Tsien RW, Smith SJ. The kinetics of synaptic vesicle recycling measured at single presynaptic boutons. *Neuron*. 1993;11(4):713–724.
280. Zhang W, Linden DJ. Neuromodulation at single presynaptic boutons of cerebellar parallel fibers is determined by bouton size and basal action potential-evoked Ca transient amplitude. *J Neurosci*. 2009;29(49):15586–15594.
281. Coombs JS, Curtis DR, Eccles JC. The generation of impulses in motoneurones. *J Physiol Lond*. 1957;139(2):232–249.
282. Palmer LM, Stuart GJ. Site of action potential initiation in layer 5 pyramidal neurons. *J Neurosci*. 2006;26(6):1854–1863.
283. Meeks JP, Mennerick S. Action potential initiation and propagation in CA3 pyramidal axons. *J Neurophysiol*. 2007;97(5):3460–3472.
284. Kole MHP, Ilschner SU, Kampa BM, Williams SR, Ruben PC, Stuart GJ. Action potential generation requires a high sodium channel density in the axon initial segment. *Nat Neurosci*. 2008;11(2):178–186.
285. Foust A, Popovic M, Zecevic D, McCormick DA. Action potentials initiate in the axon initial segment and propagate through axon collaterals reliably in cerebellar purkinje neurons. *J Neurosci*. 2010;30(20):6891–6902.
286. Grubb MS, Burrone J. Activity-dependent relocation of the axon initial segment fine-tunes neuronal excitability. *Nature*. 2010;465(7301):1070–1074.
287. Kuba H, Oichi Y, Ohmori H. Presynaptic activity regulates Na + channel distribution at the axon initial segment. *Nature*. 2010;465(7301):1075–1078.
288. Grubb MS, Shu Y, Kuba H, Rasband MN, Wimmer VC, Bender KJ. Short- and long-term plasticity at the axon initial segment. *J Neurosci*. 2011;31(45):16049–16055.
289. Grubb MS, Burrone J. Channelrhodopsin-2 localised to the axon initial segment. *PLoS One*. 2010;5(10):e13761.
290. Markram H, Lübke J, Frotscher M, Sakmann B. Regulation of synaptic efficacy by coincidence of postsynaptic APs and EPSPs. *Science*. 1997;275(5297):213–215.
291. Schoenenberger P, Schärer Y-PZ, Oertner TG. Channelrhodopsin as a tool to investigate synaptic transmission and plasticity. *Exp Physiol*. 2011;96(1):34–39.

292. Turrigiano GG, Nelson SB. Hebb and homeostasis in neuronal plasticity. *Curr Opin Neurobiol*. 2000;10(3):358–364.
293. Turrigiano GG, Leslie KR, Desai NS, Rutherford LC, Nelson SB. Activity-dependent scaling of quantal amplitude in neocortical neurons. *Nature*. 1998;391(6670):892–896.
294. Goold CP, Nicoll RA. Single-cell optogenetic excitation drives homeostatic synaptic depression. *Neuron*. 2010;68(3):512–528.

CHAPTER THIRTEEN

The Tagging and Capture Hypothesis from Synapse to Memory

Haydée Viola[*,†], **Fabricio Ballarini**[*], **María Cecilia Martínez**[*], **Diego Moncada**[*,‡]

[*]Instituto de Biología Celular y Neurociencias "Prof. E. De Robertis", Consejo Nacional de Investigaciones Científicas y Técnicas de Argentina, Facultad de Medicina, Universidad de Buenos Aires, Buenos Aires, Argentina

[†]Departamento de Fisiología, Biología Molecular y Celular, Facultad de Ciencias Exactas y Naturales, Universidad de Buenos Aires, Buenos Aires, Argentina

[‡]Neurophysiology of Learning and Memory Research Group, Leibniz-Institut für Neurobiologie, Magdeburg, Germany

Contents

Abstract

The synaptic tagging and capture theory (STC) was postulated by Frey and Morris in 1997 and provided a strong framework to explain how to achieve synaptic specificity and persistence of electrophysiological-induced plasticity changes. Ten years later, the same argument was applied on learning and memory models to explain the formation of long-term memories, resulting in the behavioral tagging hypothesis (BT). These

Progress in Molecular Biology and Translational Science, Volume 122
ISSN 1877-1173
http://dx.doi.org/10.1016/B978-0-12-420170-5.00013-1

hypotheses are able to explain how a weak event that induces transient changes in the brain can establish long-lasting phenomena through a tagging and capture process. In this framework, it was postulated that the weak event sets a tag that captures plasticity-related proteins/products (PRPs) synthesized by an independent strong event. The tagging and capture processes exhibit symmetry, and therefore, PRPs can be captured if they are synthesized either before or after the setting of the tag. In summary, the hypothesis provides a wide framework that gives a solid explanation of how lasting changes occur and how the interaction between different events leads to promotion, reinforcement, or impairment of such changes.

In this chapter, we will summarize the postulates of STC hypothesis, the common features between synaptic plasticity and memory, as well as a detailed compilation of the findings supporting the existence of BT process. At the end, we pose some questions related to BT mechanism and LTM formation, which probably will be answered in the near future.

1. INTRODUCTION

Lasting changes in synaptic plasticity strength and also in memory storage persistence are not only dependent on the characteristics of the stimuli that induce these changes. Events happening before and after these stimuli can also exert influence on the effects they have, raising an essential question about the mechanism that underlies this late-associativity phenomenon. The answer to this question came by considering that these lasting changes should occur selectively in inputs activated by the stimuli. This leads to the emergence of the synaptic tagging and capture (STC) hypothesis.[1] The STC hypothesis proposed a cellular mechanism accounting for why a stimulation that normally leads to early long-term potentiation (early-LTP) could also induce long-lasting form of long-term potentiation (late-LTP) if a separate pathway converging on an overlapping population of neurons was strongly tetanized within a specific time window. The STC hypothesis declares that LTP involves the local tagging of synapses at the moment of induction. Thus, stimuli that induce early-LTP cause the transient activation of a synaptic tag, but only those stimuli that induce late-LTP will, besides from setting a tag, also initiate protein synthesis. Those tags capture plasticity-related proteins/products (PRPs) synthesized in the soma or local dendritic domains because of the strong tetanic stimulation, and the interaction between tag and PRPs is essential for stabilization of potentiation from early-LTP to late-LTP.

In this chapter, we will discuss the involvement of synaptic changes in the long-term memory (LTM) storage processes of the mammalian brain.

Similar to those seen in synaptic plasticity assays, late-associative effects were also evidenced in learning and memory experiments in animals. It has been shown that a short-lasting memory induced by a weak training can be consolidated into a LTM if animals experience a strong event in a critical time window around the weak training. This process depends on protein synthesis induced by the strong associated experience and was originally named "behavioral tagging" (BT).[2] It has been suggested that the weak training sets a learning tag where the PRPs provided by the strong event would be captured in order to establish a persistent mnemonic trace.

Both STC and BT propose analogue mechanisms by which long-term changes induced through electrophysiological stimulations or learning processes are meant to occur. The tagging and capture dynamics is an elegant theoretical framework that is also capable to explain why the persistence of neuronal potentiation as well as the duration of memory is not only dependent on events occurring at the moment of their induction, but also on other occurring previously or subsequently to the stimulus that modifies the activity of the involved neurons. In summary it provides a wide framework capable to explain in which way lasting changes occur and how the interaction between different events leads to promotions, reinforcements, or impairments of such changes.

In this chapter, we will summarize the postulates of STC hypothesis, the features in common between synaptic plasticity and memory (SPM), as well as a detailed compilation of the findings showing the effects of the interaction between electrophysiological and behavioral stimuli drawing a line between STC and BT processes.

2. THE ABC OF TAGGING AND CAPTURE MECHANISMS

The principal idea underlying this process is that proteins and other products related to plasticity (PRPs) are used to originate long-lasting changes when they are captured by specific tags. This hypothesis, initially postulated for synaptic plasticity potentiation,[1] was demonstrated by separating the tagging stimulus from the PRPs inducing stimulus, by using a weak stimulus unable to induce PRP-dependent lasting changes and a strong independent event able to induce PRP-dependent lasting plasticity processes. The foundations of the hypothesis are based on three major points:

A. The weak stimulus that induces short-term plasticity phenomena also induces tags able to capture PRPs (induced by a strong independent

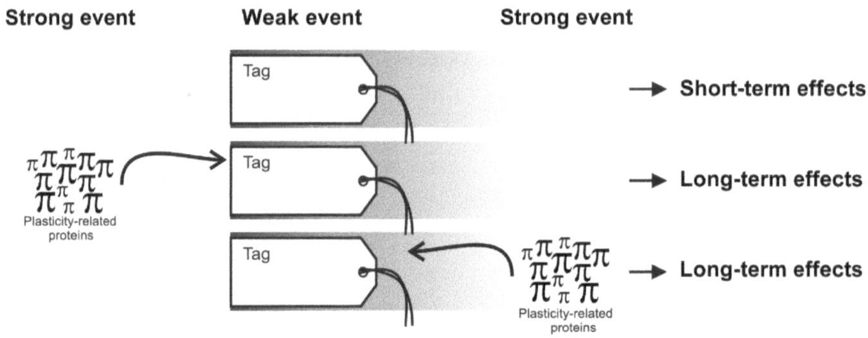

Figure 13.1 Summary of tagging and capture processes. Figure resumes how a weak event that induces transient changes in the brain can establish long-lasting phenomena through a tagging and capture process, if the tag set by the weak event captures PRPs synthesized by an independent strong event. The process exhibits symmetry and PRPs can be captured either if they are synthesized before or after the setting of the tag. In the STC model, a weak stimulation that induces early-LTP and sets a synaptic tag, in a target population of neurons, can result in late-LTP if those tags capture PRPs synthesized due to a strong stimulation applied on an independent path that targets an overlapped population of neurons at different synapses. The characteristics of the BT process are analogous to the STC model. A learning task of weak saliency only induces a short-term memory and sets a learning tag. On the other hand, a task of a stronger saliency induces protein synthesis. If both tasks are processed in same brain structures within a critical time window, the learning tags of the weak task can capture the PRPs synthesized by the strong task in order to consolidate its own lasting memory.

event occurring around the stimulus) that enables lasting plasticity processes selectively related to activated inputs (Fig. 13.1).

B. Both tag and PRPs have a transient duration.

C. In order to capture the products, tags and PRPs should be present at the same neural substrate and at the same time.

From these three key postulates, emerge three major predictions:

A. Protein synthesis inhibitors should impair the tagging and capture mechanism.

B. If PRPs arrive when the tag has already decayed, the capture mechanism should not work.

C. If PRPs are synthesized and delivered far away from the point where tags were (or will be) established, the promoting mechanism should be disrupted.

Essentially, the main requirement for the mechanism of tagging and capture to occur is the supply of PRPs at tagged sites within a critical time window around the stimulus that induces the tag.

3. WHICH CRITERIA SHOULD SATISFY A CANDIDATE FOR A TAG?

There are several criteria to be satisfied by a *synaptic tag*[3–6]: a tag can be activated by weak stimulation that induces only early-LTP; the lifetime of a tag is transient lasting less than 2 h; the activation of a tag does not require protein synthesis; a tag is induced in an input-specific manner and is relatively immobile and finally a tag must interact with (and therefore capture) PRPs for late-LTP. Using similar assumptions a *learning tag* has also been defined, where a weak training would induce a kind of mark which should fulfill these criteria[2,6,7] (but also see Ref. 8).

The current vision considers the tag as an ensemble of molecules tending to modify the morphology of the dendrite.[9–11] Among these, several protein kinases like PKA, CAMKIIα, and PKMz have been proposed as suitable candidates for the tagging machinery.[2,3,12,13]

The STC hypothesis was initially described for LTP-dependent plasticity; however, the overlap of properties for this type of synaptic plasticity and for learning and memory processes justified and fostered the search for a BT process. Below we develop these similarities.

4. MEMORY CAN BE THOUGHT OF AS CHANGES IN SYNAPTIC PLASTICITY

It is widely accepted that neural activity induced by learning triggers changes in the strength of synaptic connections within the brain. The most relevant aspect of a memory trace is that those changes in behavior, occurring as a consequence of a learning experience, persist in time. In this way, a model of synaptic plasticity where brief stimulations of a neural pathway induce long-lasting changes in the synapses could provide plausible clues of the mechanisms underlying the formation of lasting memories. These changes in the synapses' efficacy involve either up- or downregulation of synaptic strength and, when persisting for more than 1 h, they are referred as LTP or long-term depression (LTD) phenomenon, respectively. Beyond the trait of persistence, there are other three properties applicable to several forms of LTP/LTD and learning storage processes: associative induction, input-specific expression, and postinduction liability.[14] These similarities led to the proposal of the SPM hypothesis, which states that an activity-dependent synaptic plasticity is induced at the appropriate synapses during memory formation. The plastic changes must occur in those brain areas where memory is being processed and are both necessary and sufficient for the storage of the information.[15]

Considering the SPM hypothesis, one of basic criteria to meet is *detectability* (the changes in synaptic efficacy should occur and should be detectable somewhere in the nervous system of an animal displaying memory of a particular experience). In this respect, hippocampal late-LTP was recently shown to underlie LTM formation of an aversive learning task. This work showed that one-trial inhibitory avoidance (IA) learning in rats produces the same changes in hippocampal glutamate receptors as the induction of LTP with HFS, together with a spatially restricted increase in the amplitude of evoked synaptic transmission in CA1 *in vivo*. As the learning-induced synaptic potentiation occluded HFS-induced LTP, the authors conclude that IA training induces LTP in CA1.[16] Moreover, by training mice to evoke eyelid responses using a trace paradigm, it was observed that the slope of evoked field excitatory postsynaptic potentials at the CA3–CA1 synapse increased across conditioning sessions and decreased during extinction, being linearly related to the evolution of learning. These changes did not happen in control mice, when the conditioning protocol was absent, or when LTP induction was pharmacologically prevented.[17] Moreover, earlier experiments performed in brain slices obtained from previously trained rats provide a clear demonstration of a LTP-like phenomenon occurring at hippocampal or cerebral cortex areas as a result of exposure to complex environments,[18] motor learning,[19] or olfactory discrimination task.[20]

Other evidences of a relationship between activity-dependent synaptic plasticity and learning in behaving rats were obtained through an object recognition task, showing that the consolidation of this memory is accompanied by a LTP-like synaptic enhancement in the hippocampal CA3–CA1 synapse 6 h after the training session.[21] Intriguingly, recordings of the hippocampal CA1 region in freely behaving mice showed that novel spatial learning triggers endogenous LTD. Specifically, LTD was enabled when mice learned about object constellations or during a spatial object recognition (SOR) task, suggesting an important role for this type of synaptic plasticity in the creation of hippocampus-dependent memory.[22] This result is in a good agreement with the suggested role of LTD in object recognition and spatial exploration in behaving rats.[23–26]

A further criteria of the SPM hypothesis relates to the necessity and sufficiency of changes in the efficacy of synaptic plasticity for the learning and memory process.[14] Concerning the necessity criterion, different strategies were used through anterograde or retrograde interventions tending to block the mechanism that induces or expresses changes in synaptic weights, which in turn should impair new learning or affect the memory of past experiences. Moreover, if synaptic plasticity underlies memory, the behavioral parameters of learning should correlate to some parameters of synaptic plasticity. In this chapter, we focus on a recent example of accordance between LTP and learning regarding to late-associative property. As mentioned before, this was seen by registering the change on a postsynaptic response triggered by stimulation due to the action of a second spatial and temporally distant stimulation to another neuronal pathway targeting a common population of cells. In this model system, it was observed that a stimulation that normally leads to early-LTP could also induce late-LTP if a separate pathway was strongly tetanized within a specific time window.[1,27,28] A similar effect was reported if the separate pathway was stimulated with a LFS that induced LTD.[29,30] Moreover, a short-lasting LTD is capable of capturing late-phase LTD when a prolonged stimulus was applied to another input in the same population of neurons.[31] The effect of these stimulations was abolished by the application of the protein synthesis inhibitor anisomycin, showing the dependence of this phenomenon to protein synthesis. Moreover, the compartmentalization within a neuron is an important factor for the setting of clustered plasticity.[32,33]

These late-associative effects observed between two properly independent electrical stimuli also were observed in experiments combining electrical stimulation and behavioral interventions. In the following section, we will summarize these data, which are interpreted in the conceptual framework of tagging and capture. After that, we will show how these kinds of effects were also observed in learning and memory models and we will describe and discuss in detail the learning tasks, memory models, and the experiments that settled the basis of the BT hypothesis.

5. SYNAPTIC PLASTICITY WAS IMPROVED *IN VIVO* BY STRUCTURAL OR BEHAVIORAL REINFORCEMENTS

The hypothesis of STC was postulated 15 years ago using models of synaptic plasticity in hippocampal slices of rats.[1,27,34] In addition, it was recently shown that SCT also occurs in behaving animals such that the decaying early-LTP, induced by weak tetanization of the ipsilateral CA3–CA1 projection, can be converted into late-LTP by strong tetanization delivered later to the contralateral pathway.[35] Moreover, other work has showed that electrical stimulation of several structures had similar reinforcement effects on LTP duration triggered by a stimulus that induced early-LTP in the hippocampus *in vivo*. In particular, stimulation of the basolateral amygdala, medial septum, locus coeruleus (LC), and nucleus supramammillary, occurring within a specific time windows 30 min before or after a weak tetanization of the perforant path (PP), resulted in a reinforcement of early-LTP into late-LTP in the dentate gyrus (DG).[36–40] In addition, electrical activation of the ventral tegmental area (VTA), a brain region that provides major dopaminergic efferents to the CA1 region, also transformed CA1 early-LTP into late-LTP. This process depends on the functionality of D1-receptors and new protein synthesis in the hippocampus and can be observed if VTA stimulation is performed 30 min after the tetanization.[41]

In the last decade, a late-associative effect on LTP was also obtained using behavioral approaches. Before or after the induction of transient form of LTP, rats were submitted to behavioral tasks within a time window with the aim that these interventions could improve the duration of the synaptic plasticity. Several works showed that spatial exploration in an environment could modify the induction or the maintenance of LTP. Specially, the exposure of rats to a novel arena 5 min before the application of weak high-frequency stimulation in Schaffer fibers facilitates induction of LTP in CA1 region of the dorsal hippocampus. This promoting effect is dependent on hippocampal D1/D5-dopaminergic receptor functionality and on the novel nature of the environment, as the exploration to a familiar arena is unable to induce such reinforcement.[42] Similarly, exploration of a novel but not a familiar environment promotes transformation to late-LTP from a stimulus that only induces early-LTP in the DG,[43] through a β-adrenergic receptor and protein synthesis-dependent mechanism.[44] Recently, it was reported that the upregulation of synaptic strength in response to environmental enrichment *in vivo* depends on the ability of mitogen- and stress-activated kinase 1 to regulate cell surface GluA1 expression via the induction of Arc/Arg3.1.[45] Beyond this effect of novel environments on synaptic plasticity, earlier experiments showed that the motivational state of rats also affects the durability of LTP. For example, when water-deprived animals are allowed to drink water for some minutes after the induction of early-LTP in the DG, this potentiation can be reinforced into it lasting form through a mechanism dependent on β-adrenergic receptor activation and the synthesis of new proteins.[46–48] Moreover, Uzakov and coworkers[49] demonstrated that the consolidation of complex appetitive learning in a holeboard promotes the change of early-LTP induced in DG into late-LTP through a protein synthesis-dependent mechanism. Similar reinforcing effects were observed using the water maze learning task instead of the holeboard.[50]

6. LTM FORMATION WAS PROMOTED BY SYNAPTIC PLASTICITY AND BEHAVIORAL REINFORCERS

The hypothesis of SPM predicts that the interventions that affect synaptic plasticity processes should also affect memory systems in a similar fashion. So, the complementary situation where LTM formation is promoted by synaptic plasticity or behavioral reinforcers should take place. Consistent with this prediction, it was recently shown that the induction of LTD with electrical stimulation in the CA1 region of the rat hippocampus facilitated the LTM of an IA task.[26] However, a strong (but not weak) tetanic stimulation that induces LTP in the DG of the behaving animal could also promote the formation of IA–LTM after a weak training that by itself only induces short-term memory (STM). This occurs when stimulation is applied within 1 h before or after a weak IA (wIA) training and is abolished when an inhibitor of protein synthesis is infused into the hippocampus close to the tetanization.[51] Similarly, spatial object recognition (SOR) training that only induced STM can result in LTM when a strong tetanization of the PP occurs during a time window of 2 h, also in a protein synthesis-dependent way.[51]

Since 2007, several research groups around the world have shown that exploration of novel environments promotes the formation of lasting memories in rats trained in different learning tasks that normally induce only transient forms of memories. These experiments, which will be described in detail below, revealed the existence of BT process underlying the formation of LTM in a diverse set of behavioral protocols.

7. NOVELTY PROMOTES LTM FORMATION IN IA AND CONTEXTUAL FEAR TASKS

According to the BT hypothesis, a learning task that triggers both the setting of a learning tag and the induction of PRPs synthesis in the same neuronal population will be consolidated into LTM. Therefore, in this kind of training, it was impossible to interfere with either the tag setting or the PRP synthesis without having the same amnestic output in the animal behavior. For that reason, we explored the possibility of splitting these processes by using two different tasks. In that sense, a weak learning task that only induces STM does not cross through the consolidation phase and therefore removes the synthesis of PRPs from the scenario for this task. Then, if the BT and capture process exists, the learning tag set by a weak training could utilize the PRPs induced by the associated task, leading to the consolidation of the transient memory into a lasting one. To investigate this possibility we initially combined two different hippocampal-dependent behavioral tasks in rodents: the IA and the exploration of a novel open field (OF) tasks.

The OF exploration is a spatial behavioral task that, even for relatively brief periods of time, is able to trigger different interesting processes. For example, a training of 2–5 min is able to induce a LTM of habituation for this task, providing evidence for the existence of a protein synthesis-dependent consolidation that was induced by the training. This environmental novelty is also linked to the activation of the adrenergic and dopaminergic systems and to an increase in the activated levels of the transcription factor CREB and occurs specifically as a result of the detection of spatial novelty in this task.[52,53] On the contrary, prolonged exposure to the arena leads to familiarization with the environment and was associated with a decrease in pCREB levels and PKMz and subsequent lack of novelty effects.[54] All these facts, together with those findings showing that exploration of a novel arena is able to reinforce early-LTP into late forms of plasticity,[42,43,55] further justify the use of this behavioral task as a possible PRP donor for other hippocampal-dependent behavioral tags.

The IA task consists of a short single trial training, in which animals learn that stepping down from a platform located within a box results in a footshock. When animals learn and remember this situation, successive exposures to the platform dramatically increase the time to step down (latency). Therefore, an increase in the latency in the test session compared to the training session is considered an indicator of learning and memory formation, such that a longer latency indicates a stronger memory. This particular learning task has physiological and methodological advantages in searching for a BT process. Unlike multitrial learning tasks where acquisition, retrieval, and relearning overlap in all the subsequent trials after the first training session, the IA has the enormous advantage of triggering all the processes that will lead to the formation of memory after a brief, single learning session ($\sim$10 s).[56] On the other hand, the salience of the training can be easily regulated by adjusting the intensity and/or the duration of the applied foot shock (FS), without causing significant changes in the total duration of training. Thus, a weak intensity of the foot-shock results in STM and strong one in LTM.[2]

The first experiment performed in order to evaluate the BT hypothesis was aimed to analyze the effects of the exploration of a novel OF on the memory induced by a wIA training, which generates a STM but not a LTM. As the results showed that animals could learn the IA task, we reasoned that if the wIA was able to set a learning tag, then the supply of PRPs should allow the consolidation of a lasting memory. To address this possibility, some of the animals exposed to the wIA were also allowed to explore a novel OF for 5 min at different times before or after the wIA and LTM for the different groups was evaluated the next day. Remarkably, while those animals weakly trained showed only IA–STM, the others which had explored the novel arena did show a LTM for the IA task (Fig. 13.2).

This promoting effect of the novelty on the IA memory is time dependent, being restricted to a window of around 1 h before or after the wIA training but excluding peritraining time points.[2] It was suggested that the lack of efficacy of peritraining novelty exposure is due to a setting impairment of the IA-learning tag. Moreover, exposure to novelty 2 h before or after wIA training did not have any effect on IA–LTM promotion. These results are in accordance to the BT theory that postulated that the efficacy of novelty to promote LTM formation relies on the temporal coincidence of the tag with the newly synthesized PRPs, as determined by the time course of decay of the tag and the availability of PRPs (established by its synthesis/degradation dynamics and its distribution). Therefore, in order to consider this phenomenon a BT and capture process, the consolidation of the IA

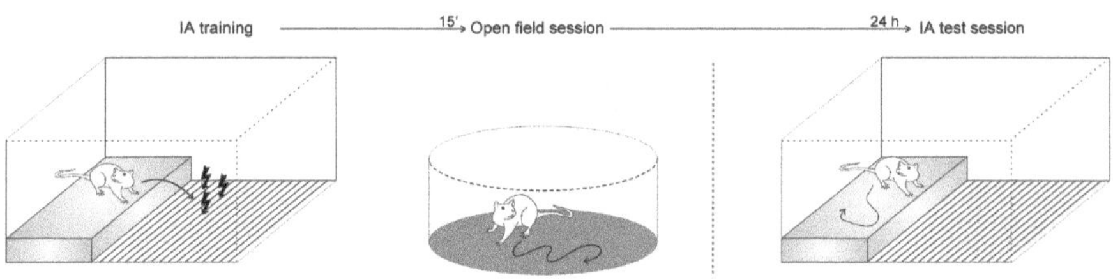

Figure 13.2 Experimental design used to study the behavioral tagging process: Inhibitory avoidance (IA) combined with the spatial exploration of a novel environment. The first experimental design performed to study the BT process involved the sequential training of rats in two hippocampus-dependent tasks: the IA and the exploration to a novel OF. A weak training in the IA normally induces a STM; however, further exploration to a novel OF 15 min after the wIA training promotes the consolidation of a LTM for this task. In the BT framework, the learning tag set by the wIA captures the PRPs derived from the OF exploration, allowing the consolidation of a lasting IA memory trace. In resemblance to the STC model, the OF exploration can promote IA–LTM whether it occurs before or after the IA training.

memory should depend on the synthesis of new proteins induced by the exploration to the novel arena. To address this possibility, cannulae were implanted in the CA1 region of rat dorsal hippocampus and either the protein synthesis inhibitor anisomycin or vehicle were infused immediately after the OF exploration. The inhibition of the synthesis of new proteins completely impaired the symmetrical promoting effect of novelty on IA–LTM. The hypothesis derived from these experiments was that if the exploration to the novel OF was effectively providing PRPs to the wIA training to allow the consolidation of its own memory, then novelty should be also able to rescue the IA memory from the amnesia induced by the infusion of anisomycin close to an strong IA training (sIA). We observed that the infusion of anisomycin previous to a sIA training impaired IA–LTM consolidation for this task. However, if animals had been previously exposed to a novel OF, this exploration could prevent this amnesic effect. Such mechanism, as well as memory promotion, was also dependent PRPs synthesis induced by the exploration of the novel arena because further inhibition by the infusion anisomycin immediately after OF exploration blocked the preventive effect of novelty on IA memory.[2] These experiments demonstrated that sIA triggers both the learning tag and the PRPs required to consolidate its own memory and also revealed that if protein synthesis was inhibited another source of these products could be used to overcome amnesia.

The experimental design described above set a starting point that boosted the study of the BT process in multiple research labs. First evidence of BT process acting in LTM formation in mice came from Bai Lu lab. They showed that novelty is able to promote an IA memory using the step-through variation of this task.[6] The following year it was shown that novelty, in this case related to the exploration of a novel object within an arena, was also able to promote step-through avoidance LTM in rats.[26] Authors proposed that promotion occurs using mechanisms associated to hippocampal-LTD.

Further evidence of BT acting in the formation of aversive LTMs came from experiments performed using a contextual fear conditioning (CFC) task. Like IA, CFC is a hippocampal-dependent learning task but with some markedly different characteristics. Here, the animal is put inside a box with metallic bars, and after a period of habituation to the environment, it receives consecutive series of FSs during a certain period of time. As a major difference to the IA task, there is nothing that the animal can do to receive or avoid receiving the FS. The shock is simply associated to being in this particular environment and leads to the formation of a fear-driven memory that can be evaluated by comparing the amount of freezing during the habituation period and the test session. An increase in freezing behavior associated with the fear to the arena is indicative of memory. In this task, shock intensity and duration can also be modulated in order to induce only short forms of memories. As observed in the IA task, a CFC–STM can be reinforced into a CFC–LTM by associating the learning experience with the exploration of a novel OF, with this promoting effect dependent on PRP-synthesis induced by the novel experience.[57] So, both operant and classic conditioning lead to the formation of LTMs through a tagging and capture process. Further experiments, analogous to the original BT ones, were performed to study memory extinction processes[8] and demonstrated that exploration of a novel arena within a critical time window of an extinction session is able to promote the long-term extinction of a CFC memory. This extinction process is dependent on gene transcription and PRPs synthesized by the novel experience. The authors propose that the extinction session is able to set a tag capable of using newly synthesized PRPs (induced by OF exploration) to induce long-term extinction. As memory extinction is indeed considered the construction of a new association and therefore a new memory that overcomes the expression of the original mnemonic trace, these results show another face of the BT process acting in LTM formation.

Another interesting property observed by different researchers is the importance of the novel nature of the associated stimulus in order to promote LTM for the weak training. We have shown that while the exploration of a novel OF 1 h before wIA training is able to promote LTM formation, the exploration of a familiar arena, that has been explored for 30 min the day before, has no promoting effect on IA–LTM.[2] In agreement, it has been shown that the exploration of a familiar environment fails to promote long-term extinction of CFC memory and that exploration of a familiar object position is unable to promote the IA–LTM.[8,26] Further, experiments performed with behavioral reinforcement of LTP showed that exploration of a novel, but not previously known, arena reinforces early-LTP into late-LTP.[42] Later in this chapter, we will refer to the novelty effects on the activation of both brain regions, the VTA and the LC, which induce the release of dopamine and norepinephrine in the hippocampus. So, it is a possibility that the detection of novelty is acting through these mechanisms to induce and to regulate the synthesis of those PRPs required for memory consolidation.

All together, these series of experiments provide an abundant body of evidence, showing that a weak learning session, unable to consolidate a lasting memory, is capable of setting a mark that can utilize newly synthesized PRPs provided by an unrelated behavioral task and thus allowing the consolidation of a lasting memory. Such a process has been termed as BT and capture, in analogy to the STC process, but whether the behavioral tag is a specific synaptic process is yet a matter of active research.

8. NOVELTY PROMOTES LTM FORMATION IN SPATIAL MEMORIES

Spatial memory is one of the most important cognitive functions in daily life. Thanks to it, it is possible to distinguish roads, places, or simply recognize an object in a certain environment.[58] In this sense, the hippocampal region that includes the CA fields, DG, and subicular complex is part of a system of anatomically related structures in the medial temporal lobe, which are important for mammalian memory.[59] Moreover, the hippocampus may be especially important for tasks that depend on relating or combining information from multiple sources, as in certain spatial memory tasks.[60]

To study BT processes in a spatial memory, rats were submitted to the hippocampus-dependent SOR task. In this kind of task, the animals should recognize that a specific object has changed its position with respect to another object.[61,62] Using a weak SOR (wSOR) training session, which only induced STM, it was evaluated if LTM could be consolidated through a relatively long-lasting temporal association with a 5-min exploration to a novel OF. A wSOR training consists in a 4-min of exploration of two identical objects located in a familiar arena. When rats were exposed to a novel OF 1 h before or 2 h after a wSOR training task, it was observed that the otherwise transient SOR memory was consolidated into a lasting form (Fig. 13.3). This promoting effect was neither observed at more distant time points nor when novelty was given very close to wSOR training. Furthermore, this permissive action of a spatial novelty on memory formation is restricted to a critical time window. These effects also depend on the novel nature of the stimulus because the exploration to a familiar environment did not possess any effect on LTM promotions. The effect of novel stimuli on LTM promotion is also dependent on protein synthesis, being its action blocked by infusion of anisomycin in CA1 region.[57]

Another example of the BT process in the field of spatial memories was provided by Wang and collaborators from Richard Morris' group, who has

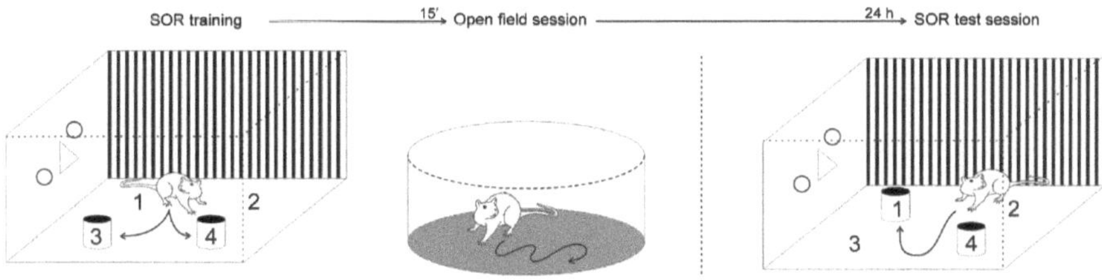

Figure 13.3 A space-related experimental design used for the study of behavioral tagging: Spatial object recognition (SOR) combined with the open field exploration. The use of a different, nonaversive, paradigm to study BT has yielded results similar to the IA paradigm, strongly supporting the BT hypothesis. A weak training in the SOR task that does not induce a long-lasting memory can be consolidated into a LTM if animals explore a novel OF 15 min after the SOR training. These two tasks depend on hippocampal processing and the promoting effect of the OF can be observed both when it is explored before or after the SOR training.

shown that an appetitive-driven spatial memory can also be promoted by effects of novelty.[63] Rats trained in an event arena for several months learned to find a food reward hidden in sand wells. After that, rats undergoing a weak encoding session to find one hidden pellet remembered the proper location of the reward for 30 min, but not 1 h after. On the contrary, a three-pellet reward encoding session induced of a 24-h lasting memory. Interestingly, animals undergoing a weak encoding training could consolidate a LTM for this task if the training was associated with the exploration of a novel OF. In concert with our observations, this promotion was dependent on the novel nature of the arena and on the synthesis of new PRPs induced by it. These findings also support that, as well as in single trial learning experiences, encoding and storage processing on everyday learning-like experience can lead to memory consolidation through a tagging and capture process.

The effects of novelty on spatial memories were also reported using the water maze task (WMT) paradigm. Almaguer-Melian and coworkers have investigated if this spatial memory could be recovered after being disrupted by FS, submitting rats to a novel OF exploration session. In the WMT learning task, a rodent is put into a small pool of water (with visual cues) that contains a hidden escape platform. When released, the subject swims around the pool searching for an exit, with the latency to find the platform used as a measure of memory. It was shown that four trials in the WMT were sufficient for rats to learn to find the platform during training. WMT–LTM trace was preserved for 24 h and FS impaired its consolidation, but preserved WMT–STM. As seen in other hippocampal-dependent learning tasks, when

OF exploration was given 15 min before or after training, the memory was improved in a protein synthesis-dependent way, overcoming the disrupting action of the FS on WM–LTM formation.[64] Similar to the results discussed earlier, when the OF was explored 4 h after WM training it lacked an effect. In addition, it has been shown that a familiar OF did not have an effect on escape latency.

As a whole, the findings reviewed to this point demonstrate that the BT process acts as a general mechanism of hippocampus-dependent LTM formation. However, this raises an important question: is this phenomenon selective just for hippocampal memories? Could a cerebral cortex-dependent memory be promoted by an appropriate associated stimulus? The conditioned taste aversion (CTA) task gave the key to respond to this query.

9. NOVELTY IMPROVES LTM FORMATION IN CTA TASK

One of the most important survival skills that animals have developed in thousands of years of evolution is taste-recognition memory. If we talk about recognition memory, we could describe it as the ability to assert the familiarity of things previously encountered.[65,66] When an animal encounters a new taste, it hesitates to eat or drink it. This behavioral response evidenced through a reduced initial consumption is known as neophobia. However, if this initial consumption does not lead to negative consequences, the factor is recognized as a safe indicator and its consumption rises again (attenuation of neophobia). But, what happens when a new taste is associated with malaise? In that case, the new taste becomes a negative signal and the animals develop a long-lasting aversion to it. This type of recognition memory is called CTA and is largely known to be dependent on the activation of the insular cortex.[67] In CTA task, rats learn to associate the consumption of flavor such as saccharine with a digestive malaise induced by intraperitoneal injection of a lithium chloride (LiCl) solution. As a consequence, there is a decrease in saccharin consumption during the test session compared to the acquisition session intake, which is used as a measure of the aversion strength.

Rats that receive a weak training in this task, by association of the consumption of saccharine with a low dose of LiCl, induced a negligible CTA–LTM but expressed a strong CTA–STM 30 min after the acquisition session.[57] Therefore, we reasoned that another gustative and novel stimulus (NaCl) that activates the same cortical region might improve the consolidation of the saccharine CTA–LTM. Particularly, strong CTA–LTM was observed when a novel taste of NaCl intake took place 1 h before or 2.5 h after the CTA training. In consonance with the observations in the IA, CFC, and SOR memories, this effect is dependent on the novel nature of the flavor (rats familiarized to drink a NaCl solution during the 3 days before training did not experience any improvement in saccharine CTA–LTM) and on the synthesis of new PRPs induced by the consumption of this novel flavor.[57] These results are in accordance with other studies demonstrating that a novel, but not a familiar, taste (NaCl) facilitates the formation of lasting memory for a weak taste by using a latent inhibition protocol.[68]

These results demonstrate that both hippocampal- and cortical-dependent learning, using aversive, nonaversive or operant tasks are able to induce a kind of mark (the learning tag) able to use PRPs to consolidate a lasting memory and suggest that the BT process acts as a general mechanism of LTM formation.

10. SPECIFIC NOVELTIES ARE REQUIRED TO PROMOTE DIFFERENT MEMORY TRACES

So far, it has been established that weak training induces a tag that has a transient lifetime and that this tag can capture proteins synthesized by novel-associated stimuli to facilitate LTM formation. The methodology used to reveal BT processes requires the integration of at least two different and separate experiences arriving at common neural substrates within a period of few hours (see item the ABC of tagging and capture). Knowing that taste induces mainly the activation of insular cortex and that spatial information is mainly related to an activation of the hippocampus, we tested if training rats in these tasks (that did not overlap their engaged neuronal population) exhibited the same promoting effects of novelty on LTM formation. In other words: is it possible to promote the formation of LTM for a cortical-dependent task like the CTA, through the exposure to a novel environment? Alternatively, can a novel taste promote the hippocampus-dependent SOR–LTM consolidation? It has been shown that rodents exposed to a novel OF 1 h before or 2.5 h after (times in which CTA–LTM was promoted by using a novel taste) a wCTA training did not express CTA–LTM. Similar results were observed when rats drank a novel taste and were trained with a wSOR. Therefore, even though the time points chosen were the ones in which a novel OF exploration promoted SOR–LTM and novel taste promoted CTA–LTM, neither the exploration to a novel arena nor the consumption of a novel taste were able to promote the appetitive nor the spatial memory, respectively.[57]

These results illustrate a specific requirement of novelty to promote different memory traces, stressing that promotion can only occur when PRPs are provided in those specific brain regions where the memory trace is being consolidated.

11. IDENTIFICATION OF TRANSMITTER SYSTEMS AND LEARNING TAG MOLECULES

The BT hypothesis implies that to allow the consolidation of LTM both the setting of the learning tag and the synthesis of new proteins have to occur. In consequence, neurotransmitter and second-messenger signaling systems should be involved in these processes. The understanding of what processes are linked to the setting of the learning tag and which ones to the synthesis of PRPs might be helpful to unveil which of them are essential to consolidate a memory and which others could be eventually replaced by the association to stimulus capable of providing PRPs required to consolidate a lasting memory.

Taken into account that a strong experience capable of inducing LTM triggers both, the setting of the learning tag and the synthesis of PRPs, the study of the mechanisms specifically involved in each of these processes might represent a challenge. According to the BT theory, the learning tag comprises a set of local processes and changes that allow the storage of information as LTM, if it captures the PRPs required for memory consolidation. Therefore, if the tag remains intact after amnesia caused by interference with the PRPs synthesis process, subsequently providing PRPs through an external stimulus should restore LTM. On the other hand, because the tag is essential to set the input specificity of the information to be stored, amnesias caused by interference with the tag should be irreparable and impossible to be prevented by association with other tasks or stimulus able to synthesize proteins. With these ideas in mind, a different series of BT experiments have been designed in order to interfere with either the PRP synthesis or the learning tag setting. In this way, pharmacological interventions performed close to a weak training might interfere with the setting of the learning tag, while interventions performed close to the strong stimulus might impair the synthesis of novelty-induced PRPs (Fig. 13.4).

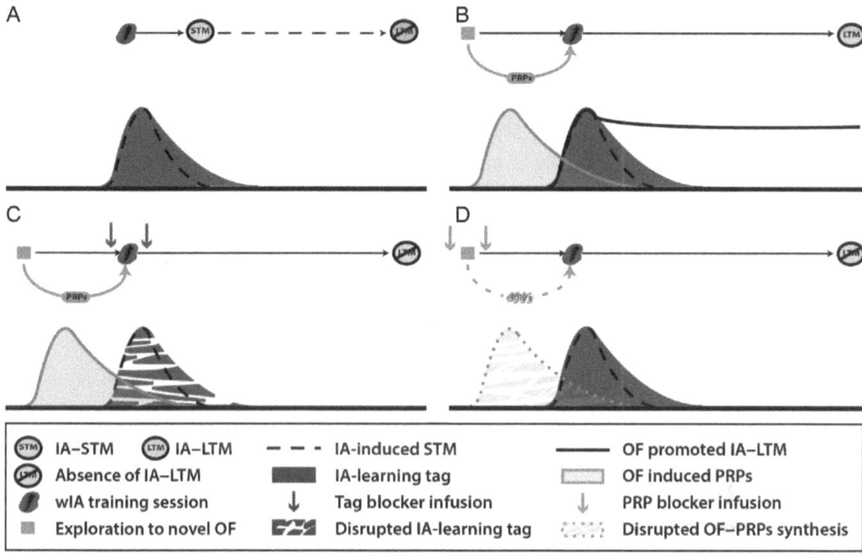

Figure 13.4 Strategies used to study the BT process in LTM formation using the inhibitory avoidance (IA) task. The figure shows different strategies that can be use to observe BT process and analyze those mechanisms associated to the setting of the learning tag or to the synthesis of PRPs, using a weak training in the IA task. In all cases, the curves represent a qualitative time course for the different processes. (A) wIA training induces short but not long-term memory and sets an IA leaning tag. (B) The exploration to a novel OF induces the synthesis of PRPs that can be used by IA-learning tag to allow IA–LTM consolidation. (C) The infusion of different drugs (i.e., CAMKIIα and PKA inhibitors, PKMz blocker, or NMDA receptor antagonist) in the hippocampus close to the wIA training can interfere with the proper setting and/or maintenance of the IA-learning tag impairing the promotion of IA–LTM. (D) The infusion of different drugs (i.e., protein synthesis inhibitors or D1/D5-dopaminergic and β-adrenergic receptors antagonists) in the hippocampus close to the OF exploration can interfere with the PRPs synthesis impairing the promotion of IA–LTM.

As has been shown by different research groups, the detection of novelty is directly associated with the activation of VTA and LC which in term releases dopamine and norepinephrine into the hippocampus, targeting mainly the CA1 and DG regions, respectively.[69–72] The activation of the D1/D5-dopaminergic and β-adrenergic receptors triggers the activation of different second-messenger cascades that can lead to gene transcription and protein translation. These facts lead directly to the question of whether these neurotransmitter systems could be responsible of triggering the synthesis of the PRPs required for memory consolidation. To investigate this issue, it was decided to apply antagonists of either the D1/D5-dopaminergic or the β-adrenergic receptors during the OF novel experience, and observe their effects on the promotion of memory in animals that were weakly trained with the IA task 60 min later. The infusion of either SHC23390 or Propranolol (D1/D5-dopaminergic and β-adrenergic receptor antagonists) in CA1 or DG, respectively, 10 min before submitting animals to a novel OF, completely blocked the promoting effect on IA–LTM formation observed in the vehicle-infused group.[7] Interestingly, the catecholaminergic dependence on memory promotion has also been observed before in experiments that combined electrophysiological and behavioral methodologies, showing that novelty-dependent LTP reinforcement also required D1/D5-dopaminergic and β-adrenergic receptors.[42,44] Moreover, the activation of these receptors was also necessary for novelty-promoted schemas memory and in the prevention of stress-induced amnesia by means of novelty-dependent BT processes.[63,64]

From these results, it is easy to think that one of the main outputs of novelty detection which induces IA–LTM formation is the release of dopamine and norepinephrine to the hippocampus. If this was really the mechanism responsible for the induction of PRPs synthesis, the direct activation of the receptors should have similar results to those observed after exposure to novelty on IA memory. This was actually observed when the exposure to the novel arena was replaced by local infusion of the dopaminergic agonist SKF33983 or adrenergic agonist Dobutamine (in CA1 and DG, respectively). In addition, the consolidation of the lasting memory induced by these agonists occurs in a temporal time window comparable to that of the OF-induced IA–LTM after a wIA training session. Remarkably, this promoting effect on IA memory is also dependent on the synthesis of new proteins triggered by the infusion of any of these agonists, ruling out other possible unspecific effects.[7] These data point directly to the idea that novelty induces, at least in the hippocampus, the synthesis of those PRPs required to consolidate the IA memory, through a mechanism that includes the release of dopamine and norepinephrine in this structure and that requires the activation of the D1/D5-dopaminergic and β-adrenergic receptors.

Further conclusions can be derived from the idea that a particular memory should need a certain pool of PRPs to be consolidated. Therefore, if the pool required by the IA memory is provided by novelty through the activation of the dopaminergic and adrenergic systems, it could be implied that these systems are indeed activated by a sIA training that induces IA–LTM and, in consequence, promotes the synthesis of the PRPs required during the consolidation process. To investigate this issue, we analyzed the role of D1/D5-dopaminergic and β-adrenergic receptors in IA memory formation by infusing their antagonists (SCH23390 or propranolol, respectively) prior to a sIA training. Remarkably, while IA–LTM was dependent on the activation of these receptors, the protein synthesis-independent IA–STM was not affected by the infusion these antagonists, strengthening the idea that they are indeed involved in triggering the synthesis of those PRPs required to establish a lasting memory. Actually, the amnesia induced by either of these two antagonists can be prevented by preexposing the animals to a novel OF, and this preventive effect is dependent on the synthesis of new proteins induced by the novel experience.[7]

This effect of the D1/D5-dopaminergic and β-adrenergic receptors acting specifically on mechanisms leading to the synthesis of PRPs is not shared by other receptors. In contrast, NMDA glutamate receptors are required to both STM and LTM formation.[7] Moreover, the amnesia induced by local infusion in the hippocampus of AP-V (an NMDA receptor antagonist) prior to an sIA training could not be prevented by the exposure to a novel OF 60 min before training, in contrast to what happened when dopaminergic or adrenergic receptors antagonists were infused.[7] Similarly the infusion of AP-V before wIA training impaired the promoting effect of novelty on IA memory. The fact that the blockade of NMDA receptors during learning induces an irreversible amnesia, or impairs memory promotion, shows that they play an essential role in the setting of the learning tag that cannot be overcome by simply providing PRPs by an external stimulus.[7] This does not mean that NMDA receptors do not play a role in the signal transduction leading to protein synthesis as well. There is a well-established role of NMDA receptors to this respect shown by several research groups around the world.[56,73–75] Indeed, by using the BT model we also observed that the blockade of NMDA receptors previous to the novelty session performed 1 h before a wIA training also impairs memory promotion, clearly suggesting a role of this receptor in PRPs synthesis as well.[7]

If NMDA receptors are involved in the setting of the IA-learning tag, they must certainly take part in registering the incoming signals and triggering the activation of the cellular machinery responsible of establishing the tag. Due to their fast activation and the speed with which they can alter the function of receptors and structural molecules, protein kinases are an interesting target in the search for components of this tagging machinery. Particularly if these kinases are usually present at the postsynaptic densities and are required for the formation of lasting memories briefly after learning.

Among these αCAMKII, PKA, and ERK 1/2 are suitable candidates because of their well-established involvement in the formation of LTMs.[56,76] To investigate the role of these kinases in LTM formation, their inhibitors were infused in the hippocampus at different times before or after wIA training with the aim to block the setting of the IA-learning tag and impair the formation of a novelty-promoted IA–LTM. The results revealed that KN-62 (αCAMKII inhibitor) and Rp-AMPc (PKA inhibitor) impaired the novelty-induced IA–LTM when infused between 10 min before and 15 min after wIA training but not 1 h latter.[7] While analyzing the role of PKMz, we observed that its blocker (Myr-Zip) had a partial influence when infused 10 min before training but its effect was still present even in animals infused 1 h after training.[77] This result suggests that both αCAMKII and PKA play an essential role in the setting of the learning tag in those moments immediately around training, but their activity is no longer required at later times. Interestingly, PKMz might be playing a role in maintaining a plausible late phase of the behavioral tag in these letter periods. Alternatively, this kinase that has been shown to be required for late maintenance of memory and plasticity[78] could be required to keep early plastic changes as well. However, PKMz role in LTM and plasticity processes is currently the center of a debate due to possible nonspecificity of its blocker Myr-zip.[79] On the other hand, neither U0126 (MEK inhibitor) nor anisomycin administered close to wIA was able to impair the promoting effect of novelty on IA–LTM formation, showing that the setting of the IA-learning tag uses already synthesized proteins and does not require the activity of ERKs 1/2.[7] By using TrkB knock-in mice, it has been demonstrated that inhibition of the receptor's kinase activity during weak training in the IA, impaired the promotion process induced by novelty. The same results were obtained in analogous *in vitro* experiments where TrkB inhibition during weak tetanization also blocked the reinforcing effect of a strong tetanization on LTP, leading them to postulate this receptor as a potential component of the behavioral and synaptic tag.[6] Interestingly, while the setting of the IA-learning tag as well as LTP- and LTD-tags are protein synthesis-independent processes, recent experiments show that the tag setting during CFC extinction learning might depend on protein synthesis.[8]

Different experiments performed using an electrophysiological model of synaptic tagging showed that the LTP-tag requires the activity of αCAMKII, PKA, and PKMz and that is a protein synthesis-independent mechanism.[12,13] Alternately, ERK 1/2 has been shown to be required for the setting of synaptic tags associated to LTD but not LTP.[80] Interestingly, using an active avoidance task, Whitlock and collaborators showed that this learning is processed through mechanism associated with LTP induction.[16] These data are consistent with the results showing that IA-tags require αCAMKII and PKA but not ERKs to be properly set.[7] Their activity is required for a certain period of time, showing that even when the training itself triggers the setting of the behavioral tag, the process is susceptible to disruption during a certain period after learning. Indeed not only pharmacological but also behavioral interventions can impair the setting or even reset the learning tag. The exploration of a novel arena close to the wIA training is unable to promote IA memory but also impairs the promoting effect of the exploration to a novel arena performed at a time in which novelty promotes memory.[77] On the other hand, the combination of two exploration sessions of different novel arenas in usually permissive time points results in a better memory than that observed by action of only one exploratory arena.[81] In fact, LTP tags are also labile and they can be reset by the application of a strong low-frequency stimulation after LTP induction,[82] suggesting that certain processes leading to LTP and LTD tag setting are driven through antagonistic mechanisms. As the exploration of a novel arena has been shown to underlie a LTD-like processing in the hippocampus,[26] this might provide an explanation of the deleterious effect observed when this novelty is experienced close to the IA training, which seems to be processed through LTP-like processing.

12. MEMORY TRACES COMPETE UNDER REGIMES OF LIMITED PROTEIN SYNTHESIS

Memories are not static entities that remain unchanged in time. Quite on the contrary, memory formation and persistence are very dynamic processes, sensitive to interventions occurring around the learning event. During our everyday life, we experience several events with multiple characteristics. However, not many of them will be stored in our LTM. Sometimes the information is coded but it is not stored or perhaps it has been effectively stored but it cannot be retrieved.

In the model of BT, one experience can promote the memory of an unrelated event, as long as the two events occur in a limited time window and they are processed in overlapping structures in the brain. Another requisite to observe this promoting effect is that one of the events triggers the synthesis of PRPs that will enable the consolidation of the trace. In this framework, the findings on BT suggest that both traces are sharing a common pool of PRPs. One possibility that could take place in such conditions is that both learning tags, each of them set by their respective learning experiences, are competing for the common pool of PRPs. Interestingly, studies in hippocampal LTP introduced the concept of "competitive maintenance."[83] Under regimes of reduced protein availability, different synapses compete for the available resources, resulting in a depotentiation of activated pathways by the influence of an independent-activated pathway. Furthermore, very recent findings provide supporting evidence for the existence of competition for PRPs by activated synapses and such effect is further increased if a protein synthesis inhibitor (which prevents the renewal of the PRPs' pool and pushes the system toward a more limited PRP availability) is applied to the bath medium.[84]

Considering the findings in LTP and competence for resources, a possible scenario is that learning tags could compete for the PRPs under limited protein resources. That is, if different tasks which are processed in overlapping networks are being consolidated into LTM, intracellular competition for PRPs will define which of the memory traces becomes stabilized. What is more, it has been postulated that during the stabilization of a memory trace, selection of the neuronal network that will be recruited into a given memory representation occurs in a competitive fashion rather than in a cell-autonomous way.[85,86] Besides, similar computational models have postulated the existence of competition while a network is being recruited for language acquisition at early stages of development.[87] All these findings lead to the idea that sharing, cross talk and capture in spine neighborhoods[88] are important mechanisms for the consolidation of plastic changes, at least at a cellular level. Thus, it is a plausible hypothesis that a similar phenomenon could be happening during the acquisition of new information.

This has been tested at behavioral level by combining wIA and novel OF, two tasks that are dependent on hippocampus processing. The results show that under regimes of reduced protein resources, but not when resources are widely available, wIA-learning task can hinder the LTM formation of OF because of their common requirement of PRPs. When rats are sequentially exposed to two different memory tasks under a regimen of limited protein synthesis, LTM for one of them is formed to the detriment of the formation of the other. It has been postulated that the amount of PRPs is insufficient to satisfy the LTM requirements of the two behavioral tasks and is what accounts for the impairment of one of the traces.[81]

Activity-regulated cytoskeletal-associated protein (Arc) has also been shown to be involved in the formation of several types of memories and has an important role in synaptic plasticity.[89–91] In particular, limiting the amount of available Arc (administering Arc mRNA antisense oligonucleotides into the dorsal hippocampus), was shown to have deleterious effects in novelty-promoted IA–LTM formation.[81] Latest research on Arc's role suggests that this protein is captured by CaMKIIβ, which induces an "inverse synaptic tagging process," recruiting Arc in the less active terminals. Arc, in turn, downregulates the amount of GluA1 at individual synapses and therefore operates as a specific sensor mediating the inactive synapse-specific control of AMPA-R clearance at weaker synapses in potentiated neurons, depending on the local history of both activity and inactivity.[92] Even though Arc is an attractive candidate as a PRP that could be competed for by different memory traces, other PRPs related to plastic changes in synaptic terminals could be considered as well.[93]

These findings suggest the existence of competition between two memory traces for their consolidation when protein resources are limited. Therefore, we suggest that interference could be explained by a mechanism of competition for protein resources, with Arc being one of the PRPs required for the consolidation of both memory traces.

Observations over a hundred years ago found that the strength of retrograde interference (RI) on learning, exerted by interpolated material, increases with the proximity between events. The degree of memory forgetting is variable and traces become less vulnerable to empirical forgetting, brain damage or retroactive interference as they consolidate with the passage of time.[94–96] Later, Skaggs[97] suggested that the interpolated task causing RI could be a mental effort per se or a similar event to be recorded, and that the RI effect is reduced when tasks are highly similar or, on the contrary, when they are markedly different. All these observations could be reinterpreted considering the BT hypothesis involving capture of PRPs by different kinds of learning tags. Thus, if the interpolated material is identical to the original, it can represent a retraining; probably reinforcing almost the same learning tags set for the original task. In such case, there would be no different kinds of learning tags capturing the PRPs. In contrast, a high dissimilarity of the material could imply its processing in different brain regions; thus, the respective learning tags would not interfere because they do not converge spatially.[11,57]

The BT model proposes a cellular mechanism to explain amnesia by retrograde and also anterograde interference, focusing on the competitive capture of proteins required for the consolidation of those memory traces. The work performed on this topic[81] provided the first evidence of molecular events underlying memory competition that could explain how some information is stored while other is lost or impaired.

13. EVIDENCE OF BT IN HUMAN

We have previously showed that rodents receiving weak training protocols that only induce STM could consolidate a LTM if the training session took place close to an unrelated novel experience. This process begins with the setting of a learning tag established by the weak training and requires synthesis of PRPs induced by novelty. Thus, an essential question arises: is this BT mechanism also acting in human LTM formation? To answer this question, we studied if a novel or familiar experience (science or music lesson) exerts a similar effect on LTM of a reading or graphical activity in the classroom. When teacher read a short story to the students, LTM improvement about its content was observed in those groups who had experienced a novel science lesson 1 h, but not 4 h, before or after the story telling. Thus, the action of novelty is symmetric and restricted to a limited time window, similar to the results previously reported in rodent models.[98] Moreover, using a visual graphical learning task (adaptation of the Rey-Osterrieth's complex figure task for children[99]) the students remembered more figure details when information was acquired 1 h before attending to a novel science lesson.[98] In contrast, a familiar lesson failed to enhance memory as there were no differences between control and experimental groups. In addition, to assess the tasks' time-specificity of this effect, students were read two different stories separated by a 3-h interval, and 1 h after the last story they attended to a novel lesson. Memory enhancement, measured the following day, was observed for the story told in the closest proximity to the novel lesson but not for the other. This suggests that the reading of the story triggers a transient process that enables the consolidation of that information by the effects of the novel experience. Taken as a whole, these results demonstrate the positive effect of novel lessons on LTM formation in schoolchildren. Memory improvement induced by novelty is not derived from an increase in the arousal state or from lowering the threshold to learn, because both types of novel experiences can exert their positive effect on memory even when they occur after the acquisition of the information to be remembered.[98]

In sum, novel lessons can exert late-associative effects on LTM formation for story telling or figure drawing in children, leading to the consolidation of these memories. Though it cannot currently be demonstrated that the effect of novelty in humans depends on protein synthesis, we put forward the hypothesis that the learning experience (story/drawing) triggers a transient process (which resembles the idea of a learning tag) that enables the consolidation of this information by the effects of the novel experience occurring around a critical time window. Therefore, taken as a whole, our findings suggest that a BT-like process could be operating in human beings.

14. CONCLUDING REMARKS

In this chapter, we have compiled findings linking synaptic plasticity with memory processes taking place in behaving animals. We then summarized experiments demonstrating that synaptic plasticity was improved *in vivo* by structural reinforcement or behavioral interventions and also experiments showing how a behavior could be improved by the induction of synaptic plasticity. Finally, the core of this review was aim to detail memory experiments, demonstrating the BT process in which one experience promotes the formation of a lasting memory for other experience.

The STC hypothesis was postulated in 1997 by Frey and Morris,[1] introducing the idea that there is a temporal window shortly after LTP induction in which PRPs are targeted selectively to activated synapses in order to establish a long-lasting form of potentiation. The most important concept of this hypothesis is the setting of a tag at the time of LTP induction whose role is to capture proteins in an input-specific manner. This tag is transient and can sequester proteins induced by any other stimulation occurring in a common neuronal population where it is set. Given that the stimulation that induces PRP synthesis can occur at times before as well as after tag induction, these late-associative effects on LTP take place in a symmetrical way. Ten years after the postulation of this hypothesis, it was demonstrated that an analogous BT process also operates in a living animal when a LTM is been formed from a weak experience.[2] In this case, a strong experience associated with a weak one helps to consolidate its LTM through a protein synthesis-dependent mechanism. Thus, STC and BT processes share some features: (1) a strong event helps to establish a persistent form of plasticity associated with another weak event, (2) this happens only when the strong event occurs in a critical time window around the weak one, (3) these events should activate a common neuronal population, (4) the mechanism is dependent on the protein synthesis induced by the strong event, and (5) is also dependent on a tag set by the weak event.

In the past 5 years, several research groups have worked on the BT process demonstrating that it was observed in operant and Pavlovian aversive paradigms, in the formation of extinction and SOR memories and in other tasks based on spatial learning.[2,6–8,26,57,63,64] Moreover, a similar phenomenon was observed also in school children who had learnt about a story or drawing, suggesting the generality of the process in long-lasting memory formation.[98] In addition, some aspects about the nature of learning tag and the identification of PRP involved in the process are beginning to be unveiled.[6–8,81] However, future studies are needed to fully understand the mechanism and the molecules involved and ultimately to visualize the PRP capture process.

In our opinion some major points remain to be addressed. First, do different learning tasks set different learning tags? In other words, does each learning experience set a characteristic mark? Are there any differences in the quality and/or quantity of learning tags between different experiences? Does a metaplasticity phenomenon that affects the duration of learning tags exist in a similar way to what has been observed in synaptic plasticity experiments?[100] Second, what are the strong experiences that are able to induce BT memory promotion by providing PRPs? Most BT experiments were performed using a novel exploration to an OF as an event of strong saliency. However, are there other events such as cognitive reinforcers or motivational situations that facilitate memory formation using BT mechanism? Finally, besides the role played by learning tags in LTM formation, are there other kinds of tags mediating other aspects of memory like the persistence of the trace? These and many questions will be probably answered in the near future.

REFERENCES

1. Frey U, Morris RG. Synaptic tagging and long-term potentiation. *Nature*. 1997;385:533–536.
2. Moncada D, Viola H. Induction of long-term memory by exposure to novelty requires protein synthesis: evidence for a behavioral tagging. *J Neurosci*. 2007;27:7476–7481.
3. Martin KC, Kosik KS. Synaptic tagging—who's it? *Nat Rev Neurosci*. 2002;3:813–820.
4. Kelleher RJ, Govindarajan A, Tonegawa S. Translational regulatory mechanisms in persistent forms of synaptic plasticity. *Neuron*. 2004;44(1):59–73.
5. Lu Y, Christian K, Lu B. BDNF: a key regulator for protein synthesis-dependent LTP and long-term memory? *Neurobiol Learn Mem*. 2008;89(3):312–323.
6. Lu Y, Ji Y, Ganesan S, et al. TrkB as a potential synaptic and behavioral tag. *J Neurosci*. 2011;31:11762–11771.
7. Moncada D, Ballarini F, Martinez MC, Frey JU, Viola H. Identification of transmitter systems and learning tag molecules involved in behavioral tagging during memory formation. *Proc Natl Acad Sci USA*. 2011;108:12931–12936.

8. de Carvalho Myskiw J, Benetti F, Izquierdo I. Behavioral tagging of extinction learning. *Proc Natl Acad Sci USA*. 2013;110(3):1071–1076.
9. Frey S, Frey JU. "Synaptic tagging" and "cross-tagging" and related associative reinforcement process of functional plasticity as the celular basis of memory formation. *Prog Brain Res*. 2008;169:117–143.
10. Ramachandran B, Frey JU. Interfering with the actin network and its effect on long-term potentiation and synaptic tagging in hippocampal CA1 neurons in slices in vitro. *J Neurosci*. 2009;29(39):12167–12173.
11. Redondo RL, Morris RGM. Making memories last: the synaptic tagging and capture hypothesis. *Nat Rev Neurosci*. 2011;12:17–30.
12. Sajikumar S, Navakkode S, Frey JU. Identification of compartment- and process-specific molecules required for "synaptic tagging" during long-term potentiation and long-term depression in hippocampal CA1. *J Neurosci*. 2007;27(19):5068–5080.
13. Redondo RL, Okuno H, Spooner PA, Frenguelli BG, Bito H, Morris RG. Synaptic tagging and capture: differential role of distinct calcium/calmodulin kinases in protein synthesis-dependent long-term potentiation. *J Neurosci*. 2010;30(14):4981–4989.
14. Martin SJ. Morris RG new life in an old idea: the synaptic plasticity and memory hypothesis revisited. *Hippocampus*. 2002;12(5):609–636.
15. Martin SJ, Grimwood PD, Morris RG. Synaptic plasticity and memory: an evaluation of the hypothesis. *Annu Rev Neurosci*. 2000;23:649–711.
16. Whitlock JR, Heynen AJ, Shuler MG, Bear MF. Learning induces long-term potentiation in the hippocampus. *Science*. 2006;313:1093–1097.
17. Gruart A, Muñoz MD, Delgado-García JM. Involvement of the CA3-CA1 synapse in the acquisition of associative learning in behaving mice. *J Neurosci*. 2006;26(4):1077–1087.
18. Green EJ, Greenough WT. Altered synaptic transmission in dentate gyrus of rats reared in complex environments: evidence from hippocampal slices maintained in vitro. *J Neurophysiol*. 1986;55(4):739–750.
19. Rioult-Pedotti MS, Friedman D, Hess G, Donoghue JP. Strengthening of horizontal cortical connections following skill learning. *Nat Neurosci*. 1998;1(3):230–234.
20. Saar D, Grossman Y, Barkai E. Reduced synaptic facilitation between pyramidal neurons in the piriform cortex after odor learning. *J Neurosci*. 1999;19(19):8616–8622.
21. Clarke JR, Cammarota M, Gruart A, Izquierdo I, Delgado-García JM. Plastic modifications induced by object recognition memory processing. *Proc Natl Acad Sci USA*. 2010;107(6):2652–2657.
22. Goh JJ, Manahan-Vaughan D. Spatial object recognition enables endogenous LTD that curtails LTP in the mouse hippocampus. *Cereb Cortex*. 2013;23(5):1118–1125. http://dx.doi.org/10.1093/cercor/bhs089.
23. Griffiths S, Scott H, Glover C, et al. Expression of long-term depression underlies visual recognition memory. *Neuron*. 2008;58(2):186–194.
24. Seoane A, Massey PV, Keen H, Bashir ZI, Brown MW. L-type voltage-dependent calcium channel antagonists impair perirhinal long-term recognition memory and plasticity processes. *J Neurosci*. 2009;29(30):9534–9544.
25. Ge Y, Dong Z, Bagot RC, et al. Hippocampal long-term depression is required for the consolidation of spatial memory. *Proc Natl Acad Sci USA*. 2010;107:16697–16702.
26. Dong Z, Gong B, Li H, et al. Mechanisms of hippocampal long-term depression are required for memory enhancement by novelty exploration. *J Neurosci*. 2012;32(35):11980–11990.
27. Frey U, Morris RG. Weak before strong: dissociating synaptic tagging and plasticity-factor accounts of late-LTP. *Neuropharmacology*. 1998;37(4–5):545–552.
28. Young JZ, Nguyen PV. Homosynaptic and heterosynaptic inhibition of synaptic tagging and capture of long-term potentiation by previous synaptic activity. *J Neurosci*. 2005;25(31):7221–7231.

29. Sajikumar S, Frey JU. Late-associativity, synaptic tagging, and the role of dopamine during LTP and LTD. *Neurobiol Learn Mem.* 2004;82(1):12–25.
30. Sajikumar S, Navakkode S, Sacktor TC, Frey JU. Synaptic tagging and cross-tagging: the role of protein kinase Mzeta in maintaining long-term potentiation but not long-term depression. *J Neurosci.* 2005;25(24):5750–5756.
31. Kauderer BS, Kandel ER. Capture of a protein synthesis-dependent component of long-term depression. *Proc Natl Acad Sci USA.* 2000;97(24):13342–13347.
32. Govindarajan A, Kelleher RJ, Tonegawa S. A clustered plasticity model of long-term memory engrams. *Nat Rev Neurosci.* 2006;7(7):575–583.
33. Sajikumar S, Korte M. Metaplasticity governs compartmentalization of synaptic tagging and capture through brain-derived neurotrophic factor (BDNF) and protein kinase Mzeta (PKMzeta). *Proc Natl Acad Sci USA.* 2011;108(6):2551–2556.
34. Frey U, Morris RG. Synaptic tagging: implications for late maintenance of hippocampal long-term potentiation. *Trends Neurosci.* 1998;21:181–188.
35. Shires KL, Da Silva BM, Hawthorne JP, Morris RG, Martin SJ. Synaptic tagging and capture in the living rat. *Nat Commun.* 2012;3:1246.
36. Frey S, Bergado-Rosado J, Seidenbecher T, Pape HC, Frey JU. Reinforcement of early long-term potentiation (early-LTP) in dentate gyrus by stimulation of the basolateral amygdala: heterosynaptic induction mechanisms of late-LTP. *J Neurosci.* 2001;21:3697–3703.
37. Straube T, Frey JU. Involvement of beta-adrenergic receptors in protein synthesis-dependent late long-term potentiation (LTP) in the dentate gyrus of freely moving rats: the critical role of the LTP induction strength. *Neuroscience.* 2003;119(2):473–479.
38. Bergado JA, Frey S, López J, Almaguer-Melian W, Frey JU. Cholinergic afferents to the locus coeruleus and noradrenergic afferents to the medial septum mediate LTP-reinforcement in the dentate gyrus by stimulation of the amygdala. *Neurobiol Learn Mem.* 2007;88(3):331–341.
39. Bergado JA, Scherf T, Almaguer-Melian W, Frey S, López J, Frey JU. Stimulation of the nucleus raphe medialis modifies basal synaptic transmission at the dentate gyrus, but not long-term potentiation or its reinforcement by stimulation of the basolateral amygdala. *Neurosci Lett.* 2009;464(3):179–183.
40. Almaguer-Melian W, Bergado JA, Martínez-Martí L, Duany-Machado C, Frey JU. Basolateral amygdala stimulation does not recruit LTP at depotentiated synapses. *Physiol Behav.* 2012;101(4):549–553.
41. Reymann KG, Frey JU. The late maintenance of hippocampal LTP: requirements, phases, 'synaptic tagging', 'late-associativity' and implications. *Neuropharmacology.* 2007;52(1):24–40.
42. Li S, Cullen WK, Anwyl R, Rowan MJ. Dopamine-dependent facilitation of LTP induction in hippocampal CA1 by exposure to spatial novelty. *Nat Neurosci.* 2003;6(5):526–531.
43. Davis CD, Jones FL, Derrick BE. Novel environments enhance the induction and maintenance of long-term potentiation in the dentate gyrus. *J Neurosci.* 2004;24(29):6497–6506.
44. Straube T, Korz V, Balschun D, Frey JU. Requirement of beta-adrenergic receptor activation and protein synthesis for LTP-reinforcement by novelty in rat dentate gyrus. *J Physiol.* 2003;552:953–960.
45. Corrêa SA, Hunter CJ, Palygin O, et al. MSK1 regulates homeostatic and experience-dependent synaptic plasticity. *J Neurosci.* 2012;32(38):13039–13051.
46. Seidenbecher T, Balschun D, Reymann KG. Drinking after water deprivation prolongs "unsaturated" LTP in the dentate gyrus of rats. *Physiol Behav.* 1995;57(5):1001–1004.
47. Seidenbecher T, Reymann KG, Balschun D. A post-tetanic time window for the reinforcement of long-term potentiation by appetitive and aversive stimuli. *Proc Natl Acad Sci USA.* 1997;94(4):1494–1499.

48. Bergado JA, Almaguer-Melian W, Kostenko S, Frey S, Frey JU. Behavioral reinforcement of long-term potentiation in rat dentate gyrus in vivo is protein synthesis-dependent. *Neurosci Lett.* 2003;351(1):56–58.
49. Uzakov S, Frey JU, Korz V. Reinforcement of rat hippocampal LTP by holeboard training. *Learn Mem.* 2005;12(2):165–171.
50. Korz V, Frey JU. Emotional and cognitive reinforcement of rat hippocampal long-term potentiation by different learning paradigms. *Neuron Glia Biol.* 2004;1(3):253–261.
51. Moncada D, Frey J. The effects of LTP induction on hippocampus-dependent memories: synaptic and behavioral tagging. In: *Society for Neuroscience Meeting November 2011, San Diego, USA*; 2011.
52. Viola H, Furman M, Izquierdo L, et al. Phosphorylated cAMP response element-binding protein as a molecular marker of memory processing in rat hippocampus: effect of novelty. *J Neurosci.* 2000;20:RC112.
53. Winograd M, Viola H. Detection of novelty, but not memory of spatial habituation, is associated with an increase in phosphorylated cAMP response element-binding protein levels in the hippocampus. *Hippocampus.* 2004;14:117–123.
54. Moncada D, Viola H. Phosphorylation state of CREB in the rat hippocampus: a molecular switch between spatial novelty and spatial familiarity? *Neurobiol Learn Mem.* 2006;86:9–18.
55. Straube T, Korz V, Frey J. Bidirectional modulation of long-term potentiation by novelty-exploration in rat dentate gyrus. *Neurosci Lett.* 2003;344:5–8.
56. Izquierdo I, Bevilaqua L, Rossato J, Bonini J, Medina J, Cammarota M. Different molecular cascades in different sites of the brain control memory consolidation. *Trends Neurosci.* 2006;29:496–505.
57. Ballarini F, Moncada D, Martinez MC, Alen N, Viola H. Behavioral tagging is a general mechanism of long-term memory formation. *Proc Natl Acad Sci USA.* 2009;106(34):14599–14604.
58. Kessels R, Jaap Kappelle L, de Haan EH, Postma A. Lateralization of spatial-memory processes: evidence on spatial span, maze learning, and memory for object locations. *Neuropsychologia.* 2000;40:1465–1473.
59. Squire L. Memory and hippocampus: a synthesis from findings with rats, monkeys, and humans. *Psychol Rev.* 1992;99:195–231.
60. O'Keefe J, Nadel L. *The Hippocampus as a Cognitive Map.* London: Oxford University Press; 1978.
61. Dix SF, Aggleton JP. Extending the spontaneous preference test of recognition: evidence of object-location and object-context recognition. *Behav Brain Res.* 1999;99(2):191–200.
62. Mumby DG, Gaskin S, Glenn MJ, Schramek TE, Lehmann H. Hippocampal damage and exploratory preferences in rats: memory for objects, places, and contexts. *Learn Mem.* 2002;9(2):49–57.
63. Wang SH, Redondo RL, Morris RGM. Relevance of synaptic tagging and capture to the persistence of long-term potentiation and everyday spatial memory. *Proc Natl Acad Sci USA.* 2010;107:19537–19542.
64. Almaguer-Melian W, Bergado-Rosado J, Pavón-Fuentes N, Alberti-Amador E, Mercerón-Martínez D, Frey U. Novelty exposure overcomes foot shock-induced spatial-memory impairment by processes of synaptic-tagging in rats. *Proc Natl Acad Sci USA.* 2012;109(3):953–958.
65. Aggleton JP, Brown MW. Episodic memory, amnesia, and the hippocampal-anterior thalamic axis. *Behav Brain Sci.* 1999;22(3):425–444.
66. Warburton EC, Brown MW. Findings from animals concerning when interactions between perirhinal cortex, hippocampus and medial prefrontal cortex are necessary for recognition memory. *Neuropsychologia.* 2010;48(8):2262–2272.

67. Berman DE, Dudai Y. Memory extinction, learning anew, and learning the new: dissociations in the molecular machinery of learning in cortex. *Science*. 2001;291:2417–2419.
68. Merhav M, Rosemblum K. Facilitation of taste memory acquisition by experiencing previous novel taste is protein-synthesis dependent. *Learn Mem*. 2008;15(7):501–507.
69. Lisman JE, Grace AA. The hippocampal-VTA loop: controlling the entry of information into long-term memory. *Neuron*. 2005;46(5):703–713.
70. Kitchigina V, Vankov A, Harley C, Sara SJ. Novelty-elicited, noradrenaline-dependent enhancement of excitability in the dentate gyrus. *Eur J Neurosci*. 1997;9(1):41–47.
71. Sara SJ, Vankov A, Hervé A. Locus coeruleus-evoked responses in behaving rats: a clue to the role of noradrenaline in memory. *Brain Res Bull*. 1994;35:457–465.
72. Vankov A, Hervé-Minvielle A, Sara SJ. Response to novelty and its rapid habituation in locus coeruleus neurons of the freely exploring rat. *Eur J Neurosci*. 1995;7:1180–1187.
73. Gao C, Gill MB, Tronson NC, et al. Hippocampal NMDA receptor subunits differentially regulate fear memory formation and neuronal signal propagation. *Hippocampus*. 2010;20(9):1072–1082.
74. Cammarota M, de Stein ML, Paratcha G, Bevilaqua LR, Izquierdo I, Medina JH. Rapid and transient learning-associated increase in NMDA NR1 subunit in the rat hippocampus. *Neurochem Res*. 2000;25(5):567–572.
75. Im HI, Nakajima A, Gong B, et al. Post-training dephosphorylation of eEF-2 promotes protein synthesis for memory consolidation. *PLoS One*. 2009;4:e7424.
76. McGaugh JL. Memory—a century of consolidation. *Science*. 2000;287(5451):248–251.
77. Moncada D, Martinez MC, Ballarini F, Alen N, Viola H. In the search for the behavioral tag. In: *Society for Neuroscience Meeting November 2008, Washington, USA*; 2008.
78. Pastalkova E, Serrano P, Pinkhasova D, Wallace E, Fenton AA, Sacktor TC. Storage of spatial information by the maintenance mechanism of LTP. *Science*. 2006;313:1141–1144.
79. Volk LJ, Bachman JL, Johnson R, Yu Y, Huganir RL. PKM-ζ is not required for hippocampal synaptic plasticity, learning and memory. *Nature*. 2013;493:420–423.
80. Navakkode S, Sajikumar S, Frey JU. Mitogen-activated protein kinase-mediated reinforcement of hippocampal early long-term depression by the type IV-specific phosphodiesterase inhibitor rolipram and its effect on synaptic tagging. *J Neurosci*. 2005;25:10664–10670.
81. Martinez MC, Alen N, Ballarini F, Moncada D, Viola H. Memory traces compete under regimes of limited Arc protein synthesis: implications for memory interference. *Neurobiol Learn Mem*. 2012;98:165–173.
82. Sajikumar S, Frey JU. Resetting of 'synaptic tags' is time- and activity-dependent in rat hippocampal CA1 in vitro. *Neuroscience*. 2004;129:503–507.
83. Fonseca R, Nägerl UV, Morris RGM, Bonhoeffer T. Competing for memory: hippocampal LTP under regimes of reduced protein synthesis. *Neuron*. 2004;44:1011–1020.
84. Govindarajan A, Israely I, Huang S-Y, Tonegawa S. The dendritic branch is the preferred integrative unit for protein synthesis-dependent LTP. *Neuron*. 2011;69:132–146.
85. Won J, Silva AJ. Molecular and cellular mechanisms of memory allocation in neuronetworks. *Neurobiol Learn Mem*. 2008;89:285–292.
86. Yasuda M, Johnson-Venkatesh E. Multiple forms of activity-dependent competition refine hippocampal circuits in vivo. *Neuron*. 2011;70:1128–1142.
87. Garagnani M, Wennekers T, Pulvermüller F. Recruitment and consolidation of cell assemblies for words by way of Hebbian learning and competition in a multi-layer neural network. *Cognit Comput*. 2009;1:160–176.
88. Bramham CR. Local protein synthesis, actin dynamics, and LTP consolidation. *Curr Opin Neurobiol*. 2008;18:524–531.
89. Tzingounis AV, Nicoll R. Arc/Arg3.1: linking gene expression to synaptic plasticity and memory. *Neuron*. 2006;52:403–407.

90. Wibrand K, Pai B, Siripornmongcolchai T, et al. MicroRNA regulation of the synaptic plasticity-related gene Arc. *PLoS One*. 2012;7:e41688.
91. Bramham CR, Alme MN, Bittins M, et al. The Arc of synaptic memory. *Exp Brain Res*. 2011;200:125–140.
92. Okuno H, Akashi K, Ishii Y, et al. Inverse synaptic tagging of inactive synapses via dynamic interaction of Arc/Arg3.1 with CaMKIIβ. *Cell*. 2012;149:886–898.
93. Barco A, Lopez de Armentia M, Alarcon JM. Synapse-specific stabilization of plasticity processes: the synaptic tagging and capture hypothesis revisited 10 years later. *Neurosci Biobehav Rev*. 2008;32:831–851.
94. Jost A. Die Assoziationsfestigkeit in ihrer Abhängigkeit von der Verteilung der Wiederholungen [The strength of associations in their dependence on the distribution of repetitions]. *Z Psychol*. 1897;14:436–472.
95. Müller GE, Pilzecker A. Experimentelle Beiträge zur Lehre vom Gedächtniss. *Z Psychol Ergänzungsband*. 1900;1:1–300.
96. Ribot T. *Les Maladies de la Memoire (Diseases of Memory)*. New York: Appleton-Century-Crofts; 1881.
97. Skaggs EB. Further studies in retroactive inhibition. *Psychol Monogr*. 1925;161:1–60.
98. Ballarini F, Martínez MC, Díaz Perez M, Moncada D, Viola H. Memory in elementary school children is improved by an unrelated novel experience. *PLoS One*. 2013;8(6): e66875.
99. Rey A. Test de copie et de reproduction de memoire de figures geometriques complexes [A test of copy and recall of a complex geometric figure]. Editions du centre de Psycholegie Appliquee. 1959.
100. Li Q, Rothkegel M, Cheng Xiao Z, Abraham W, Korte M, Sajikumar S. Making synapses strong: metaplasticity prolongs associativity of long-term memory by switching synaptic tag mechanisms. *Cereb Cortex*. 2012; doi:10.1093/cercor/bhs315.

INDEX

Note: Page numbers followed by "*f*" indicate figures and "*t*" indicate tables.

G

H

I

M

T

V

W

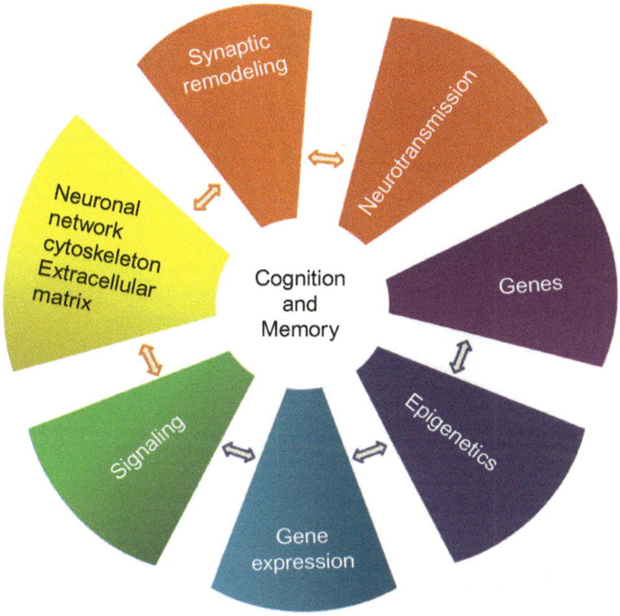

Zafar U. Khan *et al.*, Figure 1.1

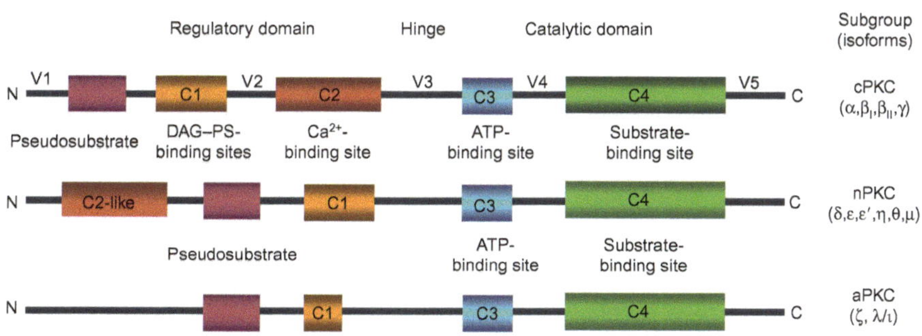

Miao-Kun Sun and Daniel L. Alkon, Figure 2.1

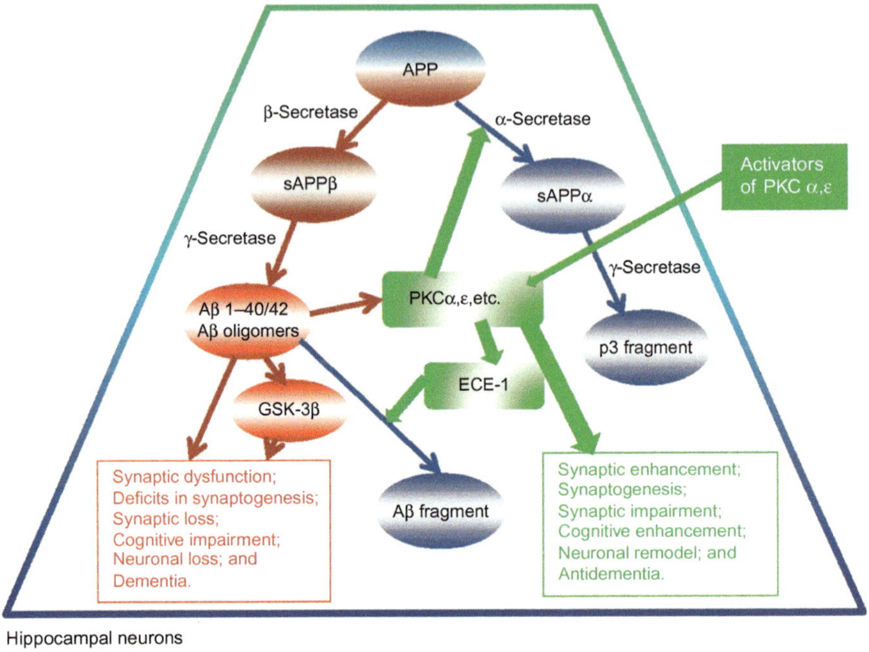

Miao-Kun Sun and Daniel L. Alkon, Figure 2.2

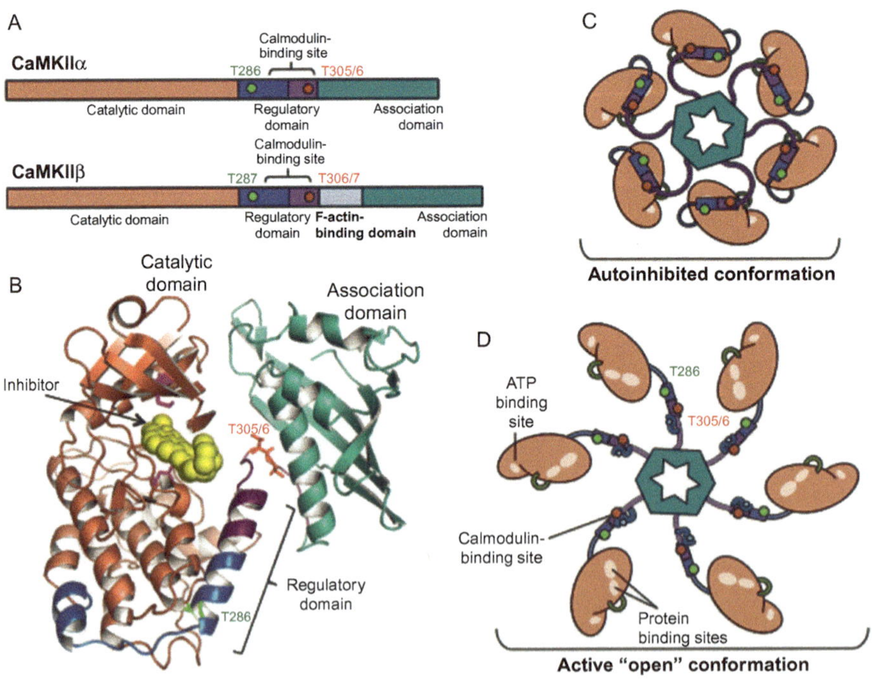

Brian C. Shonesy *et al.*, Figure 3.1

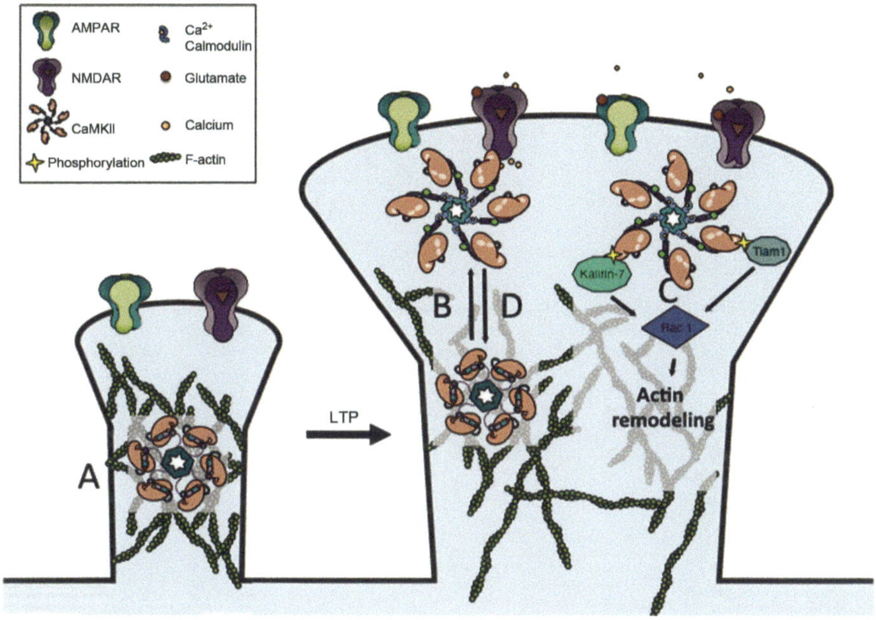

Brian C. Shonesy *et al.*, Figure 3.2

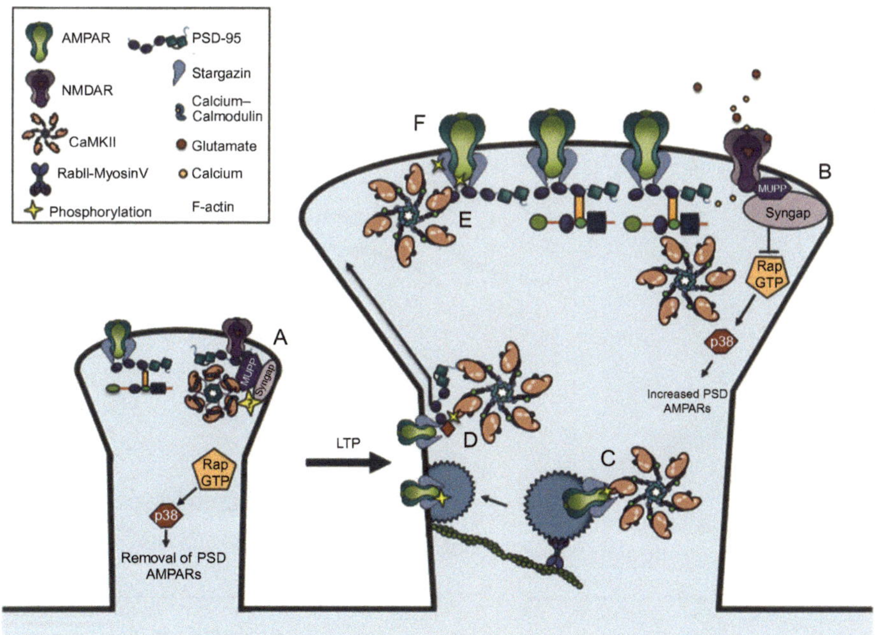

Brian C. Shonesy *et al.*, Figure 3.3

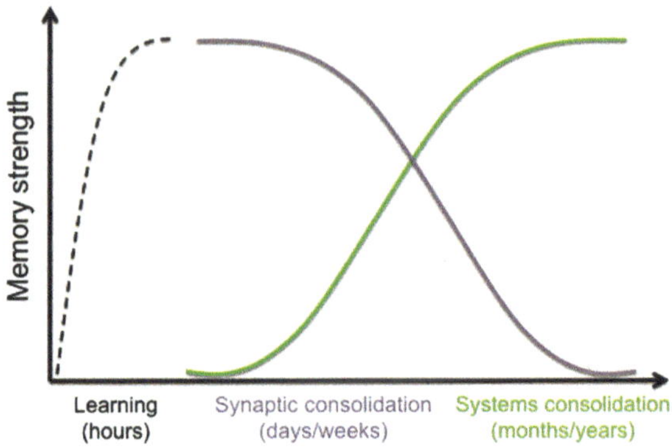

Emanuela Santini *et al.*, Figure 5.1

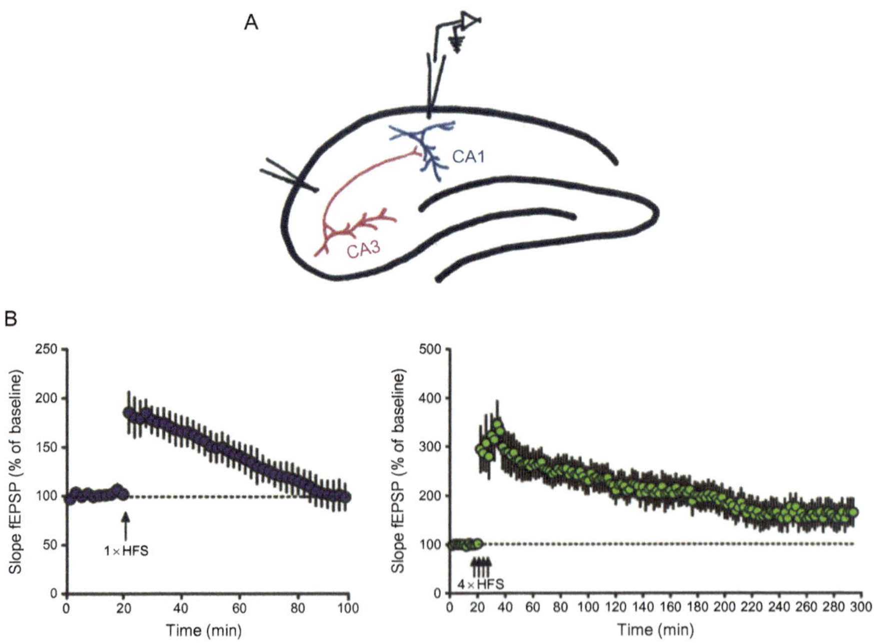

Emanuela Santini *et al.*, Figure 5.2

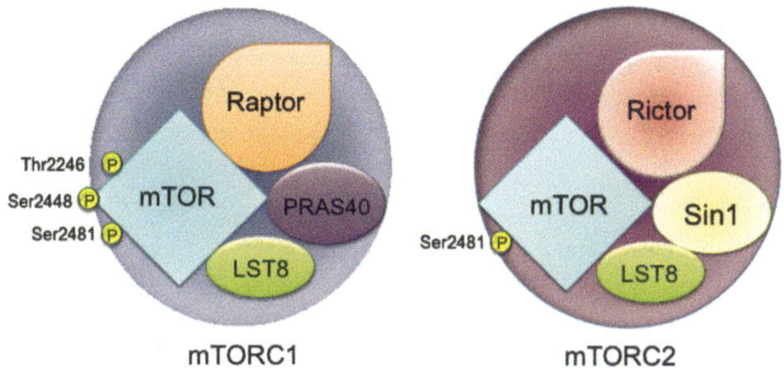

EMANUELA SANTINI *ET AL.*, FIGURE 5.3

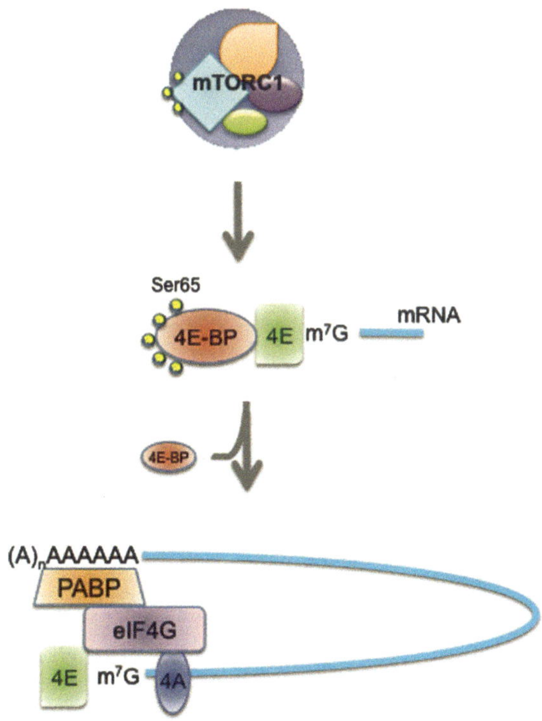

EMANUELA SANTINI *ET AL.*, FIGURE 5.4

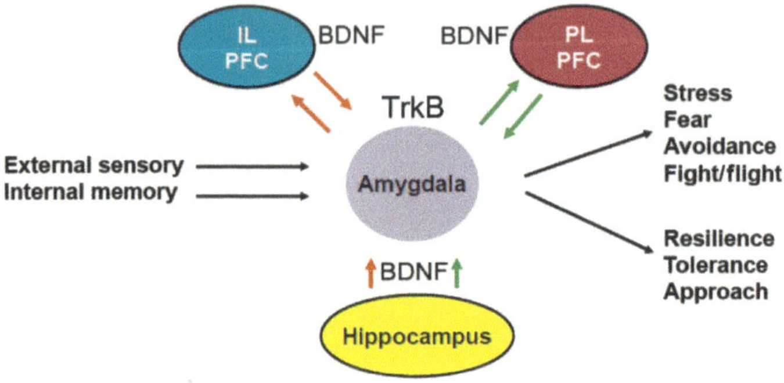

RAÜL ANDERO *ET AL.*, FIGURE 6.2

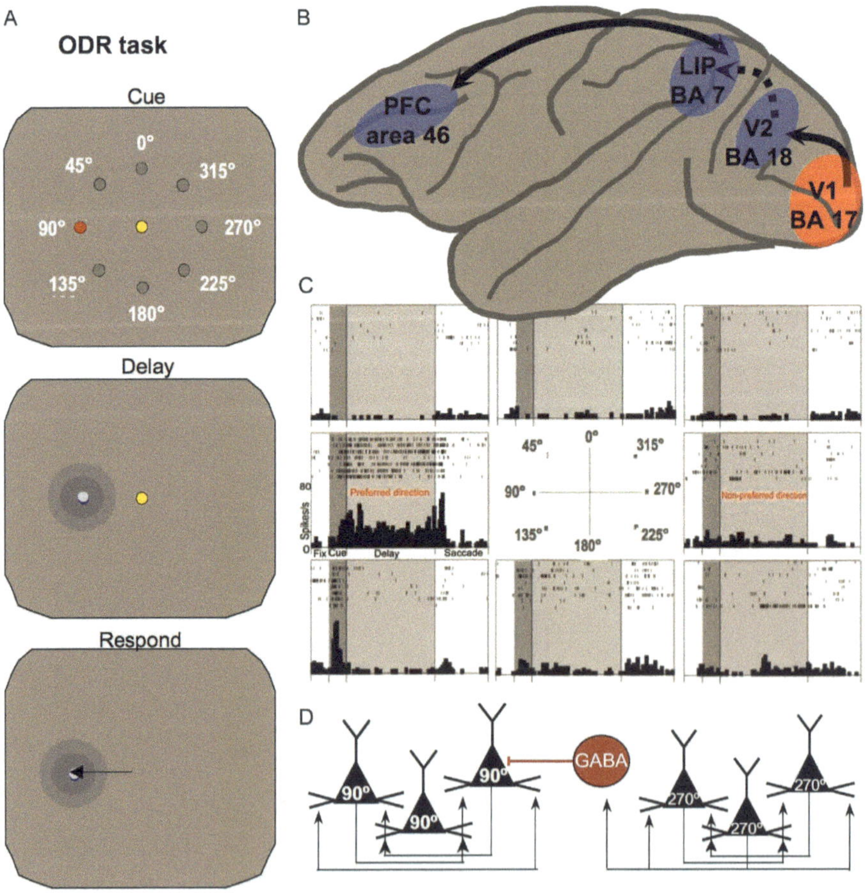

AMY F.T. ARNSTEN AND LU E. JIN, FIGURE 8.1

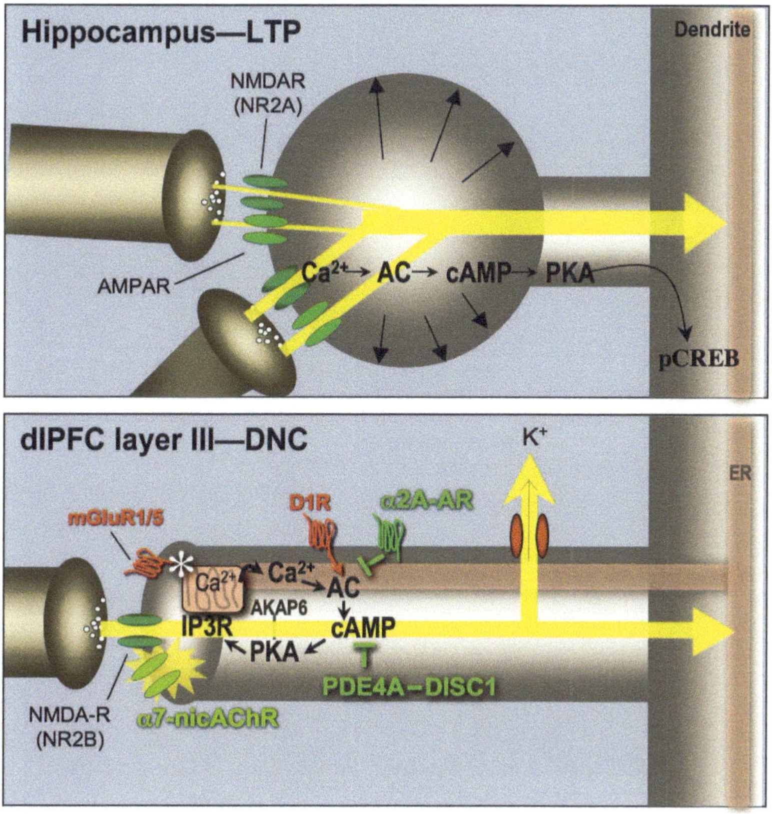

Amy F.T. Arnsten and Lu E. Jin, Figure 8.2

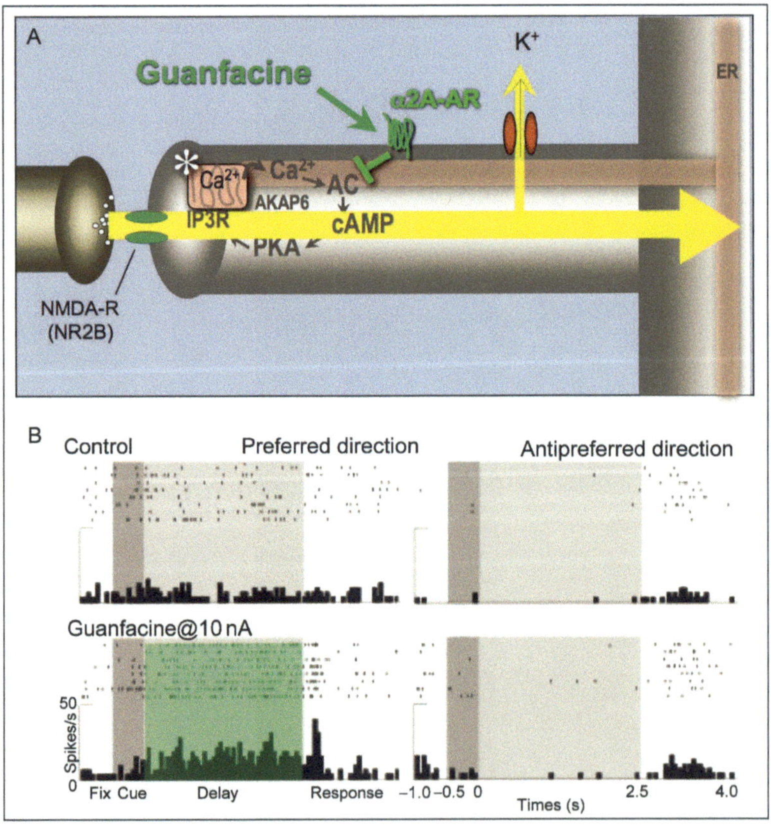

Amy F.T. Arnsten and Lu E. Jin, Figure 8.3

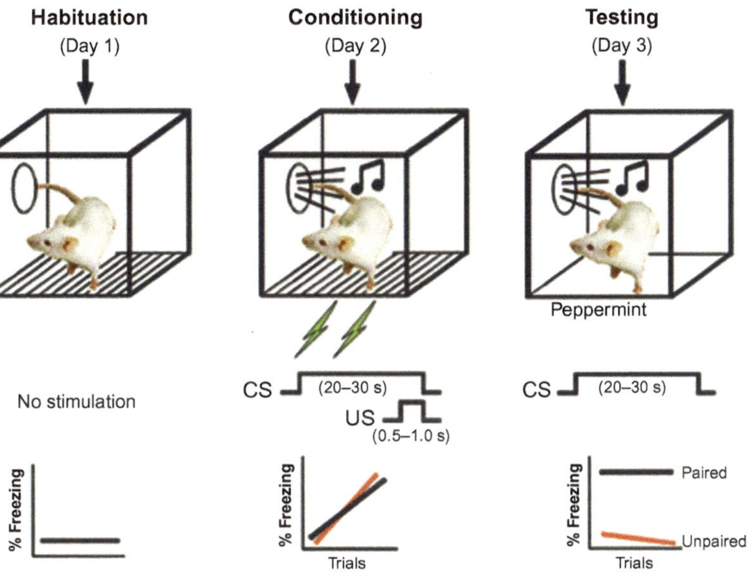

Robert M. Sears *et al.*, Figure 10.1

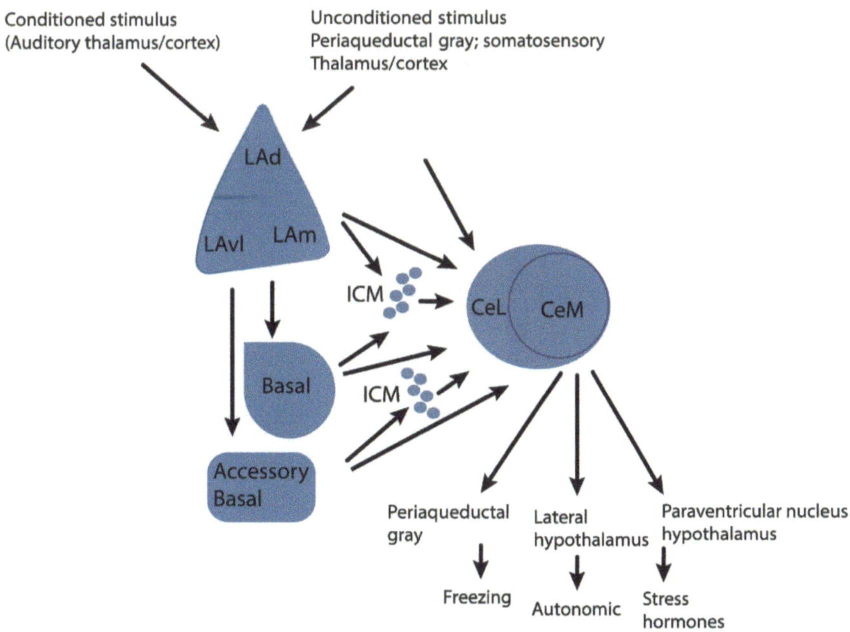

Robert M. Sears *et al.*, Figure 10.2

A Optogenetic tools used in mammalian brains

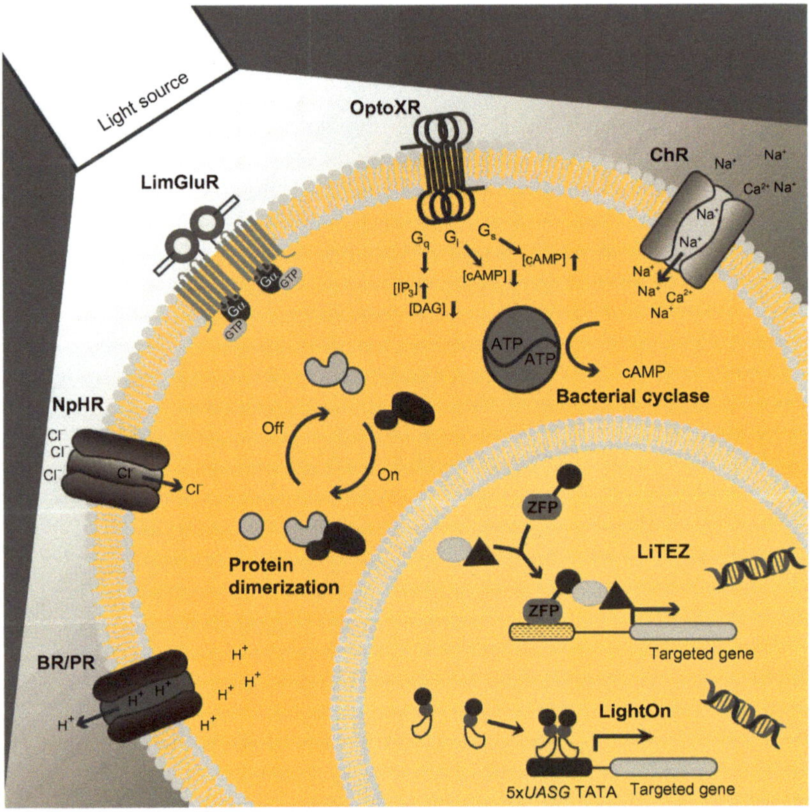

B Targeting specific cell bodies or terminals

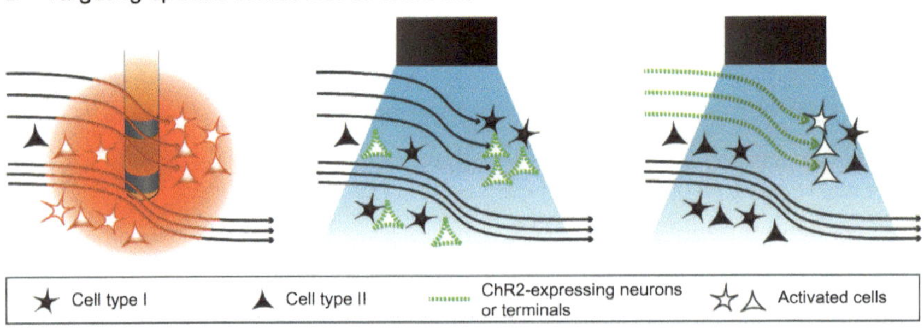

ANNA BEYELER *ET AL.*, FIGURE 12.1

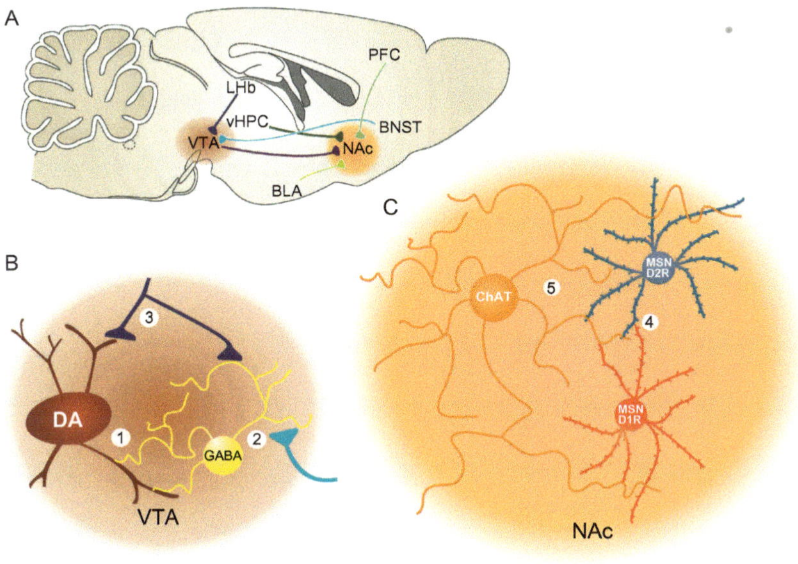

ANNA BEYELER *ET AL.*, FIGURE 12.2

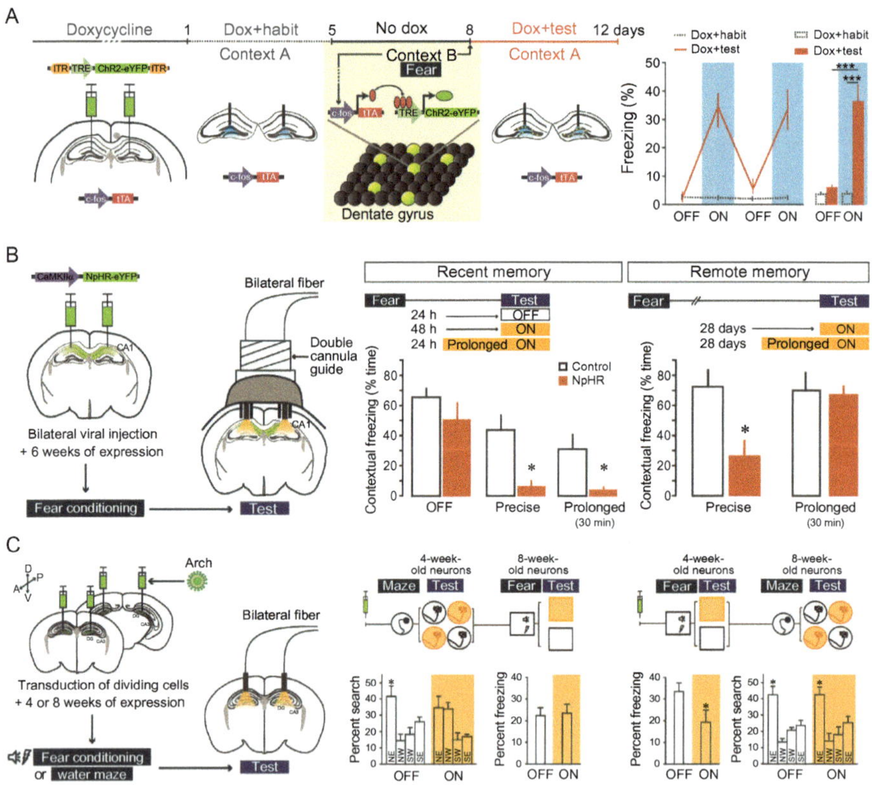

ANNA BEYELER *ET AL.*, FIGURE 12.3

FSC
www.fsc.org
MIX
Papier | Fördert gute Waldnutzung
FSC® C083411

Kolibri 360 GmbH
Ferdinand-Jühlke-Straße 7
99095 Erfurt, Deutschland
produktsicherheit@kolibri360.de